Lecture Notes in Mathematics 1524

M. Coste L. Mahé M.-F. Roy (Eds.)

Real Algebraic Geometry

Proceedings of the Conference held in
Rennes, France, June 24-28, 1991

Springer-Verlag
Berlin Heidelberg New York
London Paris Tokyo
Hong Kong Barcelona
Budapest

Editors

Michel Coste
Louis Mahé
Marie-Françoise Roy
IRMAR (CNRS, U.R.A. 305)
Université de Rennes I
Campus Beaulieu
F-35402 Rennes Cedex, France

Mathematics Subject Classification (1991): Primary: 14P
Secondary: 12D15, 32B, 32C, 58A35, 68Q40

ISBN 3-540-55992-2 Springer-Verlag Berlin Heidelberg New York
ISBN 0-387-55992-2 Springer-Verlag New York Berlin Heidelberg

Typesetting: Camera-ready by author/editor
46/3140-543210 - Printed on acid-free paper

PREFACE

The meeting on "Real Algebraic Geometry" was held in La Turballe, on the seashore not far from Rennes, from June 24 to 28, 1991. It took place ten years after the first meeting on "Géométrie Algébrique Réelle et Formes Quadratiques" (*). These Proceedings contain survey papers on some of the developements of real algebraic geometry in the last ten years, and also contributions by the participants. Every paper has been submitted to a referee, and we want to thank all of them for their collaboration.

The meeting, and the collaboration between the european teams which made it possible, received support from the Université de Rennes 1, the GDR Mathématiques-Informatique (CNRS), and the programs Réseau Européen de Laboratoires, Acces, Alliance, Actions Intégrées Franco-Espagnoles. We would like also to thank Springer-Verlag for publishing this volume, and to express our gratitude to Ms. Yvette Brunel, for her precious help for the secretary of the meeting.

Michel Coste
Louis Mahé
Marie-Françoise Roy

(*) Lecture Notes in Mathematics 959, Springer (1982)

TABLE OF CONTENTS

Survey papers

Other papers

Semialgebraic topology in the last ten years

Manfred Knebusch

Contents

§1 Brumfiel's program

Before discussing the subject named in the title it seems appropriate to outline the situation in semialgebraic topology in 1981, at the time of the first Rennes conference on real algebraic geometry.

Already in the seventies, in the long introduction to his book "Partially ordered rings and semialgebraic geometry" [B], G.W. Brumfiel had laid down a program for what we now call "semialgebraic topology". Here Brumfiel advocated a new way of handling topological problems which is closer to the spirit of algebraic geometry than traditional topology. Let me just quote the following passage:

"It thus seems to me that a true understanding of the relations between algebraic geometry and topology must stem from a deeper understanding of real algebraic geometry, or, actually, semi-algebraic geometry. Moreover, real algebraic geometry should not be studied by attempting to extend classical algebraic geometry to non-algebraically closed ground fields, nor by regarding the real field as a field with an added structure of a topology. Instead, the abstract algebraic treatment of inequalities originated by Artin and Schreier should be extended from fields to (partially ordered) algebras, with real closed fields replacing the algebraically closed fields as ground fields" [B, p.2].

In the main body of the book [B] Brumfiel develops a "real algebra" by studying partially ordered commutative rings and various sorts of convex ideals, with the perspective that this real algebra should perform a similar role in semialgebraic geometry as commutative algebra does in present day algebraic geometry. But the book does not go very far into semialgebraic topology.

§2 The two approaches

Even today not much semialgebraic topology has been done using Brumfiel's rather intricate real algebra from the seventies. Around 1979 two other approaches to semialgebraic topology emerged independently which turned out to be successful. These are the "abstract" approach by M. Coste and M.F. Roy, and the "geometric" approach by H. Delfs and M. Knebusch.

Before we get into this let me remind you of what are perhaps the two most serious difficulties which one encounters if one works over a real closed base field R different from $\mathbb{R}$.

a) R^n is totally disconnected in the strong topology (i.e. the topology coming from the ordering of R).

b) R^n has very few reasonable (i.e. geometrically relevant) compact subsets. In particular, the closed unit ball in R^n is not compact.

Let M be a semialgebraic subset of some R^n. In the abstract approach one adds to M "ideal points" which turn M into an honest (albeit not Hausdorff) topological space. More precisely, one passes from M to the corresponding constructible subset $\tilde{M}$ of the real spectrum $\operatorname{Sper} R[T_1,\ldots,T_n]$ of the polynomial ring $R[T_1,\ldots,T_n]$ (cf. [BCR, Chap. 7]). $\tilde{M}$ turns out to have only finitely many connected components, and $\tilde{M}$ is quasicompact. Thus in some sense the difficulties described above are overcome. The subspace topology of M in $\tilde{M}$ is the strong topology we started with.

One could also pass from M to the subspace $\tilde{M}^{\max}$ of closed points of $\tilde{M}$, which still contains M as a dense subset and is a compact Hausdorff space with only finitely many connected components. But although this compactification $\tilde{M}^{\max}$ of M has its merits (cf. [B$_1$]), the more interesting and more useful space is $\tilde{M}$ itself. The main reason for this is that $\tilde{M}$ is a spectral space, as defined by Hochster [Ho], and that the constructible subsets Y of $\tilde{M}$ correspond bijectively with the semialgebraic subsets N of M via the relation $Y = \tilde{N}$. A very nice consequence of this is that the "semialgebraic structure" of M is encoded to a large extent in the topology of $\tilde{M}$, since the lattice $\mathfrak{K}(\tilde{M})$ of constructible subsets of $\tilde{M}$ is by definition the boolean lattice generated by the lattice $\mathring{\mathfrak{K}}(\tilde{M})$ of quasicompact open subsets of $\tilde{M}$, and thus

$\mathfrak{K}(\tilde{M})$ is completely determined by the topology of $\tilde{M}$ (cf. [Ho]). We call the space $\tilde{M}$ the *abstraction* of the semialgebraic set M.

The wisdom of passing back and forth between the semialgebraic sets and their abstractions has been displayed well in the book [BCR] by Bochnak, Coste and Roy. Curiously another very important and fascinating aspect of the abstract approach is scarcely touched on in that book: One can study the constructible subsets of the real spectrum Sper A of *any* commutative ring A. Thus the abstract approach opens the door for an "abstract" semialgebraic topology where no base field (real closed or not) needs to be present. Coste and Roy were certainly well aware of this aspect at an early stage (cf. for example Roy's paper on abstract Nash functions [R]) but chose not to give much space to this in their book with Bochnak.

The geometric approach (cf. [DK]) relies on the following two ideas, the first one being very simple.

1° Don't consider any subset of a semialgebraic set $M \subset R^n$ which is not semialgebraic or any map $f: M \to N$ between semialgebraic sets which is not semialgebraic!

In this context a *map* f is called *semialgebraic* if the graph of f is a semialgebraic set and f is continuous with respect to the strong topologies of M and N.

2° Install on M a Grothendieck topology such that the semialgebraic functions, i.e., semialgebraic maps to R on the open semialgebraic subsets U of M (open with respect to the strong topology) form a sheaf $\mathcal{O}_M$ of R-algebras! Instead of studying M as a semialgebraic subset of R^n study the ringed space $(M, \mathcal{O}_M)$!

Let me give some comments and explanations on these ideas.

Ad 1°: The reason that this idea makes sense is Tarski's principle. It guarantees that many of the usual constructions of new sets and maps from given ones give us semialgebraic sets and maps if we start with such sets and maps. In particular, if $f: M \to N$ is a semialgebraic map between semialgebraic sets then the image $f(A)$ of a semialgebraic subset A of M is semialgebraic and the preimage $f^{-1}(B)$ of a semialgebraic subset B of N is semialgebraic. Continuity of f is not necessary for this but is appropriate since we want to do "topology".

Ad 2°: The Grothendieck topology on M is defined as follows. The underlying category is the category $\mathring{\mathfrak{S}}(M)$ of open semialgebraic subsets of M (i.e. semialgebraic subsets which are open in the strong topology), the morphisms being the inclusion mappings. An admissible open covering $(U_i | i \in I)$ of some $U \in \mathring{\mathfrak{S}}(M)$ is a family $(U_i | i \in I)$ in $\mathring{\mathfrak{S}}(M)$ with $U = \bigcup_{i \in I} U_i$, such that there exists a finite subset J of I with $U = \bigcup_{i \in J} U_i$. {Thus a property similar to quasicompactness is forced to hold.} Then the semialgebraic functions on the sets $U \in \mathring{\mathfrak{S}}(M)$ indeed form a sheaf $\mathcal{O}_M$. It turns out that a morphism from $(M, \mathcal{O}_M)$ to $(N, \mathcal{O}_N)$ is determined by the underlying map f from M to N, and that these maps f are just the semialgebraic maps from M to N as introduced above (cf. [DK, §7], by definition the morphism has to respect the R-algebra structures of the structure sheaves).

Replacing a semialgebraic set $M \subset R^n$ by the ringed space $(M, \mathcal{O}_M)$ allows us to forget the embedding $M \hookrightarrow R^n$. We call any ringed space of R-algebras which is isomorphic to such a space $(M, \mathcal{O}_M)$ an *affine semialgebraic space* over R. By abuse of notations we do not distinguish between a semialgebraic set M and the corresponding ringed space $(M, \mathcal{O}_M)$.

A *semialgebraic path* in M is a semialgebraic map from the unit interval [0,1] (which is a semialgebraic subset of R^1) to M. Having this notion of paths at hand one defines the path components of M in the obvious way. It turns out, that M has only finitely many path components $M_1, \ldots, M_r$ and that these are semialgebraic in M and closed, hence also open in the strong topology, cf. [DK]. Every M_i is "semialgebraically connected", i.e. M_i is not the union of two disjoint non empty open semialgebraic subsets, since this holds for [0,1], as is easily seen. Thus we have dealt with the first difficulty mentioned above, exploiting only idea N° 1. By the way, the abstractions $\tilde{M}_1, \ldots, \tilde{M}_r$ are the connected components of the topological space $\tilde{M}$.

In order to cope with the second difficulty one also needs idea N° 2. The category of affine semialgebraic spaces over R has fiber products. Thus we can define proper morphisms as in algebraic geometry. We call a semialgebraic map $f: M \to N$ *closed*, if the image $f(A)$ of a closed semialgebraic subset A of M is again closed. We call f *proper* if f is universally closed, i.e. for any semialgebraic map $g: N' \to N$ the cartesian square

$$\begin{array}{ccc} M \times_N N' & \xrightarrow{f'} & N' \\ g' \downarrow & & \downarrow g \\ M & \xrightarrow[f]{} & N \end{array}$$

gives us a closed semialgebraic map f'. We call an affine semialgebraic space M *complete* if the map from M to the one-point space is proper. Even more than in algebraic geometry over an algebraically closed field, it is true for many purposes, that complete spaces are the right substitute for compact spaces in topology. For example, a semialgebraic function on a complete space attains its maximum and minimum.

It turns out that there exist in abundance relevant complete affine semialgebraic spaces. Namely, the following analogue of the Heine-Borel theorem holds: A semialgebraic subset M of R^n is a complete space iff M is closed and bounded in R^n.

§3 The state of art in 1981

I give a rough sketch of the technical progress up till 1981. This is just to give an impression of the state of art at the first Rennes conference. It is not meant, of course, as a complete account of everything done up to that time.

In the geometric theory we have the following list.

1) Connected components
2) Complete affine semialgebraic spaces and the semialgebraic Heine-Borel theorem
3) Dimension theory
4) Existence of triangulations
5) Hardt's theorem
6) Semialgebraic homology

Here are some comments on these.
N° 1 and N° 2 have been described above. One may add to N° 2 that in 1981 we also had a good insight into the nature of proper maps between affine semialgebraic spaces [DK, §9 and §12].

Ad 3: The dimension $\dim M$ of a semialgebraic set can be defined as the maximal integer d such that M contains a subspace N which is isomorphic to the unit ball in R^d {[DK, §8], there a different but equivalent definition had been given}. This notion of dimension behaves very well, better than in classical topology. For example, if a partition of M into finitely many semialgebraic subsets $A_1, \dots, A_r$ is given, then $\dim M$ is the maximum of the numbers $\dim A_1, \dots, \dim A_r$.

Ad 4: If M is an affine semialgebraic space and $A_1, \dots, A_r$ are finitely many semialgebraic subsets of M then there exists a finite simplicial complex X over R and an isomorphism of spaces $\varphi: X \xrightarrow{\sim} M$ such that, for every $i \in \{1, \dots, r\}$, the set $\varphi^{-1}(A_i)$ is a subcomplex of X [DK, §2]. Here the word "simplicial complex" is used in a non classical meaning: X is the union of finitely many open simplices $\sigma_1, \dots, \sigma_t$ in some R^N such that the intersection $\bar{\sigma}_i \cap \bar{\sigma}_j$ of the closures of any two simplices σ_i, σ_j is either a common face of them or empty. Thus the closure $\bar{X}$ of X is a classical finite simplicial complex (≈ finite polyhedron), and X is obtained from $\bar{X}$ by omitting some open faces. Also "subcomplex" means just the union of some of the sets $\sigma_1, \dots, \sigma_t$. Clearly $X = \bar{X}$ iff M is complete.

In the case $R = \mathbb{R}$ the triangulation theorem has been well known since the sixties, even for semianalytic sets [L, Gi].

Ad 5: Hardt's theorem states that for every semialgebraic map $f: M \to N$ there exists a partition of N into finitely many semialgebraic subsets $N_1, \dots, N_r$ such that f is trivial over each N_j, i.e. $f^{-1}(N_j)$ is isomorphic over N_j to a direct (= cartesian) product $N_j \times F_j$, cf [DK$_1$, §6]. The theorem had been proved for $R = \mathbb{R}$ by R. Hardt around 1978 [Ha].

Ad 6: In his thesis [D] Delfs constructed homology and cohomology groups with arbitrary constant coefficients for affine semialgebraic spaces over any real closed field R. In the case $R = \mathbb{R}$ these groups coincide with the singular groups known from classical topology.

Certainly Delfs' homology theory was the most profound achievement in semialgebraic topology up till 1981. But the proofs of the triangulation theorem and of Hardt's theorem also needed new ideas beyond the known proofs for $R = \mathbb{R}$.

The triangulation theorem is the main technical tool in developing semialgebraic homology (and also semialgebraic homotopy theory, cf. §10 below). Hardt's theorem is very useful if one wants to profit from semialgebraic homology. For a good example, cf. [DK$_1$, §7]. I will say more about semialgebraic homology in the next section §4.

Remark. Only recently (1989) I learned from Gert-Martin Greuel about the unpublished dissertation of Helmut Brakhage [Bra] (Heidelberg 1954, thesis advisor F.K. Schmidt). Here Brakhage studies semialgebraic topology over an arbitrary real closed field. He exploits idea N° 1 of the geometric theory (cf. §2) to an enormous extent and obtains many of the results we had found up to 1981, in particular the triangulation theorem. The introduction to Brakhage's thesis reads very much like the talks Delfs and I used to give around 1980. He would have saved us a lot of work if we only would have known about his thesis. Brakhage is now a professor at Kaiserslautern, working mostly in applied mathematics.

It is difficult to give a good picture of the state of art in semialgebraic topology in 1981 on the abstract side, since in the abstract theory the main bias was towards algebraic problems. Topology seems to have been studied mainly as an aid for solving algebraic problems of current interest. I give the following list.

1) Connected components
2) Compactness of constructible sets
3) Specialization theory
4) Dimension theory
5) Abstract Nash functions
6) Separation of connected components by global quadratic forms

Here only N° 1 - 4 truly belong to semialgebraic topology, but N° 5 and 6 use topology in an essential way, and have also turned out to be stimulating for semialgebraic topology since 1981.

N° 1 has been discussed above, N° 2 alludes to the easily accessible but extremely important fact, that the real spectrum Sper A of any commutative ring A is compact in the constructible topology. This means that, if X is a constructible subset and $(Y_i|i \in I)$ is a family of constructible subsets of Sper A with $X \subset \bigcup_{i\in I} Y_i$, then there exists a finite subset J of I with $X \subset \bigcup_{i\in J} Y_i$. The quasicompactness of $\tilde{M}$ stated above is a rather special consequence of this.

Ad 3: If x and y are points of a topological space X then we say that y is a *specialization* of x (and x is generalization of y) if y lies in the closure of the set $\{x\}$. We write $x \succ y$ for this. N° 3 alludes to some – again simple but important – facts about specializations in a real spectrum Sper A, cf. [CR$_2$], [BCR, 7.1], [KS, III §3 and §7]. In particular, the specializations of a given point x in Sper A form a chain, i.e. if $x \succ y$ and $x \succ z$ then $y \succ z$ or $z \succ y$. Moreover if neither $x \succ y$ nor $y \succ x$ then there exist disjoint open subsets U, V in Sper A with $x \in U$ and $y \in V$.

Ad 4: The *dimension* $\dim X$ of a constructible subset X of Sper A is defined as the supremum of the lengths of the specialization chains in X. {Up till now it has been adequate to put $\dim X = \infty$ if the lengths do not have a finite bound.} The main result is that, if M is a semialgebraic set over some real closed field, then this "combinatorial" dimension $\dim \tilde{M}$ of the abstraction $\tilde{M}$ coincides with the semialgebraic dimension $\dim M$ from above, cf. [CR$_2$], [BCR].

Ad 5 and 6: One of the most important achievements in the early work of Coste and Roy is the construction of a sheaf of "abstract Nash functions" $\mathfrak{N}_A$ on the real spectrum of an arbitrary commutative ring A [R], which generalizes the sheaf of classical Nash functions for algebraic manifolds over $\mathbb{R}$. Indeed, right from the beginning they had the idea of constructing the real spectrum as a ringed space (Sper $A, \mathfrak{N}_A$) [CR], [CR$_1$], thus bringing semialgebraic geometry close to the spirit of abstract algebraic geometry in the sense of Grothendieck. The sheaf $\mathfrak{N}_A$ is more algebraic in nature than the sheaf of semialgebraic functions discussed in §2. It does not belong to semialgebraic topology, but nevertheless relies on the topological fact that every etale morphism $A \xrightarrow{\varphi} B$ induces a local homeomorphism Sper φ: Sper $B \to$ Sper A.

Building on this, Mahé was able to solve one of the main open problems of quadratic form theory from the seventies [K$_1$, Problem 16] affirmatively, namely the separation by global quadratic forms of the connected components of the set $V(\mathbb{R})$ of real points of an affine algebraic variety V, and later, together with Houdebine, also of a projective algebraic variety V over $\mathbb{R}$ [M], [HM]. In fact, they prove such a theorem over any real closed field R, and also for the real spectrum of any commutative ring.

Mahé's theorem in [M] is probably the first result which signaled to the outside world that something new in principle had happened in real algebraic geometry around 1980.

§4 Sheaves and homology

After 1981 semialgebraic topology has been dominated by two major new trends: A strong interaction between the geometric and the abstract theory, and the employment of new spaces. An important instance of the first trend is sheaf theory.

Let M be a semialgebraic set over some real closed field R. Then a (set valued) sheaf over M is essentially the same object as a sheaf over the abstraction $\tilde{M}$. Indeed, as was already known before 1981 [CR_2], [D], [De], a semialgebraic subset U of the affine semialgebraic space M is open iff the abstraction $\tilde{U}$ is open in $\tilde{M}$. Moreover, a family $(U_i|i \in I)$ of open semialgebraic subsets of M is an admissible open covering of U iff $(\tilde{U}_i|i \in I)$ is an open covering of $\tilde{U}$. The reason for this is the definition of the Grothendieck topology on M on the one hand, and the quasicompactness of $\tilde{U}$ on the other. Since the quasicompact open subsets of $\tilde{M}$ form a basis of the topology of $\tilde{M}$, all of this gives us a canonical isomorphism $\mathcal{F} \mapsto \tilde{\mathcal{F}}$ from the category of sheaves on M to the category of sheaves on $\tilde{M}$, via the rule $\mathcal{F}(U) = \tilde{\mathcal{F}}(\tilde{U})$.

Henceforth we only consider sheaves of abelian groups. Recall that M is dense in $\tilde{M}$. For $x \in M$ the stalks $\mathcal{F}_x$ and $\tilde{\mathcal{F}}_x$ are equal. It may well happen that all stalks $\mathcal{F}_x$, $x \in M$, are zero, but $\mathcal{F}$ is not zero. {An example is given in [D_2, I.1.7].} This is by no means astonishing: Of course, $\mathcal{F} \neq 0$ iff $\tilde{\mathcal{F}} \neq 0$. Then, since $\tilde{M}$ is an honest topological space, there exists some $\alpha \in \tilde{M}$ with $\tilde{\mathcal{F}}_\alpha \neq 0$. But it may happen that none of these points α lies in M.

This discussion makes it clear that most often sheaf theoretic techniques work better in the abstract setting than the geometric one. Only there one can argue "stalk by stalk" without further justification.

Now is a good moment to say something about the semialgebraic homology theory of Hans Delfs, since he has been able to simplify his theory greatly by using sheaves and the interplay back and forth between semialgebraic sets and their abstractions [D_1].

I first describe the main problem in defining homology groups $H_q(M, G)$ for a semialgebraic set M over some real closed field R and some abelian group of coefficients G. Let us assume for simplicity that M is complete. We choose a triangulation $\varphi: X \xrightarrow{\sim} M$. Here X is a finite simplicial complex in the classical sense but over R; X may be regarded as the realization $|K|_R$ over R of an abstract finite simplicial complex K, a purely combinatorial object (cf. [Spa, 3.1]; the realization is defined exactly as in the case $R = \mathbb{R}$).

It is intuitively clear that $H_q(M, G)$ should coincide, up to isomorphism, with the combinatorial homology group $H_q(K, G)$ from classical topology. To make an honest definition out of this, one has to verify that (up to natural isomorphism) the group $H_q(K,G)$ does not depend on the choice of the triangulation. The now traditional way to prove this is to define a complex $C.(M,G)$ of singular chains and to verify the seven Eilenberg-Steenrod axioms for the homology groups [ES, I §3]. Then one obtains, in a well known manner, that $H_q(C.(M,G)) \cong H_q(K,G)$ for the triangulation φ above. {One also has to consider noncomplete spaces M and the relative chain complex $C.(M,A;G)$ for A a semialgebraic subset of M. I omit these technicalities.}

We can indeed define singular chain groups $C_q(M,G)$ along classical lines, decreeing that a singular simplex is a semialgebraic map from the q-dimensional standard simplex Δ_q to M. Six of the seven Eilenberg-Steenrod axioms can be proved as in the classical theory, always using semialgebraic maps instead of continuous maps. But the excision axiom is difficult. The classical way to prove it is to make a given singular cycle Z "small" with respect to a given covering of M by two open (semialgebraic) sets U_1, U_2, by iterated barycentric subdivision of Z. This means that every singular simplex occuring in the subdivided cycle has its image either in U_1 or U_2. But if the base field R is not archimedian then this procedure completely breaks down, since then usually a given bounded semialgebraic set cannot be covered by finitely many semialgebraic sets all whose diameters are smaller than a given $\epsilon > 0$.

In his thesis [D] Delfs found the following way out of this difficulty. He defined cohomology groups $H^q(M,G)$ as the sheaf cohomology groups of the constant sheaf G_M on M, and similarly relative cohomology groups $H^q(M,A;G)$ as the sheaf cohomology groups of a suitable sheaf $G_{M,A}$ on M. {Recall that M is equipped with a Grothendieck topology.} For these groups $H^q(M,A;G)$ Delfs succeeded in verifying the Eilenberg-Steenrod axioms. Then he knew that $H^q(M,G)$ is isomorphic to the combinatorial group $H^q(K,G)$. Thus $H^q(K,G)$ is independent of the choice of the triangulation, up to natural isomorphism. From this Delfs concluded that also $H_q(K,G)$ is independent of the choice of triangulation [D].

The verification of six of the seven Eilenberg-Steenrod axioms for the groups $H^q(M,A,G)$ is straightforward, but this time the homotopy axiom causes difficulties. Delfs surmounted these difficulties in [D] by a complicated geometric procedure.

Later Delfs found an easier way [D_1]. He realized that the homotopy axiom follows from the statement that, for any sheaf $\mathcal{F}$ on M, the adjunction homomorphism $\mathcal{F} \to \pi_*\pi^*\mathcal{F}$, with π the projection from $M \times [0,1]$ to M, is an isomorphism and $R^q\pi_*(\pi^*\mathcal{F}) = 0$ for $q \geq 1$. [D_1, Prop. 4.2 and 4.4]. This then could be deduced via a stalk by stalk argument from the fact that $H^q([0,1],G) = 0$ for $q \geq 1$ and any abelian group G, which in turn can be verified in an easy geometric way. The crucial point is that one needs the fact $H^q([0,1],G) = 0$ not just over R but over the residue class fields $k(x)$ of all points $x \in \tilde{M}$. Roughly one can summarize that Delfs reduced the verification of the homotopy axiom to an easy special case using sheaf theory, at the expense of enlarging the real closed base field in many ways.

§5 Locally semialgebraic spaces

Delfs and I had already introduced *"semialgebraic spaces"* over a real closed field R before 1981 by gluing together finitely many affine semialgebraic spaces over R along open subspaces [DK, §7]. What then was still missing was a handy criterion for a semialgebraic space $M = (M, \mathcal{O}_M)$ to be again affine. Such a criterion would allow the building of semialgebraic spaces M from semialgebraic sets in an "abstract" manner, i.e. without explicitly looking at polynomials, such that M eventually turns out to be an affine space, in other words, a semialgebraic set.

In 1982 R. Robson proved his imbedding theorem [Ro] which gives such a criterion. The theorem says that a semialgebraic space M over R is affine iff M is *regular*, i.e. a point x and a closed semialgebraic subset A of M with $x \notin A$ can be separated by open semialgebraic neighbourhoods. {A subset A of M is called closed semialgebraic if the complement $M - A$ is an open semialgebraic, i.e., an admissible open subset of M.}

Robson's theorem really paved the way for the trend of employing new spaces in the geometric theory. Before I go into details about this I should say some words about covering maps.

Having semialgebraic paths at hand we may define the fundamental group $\pi_1(M, x_0)$ for M a semialgebraic space over R and $x_0 \in M$, as in the classical theory, by considering homotopy classes of semialgebraic loops with base point x_0. Of course, homotopies also have to be defined in the semialgebraic sense, starting from the unit interval [0,1] in R, cf. §10 below. It turns out that for affine M the group $\pi_1(M, x_0)$ is very respectable. It is finitely presented and coincides with the topological fundamental group in the case $R = \mathbb{R}$. {These are consequences of the two comparison theorems on homotopy sets [DK_2, III §3 and §5], to be discussed in §10 below.}

Assume since now that M is affine and path connected. The question arises whether the subgroups of $\pi_1(M, x_0)$ classify "semialgebraic covering spaces" of M, as one might expect from classical topology.

It seems clear what a semialgebraic covering map $\pi: N \to M$ has to be: N should be a semialgebraic space and π a semialgebraic map. Further there should exist an admissible open covering $(U_i | i \in I)$ of M such that π is trivial over each U_i with discrete fibers, i.e. $\pi^{-1}(U_i) \cong U_i \times F_i$ over U_i for a discrete semialgebraic space F_i. But what does it mean for a semialgebraic space F to be discrete? Reasonable answers, one can think of, are: $\dim F = 0$; the path components of F are one-point sets; the one-point sets in M are open in F; the one-point sets in M form an admissible open covering of M. – All of these properties mean the same thing, namely that the space F consists of finitely many points. We conclude that every semialgebraic covering map $\pi: N \to M$ has finite degree.

Working with path lifting techniques one verifies that the semialgebraic coverings $\pi: N \to M$ of M are indeed classified by the conjugacy classes of subgroups of $\pi_1(M, x_0)$ of finite index [K₃]. Using Robson's embedding theorem one also sees that N is again affine.

Having verified this in 1982 [DK₃], Delfs and I realized that the category of semialgebraic spaces is not broad enough. There should exist some sort of covering space N of M corresponding to any given subgroup H of $\pi_1(M, x_0)$, in particular a "universal covering", corresponding to $H = \{1\}$. This led us to introduce locally semialgebraic spaces. A *locally semialgebraic space M over R* is obtained by gluing together (maybe infinitely many) affine semialgebraic spaces over R along open semialgebraic subspaces. Of course, the gluing is meant in the sense of ringed spaces with Grothendieck topologies, cf. [DK₂, I §1].

The nice locally semialgebraic spaces are those which are *regular* (defined in the same way as above) and *paracompact*, as defined in [DK₂, I §4]. The category $LSA(R)$ of regular paracompact locally semialgebraic spaces over R contains the category of affine semialgebraic spaces over R as a full subcategory. In $LSA(R)$ we have a fully satisfactory theory of covering spaces. In particular every space $M \in LSA(R)$ has a universal covering (cf. [DK₃, §5]; a full treatment of this topic still awaits publication [K₃]).

In $LSA(R)$ there exist fibre products. There is also a good notion of subspaces. Namely, if M is a locally semialgebraic space and $(M_i | i \in I)$ is an admissible open covering of M, such that every M_i is an affine semialgebraic space, then a subset A of M is called a *subspace* if $A \cap M_i$ is semialgebraic in M_i for every $i \in I$. Indeed, collecting the affine semialgebraic spaces $A \cap M_i$ we obtain on A the structure of a locally semialgebraic space over R, which is independent of the choice of the covering $(M_i | i \in I)$. This space A is regular and paracompact if M has these properties [DK₂, I, §3 and §4].

Up to now $LSA(R)$ has proved to be the appropriate basic category for all geometric studies over R, as long as one does not pass to abstract spaces. In particular the triangulation theorem for semialgebraic sets (cf. §3 above) extends to a triangulation theorem of equal strength for these spaces (simultaneous triangulation of M and a locally finite family of subspaces of M, cf. [DK₂, Chap.II]). Also the homology theory of Delfs discussed above extends to these spaces [DK₂, Chap.III]. And we have a fairly good homotopy theory in $LSA(R)$ at hand, to be discussed below.

§6 Abstract semialgebraic functions and real closed spaces

We come back to the relationships between a semialgebraic set M over R and its abstraction $\tilde{M}$. Recall from §4 that the sheaves on the affine semialgebraic space M correspond uniquely with the sheaves on $\tilde{M}$. In particular we have a sheaf of rings

$\tilde{\mathcal{O}}_M$ on $\tilde{M}$ which corresponds to the sheaf $\mathcal{O}_M$ of semialgebraic functions on M. The question arises whether $\tilde{\mathcal{O}}_M$ generalizes in a natural way to a sheaf of rings $\mathcal{O}_X$ on any constructible subset X of any real spectrum Sper A, which then can be regarded as a sheaf of "abstract" semialgebraic functions on X.

This is indeed the case. Around 1983 G. Brumfiel [B_4] and N. Schwartz [S] gave two solutions of this problem. A (slightly "corrected", cf. [D, 1.7], [S, Example 58]) version of Brumfiel's definition runs as follows. Let p: Sper $A[T] \to$ Sper A be the natural map from the real spectrum of the polynomial ring $A[T]$ in one variable over A to Sper A, induced by the inclusion $A \hookrightarrow A[T]$. For any quasicompact open subset U of the space X the elements of $\mathcal{O}_X(U)$ are the continuous sections s of $p|p^{-1}(U): p^{-1}(U) \to U$ such that $s(U)$ is a closed constructible subset of $p^{-1}(U)$.

What does this mean? For any $x \in A$ we may identify $p^{-1}(x)$ with the real spectrum Sper $k(x)[T]$, where $k(x)$ denotes the residue class field of Sper A at x, a real closed field. This real spectrum is the abstraction of the real affine line over $k(x)$. Thus $k(x)$ injects into $p^{-1}(x)$ as a dense subset (cf. §2). For a section s as above, $s(x)$ lies in this subset and hence corresponds to an element $f(x)$ of $k(x)$, which should be regarded as the value of the abstract semialgebraic function f given by s. The section s is completely determined by the values $f(x)$ and should be regarded as the graph of f.

N. Schwartz defined an abstract semialgebraic function f on U directly as a family $(f(x)|x \in U) \in \prod_{x\in U} k(x)$ with compatibility relations between the values $f(x)$ coming from canonical valuations $\lambda_{x,y}: k(x) \to \kappa(x,y) \cup \infty$. For any pair (x,y) with $x \in U$ and y a specialization of x in U, $\kappa(x,y)$ is an overfield of $k(y)$, and $\lambda_{x,y}$ has to map $f(x)$ to $f(y) \in k(y) \subset \kappa(x,y)$. The definition of Schwartz has the advantage that here it is immediately clear that $\mathcal{O}_X(U)$ is a ring, while in Brumfiel's definition one has to work for this.

Then Delfs proved that the definitions of Brumfiel and Schwartz give the same sheaf $\mathcal{O}_X$ [D_1, §1]. The stalks of $\mathcal{O}_X$ are local rings. In the geometric case, i.e., if $A = R[T_1, \ldots, T_n]$ and $X = \tilde{M}$ with $M \subset R^n$ a semialgebraic set, we indeed have $\mathcal{O}_X = \tilde{\mathcal{O}}_M$. From now on we call the ringed space $(\tilde{M}, \tilde{\mathcal{O}}_M)$ – instead of just the topological space $\tilde{M}$ – the abstraction of the affine semialgebraic space $(M, \mathcal{O}_M)$.

In the paper [B_4] cited above Brumfiel introduced abstract semialgebraic functions as a tool to prove a vast generalization of Mahé's theorem on the separation of connected components by global quadratic forms. For every commutative ring A there is a natural homomorphism from the Witt ring $W(A)$ to the orthogonal K-group $KO_0(\text{Sper } A)$ of the real spectrum of A. Brumfiel proves that both the kernel and the cokernel of this homomorphism are 2-primary torsion groups. Thus, from our viewpoint, the localization $2^{-\infty}W(A)$ of $W(A)$ at the prime 2 is a purely topological object.

Brumfiel's paper is a bold step into the realm of abstract semialgebraic topology. A full understanding of it is a challenge even today, since some arguments are only sketched. For a discussion cf. [K, §6], and for a treatment in the geometric case cf. [BCR, 15.3].

N. Schwartz studied in [S] the spaces $(X, \mathcal{O}_X)$, with X a constructible subset of some real spectrum Sper A, for their own sake. The ring $\mathcal{O}(X)$ of global sections of $\mathcal{O}_X$ is a sort of "real closure" of the ring A. Schwartz describes how to obtain $\mathcal{O}(X)$ from the ring A in a constructive way. He further makes the important discovery that the natural map from X to Sper $\mathcal{O}(X)$ is an embedding which makes X a dense subspace of Sper $\mathcal{O}(X)$. Even more is true: the closed points and also the minimal points of Sper $\mathcal{O}(X)$ all lie in X. In the special case that X is convex in Sper A with respect to specialization, it turns out that the ringed spaces $(X, \mathcal{O}_X)$ and Spec $\mathcal{O}(X)$ are equal. In the geometric case $X = \tilde{M}$ this happens iff the semialgebraic set $M \subset R^n$ is locally closed in R^n.

Later Schwartz realized that all we have said above about $(X, \mathcal{O}_X)$ remains true if X is a *proconstructible* subset of Sper A, i.e., the intersection of an arbitrary family of constructible subsets of X [S1], [S2]. He called any ringed space isomorphic to such a space $(X, \mathcal{O}_X)$ an *affine real closed space.* He then introduced the category $\mathcal{R}$ of *real closed spaces* as a full subcategory of the category of all locally ringed spaces. The definition of a real closed space is simple: a ringed space $(X, \mathcal{O}_X)$ – always with X a genuine topological space, no Grothendieck topology – is called real closed if every point $x \in X$ has an open neighbourhood U such that $(U, \mathcal{O}_X|U)$ is an affine real closed space.

The books [S1], [S2] are both versions of Schwartz's Habilitationsschrift [S]. For the insiders they constitute a sort of bible of abstract semialgebraic topology - an incomplete bible, I should add, since more can and should be written down with the methods developed there. The shorter version [S1] is easier to read, while [S2] is closer to the original Habilitationsschrift and contains much more material.

In [S] Schwartz defined real closed spaces using as building blocks only constructible subsets of real spectra, instead of proconstructible ones. I will call these more special ringed spaces here *"abstract locally semialgebraic spaces"* and denote their category by $\mathcal{R}_0$. The analogy with locally semialgebraic spaces over a real closed field R is striking. But there is more than analogy. One can attach to any locally semialgebraic space $(M, \mathcal{O}_M)$ over R an abstract locally semialgebraic space $(\tilde{M}, \tilde{\mathcal{O}}_M)$ in a rather obvious way, starting from the abstractions of affine semialgebraic spaces discussed above. Schwartz proves that this gives an embedding of $LSA(R)$ into the category $\mathcal{R}_0$, making $LSA(R)$ a full subcategory of the category of abstract locally semialgebraic spaces over Sper R [S], [S1], [S2]. A good thing about $\mathcal{R}$ is that here more constructions - in particular more quotients - are possible than in $LSA(R)$.

In view of what has been said above about affine real closed spaces, it is clear that real closed spaces are close to schemes and many of them are in fact schemes. This constitutes a rather thorough algebraization of semialgebraic topology, since the notion of schemes originates from polynomial equalities and non-equalities ($f = 0, f \neq 0$) instead of inequalities ($f \geq 0, f > 0$).

The books [S_1], [S_2] give ample evidence that scheme theoretic notions and techniques work well in the category $\mathcal{R}$. The books are close in spirit to the foundations of Grothendieck's abstract algebraic geometry [EGA I], [EGA I*]. In particular, the transition from a locally semialgebraic space to its abstraction is fully analogous to the transition from an algebraic variety over an algebraically closed field to the associated scheme. Of course, in many respects [S_1] and [S_2] are simpler than Grothendieck's theory, since here only "topological" phenomena have to be captured. This is already reflected by the fact that no nilpotent elements occur in the structure sheaves.

It also pays well to pass from the category $\mathcal{R}_0$ to $\mathcal{R}$. For example, a finite subset of a real spectrum $\operatorname{Sper} A$ is always proconstructible but only rarely constructible. This implies that for a real closed space $(X, \mathcal{O}_X)$ every finite subset S of X gives us a "subspace" $(S, \mathcal{O}_S)$ of $(X, \mathcal{O}_X)$ which is again real closed. Especially useful are the two-point spaces $S = \{\xi, x\}$ coming from the real spectrum $\operatorname{Sper} \mathfrak{o}$ of a convex subring $\mathfrak{o}$ of a real closed field K with ξ the minimal point ("general point") and x the closed point ("special point") of $\operatorname{Sper} \mathfrak{o}$. {Recall that $\mathfrak{o}$ is a valuation ring.} These two-point spaces occur in valuative criteria for various properties of morphism between real closed spaces, cf. §13 A below. By the way, $\operatorname{Sper} \mathfrak{o} = \operatorname{Spec} \mathfrak{o}$ as a ringed space.

My student Michael Prechtel has given an interesting classification of all real closed spaces which contain only finitely many points [P].

§7 Cohomology with supports

Starting with this section, a *geometric space* means a regular paracompact locally semialgebraic space over some real closed field R, and an *abstract space* means a regular paracompact abstract locally semialgebraic space.

A word of explanation for these terms: A real closed space X is called *regular* if the specializations of a point $x \in X$ form a chain. X is called *paracompact* if X has a locally finite covering $(X_i | i \in I)$ by affine open subsets X_i.

There are several reasons why this terminology makes sense. On the one hand the abstraction $\tilde{M}$ of a locally semialgebraic space M over some real closed field R is regular if M is regular and is paracompact if M is paracompact. Thus the abstraction of a geometric space is an abstract space. On the other hand, if X is an abstract space then the subspace $X^{\max}$ of closed points of X is a paracompact topological space [D_2, Chap. I].

Abstract spaces are very amenable to sheaf cohomology with supports, as defined in classical topology (cf. e.g. [Bre]). This has been amply demonstrated by Delfs in Chapter II of his Habilitationsschrift [D_2]. Recall that if Φ is a family of supports on X (= antifilter of closed subsets), and $\mathcal{F}$ is a sheaf on X (as always, with values in abelian groups), then $\Gamma_\Phi(X,\mathcal{F})$ denotes the group of sections $s \in \mathcal{F}(X)$ with support $\operatorname{supp} s \in \Phi$, and $\mathcal{F} \mapsto H^q_\Phi(X,\mathcal{F})$ is the q-th derived functor of the left exact functor $\mathcal{F} \mapsto \Gamma_\Phi(X,\mathcal{F})$.

There is a close connection with the sheaf cohomology on the subspace $X^{\max}$, provided $X^{\max}$ is dense in X and every point of X has a specialization in $X^{\max}$ (which holds in many applications). Then we have a canonical continuous retraction $r: X \to X^{\max}$. The direct image functor r_* from the category of sheaves on X to the category of sheaves on $X^{\max}$ is easily seen to be equal to the inverse image functor i^* for i the inclusion $X^{\max} \hookrightarrow X$. For every sheaf $\mathcal{F}$ on X this gives us canonical isomorphisms $H^q_\Phi(X,\mathcal{F}) \cong H^q_\Psi(X^{\max}, i^*\mathcal{F})$, and for every sheaf $\mathcal{G}$ on X, canonical isomorphisms $H^q_\Psi(X^{\max},\mathcal{G}) \cong H^q_\Phi(X, i_*\mathcal{G})$, with Ψ denoting the set of intersections $\{A \cap X^{\max} | A \in \Phi\}$. In the case that X is affine (or, more generally, a "normal spectral space") this important observation goes back to Carral and Coste [CC].

If M is a geometric space then we have a canonical isomorphism $\mathcal{F} \mapsto \tilde{\mathcal{F}}$ from the category of sheaves on M to the category of sheaves over $\tilde{M}$, as explained above in the special case that M is affine (§4). A *family of supports* Φ *on* M is by definition an antifilter of closed subspaces of M. We can define cohomology groups $H^q_\Phi(M,\mathcal{F})$ in much the same way as in the classical theory. $\{\Gamma_\Phi(M,\mathcal{F})$ is the group of all $s \in \mathcal{F}(M)$ such that $s|M - A = 0$ for some $A \in \Phi$.$\}$ Let $\tilde{\Phi}$ denote the antifilter of closed subsets of $\tilde{M}$ generated by the abstractions $\tilde{A}$ of all $A \in \Phi$. Then it is evident that

$$H^q_\Phi(M,\mathcal{F}) = H^q_{\tilde{\Phi}}(\tilde{M},\tilde{\mathcal{F}})$$

for all $q \geq 0$.

There are many useful families of supports Φ, even more than in classical topology. In particular we can choose for Φ the family of all complete subspaces of M, "complete" being defined as in §2. We denote this family by c, suppressing its dependence on M in the notation. In the abstract setting we also have the notion of complete spaces. Here, for X an abstract space, we denote by c the antifilter of closed subsets generated by the complete subspaces of X. The groups $H^q_c(M,\mathcal{F})$ resp. $H^q_c(X,\mathcal{F})$ are the analogues of the cohomology groups with compact support in classical topology. For M a geometric space we have $H^q_c(M,\mathcal{F}) = H^q_c(\tilde{M},\tilde{\mathcal{F}})$.

In Chapter II of [D_4] Delfs develops the formal theory of sheaf cohomology with supports for geometric and abstract spaces virtually to the same extent as the classical theory (e.g. [Bre]). Some important topics are omitted, in particular cup products, but it is evident that these topics can be dealt with by the same methods. Delfs has often talked to me about such things in very clear terms.

As a sample of his theory I will now say something about the cohomology of fibers [D₂, II §8]. Let $f: M \to N$ be a morphism between geometric spaces and $\mathcal{F}$ a sheaf on M. Let $\tilde{f}: \tilde{M} \to \tilde{N}$ denote the abstraction of f. For every $\alpha \in \tilde{N}$ the fibre $\tilde{f}^{-1}(\alpha)$ is the abstraction of a geometric space over $k(\alpha)$. In particular, if $y \in N$, then $\tilde{f}^{-1}(y) = f^{-1}(y)^{\sim}$. Fix some $q \in \mathbb{N}_0$. For every $\alpha \in \tilde{N}$ we consider the cohomology group $H_c^q(\tilde{f}^{-1}(\alpha), \tilde{\mathcal{F}}|\tilde{f}^{-1}(\alpha))$ which we denote more briefly by $H_c^q(\tilde{f}^{-1}(\alpha), \tilde{\mathcal{F}})$. The question arises as to how these groups are related for varying α. A first answer is that there exists a sheaf $\mathfrak{G}$ on N, namely $\mathfrak{G} = R^q f_! \mathcal{F}$, such that, for every $\alpha \in \tilde{N}$, $H_c^q(\tilde{f}^{-1}(\alpha), \tilde{\mathcal{F}})$ is the stalk of $\tilde{\mathfrak{G}}$ at α,

$$(*) \qquad (R^q f_! \mathcal{F})_\alpha^{\sim} = H_c^q(\tilde{f}^{-1}(\alpha), \tilde{\mathcal{F}}).$$

Here $\mathcal{F} \mapsto R^q f_! \mathcal{F}$ is the q-th derived functor of a left exact functor $\mathcal{F} \mapsto f_! \mathcal{F}$ defined as follows. For U an admissible open subset of N the group $(f_! \mathcal{F})(U)$ consists of all sections s of $\mathcal{F}(f^{-1}(U))$ such that there exists some closed subspace A of $f^{-1}(U)$ such that s vanishes on $M - A$ and the restriction $f|A: A \to N$ is proper. It follows from (∗) that, for $y \in N$,

$$(**) \qquad (R^q f_! \mathcal{F})_y = H_c^q(f^{-1}(y), \mathcal{F}).$$

The equations (∗) and (∗∗) throw some light on the relationship between the geometric and the abstract setting. In the geometric theory we are interested in the groups $H_c^q(f^{-1}(y), \mathcal{F})$ for $y \in N$. But it is evident from (∗) and (∗∗) that the groups $H_c^q(\tilde{f}^{-1}(\alpha), \tilde{\mathcal{F}})$ with $\alpha \in \tilde{N}$, not necessarily $\alpha \in N$, have an influence on these groups.

§8 Borel-Moore homology

Again, let M be a geometric space and Φ a family of supports on M. Let G_M be the constant sheaf on M attached to some abelian group G. We denote the groups $H_\Phi^q(M, G_M)$ more briefly by $H_\Phi^q(M, G)$. If Φ is the set of all closed subspaces of M, henceforth denoted by cld (= "closed"), then these groups are just the semialgebraic cohomology groups $H^q(M, G)$ of Delfs discussed in §4.

It would also be desirable to have at hand homology groups $H_q^\Phi(M, G)$ with support in Φ. This is a problem even for $\Phi = \text{cld}$. It is intuitively clear that the group $H_q(M, G)$, as described in §5, should coincide with $H_q^c(M, G)$, and not with $H_q^{\text{cld}}(M, G)$, since a simplicial chain, as used there, clearly has support in a complete subspace of M. {By the way, every complete subspace of M is affine semialgebraic, cf. §13 A }.

In classical topology homology groups with closed support have been defined by Borel and Moore for locally compact spaces [BM], [Bre, Chap.V]. They used a rather complicated sheaf theoretic procedure, but their groups turned out to be very useful.

Delfs found out that, for M a locally complete geometric space, such Borel-Moore groups ${}^{BM}H_q(M,G)$ can be defined in a down-to-earth combinatorial manner as follows [D₂, Chap. III], [D₃]. We choose a triangulation $\varphi: X \xrightarrow{\sim} M$. Then X is a locally finite and locally closed simplicial complex over the base field R. This means that every open simplex of X is the face of only finitely many open simplices of X, and that X is open in $\bar{X}$. {Recall that we use the notion "simplicial complex" in a non-classical sense (§3). $\bar{X}$ is a simplicial complex in the classical sense.}

All the combinatorial technique is based on open simplices instead of closed ones. We choose a (say) total ordering of the vertices of $\bar{X}$ and then can talk about oriented open simplices. For pairwise different vertices $x_0, \dots, x_q$ of $\bar{X}$ let $\langle x_0, \dots, x_q \rangle$ denote the open oriented q-simplex with these vertices and the orientation given by the sequence $x_0, \dots, x_q$. We define a chain complex ${}^{BM}C.(X,G)$ as follows. ${}^{BM}C_q(X,G)$ consists of all formal sums $\sum_\sigma m_\sigma \sigma$ with σ running through the set of positively oriented open simplices of X and $m_\sigma \in G$. The boundary operator is given by the classical rule

$$\partial(\langle x_0, \dots, x_q \rangle) = \sum_{i=0}^{q}{}' (-1)^i \langle x_0, \dots, \widehat{x_i}, \dots x_q \rangle,$$

but omitting on the right hand side all open simplices which do not lie in X. The fact that X is locally closed guarantees that $\partial \circ \partial = 0$. ${}^{BM}H_q(M,G)$ is defined as the q-th homology group of the chain complex ${}^{BM}C.(X,G)$.

Delfs verifies that these "semialgebraic Borel-Moore homology groups" do not depend on the choice of the triangulation [D₂, III.2.1]. They have the formal properties one is used to from topological Borel-Moore homology, and in the case $R = \mathbb{R}$ they coincide with the classical Borel-Moore groups.

From now on we replace G by a principal ideal domain Λ. If M is an n-dimensional Λ-oriented paracompact locally semialgebraic manifold over R (this defined in a rather obvious way [D₂, III §3]), and $\varphi: X \xrightarrow{\sim} M$ is a triangulation, then the sum Z of all n-dimensional open suitably oriented simplices of X is an element of ${}^{BM}C_n(M,\Lambda)$ with boundary $\partial Z = 0$. The cap product with the "fundamental class" $[Z] \in {}^{BM}H_n(M,\Lambda)$ gives an isomorphism from $H^q(M,\Lambda)$ onto ${}^{BM}H_q(M,\Lambda)$ for every q, as in classical Poincaré duality theory [D, III §9].

If M is the space of real points $V(R)$ of an n-dimensional algebraic variety V over R then Delfs defines again a fundamental class $\zeta_M \in {}^{BM}H_n(M,\mathbb{Z}/2)$ in a similar way [D₃]. ζ_M is characterized by the property that its restriction to any connected component N_i of the open subset $V_{reg}(R)$ of regular points is the unique non-zero element of ${}^{BM}H_n(N_i,\mathbb{Z}/2)$. {In the case that M is complete this property had been verified before by other methods in [DK₁].} Delfs also constructs a fundamental class for $\Lambda = \mathbb{Z}$ and $M = V = V(R(\sqrt{-1}))$ with V an algebraic variety (more generally, an "isoalgebraic space") over $R(\sqrt{-1})$. Such fundamental classes have been introduced for real and complex analytic spaces by Borel and Haefliger, and used by them to establish a topological intersection theory on these spaces [BH].

Let again M be a locally complete space and Φ be a family of supports on M. In order to define homology groups with supports in Φ and, for M an oriented manifold, to establish Poincaré-duality with the cohomology groups with supports in Φ, one probably can use still combinatorial methods since we have a strong triangulation theorem at hand, but this would be more complicated than in the case Φ = cld. Instead of this Delfs returns to sheaf theoretic methods similar to those used in the classical theory [BM], [Bre]. Sheafifying the Borel-Moore chain complex ${}^{BM}C.(M,\Lambda)$ he obtains a complex $\Delta.$ of sheaves of simplicial chains. For $\mathcal{F}$ a sheaf of Λ-modules he defines $H_q^\Phi(M,\mathcal{F})$ as the q-th homology groups of $\Gamma_\Phi(M,\Delta.\otimes_\Lambda\mathcal{F})$ [D_2, Chap. III]. For G a Λ-module and $\mathcal{F}$ the constant sheaf G_M this group coincides with ${}^{BM}H_q(M,G)$ if Φ = cld, and with the semialgebraic homology group $H_q(M,G)$ if $\Phi = c$.

Being able to vary Φ and $\mathcal{F}$, Delfs has enough flexibility to establish Poincaré-duality and semialgebraic intersection theory, the latter for $M = V(R)$, $\Lambda = \mathbb{Z}/2$, resp. $M = V(R(\sqrt{-1}))$, $\Lambda = \mathbb{Z}$, with V an algebraic manifold over R or $R(\sqrt{-1})$ respectively (even an isoalgebraic manifold in the latter case) [D_2].

In particular Delfs obtains an important result about algebraic intersection numbers. Let C be an algebraically closed field of characteristic zero. Let V be a smooth algebraic variety over C, and let X_1, X_2 be irreducible subvarieties of V. Let Y be an irreducible component of $X_1 \cap X_2$ with codim Y = codim X_1 + codim X_2. Then the algebraic intersection number $i(X_1 \cdot X_2, Y)$ is a well defined natural number. Now choose a real closed field R with $C = R(\sqrt{-1})$, which can of course usually be done in very many different ways. Then $V = V(C)$ is also a geometric space over R, and Borel-Moore homology with $\Lambda = \mathbb{Z}$ gives us a semialgebraic intersection number $i_R(X_1 \cdot X_2, Y)$. Delfs proves that this number coincides with $i(X_1 \cdot X_2, Y)$ for every choice of R.

§9 Base field extension and comparison theorems

We fix a real closed base field R and a real closed field extension K of R. Let M be a semialgebraic subset of R^n. Then we obtain from M a semialgebraic subset $M(K)$ of K^n by "base field extension" as follows. We write

$$M = \bigcup_{i=1}^{r} \{x \in R^n | f_{ij}(x) > 0,\ j = 1, \dots, s(i);\ g_{ik}(x) \geq 0,\ k = 1, \dots, t(i)\}$$

with some polynomials f_{ij}, g_{ik} in n variables over R. Then we define

$$M(K) := \bigcup_{i=1}^{r} \{x \in K^n | f_{ij}(x) > 0,\ j = 1, \dots, s(i);\ g_{ik}(x) \geq 0,\ k = 1, \dots, t(i)\}.$$

It is an immediate consequence of Tarski's principle that $M(K)$ does not depend on the choice of the description of M above.

If $f: M \to N$ is a semialgebraic map between semialgebraic sets over R then, again by Tarski's principle, there exists a well defined semialgebraic map $f_K: M(K) \to N(K)$ whose graph $\Gamma(f_K)$ is obtained from $\Gamma(f)$ by base field extension, $\Gamma(f_K) = \Gamma(f)(K)$.

Thus we have a functor $M \mapsto M(K)$ from the category of affine semialgebraic spaces over R to the analogous category over K. This functor extends in a natural way to a functor $M \mapsto M(K)$ from the category $LSA(R)$ of geometric (= regular paracompact locally semialgebraic) spaces over R to $LSA(K)$ [DK$_2$, Chap. I].

This functor "base field extension from R to K" has an illuminating interpretation in the abstract setting. Let $\mathfrak{S}$ denote the category of abstract spaces (a full subcategory of the category $\mathcal{R}_0$ considered in §6). Recall that the abstraction functor $M \mapsto \tilde{M}$ embeds $LSA(R)$ into $\mathfrak{S}$. In the same way, $LSA(K)$ is a subcategory of $\mathfrak{S}$. In particular $\mathfrak{S}$ contains the one-point ringed spaces Sper R and Sper K. The inclusion $R \hookrightarrow K$ gives us a morphism Sper $K \to$ Sper R. {By the way, Sper R = Spec R and Sper K = Spec K.} In $\mathfrak{S}$ we have fibre products [S$_1$], [S$_2$], and

$$\widetilde{M(K)} = \tilde{M} \times_{\operatorname{Sper} R} \operatorname{Sper} K.$$

For semialgebraic homology $H_*(-, G)$ with values in some abelian group G, the following comparison theorems are rather evident from the simplicial definition of $H_*(-, G)$ (cf. [DK$_2$, III §7] for a more precise version).

First Comparison Theorem. There is a canonical isomorphism

$$H_q(M, G) \xrightarrow{\sim} H_q(M(K), G).$$

Second Comparison Theorem. If $R = \mathbb{R}$ there is a canonical isomorphism

$$H_q(M, G) \xrightarrow{\sim} H_q(M_{\text{top}}, G).$$

Here M_{top} denotes the set M equipped with the topology which has the admissible open subsets of M as a basis, the so called "strong topology". Notice that this is the subspace topology of M in $\tilde{M}$. The groups $H_q(M_{\text{top}}, G)$ are the singular homology groups. There is an important case, namely if M is "partially complete" (cf. §13 A below for this), in which M_{top} coincides with the space $M^{\max}$ of closed points in $\tilde{M}$.

We have similar comparison theorems in sheaf cohomology due to Delfs [D$_2$, Chap. II]. Let Φ be a family of supports on M. It gives us a family of supports $\Phi(K)$ on $M(K)$, namely the antifilter of closed subspaces generated by $\{A(K) | A \in \Phi\}$. In the case $R = \mathbb{R}$ it also gives us a family of supports Φ_{top} on M_{top}, namely the antifilter of closed subsets generated by Φ.

Let $\mathcal{F}$ be a sheaf on M. It gives us a sheaf $\mathcal{F}(K)$ on $M(K)$ by the rule $\mathcal{F}(K)^{\sim} = \pi^*(\tilde{\mathcal{F}})$, with π the natural projection from $M(S)^{\sim} = \tilde{M} \times_{\mathrm{Sper}\, R} \mathrm{Sper}\, K$ to $\tilde{M}$. In the case $R = \mathbb{R}$ we also obtain a sheaf $\mathcal{F}_{top}$ on M_{top} by the rule $\mathcal{F}_{top} = i^*(\tilde{\mathcal{F}})$ with i the inclusion $M_{top} \hookrightarrow \tilde{M}$.

First Comparison Theorem [D2, II.6.1]. There are canonical isomorphisms

$$H^q_\Phi(M, \mathcal{F}) \xrightarrow{\sim} H^q_{\Phi(K)}(M(K), \mathcal{F}(K)).$$

Second Comparison Theorem [D2, II §5]. Let $R = \mathbb{R}$. Assume either that $\mathcal{F}$ is locally constant or that M is partially complete. Then there are canonical isomorphisms

$$H^q_\Phi(M, \mathcal{F}) \xrightarrow{\sim} H^q_{\Phi_{top}}(M_{top}, \mathcal{F}_{top}).$$

Delfs points out that some restrictive hypothesis in the second theorem is indeed necessary [D2, II.5.6]. One should add that the first theorem also does not always give the result one would like to have. For example, if Φ is the set of all complete subspaces of M, denoted previously by c, then $\Phi(K)$ may be smaller than the set of all complete subspaces of $M(K)$. Nevertheless Delfs proves that there exist canonical isomorphisms

$$H^q_c(M, \mathcal{F}) \xrightarrow{\sim} H^q_c(M(K), \mathcal{F}(K))$$

if M is locally complete [D2, II.6.10].

If M is locally complete then $M(K)$ is locally complete, and, in the case $R = \mathbb{R}$, the space M_{top} is locally compact. Thus one may also look for comparison theorems in Borel-Moore homology with supports and with sheaves as coefficients. Delfs proves such theorems, again with a necessary restriction in the assumptions of the second comparison theorem [D2, III §10 and §11].

§10 Homotopy sets

We fix a real closed base field R. A pair of geometric spaces (M, A) over R consists of a geometric space M and a subspace A of M. A morphism f from (M, A) to another pair (N, B) is a morphism $f: M \to N$ with $f(A) \subset B$. {Then the restriction $f|A: A \to B$ is again a morphism [DK2, I.3.2].}

It is now clear how to define a homotopy F between two morphisms $f, g: (M, A) \rightrightarrows (N, B)$; namely, F is a morphism from $(M \times [0,1], A \times [0,1])$ to (N, B) with $F(x, 0) = f(x)$ and $F(x, 1) = g(x)$ for every $x \in M$. More generally all the basic notions of the elementary classical homotopy theory make sense in the category $LSA(R)$. But notice that we do not have something like spaces of morphisms at our disposal.

Let $[(M,A),(N,B)]$ denote the set of homotopy classes of morphisms from (M,A) to (N,B). One would like to know as much as possible about it. In the case that $B = \{x\}$ is a one-point set and $(M,A) = (S^q(R),\{\infty\})$, with $S^q(R)$ the unit sphere in R^{q+1} and ∞ its north pole, we equip this set with a group structure, in the classical way which is of semialgebraic nature. This gives us the homotopy group $\pi_q(N,x)$. It is abelian for $q \geq 2$.

For homotopy sets we have the following two comparison theorems.

First Comparison Theorem. If K is a real closed overfield of R then the natural map $[f] \to [f_K]$ from $[(M,A),(N,B)]$ to $[(M(K),A(K)),(N(K),B(K))]$ is bijective.

Second Comparison Theorem. If $R = \mathbb{R}$ then the natural map from $[(M,A),(N,B)]$ to the set of homotopy classes of continuous maps $[(M_{\text{top}},A_{\text{top}}),(N_{\text{top}},B_{\text{top}})]$ is bijective.

All this can be proved by using Tarski's principle, triangulations, and simplicial approximation techniques [DK$_2$, Chap. III]. More generally such theorems hold for relative homotopy classes, where a subspace of M has to be fixed pointwise by the homotopies [loc. cit.].

If $B = \{x\}$ and $(M,A) = (S^q(R),\{\infty\})$ then the bijections in the theorems are, of course, group isomorphisms. The theorems now tell us what the homotopy groups of a pointed geometric space (N,x) over R look like "in principle". To be more precise, we choose a triangulation $\varphi: X \xrightarrow{\sim} N$ such that $\varphi^{-1}(x)$ is a vertex e of X. Let us assume for simplicity that N is "partially complete" (cf. §13 A below), a property which forces the simplicial complex to coincide with its closure $\bar{X}$. Then X is the realization $|L|_R$ of a locally finite abstract simplicial complex L (in the classical sense [Spa, Chap. 3]). Let R_0 denote the real closure of $\mathbb{Q}$ with respect to its unique ordering. This field has a unique embedding into any other real closed field. Of course, $|L|_R = |L|_{R_0}(R)$. Now the isomorphism φ and the two comparison theorems give us group isomorphisms

$$\pi_q(N,x) \cong \pi_q(|L|_R,e) \cong \pi_q(|L|_{R_0},e) \cong \pi_q(|L|_{\mathbb{R}},e) \cong \pi_q(|L|_{\mathbb{R},\text{top}},e).$$

The last group, a classical homotopy group, can be written - somewhat tautologically - as the homotopy group $\pi_q(L,e)$ of the abstract simplicial complex L.

§11 Weakly semialgebraic spaces, and quotients

The comparison theorems stated in the last section, and analogous theorems for triples, quadruples, etc. of geometric spaces instead of pairs, allow us to transfer a considerable amount of classical homotopy theory to the category of geometric spaces $LSA(R)$, cf. [DK$_2$, III §6] for examples.

Nevertheless, homotopy theory in $LSA(R)$ is hampered by the fact that many of the constructions used in the classical theory do not have counterparts here. In particular, quotients by seemingly innocent equivalence relations can cause problems. For example, CW-complexes exist in $LSA(R)$ only to a limited extent. Even the (reduced) suspension SM of a pointed geometric space (M, x_0) usually does not exist, since it may be impossible to contract the closed subspace $M \times \{0\} \cup M \times \{1\} \cup \{x_0\} \times [0,1]$ of $M \times [0,1]$ to a point in some universal way. Thus we cannot start stable homotopy theory in $LSA(R)$.

Fortunately there is a way out of many of these difficulties. As has been shown in [K$_2$], we can embed $LSA(R)$ as full subcategory in the category $WSA(R)$ of "weakly semialgebraic spaces" over R, where homotopy theory is more pleasant.

A *weakly semialgebraic space* $M = (M, \mathcal{O}_M)$ over R is a set M equipped with a Grothendieck topology, consisting of admissible open subsets and admissible open coverings, and a structure sheaf $\mathcal{O}_M$, consisting of R-valued functions, such that M arises from affine semialgebraic spaces by gluing along *closed* subspaces carefully observing certain rules, cf. [K$_2$, p.3 f]. As for locally semialgebraic spaces, a morphism $(M, \mathcal{O}_M) \to (N, \mathcal{O}_N)$ between such ringed spaces is determined by the underlying map $f: M \to N$, and will henceforth be identified with this map (cf. [K$_2$, IV §1 for the definition of morphisms).

In $WSA(R)$ there exist fibre products. Thus notions like proper morphisms and complete spaces make sense here. It turns out that every complete weakly semialgebraic space is a complete affine semialgebraic space, hence isomorphic to a polytope (= finite polyhedron), [K$_2$, p. 44] and [DK$_2$, p. 59 f].

Of particular importance for homotopy theory are the weakly semialgebraic spaces which can be obtained by gluing *complete* affine spaces along closed subspaces. They are called *weak polytopes.* Every weakly semialgebraic space is homotopy equivalent to a weak polytope [K$_2$, V §4].

Let T be an equivalence relation on a weakly semialgebraic space M. Under which conditions can we expect that the quotient M/T exists in a reasonable sense? Certainly we should assume that T is a closed subspace of $M \times M$. Let $p_T: T \to M$ denote the natural projection from T to the first factor in $M \times M$. For M affine semialgebraic and p_T proper ("proper equivalence relation") Brumfiel has proved the important theorem that a "strong" quotient M/T exists [B_5]: the set theoretic quotient M/T can be equipped in a unique way with the structure of an affine semialgebraic space such that the projection map $\pi_T: M \to M/T$ is "identifying" in a strong sense. This implies, in particular, that π_T is the categorical quotient of M by T in the category of affine semialgebraic spaces.

In $WSA(R)$ we can do even better. As an easy consequence of Brumfiel's theorem such a strong quotient $\pi_T: M \to M/T$ exists if $p: T \to M$ is *partially proper*, i.e. the restriction of p to any closed semialgebraic subspace of T is proper [K_2, IV §11].

At first glance, partially proper equivalence relations might not look much more general than proper ones. But in fact quotients by partially proper equivalence relations suffice for many important topics in homotopy theory. For instance, if A is a closed subspace of a weakly semialgebraic space M, and if $f: A \to N$ is a partially proper map, then we have a space $M \cup_{A,f} N$ at our disposal, obtained by gluing M to N along A via f.

In particular take M to be a direct sum of closed balls, and A the boundary ∂M, which is a direct sum of spheres. This is the essential step to build CW-complexes. Notice that now A is a weak polytope, and every morphism from a weak polytope to a weakly semialgebraic space is partially proper.

As a result we can construct CW-complexes and, more generally, relative CW-complexes in $WSA(R)$ in much the same way and for much the same purposes (e.g. killing of homotopy groups) as one is used to in classical topology.

If (M, x_0) is a pointed weak polytope then we obtain the suspension SM by choosing $A = M \times \{0\} \cup M \times \{1\} \cup \{x_0\} \times [0,1]$ and f as the morphism from A to the one point space $*$. Thus we can start stable homotopy in the full subcategory $\mathcal{P}(R)$ of $WSA(R)$ whose objects are the weak polytopes over R.

If M is not a weak polytope then also A is not a weak polytope, and this means that the morphism $A \to *$ is not partially proper. It turns out, that then SM does not exist as a strong quotient of $M \times [0,1]$ in the sense indicated above. One can conclude from [Sch, §3] that SM does not exist even as a categorical quotient, at least if M is a geometric space.

The existence of strong quotients by partially proper equivalence relations seems to be a best possible result. Indeed, C. Scheiderer has proved a "converse" to Brumfiel's theorem [Sch]: If T is a closed semialgebraic equivalence relation on a locally complete semialgebraic space M, and the strong quotient M/T exists, then, up to obvious modifications, T is proper. Scheiderer's method is very interesting. He makes essential use of abstract spaces and applies an extension of Tarski's principle to real closed fields with convex valuation rings [loc. cit.].

Weakly semialgebraic spaces often cannot be triangulated, but "patch techniques" have been developed to work with them nearly as easily as with triangulable spaces, cf. [K_2, Chap. V].

For every real closed overfield K of R we have a natural base field extension functor from $WSA(R)$ to $WSA(K)$ which extends the base field extension functor from $LSA(R)$ to $LSA(K)$ discussed above (§9). It turns out that the two comparison theorems for homotopy sets from above (§10) remain true for weakly semialgebraic spaces.

Thus nothing is lost and much is gained in homotopy theory by passing from geometric spaces to weakly semialgebraic spaces.

In classical homotopy theory one often works in the category of topological spaces which are homotopy equivalent to CW-complexes, or variants of this category, which all look a little artificial, since a topological space may carry many structures of a CW-complex (cell decompositions), none of which is given in an intrinsic way. A nice fact about $WSA(R)$ is that every weak polytope is homotopy equivalent to a CW-complex [K_2, V §7]. Thus here the category $\mathcal{P}(R)$ is a satisfactory basic category for homotopy theory, the notion of a weak polytope being intrinsic and quite natural, cf. §13 A below.

What are the "abstractions" of weakly semialgebraic spaces? Do they exist? Recently N. Schwartz gave a fascinating solution to this problem by inventing his *"inverse real closed spaces"* [S_3]. In some sense the approach of Schwartz to inverse real closed spaces is simpler than my approach to weakly semialgebraic spaces in [K_2], and thus throws new light on them. Inverse real closed spaces have passed early tests of usefulness successfully. For example they are instrumental in the homotopy classification of vector bundles and the definition of Stiefel-Whitney classes on real closed spaces (which helps to understand the old paper [B_4] of Brumfiel, in particular) [Schwartz, talk at the Ragsquad seminar in Berkeley, October 1990]. I refrain here from a discussion of inverse real closed spaces, since this subject is in a less mature state than the other topics dealt with in this article. Instead I refer the reader to the paper [S_3].

§12 Generalized homology

Let $\mathcal{P}^*(R)$ denote the category of weak polytopes with base points over R. In this category the direct sum of an arbitrary family of objects $(M_\lambda, \lambda \in \Lambda)$ is the wedge $\bigvee_{\lambda\in\Lambda} M_\lambda$, obtained from the disjoint union $\bigsqcup_{\lambda\in\Lambda} M_\lambda$ by identifying the base points of all the M_λ. We also have the suspension functor $S: \mathcal{P}^*(R) \to \mathcal{P}^*(R)$. We pass from $\mathcal{P}^*(R)$ to the associated homotopy category $H\mathcal{P}^*(R)$. It has the same objects as $\mathcal{P}^*(R)$, but the morphisms are the homotopy classes of morphisms in $\mathcal{P}^*(R)$. Now we have all the tools we need to define generalized homology theories as one does in classical topology. There one uses the homotopy category $H\mathfrak{W}^*$ of pointed topological CW-complexes instead of $H\mathcal{P}^*(R)$.

Definition (cf. [Sw, p. 109 f]). A *reduced semialgebraic homology theory* over R is a family $(k_n | n \in \mathbb{Z})$ of covariant functors k_n from $H\mathcal{P}^*(R)$ to the category Ab of abelian groups, together with a family $(\sigma_n | n \in \mathbb{Z})$ of natural equivalences $\sigma_n: k_n \xrightarrow{\sim} k_{n+1} \circ S$, such that the following two axioms hold:

Exactness Axiom. For every pair (M, A) of pointed weak polytopes over R, the sequence

$$k_n(A) \longrightarrow k_n(M) \longrightarrow k_n(M/A),$$

induced by the inclusion $A \hookrightarrow M$ and the projection $M \to M/A$, is exact.

Wedge Axiom. For every family $(M_\lambda | \lambda \in \Lambda)$ of pointed weak polytopes, and every $n \in \mathbb{Z}$, the map

$$\bigoplus_{\lambda \in \Lambda} k_n(M_\lambda) \longrightarrow k_n(\bigvee_{\lambda \in \Lambda} M_\lambda),$$

induced by the inclusions $M_\lambda \hookrightarrow \bigvee_{\mu \in \Lambda} M_\mu$, is an isomorphism.

Applying the homotopy theory developed in [K$_2$] one proves that every reduced homology over R "extends" in an unique way to a non reduced homology theory $(h_n | n \in \mathbb{Z})$ on the category of all pairs of weakly semialgebraic spaces over R. Such unreduced homology theories are again defined in an axiomatic way analogous to classical topology, in other words, the Eilenberg-Steenrod axioms [ES, I §3] - with the exception of the dimension axiom - have to hold, cf. [K$_2$, VI §4]. The relationship between h_n and k_n is given by $h_n(M, A) = k_n(M/A)$ for any pair (M/A) of weak polytopes, the space M/A being equipped with its natural base point A/A. {Here I suppress a consideration of the boundary maps $\partial_n: h_n(M, A) \to h_{n-1}(A, \emptyset)$, which are present in the axioms of an unreduced homology theory.} It turns out that the functors h_n satisfy a strong excision property [K$_2$, VI.6.10], which is even better than excision in classical topology.

A main result in [K$_2$] about generalized homology is the following: Given a reduced homology theory $(k_n^{\text{top}} | n \in \mathbb{Z})$ on the homotopy category $H\mathfrak{W}^*$ of topological CW-complexes, there exists, for every real closed field R, a reduced homology theory $(k_n^R | n \in \mathbb{Z})$ on $H\mathcal{P}^*(R)$ such that two comparison theorems analogous to those stated in §9 are true (cf. [K$_2$, VI §3] for a more precise statement). One then also has comparison theorems for the associated unreduced homology theories [K$_2$, VI.5.4].

Running parallel to all of this, there is a theory of reduced and unreduced cohomology with results analogous to those just stated.

To appreciate the content of these results let us choose for (k_n^{top}) the reduced singular homology theory with coefficients in some abelian group G. Then we obtain a homology theory (h_n^R) for pairs of weakly semialgebraic spaces over R which satisfies all the Eilenberg-Steenrod axioms. Given, say, a complete geometric space M over R and a triangulation $|L|_R \xrightarrow{\sim} M$, with L a finite abstract simplicial complex, one verifies in the classical way that $h_q^R(M) \cong H_q(L, G)$. Thus we have obtained anew the result of Delfs that (up to canonical isomorphism) the combinatorial homology group $H_q(L, G)$ does not depend on the choice of the triangulation.

In some sense, this new proof of Delfs' result is considerably easier than the two original proofs discussed in §4. Once one knows that weakly semialgebraic spaces exist with suitable formal properties, in particular the two comparison theorems for homotopy sets, the proof works by straightforward homotopy methods, cf. [K_2]. But notice that the new proof gives something less than the old ones: a connection with the sheaf cohomology groups $H^q(M, G_M)$ is missing.

From now on I call the groups $H_q(M, A; G)$ which come from the topological singular homology theory *ordinary homology groups.* They coincide with the semialgebraic homology groups of Delfs discussed in §4 and later.

There exist many prominent homology and cohomology theories in topology. For example, let $(h^n | n \in \mathbb{Z})$ be one of the classical cobordism theories [Sw, 12.24]. Then, by the results above, we have associated groups $h^n(M, A)$ for every pair of weakly semialgebraic spaces (M, A) over some real closed field. {From here on I omit the subscripts R and "top".} Many real algebraic geometers will be interested in these groups only for (M, A) a pair of semialgebraic sets. But it seems to be difficult to understand the groups $h^n(M, A)$ working only with semialgebraic sets or, more generally, with geometric spaces. So weakly semialgebraic spaces serve us well, even if we have no geometric interest in them.

There remains the task of giving a *geometric* interpretation of the groups $h^n(M, A)$ for, say, semialgebraic sets. They can probably be described using some semialgebraic differential topology, of the sort that appears in the *definition* of the topological cobordism groups. The theory in [K_2] does not give such an interpretation. To the best of my knowledge, nobody has tackled this task up to now.

In the category $\mathcal{P}(R)$ of weak polytopes over R there exists the realization $|L|_R$ of any simplicial set L (= "semisimplicial set" in older terminology) [K_2, Chap. VII], with the formal properties one expects from classical topology [Mi], [May]. This makes it possible to solve a problem left open in Delfs' semialgebraic homology theory discussed in §4, namely the question of whether the semialgebraic singular chain complexes give us the ordinary homology groups. For M a weakly semialgebraic space over R let Sin$M$ denote the singular set of M. This is the simplicial set consisting of the semialgebraic maps from the standard simplices over R to M. As in topology one has an obvious morphism $j_M: |\mathrm{Sin}M|_R \longrightarrow M$. One can prove directly that j_M is a homotopy equivalence [K_2, VII §7]. {In topology the analogous map is only a "weak" homotopy equivalence [Mi].} From this it follows immediately that the homology groups of the singular chain complex of M with coefficients in G are the ordinary homology groups $H_q(M,G)$ [loc. cit.].

In particular, we now know that semialgebraic singular homology obeys the excision axiom. It would be desirable to have a more elementary proof of this, say, by nonlinear subdivision of singular chains. Certainly we would learn a lot from this about the geometry behind ordinary homology.

§13 Novel features of semialgebraic topology: three examples.

Much of what has been said up to now gives the impression that semialgebraic topology, at least in the geometric setting, is very similar to traditional topology. The definition of spaces, morphisms, etc. may look a little exotic to the classical topologist, but then the results in homology and homotopy are as he or she is used to.

In fact semialgebraic topology has also features which may be unexpected from the classical viewpoint, even if the base field R is the field $\mathbf{R}$ of real numbers. In the following I give three examples.

A. Proper and partially proper maps

I start with some definitions. Let M be a locally semialgebraic space over R. A subset A of M can carry at most one structure of a subspace of M (cf. §5 above and [DK_2, I §3] for the definition of subspace). If it does we call A a *locally semialgebraic subset* of M. Locally semialgebraic subsets are the only subsets of M which have a geometric meaning. If A carries the structure of a subspace which is even a semialgebraic space we call A a *semialgebraic subset* of M.

If A and B are locally semialgebraic subsets of M with $B \subset A$ and A is semialgebraic then also B is semialgebraic. The image $f(A)$ of a semialgebraic subset A of M under any morphism $f: M \to N$ is a semialgebraic subset of N. All this justifies the idea that "semialgebraic" is something like "small". This notion of smallness is alien to classical topology.

If $f: M \to N$ is a morphism and B is a locally semialgebraic subset of N, then the preimage $f^{-1}(B)$ is a locally semialgebraic subset of M. {It is not true in general that images of locally semialgebraic subsets of M are locally semialgebraic in N}. We call the morphism f *semialgebraic* if the preimages of semialgebraic subsets of

N are semialgebraic, and we call f *affine semialgebraic*, if the preimages of affine semialgebraic subsets of N are affine semialgebraic.

Theorem 1. Every proper morphism $f: M \to N$ is affine semialgebraic.

This theorem has an interesting history. Delfs and I gave an argument which proves that f is semialgebraic [DK$_2$, p. 59 ff]. This reduces the proof of the theorem to the case that N is affine semialgebraic and M is semialgebraic. In this case the theorem is due to N. Schwartz, who proved it by transition to abstract spaces [S], [S$_1$], [S$_2$]. It was one of the really remarkable early applications of his theory of real closed spaces. Then Robson gave a proof within the geometric setting in the special case that N is the one-point space, applying his embedding theorem, cf. [DK$_2$, p. 60]. It was not until much later that R. Huber found a proof in general which stays in the geometric setting. This proof is contained in a joint paper by Huber and Scheiderer [HuS], which contains related results about "locally proper maps", and which, in my opinion, also gives a good feeling for the interplay between the geometric and the abstract setting for such questions. In addition it contains a deepening of Schwartz's result on abstract spaces [S$_1$, p. 83], [S$_2$, V.5.7] which is behind his proof of Theorem 1 for M and N semialgebraic.

If one views a morphism $f: M \to N$ as a "family of spaces" then Theorem 1 may be regarded as a negative result. Properness is commonly regarded as a condition which is often needed to ensure good behaviour in families of spaces. Theorem 1 tells us that we do not get anything new in essence if we study proper morphisms between geometric spaces instead of just affine semialgebraic spaces.

Fortunately there is another notion, alien to classical topology, but related to properness, which still gives us good families of geometric spaces which are not necessarily semialgebraic. A morphism $f: M \to N$ is called *partially proper* if the restriction $f|A: A \to N$ to any closed semialgebraic subset A of M is a proper morphism. A geometric space M is called *partially complete* if the morphism from M to the one-point space is partially proper.

These notions are quite natural, as is illustrated by the following "relative path completion criterion" [DK$_2$, I §6]:

Theorem 2. A morphism $f: M \to N$ between geometric spaces is partially proper iff for every commuting square of morphisms (solid arrows)

$$\begin{array}{ccc}]0,1] & \xrightarrow{\alpha} & M \\ i \downarrow & \overset{\tilde{\alpha}}{\nearrow} & \downarrow f \\ [0,1] & \xrightarrow[\beta]{} & N \end{array}$$

with i the inclusion map, there exists a path $\bar{\alpha}$ (dotted arrow) which extends α. {N.B. Since geometric spaces are "separated" there can exist at most one such path $\bar{\alpha}$, and $\bar{\alpha}$ lies over β, $f \circ \bar{\alpha} = \beta$.}

In the special case that N is the one-point space the theorem tells us that a geometric space M is partially complete iff every "incomplete path" $\alpha\colon]0,1] \to M$ can be completed. Here the classical notion of sequential compactness comes to our mind, but sequences are a blunt instrument in semialgebraic topology. {There even exist real closed fields which do not contain any non-trivial zero sequences.} Incomplete paths may be regarded as the right substitute of sequences in the geometric theory. From this viewpoint partial completeness, and not completeness, is the right analogue of compactness.

By the way, all the definitions in this subsection A make sense more generally for weakly semialgebraic spaces, and both theorems 1 and 2 remain true for them. The partially complete weakly semialgebraic spaces are precisely the weak polytopes discussed in §11.

In the abstract setting the counterparts of path completion and path lifting criteria like Theorem 2 are *valuative criteria* similar to the ones in algebraic geometry. The books of Schwartz [S1], [S2], his paper [S4], and the paper [HuS] by Huber and Scheiderer contain several such criteria. Here the unit interval $[0,1]$ is replaced by the two point subspace $\{\xi, x\}$ of the real spectrum $\mathrm{Sper}\,\mathfrak{o} = \mathrm{Spec}\,\mathfrak{o}$ of a convex subring $\mathfrak{o}$ of some real closed field, with ξ the generic and x the closed point of $\mathrm{Sper}\,\mathfrak{o}$, while $]0,1]$ is replaced by $\{\xi\}$.

B. Different semialgebraic structures on the same classical space

Let M be a locally closed semialgebraic subset of R^n. Then we have on the set M the structure of a locally complete affine semialgebraic space, which we denote again by M.

Let $c(M)$ denote the direct system of all complete semialgebraic subsets K of M. For every $K \in c(M)$ the interior $\mathring{K}$ is again semialgebraic, thus both K and $\mathring{K}$ are affine semialgebraic spaces. We can form the direct limit

$$M_{\mathrm{loc}} = \varinjlim_{K \in c(M)} \mathring{K}$$

in the category $LSA(R)$ [DK2, I.2.6]. M_{loc} is also the direct limit of the spaces K [DK2, I.7.8]. It is a partially complete geometric space over R which has the same underlying set as M. Even $M_{\mathrm{top}} = (M_{\mathrm{loc}})_{\mathrm{top}}$.

Assume that M is not complete. The identity map $M_{\mathrm{loc}} \to M$ is a morphism but by no means an isomorphism. Indeed, a subset A of M is semialgebraic in M_{loc} iff A is semialgebraic and bounded in $\mathbf{R}^n$ and A has a positive distance from the boundary $\bar{M} - M$ of M, provided this boundary is not empty. In particular the space M_{loc} itself is not semialgebraic.

A function $f\colon M \to \mathbf{R}$ is a global section of the structure sheaf of M_{loc} iff f is continuous and, for every $x \in M$, the restriction of f to a suitable semialgebraic neighbourhood of x in M is semialgebraic ("locally semialgebraic functions"). Thus $\mathcal{O}(M_{\text{loc}})$ is a bigger ring than $\mathcal{O}(M)$.

The different nature of M and M_{loc} is perhaps best illustrated by looking at triangulations. For instance, if the set M is an open n-simplex in $\mathbf{R}^n$, then the space M has the tautological triangulation id: $M \to M$, while we obtain a triangulation of M_{loc} by subdividing the set M into infinitely many closed simplices.

If M is an *arbitrary* semialgebraic subset of $\mathbf{R}^n$ then we can still form the inductive limit of the spaces K with $K \in c(M)$ in the category $WSA(\mathbf{R})$ of weakly semialgebraic spaces, and we obtain a weak polytope $P(M)$. If M is locally closed in $\mathbf{R}^n$ the ringed space $P(M)$ equals M_{loc}.

The space $P(M)$ has the same underlying set as M, and the identity map $j_M\colon P(M) \to M$ is a morphism in $WSA(\mathbf{R})$. It has the following universal property: any morphism from a weak polytope to M factors through j_M in a unique way [K$_2$, IV §9].

It is evident from this that j_M induces isomorphisms between the homotopy groups of $P(M)$ and M. By a semialgebraic version of the Whitehead theorem [K$_2$, V.6.10] it follows that j_M is a homotopy equivalence.

Intuitively, replacing M by $P(M)$ means forgetting what happens in M "at infinity".

Everything said in this subsection remains true if we replace $\mathbf{R}$ by a *"sequential"* real closed base field R, i.e. a real closed field which contains at least one non trivial zero sequence. Fortunately, most real closed fields occuring in practice are sequential.

C. Fixed points

Let R be an *arbitrary* real closed field and M a locally closed semialgebraic subset of R^n. Then the direct limit M_{loc} of the spaces $\mathring{K}$ with $K \in c(M)$ still exists in the category of all locally semialgebraic spaces over R. {M_{loc} is regular but perhaps not paracompact.}

It turns out that the abstraction $\tilde{M}_{\text{loc}}$ of M_{loc}, as a locally ringed space, is an open subspace of $\tilde{M}$, namely the union of the open subspaces $(\mathring{K})^\sim$ of $\tilde{M}$ with K running through $c(M)$. It follows that $\tilde{M}_{\text{loc}}$ (as a set) is also the union of the closed subsets $\tilde{K}$ of $\tilde{M}$.

A good understanding of $\tilde{M}_{\text{loc}}$ as a subset of $\tilde{M}$ is possible using ultrafilters of semialgebraic sets. Let $\mathfrak{S}(M)$ denote the boolean lattice of all semialgebraic subsets of M. As is well known, the points of $\tilde{M}$ can be identified with the ultrafilters in $\mathfrak{S}(M)$ [Br, §4], [BCR, 7.2.4], [KS, III §5], namely the ultrafilter $F(\alpha)$ corresponding to a point $\alpha \in \tilde{M}$ consists of all $S \in \mathfrak{S}(M)$ with $\alpha \in \tilde{S}$. In this interpretation $\tilde{M}_{\text{loc}}$ is the set of all $\alpha \in \tilde{M}$ with $K \in F(\alpha)$ for at least one $K \in c(M)$.

Definition. The *fringe* $\text{Frin}(M)$ of M is the set $\tilde{M} - \tilde{M}_{\text{loc}}$.

Intuitively, $\text{Frin}(M)$ consists of the points in $\tilde{M}$ "at infinity".

We now turn to fixed points. Let $f: M \to M$ be a semialgebraic map. It induces $\mathbb{Q}$-linear maps $H_q(f): H_q(M, \mathbb{Q}) \to H_q(M, \mathbb{Q})$ in ordinary (= semialgebraic) homology with coefficients in $\mathbb{Q}$. The trace $Tr H_q(f)$ of $H_q(f)$ is well defined and an integer, since $H_q(f)$ comes from an endomorphism of the finitely generated abelian group $H_q(M, \mathbb{Z})$. We define the *Lefschetz number* $\Lambda(f)$ as usual,

$$\Lambda(f) := \sum_{i=0}^{n} (-1)^i \, Tr H_i(f).$$

Brumfiel has proved the following remarkable theorem ([B$_3$], cf. also [B$_2$]):

Theorem. Assume that $\Lambda(f) \neq 0$. Then the abstraction $\tilde{f}: \tilde{M} \to \tilde{M}$ of f has a fixed point. More precisely, either $f: M \to M$ has a fixed point or $\tilde{f}$ has a closed fixed point in $\text{Frin}(M)$.

Example. Let $R = \mathbb{R}$, $M = \mathbb{R}^2$, $f(x, y) = (x + 1, xy)$. Certainly $f: M \to M$ has no fixed point. But $\Lambda(f) = 1$, since M is contractible. Thus $\tilde{f}$ must have at least one closed fixed point in $\text{Frin}(\mathbb{R}^2)$. Such a fixed point is provided by the graph of the gamma function $\Gamma(x)$. Let F denote the set of all $A \in \mathfrak{S}(M)$ such that A contains the set $\{(x, \Gamma(x)) | x \geq c\}$ for some $c > 0$. This is an ultrafilter which gives us a closed point $\alpha \in \text{Frin}\,(M)$ fixed under $\tilde{f}$.

Brumfiel's fixed point theorem is important for his "real spectrum compactification" $\widehat{M_g}$ of the Teichmüller space M_g of compact Riemann surfaces of genus g [B$_1$], since it implies that every element of the Teichmüller modular group has a fixed point in $\widehat{M_g}$.

§14 An outlook

What can be said about the situation in semialgebraic topology now, in the year 1991, and about prospects for the future, without too much speculation?

Certainly abstract spaces contain a rich potential for further research. They should be "superior" to geometric spaces. But up to now nearly all the deeper results in homology and homotopy theory rely heavily on arguments in the geometric category, using triangulations, simplicial approximations, the homotopy extension property and the like.

It would be highly desirable to have a homotopy theory for real closed spaces. There can be no doubt that we have found the "right" homotopy groups in the geometric setting. But all this does not tell us how to proceed in the abstract setting. It is only clear that an abstract homotopy theory should give us the geometric theory if we apply it to the abstractions of geometric spaces.

For abstract cohomology the analogous problem seems to be easier. There exists an approach by N. Schwartz using inverse real closed spaces which looks promising. Schwartz talked about this at Luminy in October 1989, and then in the Ragsquad seminar at Berkeley in October 1990.

In abstract homotopy one could proceed along the lines given in chapter II of Baues' book [Ba], where the homotopy theory of a "cofibration category" has been developed. Or one could try to associate with a real closed space X (perhaps fulfilling some conditions like regularity or "tautness" [D_2, I §3]) a simplicial set or, perhaps better, something like a simplicial space VX (in the classical sense, with Hausdorff topology), such that the homotopy type of VX encodes the homotopy information about X.

In his talk at Oberwolfach in June 1990 C. Scheiderer proposed a simplicial space VX (together with a "quasiaugmentation" $VX \to X$) which looks promising. The definition of VX employs chains of real valuations of commutative rings. Scheiderer was able to verify that, for any sheaf $\mathcal{F}$ on X, the cohomology groups $H^q(X, \mathcal{F})$ can be computed from VX. Also, for X the abstraction of a geometric space M, VX represents the homotopy type of M. These are hints that VX gives the right homotopy type.

In the last years the interest in abstract spaces also has been enhanced by the discovery of the Spanish school (Andradas, Ruiz, ...), that questions in semianalytic geometry can be treated by considering real spectra. I just mention two basic observations: If A is a local ring of real analytic functions then the constructible subsets of Sper A are in natural one-to-one correspondence with the germs of semianalytic sets [Rz]. If A is the ring of global analytic functions on a compact real analytic manifold M then the constructible subsets of Sper A correspond bijectively with the globally semianalytic subsets of M [Rz_1]. The homotopy or homology of such a constructible set (which is a real closed space) should have a close relation to the homotopy resp. homology of the corresponding semianalytic object.

Concerning the two trends in the last ten years mentioned above (§4) we can safely say that the first one has run its course. Interaction between the geometric and the abstract theory is no longer a "trend" but a well established and widely used technique. However the first trend may continue in a wider context if we allow other meanings for the word "geometric".

Geometric spaces might be built, for instance, from semianalytic sets, or perhaps even from subanalytic sets. J. Denef and L. van den Dries have found a new approach to subanalytic sets [DvdD] which gives much hope that subanalytic sets are amenable to techniques similar to the ones we now have in the geometric semialgebraic topology. Subanalytic sets (more precisely, subsets of $\mathbf{R}^n$ which are subanalytic in $\mathbf{P}^n(\mathbf{R})$) and semialgebraic sets can both be subsumed under *o-minimal structures*, a notion stemming from model theory, cf. [PiS]. In his talk in the Ragsquad seminar at Berkeley, April 1991, van den Dries has outlined a "tame topology of o-minimal structures" with many of the features we are used to for semialgebraic sets, in particular N° 1 - N° 5 of the list in §3 for the geometric theory (cf. his forthcoming book [vdD], as soon as it appears). Thus o-minimal structures seem to provide a good framework for highly interesting new geometric spaces in the years to come.

In 1990 R. Huber introduced "semirigid functions" [Hu_1] as an offspring of his abstract approach to rigid analytic geometry [Hu]. They give us semirigid sets, which are vaguely analoguous to semianalytic sets in real analytic geometry. To understand these sets well, one definitely needs abstract spaces which are derivates of the *real valuation spectra* of commutative rings [Hu_1].

This, and the frequent occurrence of valuations in the theory of real closed spaces, are hints that valuation spectra will play a role in a further development of semialgebraic topology, for which the word "semialgebraic" is probably no longer appropriate. An introduction to valuation spectra has been given in [Hu, Chap. I] and [HuK].

From all of this it is pretty clear that the second trend mentioned in §4, i.e., the employment of new spaces - both geometric and abstract, will persist.

Acknowledgements. I thank Roland Huber, Claus Scheiderer, Colm Mulcahy, and Victoria Powers for helpful criticism, both in mathematics (R.H., C.S.) and in language (C.M., V.P.). I further thank Michel Coste, Louis Mahé, and Marie-Francoise Roy for organizing the conference and thus providing - among many other things - a very pleasant opportunity to give a talk, of which this article is an expanded version.

References

[Ba] **H.J. Baues**, "Algebraic homotopy". Cambridge studies advanced math. 15, Cambridge Univ. Press 1988.

[BCR] **J. Bochnak, M. Coste, M.-F. Roy**, "Géométrie algébrique réelle". Ergebnisse Math. Grenzgeb. 3. Folge Band 12, Springer 1987.

[BH] **A. Borel, A. Haefliger**, La classe d'homologie fondamentale d'un espace analytique, Bull. Soc. Math. France 89, 461 - 513 (1961).

[BM] **A. Borel, J.C. Moore**, Homology for locally compact spaces. Mich. Math. J. 7, 137 - 159 (1960).

[Bra] **H. Brakhage**, "Topologische Eigenschaften algebraischer Gebilde über einem beliebigen reell-abgeschlossenen Grundkörper". Dissertation Heidelberg 1954.

[Bre] **G. Bredon**, "Sheaf theory". McGraw Hill 1967.

[Br] **L. Bröcker**, Real spectra and distribution of signatures. In: "Géométrie algébrique réelle et formes quadratiques" (Ed. J.-L. Colliot-Thélène, M. Coste, L. Mahé, M.F. Roy), pp. 249 - 272. Lecture Notes Math. 959, Springer 1982.

[B] **G.W. Brumfiel**, "Partially ordered rings and semialgebraic geometry". London Math. Soc. Lecture Notes 37, Cambridge Univ. Press 1979.

[B_1] ——, The real spectrum compactification of Teichmüller spaces. In: "Geometry of group representations" (Ed. W.M. Goldman, A.R. Magid), Contemporary Math. 74, A.M.S. 1988, pp 51 - 75.

[B_2] ——, A semi-algebraic Brouwer fixed point theorem for real affine spaces, ibid., pp 77 - 82.

[B_3] ——, A Hopf fixed point theorem for semialgebraic maps. Preprint Stanford University.

[B_4] ——, Wittrings and K-theory. Rocky Mountain J. Math. 14, 733 - 765 (1984).

[B_5] ——, Quotient spaces for semialgebraic equivalence relations. Math. Z. 195, 69 - 78 (1987).

[CC] **M. Carral, M. Coste**, Normal spaces and their dimensions. J. Pure Appl. Algebra 30, 227 - 235 (1983).

[CR] **M. Coste, M.-F. Coste-Roy**, Le spectre réel d'un anneau est spatial. C.R.A.S. Paris 290, 91 - 94 (1980).

[CR_1] ——, ——, Le topos étale réel d'un anneau. In: 3e colloque sur les catégories dédier a Charles Ehresmann, Cahiers de topologie et géom. diff. 22-1, 19 - 24 (1981).

[CR_2] ——, ——, La topologie du spectre réel. In: "Ordered fields and real algebraic geometry" (Ed. D.W. Dubois, T. Recio), Contemp. Math. 8, 27 - 59 (1982).

[D] **H. Delfs**, Kohomologie affiner semialgebraischer Räume. Dissertation Regensburg 1980.

[D_1] ——, The homotopy axiom in semialgebraic cohomology. J. reine u. angew. Math. 355, 108 - 128 (1985).

[D_2] ——, Homology of locally semialgebraic spaces. Lecture Notes in Math. 1484. Springer, to appear (≈ Habilitationsschrift Regensburg 1984).

[D_3] ——, Semialgebraic Borel-Moore homology, Rocky Mountain J. Math. 14, 987 - 990 (1984).

[DK] **H. Delfs, M. Knebusch**, Semialgebraic topology over a real closed field II: Basic theory of semialgebraic spaces. Math. Z. 178, 175 - 213 (1981).

[DK_1] ——, ——, On the homology of algebraic varieties over real closed fields. J. reine u. angew. Math. 335, 122 - 163 (1982).

[DK_2] ——, ——, "Locally semialgebraic spaces". Lecture Notes Math. 1173 (1985).

[DK_3] ——, ——, An introduction to locally semialgebraic spaces. Rocky Mountain J. Math. 14, 945 - 963 (1984).

[De] **C.N. Delzell**, A finiteness theorem for open semialgebraic sets, with applications to Hilbert's 17th problem. In: "Ordered fields and real algebraic geometry" (Ed. D.W. Dubois, T. Recio), Contemporary Math. 8, 79 - 97 (1982).

[DvdD] **J. Denef, L. van den Dries**, p-adic and real subanalytic sets. Ann. Math. 128, 79 - 138 (1988).

[ES] **S. Eilenberg, N. Steenrod**, "Foundations of algebraic topology". Princeton Univ. Press 1952.

[Gi] **B. Giesecke**, Simpliziale Zerlegung abzählbarer analytischer Räume. Math. Z. 83, 177 - 213 (1964).

[EGA I] **A. Grothendieck, J. Dieudonné**, "Élcments dc géométric algébriquc I: Le language des schémas. Publ. Math. N° 4, Inst. Hautes Etudes Sci. 1960.

[EGA I*] ——, ——, "Élements de géométrie algébrique I". Grundlehren math. Wiss. 166, Springer 1971.

[Ha] **R. Hardt**, Semialgebraic local triviality in semi-algebraic mappings. Amer. J. Math. 102, 291 - 302 (1980).

[Ho] **M. Hochster**, Prime ideal structure in commutative rings. Trans. Amer. Math. Soc. 142, 43 - 60 (1969).

[HM] **J. Houdebine, L. Mahé**, Séparation des composantes connexes réelles dans le cas des variétés projectives. In: "Géometrie algébrique réelle et formes quadratiques (Ed. J.-L. Colliot-Thélène, M. Coste, L. Mahé, M.F. Roy), pp 358 - 370. Lecture Notes Math. 959, Springer 1982.

[Hu] **R. Huber**, "Bewertungsspektrum und rigide Geometrie". Habilitationsschrift, Regensburg 1990.

[Hu_1] ——, Semirigide Funktionen. Preprint Regensburg 1990.

[HuK] **R. Huber, M. Knebusch**, On valuation spectra. To appear in the proceedings of the RAGSQUAD-seminar Berkeley 1990/91, Amer. Math. Soc.

[HuS] **R. Huber, C. Scheiderer**, A relative notion of local completeness in semialgebraic geometry. Arch. Math. 53, 571 - 584 (1989).

[K] **M. Knebusch**, An invitation to real spectra. In: "Quadratic and hermitian forms" (Ed. C.R. Riehm, I. Hambleton), Canadian Math. Soc. Conference Proceedings 4, 51 - 105 (1984).

[K_1] ——, Some open problems. In: "Conference on quadratic forms - 1976" (Ed. G. Orzech). Queen's Papers Pure Appl. Math. 46, 361 - 370 (1977).

[K_2] ——, "Weakly semialgebraic spaces". Lecture Notes Math. 1367, Springer 1989.

[K_3] ——, "Semialgebraic fibrations and covering maps". In preparation.

[KS] **M. Knebusch, C. Scheiderer**, "Einführung in die reelle Algebra", Vieweg, Braunschweig-Wiesbaden 1989.

[L] **S. Lojasiewicz**, Triangulation of semi-analytic sets. Ann. Scuola Norm. Sup. Pisa (3) 18, 449 - 474 (1964).

[M] **L. Mahé**, Signatures et composantes connexes. Math. Ann. 260, 191 - 210 (1982).

[May] **J.P. May**, "Simplicial objects in algebraic topology". Van Nostrand Math. Studies 11, Van Nostrand 1967.

[Mi] **J. Milnor**, The geometric realization of a semi-simplicial complex. Annals of Math. 65, 357 - 362 (1957).

[P] **M. Prechtel**, "Endliche semialgebraische Räume". Diplomarbeit Regensburg 1988.

[PiS] **A. Pillay, C. Steinhorn**, Definable sets in ordered structures, I. Trans. Amer. Math. Soc. 295, 565 - 592 (1986).

[Ro] **R. Robson**, Embedding semialgebraic spaces. Math. Z. 183, 365 - 370 (1983).

[R] **M.-F. Roy**, Faisceau structural sur le spectre réel et fonctions de Nash. In: "Géométrie algébrique réelle et formes quadratiques", pp. 406 - 432. Lecture Notes Math. 959, Springer 1982.

[Rz] **J.M. Ruiz**, Basic properties of real analytic and semianalytic germs, Publ. Inst. Recherche Math. Rennes 4, 29 - 51 (1986).

[Rz_1] ——, On the real spectrum of a ring of global analytic functions. ibid., 84 - 95.

[Sch] **C. Scheiderer**, Quotients of semi-algebraic spaces. Math. Z. 201, 249 - 271 (1989).

[S] **N. Schwartz**, "Real closed spaces". Habilitationsschrift Ludwig-Maximilians-Universität München 1984.

[S_1] ——, "The basic theory of real closed spaces". Memoirs Amer. Math. Soc. 77, N° 397 (1989).

[S_2] ——, "The basic theory of real closed spaces". Regensburger Mathematische Schriften 15, Fakultät für Mathematik der Universität Regensburg 1987.

[S_3] ——, Inverse real closed spaces. Illinois J. Math., to appear soon.

[S_4] ——, Open morphisms of real closed spaces. Rocky Mountain J. Math. 19, 913 - 939 (1989).

[Spa] **E.H. Spanier**, "Algebraic topology". McGraw-Hill 1966.

[Sw] **R.M. Switzer**, "Algebraic topology - Homotopy and homology". Grundlehren math. Wiss. 212, Springer 1975.

Address: Fakultät für Mathematik, Universität Regensburg,
Universitätsstr. 31, D-8400 Regensburg

Algebraic Geometric Methods in Real Algebraic Geometry

R. Parimala

This article is based on the talk given at the conference. It is a brief overview of certain recent developments in the study of the topology of a real algebraic variety through a combination of techniques from algebraic geometry and the theory of quadratic forms. Our underlying theme is to illustrate how the higher dimensional variations of some basic theorems of Witt ([W]) for real curves have encompassed a wide range of mathematical activities in the last decade. Thus, this article is in no sense a comprehensive survey of the results linking algebraic geometry and real algebraic geometry. For instance, an area left out is the construction of real moduli spaces.

Let X be a real variety and $X(\mathbb{R})$ its set of $\mathbb{R}$-rational points, equipped with the strong (Euclidean) topology. We consider principally two classes of maps which relate geometric invariants of X to topological invariants of $X(\mathbb{R})$.

I. Suppose X is smooth and $X(\mathbb{R})$ compact. We discuss the cycle map from the Chow group of X to the homology groups of $X(\mathbb{R})$ with $\mathbb{Z}/2$ coefficients, its kernel and image.

II. We discuss two kinds of fibre maps:

1) The signature map from the Witt group of X to the group of $\mathbb{Z}$-valued continuous functions on $X(\mathbb{R})$.

2) The "mod-2" signature map from the unramified étale cohomology of X to the cohomology of $X(\mathbb{R})$.

The study of these maps for real curves is contained in the above mentioned paper of Witt.

We fix the following notation for the rest of this article. By a *variety* X over a field k, we mean a reduced irreducible scheme over k and by a *subvariety* a closed subscheme which is reduced and irreducible.

1 The cycle map

Let X be a nonsingular variety over $\mathbb{R}$ such that $X(\mathbb{R})$ is compact. Let $Z_k(X)$ denote the group of k-cycles on X; namely the free abelian group on k-dimensional sub-varieties of X. Let $P_k(X)$ denote the subgroup generated by cycles which are rationally equivalent to zero. Let $Z_k^{th}(X)$ denote the subgroup of $Z_k(X)$ generated by "*thin cycles*", i.e. $Z_k^{th}(X)$ is generated by k-dimensional subvarieties Y of X with $Y(\mathbb{R})$ not Zariski dense in Y. The cycle map $cl_k : Z_k(X) \to H_k(X(\mathbb{R}), \mathbb{Z}/2)$ is defined as follows: Let Y be a k-dimensional subvariety of X. If Y is thin, then we define $cl_k(Y) = 0$. Otherwise, let $\sigma_Y \in H_k(Y(\mathbb{R}), \mathbb{Z}/2)$ be the fundamental class of $Y(\mathbb{R})$. Then $cl_k(Y) = i(\sigma_Y)$, where $i : H_k(Y(\mathbb{R}), \mathbb{Z}/2) \to H_k(X(\mathbb{R}), \mathbb{Z}/2)$ is induced by the inclusion $Y(\mathbb{R}) \to X(\mathbb{R})$ (cf. [Bor-H] 5.12). The image of the cycle map is called the *group of algebraic homology classes*, denoted $H_*^{alg}(X(\mathbb{R}), \mathbb{Z}/2)$. If $\rho : H^*(X(\mathbb{R}), \mathbb{Z}/2) \to H_*(X(\mathbb{R}), \mathbb{Z}/2)$ denotes the Poincaré duality isomorphism, then $\rho^{-1}(H_*^{alg}(X(\mathbb{R}), \mathbb{Z}/2)) = H_{alg}^*(X(\mathbb{R}), \mathbb{Z}/2)$ is called the *group of algebraic cohomology classes*.

We first look at the kernel of the cycle map. This map respects rational equivalence ([Bor-H] 5.13) so that $cl_k(P_k(X)) = 0$. Further, by definition, $cl_k(Z_k^{th}(X)) = 0$. The question whether the kernel of the cycle map is precisely the group $P_k(X) + Z_k^{th}(X)$ for smooth projective varieties X over $\mathbb{R}$ remained open until recently. We first indicate how the results of Witt imply that for a smooth projective real curve X, kernel of cl_0 is equal to $P_0(X) + Z_0^{th}(X)$. Let $X(\mathbb{R}) = \bigcup C_i, \{C_i\}$ denoting the set of connected components of $X(\mathbb{R})$. Each C_i is topologically a circle. Suppose we are given an even number of points $\{P_1, P_2, \ldots, P_{2n}\}$ on some C_i. Witt proves that there is a rational function $f \in \mathbb{R}(X)$, regular on $X(\mathbb{R})$ which changes sign on C_i exactly at the points P_i. Knebusch ([K]$_1$ 4.5 a, p. 61) refines this into the following theorem: There exists a rational function $f \in \mathbb{R}(X)$, regular on $X(\mathbb{R})$ such that

$$\operatorname{div} f = \sum_{1\leq i\leq 2n} P_i + D,$$

where D is a divisor on X with $(\operatorname{supp} D)\cap X(\mathbb{R}) = \phi$ (Indeed one may choose supp D to be any given complex point of X). This implies that kernel $cl_0 = Z_0^{th}(X) + P_0(X)$.

Colliot-Thélène and Ischebeck [C-I] prove that ker $cl_0 = Z_0^{th}(X)+P_0(X)$ for varieties X of arbitrary dimension and their method of proof is to reduce to the case of curves. Bröcker [Br]$_1$ proves the same result for the case of divisors, namely, $k = \dim X - 1$. Ischebeck and Schülting [I-Sc] prove more generally that ker $cl_k = Z_k^{th}(X) + P_k(X)$ for all varieties X proper over $\mathbb{R}$. Their proof has two main steps : 1) If $f : Z \to X$ is a morphism of smooth projective varieties of dimensions k and n respectively, with $f^{\mathbb{R}} : Z(\mathbb{R}) \to X(\mathbb{R})$ bordant, then $f_*([Z])$ is rationally equivalent to a thin cycle. 2) If $z \in Z_k(X)$ with $cl_k(z) = 0$, then there is a morphism $f : Z \to X$ of smooth projective varieties with $f^{\mathbb{R}} : Z(\mathbb{R}) \to X(\mathbb{R})$ bordant and $f_*([Z]) = z + z', z' \in Z_k^{th}(X)$. This is achieved by relating $H_k^{alg}(X(\mathbb{R}), \mathbb{Z}/2)$ to the algebraic part of the k-bordism group.

Schülting [Sc] studies the behaviour of the cokernel of the cycle map under birational transformations of smooth projective varieties. Let $T_k(X) = H_k(X(\mathbb{R}), \mathbb{Z}/2)/H_k^{alg}(X(\mathbb{R}), Z/2)$. If X is a smooth projective variety of dimension n and $\pi : Y \to X$ is a blow-up with smooth center Z of dimension $n-d$, then he shows that

$$T_k(Y) \simeq T_k(X) \bigoplus \bigoplus_{k-d+1\leq j\leq k-1} T_j(Z).$$

In particular, $T_1(X) = H_1(X(\mathbb{R}), \mathbb{Z}/2)/H_1^{alg}(X(\mathbb{R}), \mathbb{Z}/2)$ is a birational invariant for smooth projective varieties. If $\dim X = n, T_n(X) \simeq (\mathbb{Z}/2)^{s-1}$, where s is the number of connected components of $X(\mathbb{R})$, which is also a birational invariant for smooth projective varieties over $\mathbb{R}$ (cf [Bo-C-R] Cor 10.4.6. p. 227 and [D-K] §13). Thus, for a smooth projective surface $X, T_*(X)$ is a birational invariant of X. Silhol ([Si] Th. 5) proves that for real rational surfaces (i.e., surfaces for which $X_{\mathbb{C}}$ is birational to $\mathbb{P}^2_{\mathbb{C}}$), $T_1(X) = 0$.

2 Image of the cycle map; Connections with K-Theory

Let X be an affine algebraic variety over $\mathbb{R}$ with $X(\mathbb{R})$ compact. Let $K_F(X)$ denote the Grothendieck group of topological F-vector bundles on $X(\mathbb{R}), F$ being $\mathbb{R}, \mathbb{C}$ or $\mathbb{H}$. By a theorem of Swan ([Sw]$_1$ p. 268), $K_F(X) \simeq K_0(F \otimes \mathcal{C}(X))$, the Grothendieck group on finitely generated projective $F \otimes \mathcal{C}(X)$-modules, $\mathcal{C}(X)$ denoting the ring of continuous functions on $X(\mathbb{R})$. We define a certain algebraic part $K_{alg}(X)$ of $K_{\mathbb{R}}(X)$ which, under the Stiefel-Whitney map

$$w : K_{\mathbb{R}}(X) \to H^*(X(\mathbb{R}), \mathbb{Z}/2),$$

maps into $H^*_{alg}(X(\mathbb{R}), \mathbb{Z}/2)$. Let $\mathcal{R}(X)$ denote the ring of regular functions on X; i.e. rational functions $f/g \in \mathbb{R}(X)$ such that g is non-vanishing on $X(\mathbb{R})$. A continuous vector bundle ξ on $X(\mathbb{R})$ is said to be *strongly algebraic* if its sheaf of sections is associated to a projective module over $\mathcal{R}(X)$. The group $K_{F-alg}(\mathcal{C}(X))$ is simply the image of the natural homomorphism $K_0(F \otimes \mathcal{R}(X)) \to K_F(\mathcal{C}(X))$. The Stiefel Whitney map w maps $K_{\mathbb{R}-alg}(X)$ into $H^*_{alg}(X(\mathbb{R}), \mathbb{Z}/2)$ ([Be-T]$_2$ 2.4) and the two groups $K_{\mathbb{R}-alg}(X)$ and $H^*_{alg}(X(\mathbb{R}), \mathbb{Z}/2)$ are objects of extensive study for compact real varieties X.

By a theorem of Swan ([Sw]$_2$ Th. 2.2) for compact varieties X, $K_0(\mathcal{R}(X)) \to K_{\mathbb{R}}(X)$ is injective. Further, a vector bundle ξ on $X(\mathbb{R})$ is strongly algebraic if and only if its class in $K_{\mathbb{R}}(X)$ belongs to $K_{\mathbb{R}-alg}(X)$. Thus, we may identify the two groups $K_0(\mathcal{R}(X))$ and $K_{\mathbb{R}-alg}(X)$. One of the basic problems is to characterise strongly algebraic vector bundles on $X(\mathbb{R})$. Typical examples of varieties X for which $K_{\mathbb{R}-alg}(X) = K_{\mathbb{R}}(X)$ are spheres ([F]) and projective spaces ([G-R]). Swan shows [Sw]$_3$ that for an n-dimensional sphere X over $\mathbb{R}, K_0(X) \simeq K_0(\mathcal{R}(X)) \simeq K_{\mathbb{R}}(X)$. Benedetti and Tognoli [Be-T]$_1$ prove that if X is a smooth affine variety over $\mathbb{R}$, homeomorphic to $S^n, K_{\mathbb{R}-alg}(X) = K_{\mathbb{R}}(X)$. In [Bo-K]$_3$, there is a characterization of strongly algebraic vector bundles on $X(\mathbb{R})$ in terms of Stiefel-Whitney invariants, for low dimensional varieties. More precisely, Bochnak and Kucharz prove that if dim $X \leq 3$, a continuous vector bundle ξ of constant rank on $X(\mathbb{R})$ is strongly algebraic, if and only if its Stiefel-Whitney classes are algebraic.

Bochnak, Büchner and Kucharz ([Bo-B-K] Th.1.1) exhibit a large family of real varieties X for which the group $\widetilde{K}_{F-alg}(X)$ is rather small. They show that if $X \hookrightarrow \mathbb{R}P^n$ is a real variety of dimension at least 1 such that $X(\mathbb{R})$ is connected and $X_{\mathbb{C}}$ is a non-singular complete intersection, $\widetilde{K}_{F-alg}(X)$ is either finite or a direct sum of a finite group and $\mathbb{Z}$. If dim X is odd and $H^{even}(X,\mathbb{Z})$ is torsion free, they show that $\widetilde{K}_{\mathbb{C}-alg}(X)=0$. However, the groups $\widetilde{K}_F(X)$ may generally have arbitrarily large rank (cf. [Bo-B-K]).

3 Certain types of algebraic models for compact C^∞-manifolds

After the solution of the Nash conjecture by Tognoli [T], the following natural questions, related to approximations of algebraic models by non-singular algebraic sets, were raised (cf. [Bo]). Let M be a compact (connected) C^∞-manifold. An *algebraic model* of M is a non-singular affine scheme X with $X(\mathbb{R})$ diffeomorphic to M.

Q1) Given M as above, does there exist an algebraic model X with $H^*_{alg}(X(\mathbb{R}),\mathbb{Z}/2) = H^*(X(\mathbb{R}), Z/2)$?

Q2) Characterise all compact C^∞-manifolds M satisfying the following condition: For every algebraic model X of M, $H^*_{alg}(X,\mathbb{Z}/2) = H^*(X(\mathbb{R}),\mathbb{Z}/2)$.

It was shown in [A-K] and [Be-T]$_2$ that there exists an algebraic model X for M with $H^1_{alg}(X(\mathbb{R}),\mathbb{Z}/2) = H^1(X(\mathbb{R}),\mathbb{Z}/2)$. In general,however, Q1 has a negative answer. Benedetti and Dedo ([Be-D]) construct a 11-dimensional compact C^∞-manifold M with $H^2_{alg}(X(\mathbb{R}),\mathbb{Z}/2) \neq H^2(X(\mathbb{R}),\mathbb{Z}/2)$ for any algebraic model X of M. Regarding $Q2$, $\mathbb{R}P^{2n}$ are examples of such manifolds. For any algebraic model X of $\mathbb{R}P^{2n}$, $H^1(X(\mathbb{R}),\mathbb{Z}/2) \simeq \mathbb{Z}/2$, with the non-trivial element given by the first Stiefel-Whitney class of the determinant of the tangent bundle of $\mathbb{R}P^{2n}$. Thus the non-orientability of $\mathbb{R}P^{2n}$ implies that $H^i_{alg}(X(\mathbb{R}),\mathbb{Z}/2) = H^i(X(\mathbb{R}),\mathbb{Z}/2)$. However, until recently, it remained open whether $\mathbb{R}P^3$ has this property. In ([Bo-B-K]), examples of 3-dimensional varieties X, diffeomorphic to $\mathbb{RP}^3$ were constructed, which have the property that $\widetilde{K}_{\mathbb{C}-alg}(X) = 0$. In fact, for such an X, $H^1_{alg}(X(\mathbb{R}), \mathbb{Z}/2) = 0$ (cf. [Bo-B-K], Prop. 7.8). Thus $\mathbb{RP}^3$ does not belong to the class defined in Q2.

The question, given a subgroup $G \subseteq H^1(M, \mathbb{Z}/2), M$ a compact C^∞-manifold, whether there exists an algebraic model X with $H^1_{alg}(X(\mathbb{R}), \mathbb{Z}/2) = G$, has been studied by Bochnak and Kucharz in [Bo-K]$_4$. A necessary condition to be satisfied by G, for non-orientable M, is that it should contain the first Stiefel-Whitney class of M. It is shown in [Bo-K]$_4$ that if M is a compact connected orientable C^∞-manifold of dimension ≥ 3, for any $G \subseteq H^1(M, \mathbb{Z}/2)$, there exists an algebraic model X such that $H^1_{alg}(X, \mathbb{Z}/2) = G$. The same conclusion holds for non-orientable M, provided the first Stiefel-Whitney class of M belongs to G.

For a detailed bibliography on applications of the analysis of the groups H^1_{alg} to problems in various branches of mathematics, we refer to [Bo-K]$_4$. We mention that Bochnak and Kucharz use their computations of $K_{F-alg}(X)$, (cf. §2) to construct algebraic models $X \hookrightarrow \mathbb{R}^{2k+1}$ for a compact C^∞ hypersurface in $\mathbb{R}^{2k+1}$ with the property that every regular mapping from X into S^{2k} is null-homotopic [Bo-K]$_2$. They contrast this result with the odd dimensional case, where for any real algebraic set of dimension $2k+1$ in $\mathbb{R}^n$, compact, connected and orientable, either every smooth mapping into S^{2k+1} is homotopic to a regular map or precisely those maps of even topological degree have this property [Bo-K]$_1$.

4 Witt groups of real varieties and signatures

Let X be any variety over a real closed field R and let $\mathrm{Spec}_r X$ denote the real spectrum of X (cf. [Bo-C-R], §7.1). We recall that the points of $\mathrm{Spec}_r X$ are pairs (x, ν), where $x \in X$ and ν is an ordering of $R(x)$. Let $W(X)$ denote the Witt group of X. The total signature homomorphism $\mathrm{sgn} : W(X) \to \mathcal{C}(\mathrm{Spec}_r X, \mathbb{Z})$ assigns to any quadratic bundle q on X, the function $\mathrm{sgn}(q) : \mathrm{Spec}_r X \to \mathbb{Z}$, defined by $\mathrm{sgn}(q)(x, \nu) =$ signature of the restriction of q to the fibre at x, with respect to the ordering ν. Let s denote the number of semi-algebraic components of $X(R)$. Then $\mathcal{C}(\mathrm{Spec}_r X, \mathbb{Z}) = \mathbb{Z}^s$ and in the case $R = \mathbb{R}$, s is the number of connected components of $X(\mathbb{R})$. The question whether the signatures separate semi-algebraic components of $X(R)$ was raised by Knebusch in $[K]_3$. The separation of connected components of curves over $\mathbb{R}$ by signatures is due to Witt. Knebusch himself proves $[K]_1$ that the signatures separate semi-algebraic components of curves over real closed fields, using Geyer's algebraisation of Witt's results $[Ge]_1$. In the beautiful paper $[M]_1$, Mahé gives an affirmative answer to this question in the most general setting of any affine scheme Spec$A$ with 2 invertible in A. He proves in fact that given an open and closed subset $U \subset \mathrm{Spec}_r A$, there exists a quadratic bundle q over A and an integer $N > 0$ such that $\mathrm{sgn}(q) \mid U = 2^N$ and $\mathrm{sgn}(q) \mid \mathrm{Spec}_r A \setminus U = 0$. The separation of semi-algebraic components of a smooth projective variety over a real closed field was proved in Houdebine-Mahé [H-M]. Thus the cokernel of the total signature map is 2-primary torsion.

A natural question is whether the exponent 2^N of the cokernel of the signature map is bounded by a function of the dimension of X. Mahé $[M]_1$ proved that N can be bounded by a function $f(s,m)$, where s is the number of inequalities necessary to define a semi-algebraic basic open subset of $X(R)$, and m is the least integer such that every totally positive element of $R[X]$ divides an element of the form $1+\sum_{1\leq i\leq m} x_i^2, x_i \in R[X]$. Bröcker ($[Br]_2$ and $[Br]_3$) proved that s is bounded by a function of the dimension of X. Mahé $[M]_2$ proves that $m \leq n-1+2^{n+1}$, where $n = \dim X$. This settles the above question in the affirmative.

It would be interesting to have more precise bounds for this exponent in terms of the dimension of X. For instance, one could ask whether this exponent is bounded by 2^{n+1}, for smooth varieties of dimension n over $\mathbb{R}$. This is true for curves, which is due to Witt and for smooth projective surfaces and abelian varieties, which is due to Colliot-Thélène and Sansuc ([C-S], Th. 4.1 and Th. 5.4). They use Geyer's duality theorem for abelian varieties X over $\mathbb{R}$ $[Ge]_2$ which gives the bound $s \leq 2^g$ where g is the dimension of X and s denotes the number of connected components of $X(\mathbb{R})$.

We now mention a refinement of Mahé's result due to Brumfiel [Bru], which compares the Witt group $W(X)$ and the topological Witt group $W(X(R))$. Let $S^\circ(X)$ denote the ring of semi-algebraic continuous functions on $X(R)$. Since every semi-algebraic bundle on $X(R)$ has a unique positive definite structure upto isometry (cf. [Bo-C-R] 15.1.11, p.331), there is an isomorphism $K_0(\mathcal{S}^\circ(X)) \simeq W(\mathrm{Spec}_r X)$. The natural inclusion $R[X] \hookrightarrow \mathcal{S}^\circ(X)$ induces a homomorphism $i : W(X) \to W(\mathcal{S}^\circ(X)) = W(\mathrm{Spec}_r X) = K_0(\mathcal{S}^\circ(X))$ and we have a commutative diagram

$$\begin{array}{ccc} W(X) & \xrightarrow{i} & K_0(\mathcal{S}^\circ(X)) \\ & {\scriptstyle sgn}\searrow \quad \swarrow & \\ & \mathcal{C}(\mathrm{Spec}_r X, \mathbb{Z}) & \end{array}$$

Brumfiel proves that the cokernel of i is 2-primary torsion. If $R = \mathbb{R}$, we have $K_0(\mathcal{S}^\circ(X)) \simeq K_0(\mathcal{C}(X)) = K_{\mathbb{R}}(X)$, ([Bo-C-R], 12.7.8, p. 287). Thus the theorem above compares the Witt group of X with the K-theory of $X(\mathbb{R})$ and asserts that the cokernel of this map is 2-primary torsion.

We record a few computations for the Witt groups of real varieties. The computation of the Witt groups of real curves, or more generally of curves over real closed fields is due to Knebusch $[\mathrm{K}]_2$. The Witt group is a birational invariant for smooth projective surfaces. In [Su], Sujatha computes the Witt group of a real projective surface X in terms of certain birational invariants of X. For smooth projective real rational surfaces X (that is $X_{\mathbb{C}}$ is birational to $\mathbb{P}^2_{\mathbb{C}}$), the number of connected components s of $X(\mathbb{R})$ determines $W(X)$ completely: if $X(\mathbb{R})$ is not empty, $W(X) \simeq \mathbb{Z}^s \oplus (\mathbb{Z}/2)^{(s-1)}$ and if $X(\mathbb{R})$ is empty, $W(X) \simeq \mathbb{Z}/4$, being generated by $< 1 >$ (cf. [Su] § 4), noting that Theorem 4.2 needs a slight modification).

We also mention here some computations of the symplectic Witt group of a real variety in terms of algebraic cohomology classes. Let X be a compact affine real variety of dimension at most 3. Let $WSp(X)$ denote the symplectic Witt group of X. An element $[\xi] \in WSp(X)$ is represented by a rank 2 symplectic bundle ξ on X, since every symplectic bundle of rank ≥ 4 on X has a non-vanishing section and hence splits off a hyperbolic plane ([B], §4.12). Let $\alpha(\xi) \in H^2_{alg}(X(\mathbb{R}), \mathbb{Z})$ denote the Euler class of ξ. Indeed, the image of $\alpha(\xi)$ in $H^2(X(\mathbb{R}), \mathbb{Z}/2)$ is simply the second Stiefel-Whitney class $w_2(\xi)$ of ξ. Let $\beta : H^1_{alg}(X(\mathbb{R}), \mathbb{Z}/2) \to H^2_{alg}(X(\mathbb{R}), \mathbb{Z})$ denote the Bockstein homomorphism. Then there is a homomorphism

$$\tilde{\alpha} : WSp(X) \to H^2_{alg}(X(\mathbb{R}), \mathbb{Z})/\beta(H^1_{alg}(X(\mathbb{R}), \mathbb{Z}/2))$$

induced by α. Barge and Ojanguren [Ba-O] show that if $\dim X \leq 2$, $\tilde{\alpha}$ is an isomorphism, observing that in dimension 2, $H^2_{alg} = H^2$. The mod 2 version of this result, namely that $WSp(X)/2 \simeq (\mathbb{Z}/2)^s$, s denoting the number of connected components of $X(\mathbb{R})$, is due to Pardon ([P], Th.B) and a refinement of Pardon's result through different methods is in [O-P-S]. Bochnak and Kucharz ([Bo-K]$_3$ Prop. 2.3) show that for dim $X \leq 3$, α induces an isomorphism $\tilde{\alpha} : WSp(\mathcal{R}(X)) \simeq H^2_{alg}(X(\mathbb{R}), \mathbb{Z})/\beta H^1_{alg}(X(\mathbb{R}), \mathbb{Z}/2)$, where $\mathcal{R}(X)$ denotes the ring of regular functions on X.

5 Etale cohomology and real cohomology

Let X be a smooth quasi-projective curve over $\mathbb{R}$. Let $\mathrm{Br}(X)$ denote the Brauer group of X. Given an element $\zeta \in \mathrm{Br}(X)$, represented by an Azumaya algebra A on X, for any rational point $x \in X(\mathbb{R})$, the restriction of A to the fibre at x defines a central simple algebra over $\mathbb{R}$ which is the class of $\mathbb{H}$ or zero in the Brauer group of $\mathbb{R}$. This defines a map

$$\mathrm{Br}(X) \to \mathcal{C}(X(\mathbb{R}), \mathbb{Z}/2).$$

Witt's construction of regular functions on X with prescribed signs on each connected component of $X(\mathbb{R})$ shows that this map is surjective. We now explain a higher dimensional analogue of this theorem. Let X be any quasi-projective scheme. We define (cf. [C-P]) a fibre map, called the *mod-2 signature map*,

$$h_n : H^n_{et}(X, \mu_2) \to \mathcal{C}(\mathrm{Spec}_r X, \mathbb{Z}/2)$$

as follows: Let $(x, \nu) \in Spec_r X$, $\nu : k(x) \hookrightarrow L$ being an imbedding in a real closure. Let $\psi_{(x,\nu)} : H^n_{et}(X, \mu_2) \to H^n(k(x), \mu_2) \xrightarrow{\nu} H^n(L, \mu_2) \simeq \mathbb{Z}/2$ be the restriction to the fibre. We define $h_n(\xi)(x, \nu) = \psi_{(x,\nu)}(\xi)$. Let X be a smooth quasi-projective variety over a real closed field R. Let $\mathcal{H}^n$ denote the Zariski sheaf associated to the presheaf $U \mapsto H^n_{et}(U, \mu_2)$. Since every section of $\mathcal{H}^n$ extends uniquely, at a point $x \in X$, to an étale cohomology class, the naturality of h_n yields a homomorphism

$$h_n : H^o(X, \mathcal{H}^n) \to \mathcal{C}(\mathrm{Spec}_r X, \mathbb{Z}/2).$$

It is shown in [C-P] that, if X is a smooth quasi-projective variety of dimension d over a real closed field, then for $n \geq d+1$, the maps h_n are isomorphisms. In particular for $n \geq d+1$ the groups $H^{\circ}(X, \mathcal{H}^n)$ are isomorphic to $(\mathbb{Z}/2)^s$, s denoting the number of semi-algebraic components of $X(R)$, and hence they are finite.

We point out a consequence of this result for quadratic forms. Knebusch $[\mathrm{K}]_3$ raised the following question: "Is the Witt group of a smooth affine algebraic variety over $\mathbb{R}$ finitely generated?" The answer to this question is positive for $dim X \leq 2$, the case of curves being treated by Knebusch $[\mathrm{K}]_2$, and the dimension 2 case by Ayoub [Ay]. It was shown in [Pa] that for a smooth affine threefold X over $\mathbb{R}$, $W(X)$ is finitely generated if and only if $CH^2(X)/2$ is finite, CH^2 denoting the group of codimension two cycles, modulo rational equivalence. The finiteness of $CH^2(X)/2$ is a wide open question in dimension 3 and the above theorem, which establishes the equivalence of this problem in geometry to one on quadratic forms, is proved via real algebraic geometry, namely through the finiteness of $H^0(X, \mathcal{H}^4)$.

We describe an interpretation of the map h_n, due to Scheiderer, which leads to calculations of the higher cohomology groups of the sheaf $\mathcal{H}^n$ in terms of the semi-algebraic cohomology of $X(R)$. Let $\varphi : \mathrm{Spec}_r X \to X$ be the natural map, $\varphi(x,v) = x$. The maps $h_n : H^n_{et}(X, \mu_2) \to \mathcal{C}\,(\mathrm{Spec}_r X, \mathbb{Z}/2)$ are functorial and yield a sheaf morphism $h_n : \mathcal{H}^n \to \phi_*(\mathbb{Z}/2)$. The theorem of [C-P] can be restated as an isomorphism $h_n : \mathcal{H}^n \simeq \varphi_*(\mathbb{Z}/2)$ of sheaves for $n \geq d+1$, where X is a smooth quasi-projective variety of dimension d over a real closed field R. In particular, for $i \geq 0, n \geq d+1, H^i(X, \mathcal{H}^n) \simeq H^i(X, \varphi_*\mathbb{Z}/2)$. The beautiful remark of Scheiderer (cf $[\mathrm{S}]_1$) is, that for any variety X over a real closed field R, even possibly singular, $R^q\varphi_*(\mathbb{Z}/2) = 0$, so that $H^i(X, \varphi_*\mathbb{Z}/2) \simeq H^i\,(\mathrm{Spec}_r X, \mathbb{Z}/2) = H^i_{sa}(X(R), \mathbb{Z}/2)$ (cf [D-K]). Specializing R to the field of real numbers, we obtain an isomorphism

$$h_n : H^i(X, \mathcal{H}^n) \simeq H^i(X(\mathbb{R}), \mathbb{Z}/2),$$

for $n \geq d+1, i \geq 0$. In particular, if X is a smooth surface over $\mathbb{R}$ with $X(\mathbb{R})$ compact, and s is the number of connected components of $X(\mathbb{R})$, the degeneracy of the Bloch-Ogus spectral sequence ([C-P]) gives, for $i \geq 4$,

$$\begin{aligned} H^i(X) &\simeq H^0(X, \mathcal{H}^4) \oplus H^1(X, \mathcal{H}^3) \oplus H^2(X, \mathcal{H}^2) \\ &\simeq (\mathbb{Z}/2)^s \oplus H^1(X(\mathbb{R}), \mathbb{Z}/2) \oplus (\mathbb{Z}/2)^s \end{aligned}$$

(cf [C-P] 3.2.1 and $[\mathrm{S}]_1$).

The following observation of Colliot-Thélène [C] allows us to dispense with the smoothness assumption on X in the above theorems. Cox in [Co] proves that, for any variety X over $\mathbb{R}$, there is a natural isomorphism

$$H^m_{et}(X, \mathbb{Z}/2) \simeq \bigoplus_{0 \le i \le d} H^i(X(\mathbb{R}), \mathbb{Z}/2),$$

for $m \geq 2d+1$. This yields an isomorphism $\mathcal{H}^m \simeq \varphi_* \mathbb{Z}/2$, for $m \geq 2d+1$, in view of the above mentioned remarks of Scheiderer. Since for an affine scheme X, and for $m \geq d+1, H^m_{et}(X_{\mathbb{C}}, \mathbb{Z}/2) = 0$, the Kummer sequence yields a long exact sequence in etale cohomology:

$$H^m(X_{\mathbb{C}}, \mathbb{Z}/2) \to H^m(X, \mathbb{Z}/2) \overset{\cup x_{-1}}{\to} H^{m+1}(X, \mathbb{Z}/2) \to H^{m+1}(X_{\mathbb{C}}, \mathbb{Z}/2),$$

which, for $m \geq d+1$, yields an isomorphism $\mathcal{H}^m \overset{\sim}{\to} \mathcal{H}^{m+1}$ of sheaves and which commutes with h, so that one has $h_m : \mathcal{H}^m \overset{\sim}{\to} \varphi_* \mathbb{Z}/2$, for $m \geq d+1$.

In $[S]_2$, Scheiderer tries to interpret the results of Cox in the general setting of the real spectrum of any scheme. This leads to a natural isomorphism

$$\lim_{\to n} H^n_{et}(X, \mathbb{Z}) \simeq \bigoplus_{i \geq 0} H^i(\mathrm{Spec}_r X, \mathbb{Z}/2),$$

for any affine scheme X in which 2 is invertible. This places in an abstract setting, the results of Arason-Elman-Jacob for fields ([A-E-J], 2.4). One of the interesting consequences of this analysis is that the connected components of the real spectrum of any commutative ring are separated by the étale cohomology classes.

Acknowledgements. I thank J.-L. Colliot-Thélène for his help in the preparation of this talk. I also thank M.A. Knus and J. Barge for critically reading the manuscript.

References

[A-K] S. Akbulut, H. King, A relative Nash theorem, Trans. Amer. Math. Soc. 267 (2), 465-481 (1981).

r-E-J] J.Kr.Arason, R.Elman, B.Jacob, The graded Witt ring and Galois cohomology, CMS.Conf. Proc. Vol.4, Providence: AMS, 1984, 17-50.

[Ay] G. Ayoub, Le groupe de Witt d'une surface réelle. Comment. Math. Helv. 62, 74-105 (1987).

[B] H. Bass, Unitary algebraic K-theory, Hermitian K-theory and Geometric Applications, Battelle Institute Conference 1972. Algebraic K-theory III, SLN 343, 1973, pp. 57-265.

[Ba-O] J. Barge, M. Ojanguren. Fibrés algébriques sur une surface réelle. Comment. Math. Helv. 62, 616-629 (1987).

[Be-D] R, Benedetti, M. Dedo, Counterexamples to representing homology classes by real algebraic subvarities up to homeomorphism. Compositio Math. 53, 143-151 (1984).

[Be-T]$_1$ R. Benedetti, A. Tognoli, Sur les fibrés vectoriels algébriques réels. C.R. Acad. Sc. Paris, t. 287, Série A-831-833 (1978).

[Be-T]$_2$ R. Benedetti, A. Tognoli, Remarks and counterexamples in the theory of real algebraic vector bundles and cycles, Géométrie algébrique réelle et formes quadratiques, Rennes 1981, SLN 959, pp. 198-211.

[Bo] J. Bochnak, Topology of real analytic sets - Some open problems. Rennes 1981, SLN 959, pp. 212-217.

[Bo-B-K] J. Bochnak, M. Buchner and W. Kucharz, Vector bundles on real algebraic varieties, K-theory, 3, 271-298 (1990).

[Bo-C-R] J. Bochnak, M. Coste, M-F.Roy, Géométrie algébrique réelle, Ergeb. der Math. 3. Folge, Bd. 12, Springer 1987.

[Bo-K]$_1$ J. Bochnak, W. Kucharz, Representation of homotopy classes by algebraic mappings, J. Reine. Angew. Math. 377, 159-169 (1987).

[Bo-K]$_2$ J. Bochnak, W. Kucharz, On real algebraic morphisms into even dimensional spheres, Ann. of Math. 128, 415-433 (1988).

[Bo-K]$_3$ J. Bochnak, W. Kurcharz, K-theory of real algebraic surfaces and threefolds, Math. Proc. Camb. Phil. Soc. 106, 471-480 (1989).

[Bo-K]$_4$ J. Bochnak, W. Kucharz, Algebraic models of smooth manifolds, Invent. Math. 588-611 (1989).

[Bor-H] A. Borel, A. Haefliger, La classe d' homologie fondamentale d'un espace analytique. Bull. Soc. Math. France, 89, 461-513 (1961).

[Br]$_1$ L. Bröcker, Reelle Divisoren, Arch. Math. 35, 140-143 (1980).

[Br]$_2$ L. Bröcker, Minimale Erzeugung von Positivbereich. Geom. Dedication 16. 335-350 (1984).

[Br]$_3$ L. Bröcker, Spaces of orderings and semialgebraic sets, Quadratic and hermitian forms, CMS Conf. Proc. Vol. 4, Providence: AMS, 1984, pp. 231-248.

[Bru] G.W. Brumfiel, Witt rings and K-theory. Rocky Mountain J. of Math. 14, 733-765 (1984).

[C] J.-L. Colliot-Thélène, Abstract of a talk at Oberwolfach, 1990.

[C-I] J.-L. Colliot-Thélène, Ischebeck, L' équivalence rationnelle sur les cycles de dimension zéro des variétés algébriques réelles. C.R. Acad. Sci. Paris, Serie I, 292, 723-725 (1981).

[C-P] J.-L. Colliot-Thélène, R. Parimala, Real components of algebraic varieties and étale cohomology, Invent. Math. 101, 81-99 (1990).

[C-S] J.-L. Colliot-Thélène, J.-J. Sansuc, Fibrés quadratiques et composantes connexes réelles, Math. Ann. 244, 105-134 (1979).

[Co] D.A. Cox, The étale homotopy type of varieties over $\mathbb{R}$, Proc. Amer. Math. Soc. 76, 17-22 (1979).

[D-K] H. Delfs, M. Knebusch, Semialgebraic topology over a real closed field II: Basic theory of semialgebraic spaces. Math. Z. 178, 175-213 (1981).

[F] R. Fossum, Vector bundles over spheres are algebraic. Invent. Math. 8, 222-225 (1969).

[G-R] A.V. Geramita, L.G. Roberts, Algebraic vector bundles on projective space. Invent. Math. 10, 298-304 (1970).

[Ge]$_1$ W.D. Geyer, Ein algebraischer Beweis des Satzes von Weichold über reelle algebraische Funktionenkörper. In : Algebraische Zahlentheorie, Oberwolfach, 1964.

[Ge]$_2$ W.D. Geyer, Dualität bei abelschen Varietäten über reelle abgeschlossen Körpern, J. reine. angew. Math. 293-294, 62-66 (1977).

[H-M] J. Houdebine, L. Mahé, Séparation des composantes connexs réelles dans le cas des varietés projectives, Géométrie algébrique réelle et formes quadratiques, Rennes 1981, SLN 959, p. 358-370.

[I-Sc] F. Ischebeck, H-W. Schülting, Rational and homological equivalence for real cycles, Invent. Math. 94, 307-316 (1988).

[K]$_1$ M. Knebusch, On algebraic curves over real closed fields I. Math. Z. 150, 49-70 (1976).

[K]$_2$ M. Knebusch, On algebraic curves over real closed fields II. Math. Z. 151, 189-205 (1976).

[K]$_3$ M. Knebusch, Some open problems, Queen's Papers in Pure and Appl. Math. 46, 361-370 (1977).

[M]$_1$ L. Mahé, Signatures et componentes connexes. Math. Ann. 260, 191-210 (1982).

[M]$_2$ L. Mahé, Théorème de Pfister pour les variétés et anneaux de Witt réduits, Invent. Math. 85, 53-72 (1986).

[O-P-S] M. Ojanguren, R. Parimala and R. Sridharan, Symplectic bundles over affine surfaces, Comment. Math. Helv. 61, 491-500 (1986).

[P] W. Pardon, A relation between Witt groups and zero cycles in a regular ring, Algebraic K-theory, Geometry and Analysis, SLN 1046, 1984, pp. 261-328.

[Pa] R. Parimala, Witt groups of affine threefolds, Duke Math. J. 57, 947-954 (1988).

[S]$_1$ C. Scheiderer, A remark on the paper, Real components of algebraic varieties and étale cohomology by J.-L. Colliot-Thélène, R. Parimala, Preprint, 1990.

[S]$_2$ C. Scheiderer, Real and étale cohomology, Preprint, 1991.

[Sc] H.W. Schülting, Algebraische und topologische reelle Zykeln unter birationalen Transformationen, Math. Ann. 272, 441-448 (1985).

[Si] R. Silhol, Cohomologie de Galois et cohomologie des variétés algébriques réelles; Applications aux surfaces rationnelles, Bull. Soc. Math. France 115, 107-125 (1987).

[Su] R. Sujatha, Witt group of real projective surfaces, Math. Ann. 288, 89-101 (1990).

[Sw]$_1$ R. Swan, Vector bundles and projective modules. Trans. Amer. Math. Soc. 105, 264-277 (1962).

[Sw]$_2$ R. Swan, Topological examples of projective modules, Trans. Amer. Math. Soc. 230, 201-234 (1977).

[Sw]$_3$ R. Swan, K-theory of quadric hypersurfaces, Ann. of Math. 122, 113-154 (1985).

[T] A. Tognoli, Su una congettura di Nash. Ann. Scuola Norm. Sup. Pisa 27, 167-185 (1973).

[W] E. Witt, Zerlegung reeller algebraischer Funktionen in Quadrate, Schiefkörper über reelle Funktionenkörper. J. Reine Angew. Math. 171, 4-11 (1934).

School of Mathematics
Tata Institute of Fundamental Research
Bombay - 400005
India.
E-mail : PARIMALA@TIFRVAX.BITNET

ON THE CLASSIFICATION OF DECOMPOSING PLANE ALGEBRAIC CURVES

G. M. Polotovskiĭ

(Nizhniĭ Novgorod, USSR)

This report represents a survey of results on the arrangement of curves which are decomposed into product of curves "in general position" in real projective plane RP^2. The problem under consideration is in close connection with the classical question of the Hilbert 16-th problem on the topology and arrangement of non-singular algebraic varieties. Note that in some cases we receive, in fact, the classification of the non-singular affine plane algebraic curves (see sections III - VI below).

Let's denote the non-singular plane real curve of degree m by symbol C_m, and the set of points of this curve in RP^2 by the same symbol. Let's denote the union of transversal intersecting curves C_m and $\hat{C}_n$ $(m \le n)$ by $C_m \sqcup \hat{C}_n$ *) . Two of such curves $C_m \sqcup \hat{C}_n$ and $C'_m \sqcup \hat{C}'_n$ are *equivalent* if topological triples $(RP^2, C_m \sqcup \hat{C}_n, C_m)$ and $(RP^2, C'_m \sqcup \hat{C}'_n, C'_m)$ are homeomorphic, or $m=n$ and $C_m = \hat{C}'_n$ and $C'_m = \hat{C}_n$ (i.e. curves differ only by the sequence of writing the factors).

Definition. The curve $C_m \sqcup \hat{C}_n$ is called the *(M-i)-decomposing curve* if C_m is $(M-p)$-curve, $\hat{C}_n$ is $(M-q)$-curve, the number of common real points of C_m and $\hat{C}_n$ is equal to $mn-r$ and $p+q+r=i$.

The similar definitions are used for the case $C_m \sqcup \hat{C}_n \sqcup \hat{\hat{C}}_k$; in particular, these curves have only non-degenerate two-fold singular points.

The problem of classification of decomposing curves is (as in the case of non-singular curves) most interesting and complicated

---

*) The notations of the factors are distinguished in the case m=n by "the hat".

in "maximalities cases". We shall discuss M-, $(M-1)$- and $(M-2)$-cases mainly.

I. $C_m \sqcup \hat{C}_n$ in cases $m+n<6$ and $C_m \sqcup \hat{C}_n \sqcup \hat{C}_k$ in cases $m+n+k<7$

In this cases the classification is not a problem. For example, restrictions arising from Harnack and Bezout theorems allow only 7 types of topological models for M-decomposing curves $C_2 \sqcup C_3$ - see Fig. 1 where the last model is non-realizable.

Definition. If topological model satisfies the restrictions arising from Harnack and Bezout theorems it is called *admissible*.

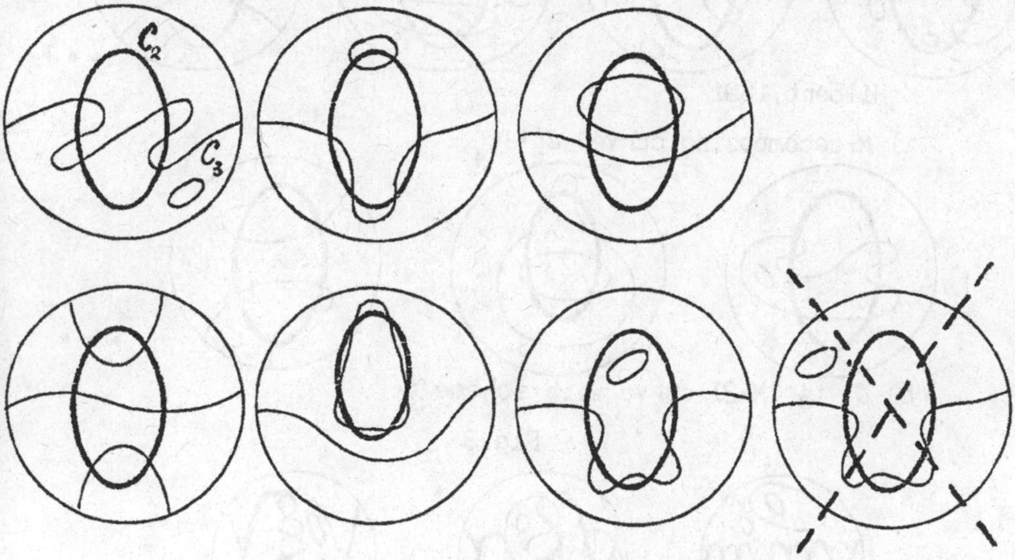

Fig. 1

II. The decomposing curves of 6-th degree

The classification of $C_m \sqcup \hat{C}_n$ where $m+n=6$ is the first non-trivial case. This problem has been solved in my papers [1]-[3]. If all common points of C_1 and C_5 belong to odd branch of C_5 then all realizable types of M-decomposing curves $C_1 \sqcup C_5$ are shown in Fig. 2. Two of these types were unknown earlier.

For example, some realized types of M- and $(M-2)$-decomposing curves $C_2 \sqcup C_4$ are shown in Fig. 3 (the first and the second curves are Hilbert ones; the symbol $\frac{1}{1}$ denotes "one oval in the other").

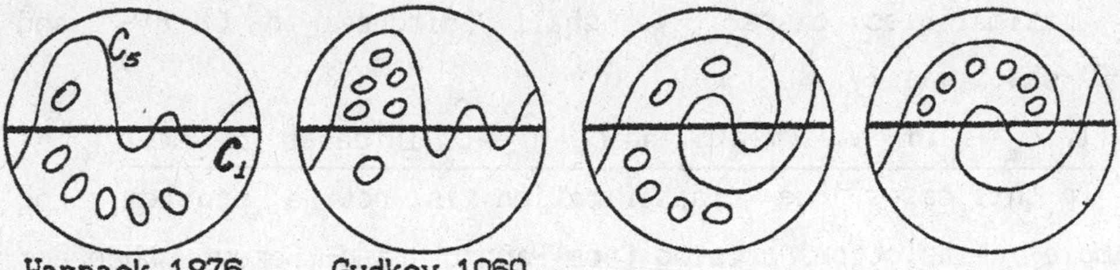

Harnack, 1876 Gudkov, 1969

Fig. 2

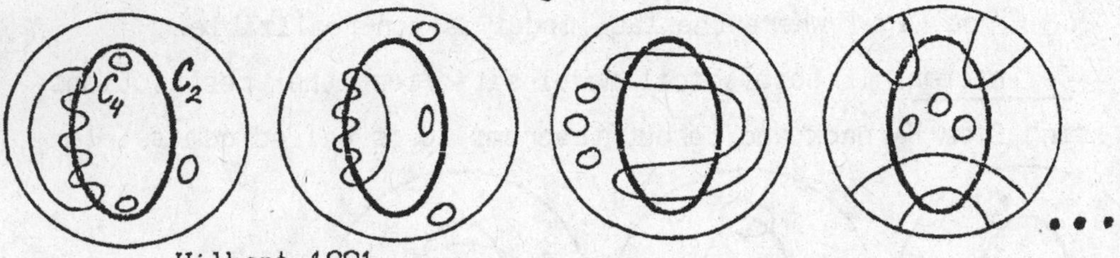

Hilbert, 1891

a) M-decomposing curves $C_2 \sqcup C_4$

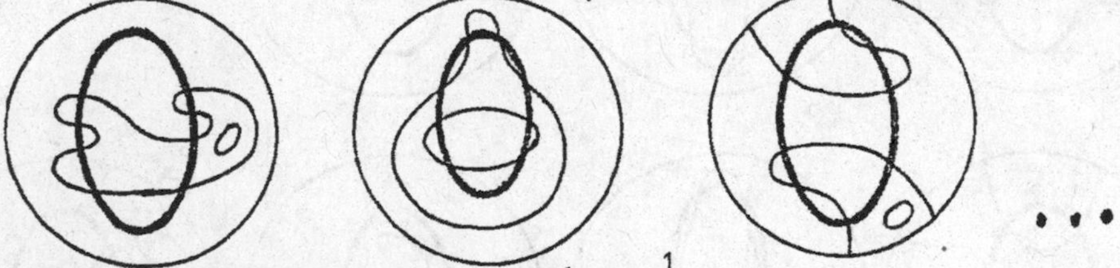

b) C_4 is $(M-2)$-curve with scheme $\frac{1}{1}$

Fig. 3

Fig. 4

When all common points of C_3 and $\hat{C}_3$ belong to odd branches of these curves all realized types of *M*-decomposing curves $C_3 \sqcup \hat{C}_3$ are shown in Fig. 4.

The total statistics of results for the case $m+n=6$ is given in Table 1. I shall remark that 156 models are admissible but non-realizable - examples are shown in Fig. 5. I shall remark also that all models of Fig. 5 and some others satisfy all known by 1980 restrictions (Petrovskii inequality, Kharlamov and Viro inequalities and so on) which were applicable in our situation.

Table 1

$C_m \sqcup \hat{C}_n$	The number of realizable models	
	total	*M*-decomposing
$C_1 \sqcup C_5$	149	15
$C_2 \sqcup C_4$	321	42
$C_3 \sqcup \hat{C}_3$	170	46
total	640	103

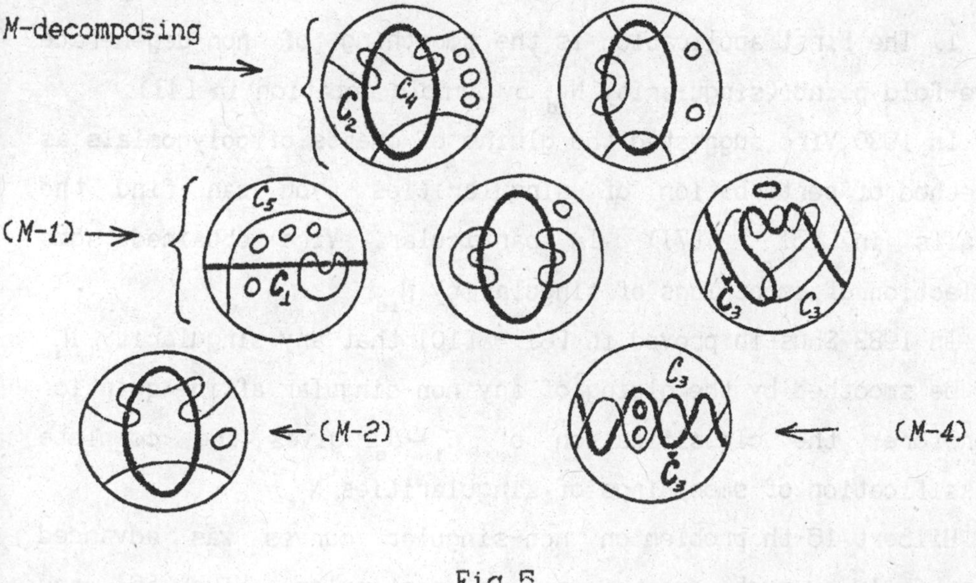

Fig. 5

To obtain the classification of the 6-th degree decomposing curves I used the following techniques: 1) "Geometric combinatorics" for the enumeration of admissible models; 2) Quadratic transformation $RP^2 \to RP^2$ defined by formulae $y_0=x_1x_2, y_1=x_0x_2, y_2=x_0x_1$; 3) Hilbert-Rohn method, i.e. study of bifurcation of curves during the motion in the space RP^{27} of real sextics.

III. The applications of the decomposing curves of 6-th degree

Recall that the classification of $C_1 \sqcup C_n$ is the classification of non-singular affine curves of degree n: take C_1 as a line at infinity in affine plane, see Fig. 6.

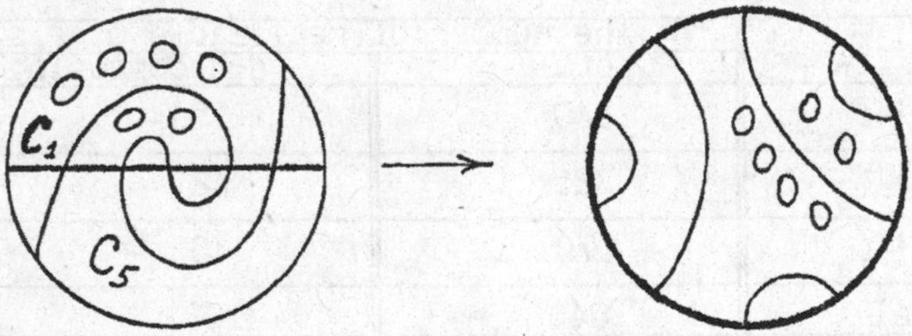

Fig. 6

1) The first application is the smoothing of non-degenerate five-fold point (singularity N_{16} by Arnold notation in [4]).

In 1980 Viro suggested the gluing of charts of polynomials as a method of perturbation of singularities (you can find the details in [5] - [7]). In particular, Viro obtained some collection of smoothings of singularity N_{16}.

In 1983 Shustin proved in [8] - [10] that any singularity N_{16} may be smoothed by the gluing of any non-singular affine quintic. Therefore, the classification of $C_1 \sqcup C_5$ gives the complete classification of smoothings of singularities N_{16}.

Hilbert 16-th problem on non-singular curves was advanced essentially by such smoothings. In particular, Viro [6] and Shustin [11] constructed M-curves, and I [12] - [14] constructed (M-1)-, (M-2)- and (M-3)-curves of degree 8 by this method. You can see the example of such construction in Fig. 7. The curve C_8 with singularity N_{16} was constructed by the hyperbolism from the Gudkov quintic from paper [15], and then by smoothing of singularity J_{10} by Viro method. In these pictures and below I denote the number of empty ovals by Greek letters in the corresponding regions.

some realized values

α	β	γ
1	3	6
5	3	2

realizable values by [1]

δ	ε
2	4
6	0

hyperbolism+smoothing of J_{10} by Viro method [5]

gluing by Viro [5]

The result: the curve $\widetilde{C}_8$ with the scheme
$$\frac{\alpha}{1} + \frac{\delta}{1} + \frac{\beta+1}{1} + (\gamma+\varepsilon)''$$

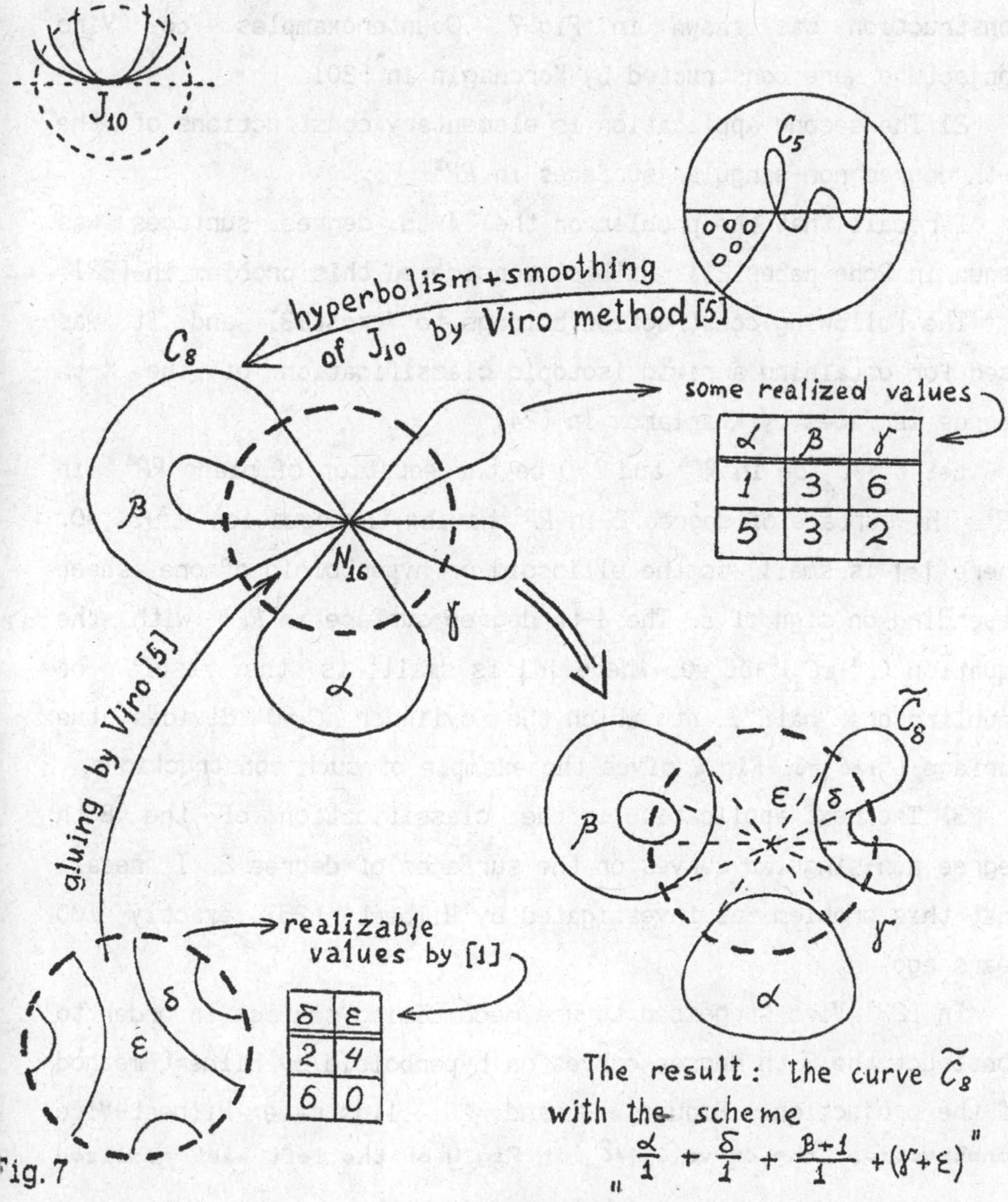

Fig. 7

By analyzing the results of such constructions I found in 1981 (see [16]) that Viro conjecture on M-schemes of degree 8 with the deep nest (see [7]) and Rohlin conjecture on schemes of type I (see [17]) contradict of each other. Later it was found that both conjectures were false. Counterexamples to Rohlin conjecture were constructed in [18], [19]. One variant of the construction was shown in Fig.7. Counterexamples to Viro conjecture were constructed by Korchagin in [20].

2) The second application is elementary constructions of the 4-th degree non-singular surfaces in RP^3.

I recall that the problem on the 4-th degree surfaces was begun in Rohn paper[21]. Hilbert considered this problem in [22].

The following construction belongs to Viro [23] and it was used for obtaining a rigid isotopic classification of the 4-th degree surfaces by Kharlamov in [24].

Let $C_2 \sqcup C_4$ be in RP^2 and L=0 be the equation of plane RP^2 in RP^3. The surface of degree 2 in RP^3 having the equation $L^2+\varepsilon C_2=0$, where $|\varepsilon|$ is small, is the ellipsoid or hyperboloid of one sheet depending on sign of ε. The 4-th degree surface in RP^3 with the equation $(L^2+\varepsilon C_2)^2+\delta C_4=0$, where $|\delta|$ is small, is the result of doubling one "half", into which the cylinder $C_4=0$ divides the surface $L^2+\varepsilon C_2=0$. Fig.8 gives the example of such constructions.

3) The next application is the classification of the 8-th degree non-singular curves on the surfaces of degree 2. I recall that this problem was investigated by Hilbert [25] exactly 100 years ago.

In [23] Viro suggested to use decomposing sextics in order to construct the 8-th degree curves on hyperboloid by Hilbert method of the projection. Figures 9 and 10 illustrate Hilbert-Viro construction: the curve $C_3 \sqcup \hat{C}_3$ in Fig.9 on the left was realized

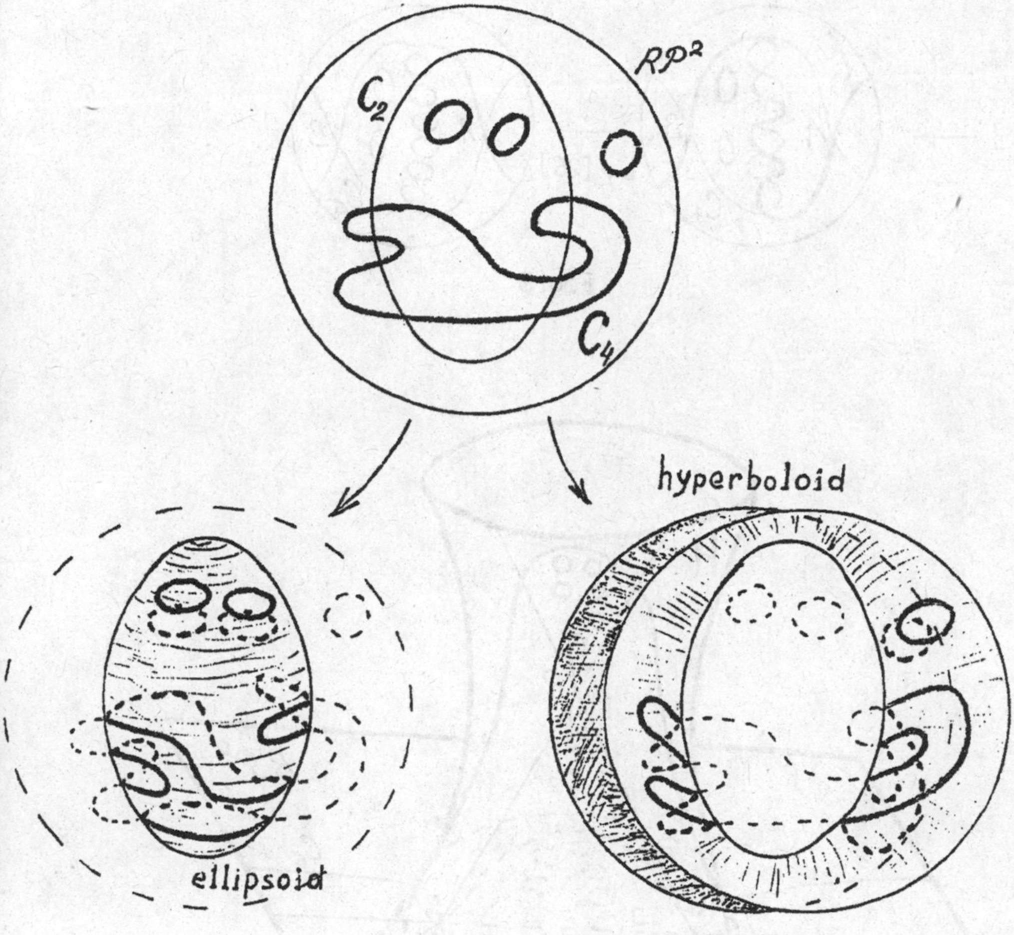

Fig. 8

by me in [1]; we obtain the sextic in Fig. 9 on the right by smoothings of 7 knots. It is possible by Brusotti theorem [26]. Now we make the projection of C_6 with knots A and B on the hyperboloid H intersecting plane RP^2 along the conic which passes through points A and B. The center of projection is the point P of rectilinear generators passing through A and B. As a result we obtain M-curve of degree 8 on H.

By this method, Gudkov [27] obtained the classification of the 8-th degree non-singular curves on hyperboloid, and together

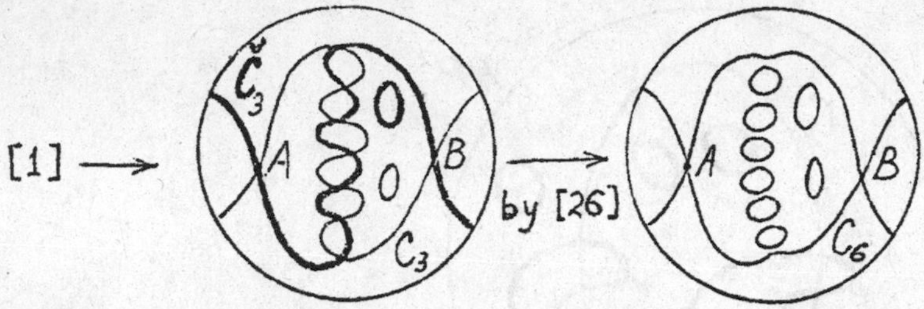

Fig. 9

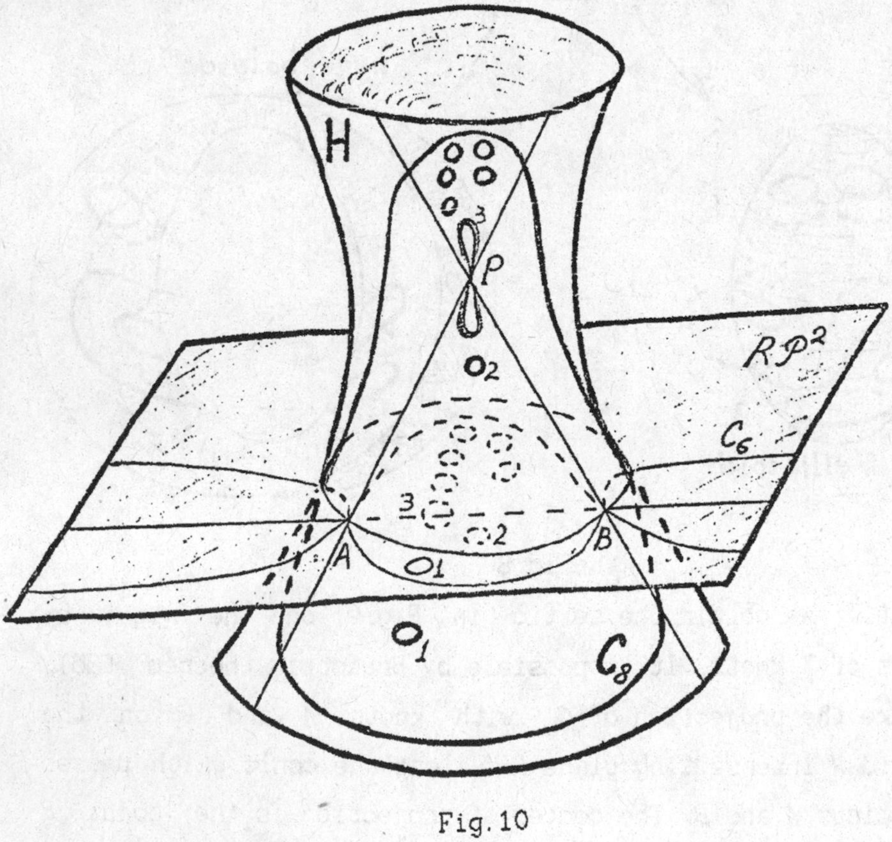

Fig. 10

with Shustin [28] - on ellipsoid.

4) By analysis of classification of decomposing sextics, I found a formulation of the below theorem on complex orientations.

Let P be point of intersection of branches A and B which are provided with orientations. The smoothing of P in Fig. 11a) is called *consented* with orientation, and in Fig. 11b) – *non-consented*.

Fig. 11

Let curves C_m and $\hat{C}_n$ be provided with orientations, and let C_{m+n} be non-singular curve obtained by the consented smoothings of knots of $C_m \sqcup \hat{C}_n$. Let $\tilde{C}_{m+n}$ be obtained also if we preliminarily replace the orientation of one of curves C_m or $\hat{C}_n$ by the opposite one.

(I recall that according to Rohlin [17], the curve is of type I (resp. II) if it divides (resp. not divides) its own complexification; the complex orientation of the curve of type I is the induced border orientation of one of half of its complexification).

Theorem A. Let C_m and $\hat{C}_n$ be of type I and let the number of common real points of C_m and $\hat{C}_n$ be $mn - 2$. Let C'_{m+n} be obtained from $C_m \sqcup \hat{C}_n$ by smoothings of all knots. Then C'_{m+n} is of type I iff $C'_{m+n} = C_{m+n}$ and $\tilde{C}_{m+n}$ is of type II or, vice versa, $C'_{m+n} = \tilde{C}_{m+n}$ and C_{m+n} is of type II. Moreover, the arising orientation is the complex orientation of C'_{m+n}.

The proof of this theorem was reported to me by Fiedler and by Makeev independently. Earlier Fiedler [29] and Marin [30] have obtained similar results in the case when the number of real common points of C_m and C_n equals mn, and later Viro and Zvonilov

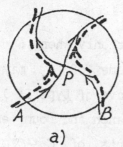

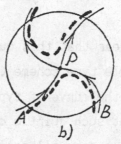

(see [7]) have found the formulation which contains all cases.

IV. The problem on a type C_5 if $C_1 \sqcup C_5$ is given

Let the curve IV. Now I shall consider the following problem: let there is $C_1 \sqcup C_5$ be given. What type (I or II) can the curve C_5 be of ?

It is clear that this question has a sense only when C_5 has the *indefinite* type scheme (i.e. the real scheme of C_5 may be realized by the curve of type I and by the curve of type II). It is known that such quintic consists of odd branch and four empty ovals. In this case all realizable arrangements of $C_1 \sqcup C_5$ are shown in Fig. 12 where the number of distinct situations is equal to 30.

$A_\alpha \ (\alpha + \beta = 4)$　$B_\alpha \ (\alpha + \beta = 4)$　$C_\alpha \ (\alpha + \beta = 4)$　$D_\alpha \ (\alpha + \beta = 3)$

$E_\alpha \ (\alpha + \beta = 3)$　$F_\alpha \ (\alpha + \beta = 2; \alpha \leq \beta)$　$G_\alpha \ (\alpha + \beta = 3; \alpha < \beta)$　$H_\alpha \ (\alpha + \beta = 4; \alpha \leq \beta)$

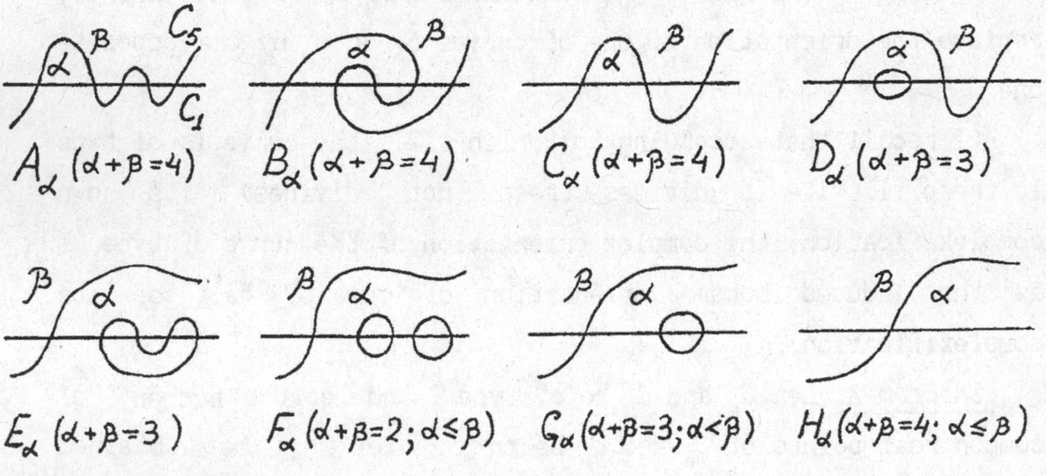

Fig. 12

The following theorem [31] gives the answer to the question above.

Theorem B. The curve C_5 1)can not be of type II in cases A_2, B_3, E_2; 2)can not be of type I in cases $A_1, A_3, A_4, B_0, B_2, B_4, C_4, D_3, E_1, E_3$; 3)may be of an arbitrarily type (I or II, both situations are realized) in all other 17 cases of Fig. 12.

Corollary. Pairs $(C_1 \sqcup C_6, C_1)$ make up $\geq$ 149+17=166 classes with respect to the rigid isotopy in RP^2.

Techniques used for the proof of theorem *B* are the following: quadratic transformations and theorem *A* mentioned above.

Remark. Later in [32], Kharlamov and Viro obtained (by another way) the result from which the part of theorem *B* follows. But authors remarked in [32] that my classification of decomposing sextics was one of the initial points for their paper.

I also remark that some of my curves were used as examples for proving the exactness of some inequalities. See, for example, Zvonilov paper [33]. There are some other applications.

So, the decomposing sextics classification was useful in Hilbert 16-th problem, and it is naturally to investigate decomposing curves of degree 7.

Y. The curves $C_1 \sqcup C_6$

a) In case when all points of intersection of C_1 and C_6 belong to one of the ovals of C_6, *M*-decomposing curves were investigated by Korchagin and Shustin [34]. The Table 2 gives the statistics of the results. In the first column the arrangements of intersecting branches are shown. The topological models for which the question on realizability is open are listed in Fig. 13.

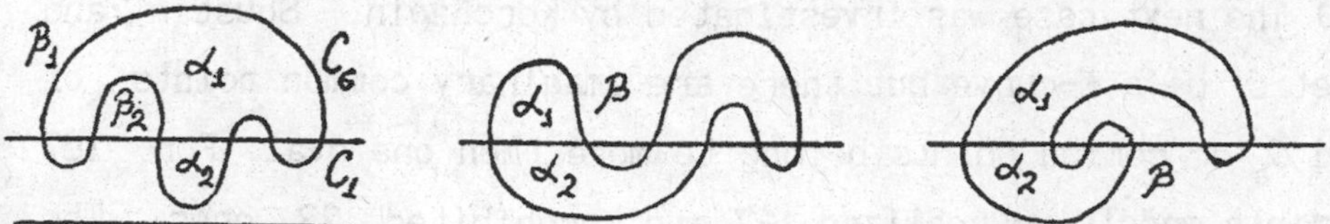

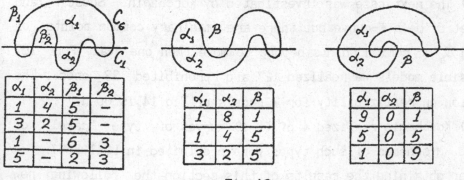

α_1	α_2	β_1	β_2
1	4	5	–
3	2	5	–
1	–	6	3
5	–	2	3

α_1	α_2	β
1	8	1
1	4	5
3	2	5

α_1	α_2	β
9	0	1
5	0	5
1	0	9

Fig. 13

Table 2

The arrangement of intersecting branches	The number of admissible models			
	total	realized	prohibited	the question is open
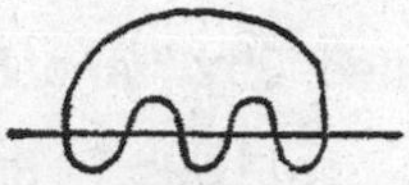 "comb"	51	17	30	4
 "serpent"	27	9	15	3
 "snail"	33	7	23	3
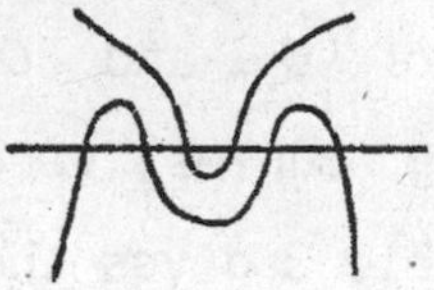 "camel"	18	0	18	0
total	129	33	86	10

b) The next case was investigated by Korchagin, Shustin and me. Let C_6 be a *M*-curve but there are imaginary common points of C_1 and C_6 or common points belong to more then one oval. For 167 admissible models we realized 127 and prohibited 33 ones. The question on realizability for 7 models of Fig. 14 is open.

c) Korchagin realized 4 of (*M-1*)-curves of type "camel" of Fig. 15. *M*-schemes of such types were prohibited in [34], [35].

For obtaining the results of this section the following new techniques were used. First, the method of "reduction of singularities" which was found by Korchagin [36] and Shevaillier [37] independently. Second, the complex orientation theory (see [17] and [38]). Finally, homological Viro method. (You can find details in [34],[35].)

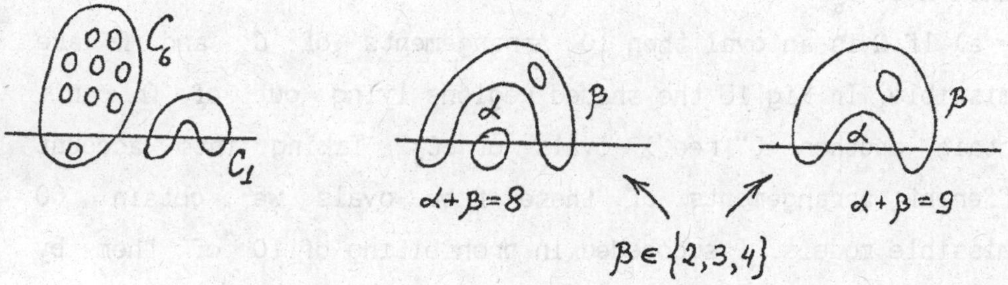

Fig. 14

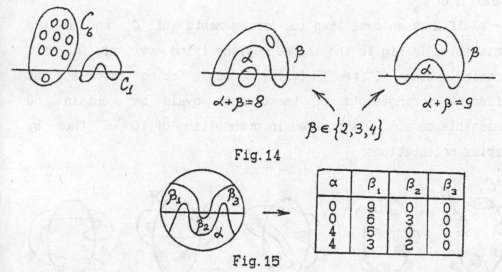

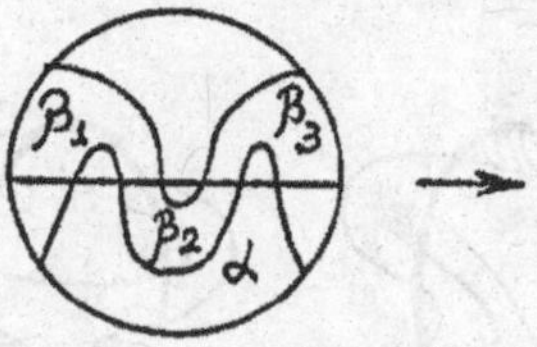

α	β_1	β_2	β_3
0	9	0	0
0	6	3	0
4	5	0	0
4	3	2	0

Fig. 15

Affine sextics were used in [12] for constructions of the 8-th degree (*M-1*)-curves by smoothings of six-fold points.

The next sections of my report have not been published because they are not finished.

VI. The problem on a type C_8 if $C_1 \sqcup C_8$ is given

Let $C_1 \sqcup C_8$ be given. What type (I or II) can the curve C_8 be of?

This question is similar to the problem of section IV above and has a sense only when C_8 has the indefinite type scheme. By Rohlin [17] then C_8 has one of the following schemes:

$$\frac{8}{1},\ \frac{4}{4}4,\ 9,\ \frac{5}{1}1,\ \frac{3}{1}3,\ \frac{1}{1}5,\ \frac{4}{1},\ \frac{2}{1}2.$$

Now I know that 361 models must be investigated. For 80 models I prove that if the model is realizable then C_8 belongs to type II. 26 models are realized by curves of type I. There exist arrangements in which C_6 can be of an arbitrary type (I or II).

VII. The curves $C_2 \sqcup C_5$

In this section I consider only M-decomposing curves such that all 10 common real points of C_2 and C_5 belong to the same branch Ω of C_5.

a) If Ω is an oval then 10 arrangements of C_2 and Ω are admissible. In Fig. 16 the shaded regions lying out of Ω can't contain another ("free") ovals of C_5. Taking into account different arrangements of these free ovals we obtain 60 admissible models. I succeeded in prohibiting of 10 of them by complex orientations.

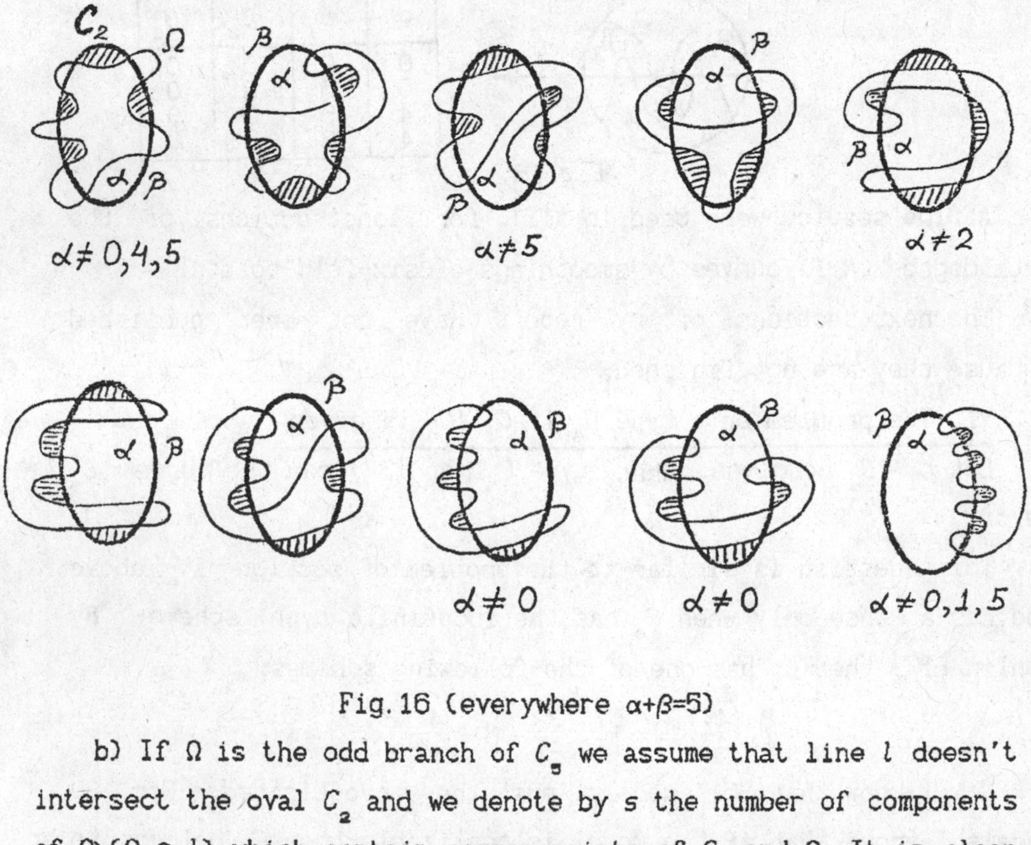

Fig. 16 (everywhere $\alpha+\beta=5$)

b) If Ω is the odd branch of C_5 we assume that line l doesn't intersect the oval C_2 and we denote by s the number of components of $\Omega \setminus (\Omega \cap l)$ which contain common points of C_2 and Ω. It is clear that s may be equal to 1, or 3, or 5. Examples of admissible models you can see in Fig. 17.

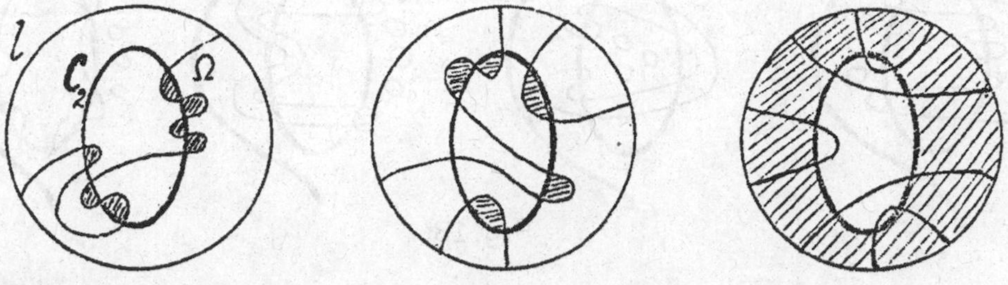

s=1 s=3 s=5

Fig. 17

The results which I succeeded in obtaining at present time are reflected in Table 3.

Table 3

Ω	s	The number of admissible models				
		$C_2 \sqcup \Omega$	$C_2 \sqcup C_5$	prohibited	realized	the question is open
oval	-	10	60	10	-	50
odd	1	33	2223	1449	19	705
branch	3	15	1375	1075	43	257
	5	2	14	8	4	2
Total		60	3672	2592	66	1014

VIII. *M*-decomposing curves $C_3 \sqcup C_4$

In this section I consider only two cases:

a) All common points of C_3 and C_4 belong to the oval of C_3 and to one of the ovals of C_4. In this case 16 arrangements of intersecting ovals are admissible. Taking into account free ovals of C_4 we obtain 64 possibilities in which I prohibited 19 models by complex orientations and realized 4 models which you can see in Fig. 18. The question on realizability of 41 others is open.

b) All common points of C_3 and C_4 belong to odd branch of C_3 and to one of the ovals of C_4, and this oval is *convex*. There are 3 admissible models for intersecting branches, see Fig 19. Taking into account free ovals we have again 64 arrangements in which I

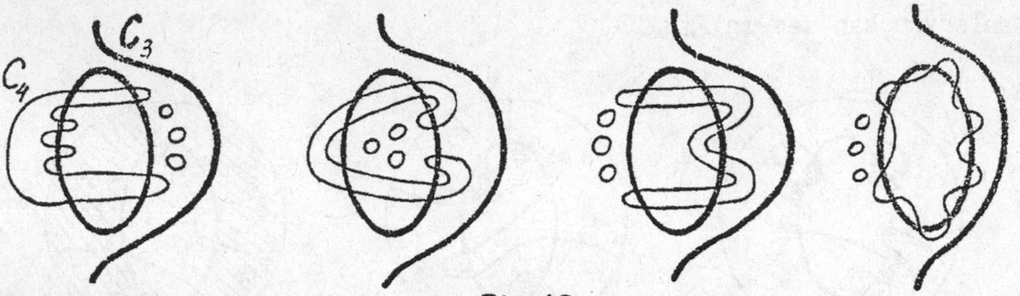

Fig. 18

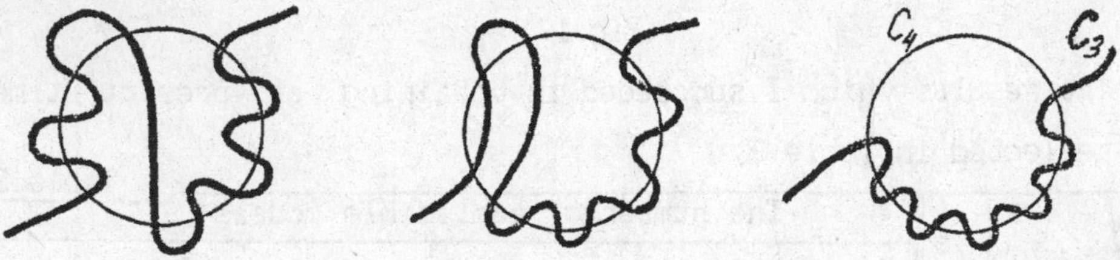

Fig. 19

prohibited 14 by complex orientations.

IX. *M*-decomposing curves $C_5 \sqcup C_1 \sqcup \hat{C}_1$ where all 10 common points of C_5 and $C_1 \sqcup \hat{C}_1$ belong to odd branch of C_5

Such curves appeared in Brusotti paper [39] (so-called "bifrontal generating") and were used for constructions of the 10-th degree *M*-curves in [40].

The curve $C_1 \sqcup \hat{C}_1$ separates RP^2 in two disks *A* and *B* (Fig. 20).

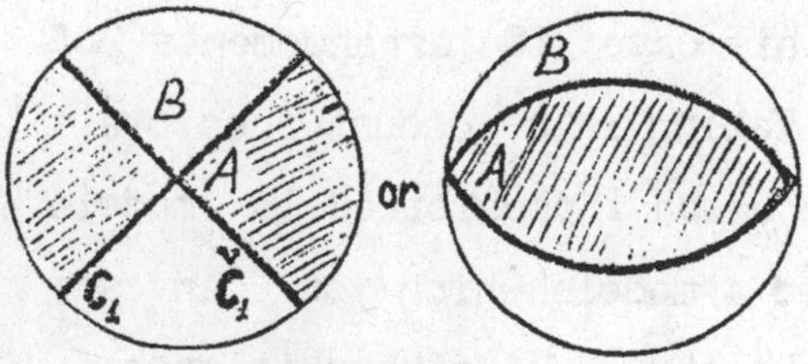

Fig. 20

Let *s* (resp. *t*) be the number of arcs of Ω, connecting C_1 and $\hat{C}_1$ in the disc *A* (resp. *B*). Since *A* and *B* are equivalent then it is sufficient to consider only $s \geq t$. It is evident that $s+t$ is even and $2 \leq s+t \leq 8$. You can see examples of arrangements of $C_1, \hat{C}_1$ and Ω for different values of (s, t) in Fig. 21.

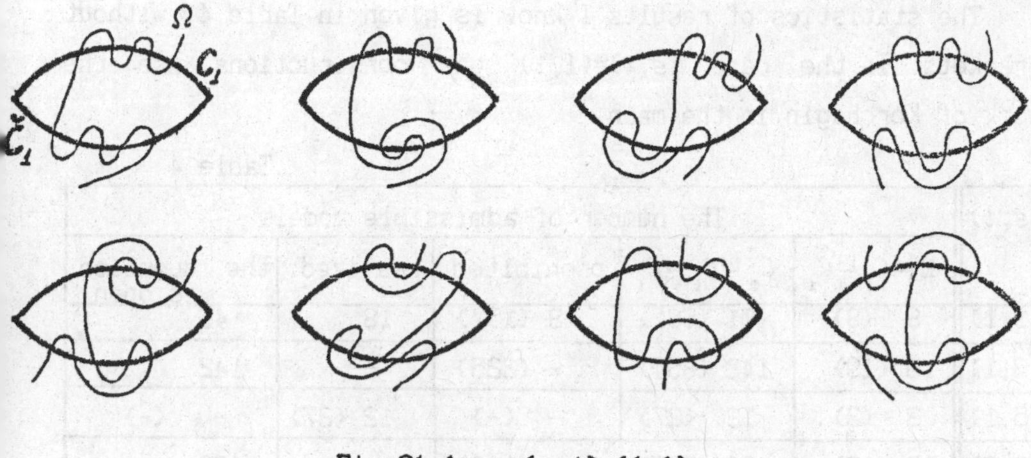

Fig. 21.1 $(s,t)=(1,1)$

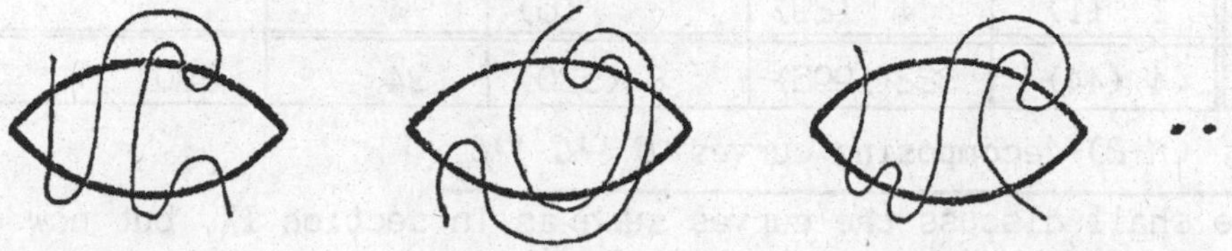

Fig. 21.2 $(s,t)=(3,1)$ (Total 19 pictures)

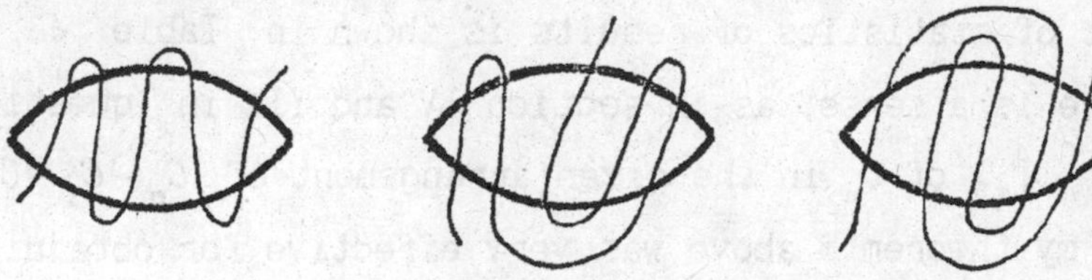

Fig. 21.3 $(s,t)=(5,1)$

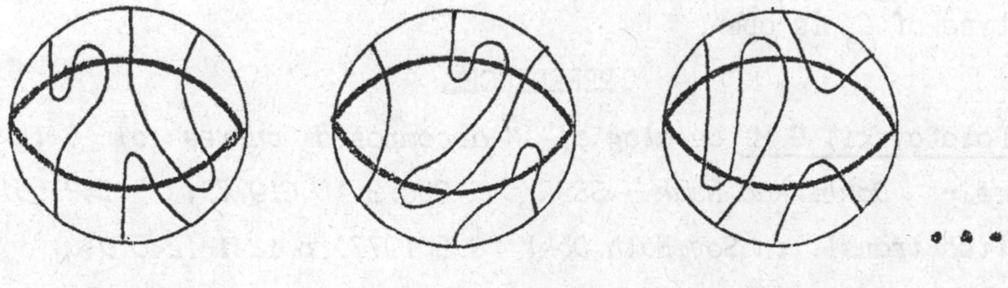

Fig. 21.4 $(s,t)=(3,3)$ (Total 13 pictures)

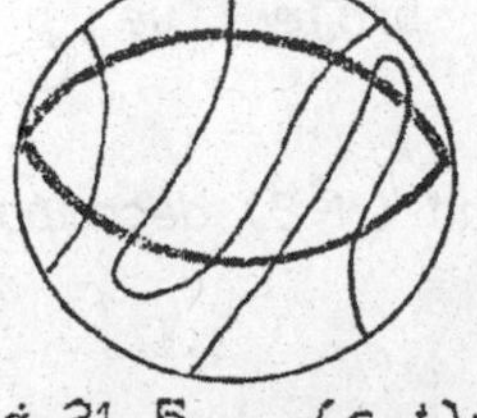

Fig. 21.5 $(s,t)=(5,3)$

The statistics of results I know is given in Table 4 without brackets. In the case $(s,t)=(1,1)$ the constructions are the work of Korchagin in the main.

Table 4

(s,t)	The number of admissible models				
	$\Omega \sqcup C_1 \sqcup \hat{C}_1$	$C_5 \sqcup C_1 \sqcup \hat{C}_1$	prohibited	realized	the question is open
(1,1)	8 {8}	71 {561}	8 {157}	18	45
(3,1)	19 {19}	142 {854}	- {225}	-	142
(5,1)	3 {3}	12 {27}	- {-}	12 {27}	- {-}
(3,3)	13 {13}	93 {525}	- {142}	-	93
(5,3)	1 {1}	4 {29}	- {6}	4	-
Total	44 {44}	322{1996}	8 {530}	34	280

X. $(M-2)$-decomposing curves $C_5 \sqcup C_1 \sqcup \hat{C}_1$

We shall discuss the curves such as in section IX, but now C_5 is $(M-2)$-curve, i.e. consists of odd branch and four empty ovals. The part of statistics of results is shown in Table 4 in brackets. There is a sense, as in section IV and YI, in question on type (I or II) of C_5 in the given arrangement of $C_5 \sqcup C_1 \sqcup \hat{C}_1$. I remark that my theorem B above was very effective for obtaining of these results. In particular, only in 55 cases the question on the type of C_5 is open.

REFERENCES

1. Polotovskii G.M. *Catalog of M-decomposed curves of 6-th degree.- Dokl. Akad. Nauk SSSR, 236:3 (1977), 548-551. (English transl. in Sov. Math. Dokl 18:5(1977), p. 1241-1245.)*

2. Polotovskii G.M. *Complete classification of M-decomposing curves of 6-th degree in real projective plane.- deposited in VINITI N1349-78Dep, 1-103.*

3. Polotovskii G.M. *(M-1)- and (M-2)-decomposing curves of 6-th*

degree.-*Methods of the Qualitative Theory of Diff. Equations, Gos. Univ., Gorky (1978), 130-148.*

4. Arnold V.I., Varchenko A.N., Gusein-Zade S.M. *Singularities of differentiable mappings, Vol. I, "Nauka", Moscow, 1982 (English transl., Birkhäuser, 1985.)*

5. Viro O.Y. *Gluing of plane real algebraic curves and constructions of curves degrees 6 and 7.- Lect.N.Math., Vol. 1960 (1984), 187-200.*

6. Viro O.Y. *Gluing algebraic hypersurfaces, smoothings of singularities and constructions of curves.- Trudy Leningr. Mezhdunar. Topol. Confer., Leningrad, 1983, 149-197.*

7. Viro O.Y. *Progress in topology of real algebraic varieties over the last six years.- Usp.Mat.Nauk., 41:3 (1986), 45-67.*

8. Shustin E.I. *The Hilbert-Rohn method and smoothings of complicated singular points of curves of 8-th degree.- Usp.Mat.Nauk., 38:6 (1983), 157-158.*

9. Shustin E.I. *Real smoothings of simple 5-fold singular point.-deposited in VINITI №6872-83Dep, 1-21.*

10. Shustin E.I. *The Hilbert-Rohn method and smoothing of singular points of real algebraic curves.- Dokl.Akad.Nauk SSSR, 281:1 (1985), 33-36.*

11. Shustin E.I. *Independence of smoothings of singular points and new M-curves of degree 8.- Usp.Mat.Nauk., 40:5 (1985), 212.*

12. Goryacheva T.V., Polotovskiĭ G.M. *Constructions of (M-1)-curves of 8-th degree.- deposited in VINITI №4441 85Dep, 1-30.*

13. Polotovskiĭ G.M. *(M-2)-curves of 8-th degree: constructions, open problems.- deposited in VINITI N1185-85Dep, 1-194.*

14. Polotovskiĭ G.M., Tscherbakova A.V. *On the construction of (M-3)-curves of 8-th degree.- deposited in VINITI №4440-85dep, 1-23.*

15. Gudkov D.A. *Construction of new series of M-curves.- Dokl.Akad.Nauk SSSR, 200:6 (1971), 1269-1272.*

16. Polotovskii G.M. *On the classification of (M-2)-curves of order 8.- Selecta Math. Sovietica 9:4 (1990),403-409.*

17. Rohlin V.A. *Complex topological characteristics of real algebraic curves.- Usp.Mat.Nauk.,33:5 (1978), 77-89 (English transl. in Russian Math.Surveys, 33 (1978) .)*

18. Shustin E.I. *Counterexamples to Rohlin conjecture.- Funkt. Anal.Appl., 19:2 (1985), 94-95.*

19. Polotovskii G.M.,Shustin E.I. *Construction of counterexamples to Rohlin conjecture.- Usp.Mat.Nauk 39:4 (1984), 113.*

20. Korchagin A.B. *New M-curves of degrees 8 and 9.- Dokl. Akad.Nauk SSSR, 306:5 (1989), 1038-1041 (English transl. in Sov. Math. Dokl., 39:3(1989),p.569-572.)*

21. Rohn K. *Die Fläche 4 Ordnung hinsichtlich ihrer Knotenpunkte und ihrer Gestaltung.- Preisschrift der Fürstl. Jablonowskischen Geselschaft, Leipzig,1886.*

22. Hilbert D. *Uber die Gestalt einer Fläche vierter Ordnung.- Gött.Nachrichten (1909), 308-313.*

23. Viro O.Y. *Construction of multicomponent real algebraic surfaces.- Dokl.Akad.Nauk SSSR, 248:2 (1979), 279-282.*

24. Kharlamov V.M. *On the classification of non-singular surfaces of degree 4 in RP^3 with respect to rigid isotopies.- Funkt.Anal.Appl., 18:1 (1984), 49-56.*

25. Hilbert D. *Ueber die reellen Züge algebraischer Kurven.- Math.Ann., 38 (1891), 115-138.*

26. Brusotti L. *Sulla "piccola variazione" di una curva piana algebrica reali.- Rend.Rom.Acc.Linc.(5) 30 (1921), 375-379.*

27. Gudkov D.A. *On the topology of algebraic curves on hyperboloid.- Usp.Mat.Nauk.,34:6 (1979), 26-32.*

28. Gudkov D.A., Shustin E.I. *The classification of non-singular curves of 8-th degree on ellipsoid.- Methods of the Qualitative Theory of Diff. Equations, Gos. Univ., Gorky (1980), 104-107.*

29. Fiedler T. *Eine Beschränkung fur die Lage von reellen ebenen algebraischen Kurven.- Beiträge zur Algebra und Geometrie, 11 (1981), 7-19.*

30. Marin A. *Quelques remarques sur les courber algébriques planes reelles.-Publ. Math. Univ. Paris VII, 9 (1980), 51-68.*

31. Polotovskii G.M. *The connection between rigid isotopic class of nonsingular curve of 5-th degree in RP^2 and its arrangement with respect to line.- Funkt. Anal. Appl., 20:4 (1986), 87-88.*

32. Kharlamov V.M., Viro O.Y. *Extensions of the Gudkov-Rohlin congruence.- Lect. N. Math., 1346 (1988), 357-406.*

33. Zvonilov V.I. *Strengthened Petrovskiî and Arnold inequalities for curves of odd degree.- Funkt. Anal. Appl., 13:4 (1979), 32-39.*

34. Korchagin A.B., Shustin E.I. *Affine curves of degree 6 and smoothings of a non-degenerate sixth order singular point.- Isv. Akad. Nauk SSSR, Ser. Matem., 52:6 (1988), 1181-1199. (English transl. in Math. USSR Isvestiya, 33:3 (1989), p. 501-520.)*

35. Shustin E.I. *On the isotopic classification of affine M-curves of degree 6.-Methods of the Qualitative Theory of Diff. Equations, Gos. Univ., Gorky (1988), 97-105.*

36. Korchagin A.B. *On the reduction of singularities and the classification of nonsingular affine sextics.- deposited in VINITI N1107-B (1986), 1-16.*

37. Chevaillier B. *A propos des courbes de degree 6.- C.R. Acad. Sci. Paris, Ser. A 302:1 (1986), 33-38.*

38. Fiedler T. *Pencils of lines and the topology of real algebraic curves.- Isv. Akad. Nauk SSSR, Ser. Matem., 46:4 (1982), 853-863. (English transl. in Math. USSR Isvestiya, 21 (1983).)*

39. Brusotti L. *Nuovi metodi costruttivi di curve piano d'ordine assegnato, dotate del massimo numero di circuiti.-* Rend. Ist. Lomb., Ser. II^a, 47(1914)-49(1916).

40. Chislenko Y.S. *M-curves of degree ten.-* Zap. nauchn. semin. nLOMI, v (1982), 146-161. (English transl. in J. Soviet Math., 26 (1984), no. 1)

Real Algebra and Its Applications to Geometry In the Last Ten Years: Some Major Developments and Results

Claus Scheiderer
Fakultät für Mathematik, Universität Regensburg, 8400 Regensburg, Germany

This is the slightly expanded version of a talk I have given at the conference. Its object was to report on some progress obtained in real algebra and geometry during the past decade. I want to stress right at the beginning that no attempts have been made to give a survey of encyclopedic character. This cannot and will not be a comprehensive account; rather choices had to be made about which topics to include. They reflect the personal taste of the organizers and of myself, and are not meant to imply any kind of ranking.

The presentation is divided into three parts. The first is concerned with problems on sums of squares in fields and rings, both qualitative and quantitative ones. The second assembles results on sums of higher (even) powers, together with related subjects like real valuations and real holomorphy ring. The last part presents some applications of real algebra techniques to geometric questions. Here I had to be even more selective, since geometric problems have always been a strong incentive for the development of algebraic methods, and so the applications abound. On the other hand, it is a pleasure and sometimes a surprise to see how this can be turned around, so that geometry is used to solve purely algebraic problems. We shall meet this phenomenon in several instances.

Among the topics which had to be excluded are pure quadratic form theory (although quadratic forms are very much present, at least behind the scenes), Galois theory and Galois cohomology of fields in connection with reality phenomena, and the extensive work on semialgebraic topology and its abstract generalizations in the real spectrum.

It is a pleasure for me to express my gratitude to D. Gondard, M. Knebusch, T.Y. Lam, L. Mahé, A. Prestel and D. Shapiro for their generous help with the preparation of this talk.

1. Sums of squares

As my point of departure I want to take the famous 17th Hilbert problem. It is well known that through its solution by Emil Artin it catalysed the development of specific real algebraic techniques. Though Artin solved it completely in its original form, both the problem and its solution lead naturally to a variety of related questions, many of which are still unsolved.

Hilbert originally had asked if a rational function $f \in \mathbb{Q}(X)$, $X := (X_1, \ldots, X_n)$, which takes non-negative values where it is defined, is a sum of squares of rational functions. Artin's proof showed much more generally, for any ordered field (k, P), that every rational function in $k(X)$ which takes non-negative values over the *real closure* $\bar{k}_P$ can be written $f = \sum_i a_i g_i^2$ with $a_i \in P$ and $g_i \in k(X)$. Thus, a natural way to generalize the question to arbitrary fields k (with possibly many orderings) is to ask: If k is a field and T is a preordering of k, when is it true that every rational function $f \in k(X)$, which over k takes values in T, can be written $f = \sum_i t_i g_i^2$ with $t_i \in T$ and $g_i \in k(X)$? This has been called the weak Hilbert property for (k, T). In the cases where T is an ordering, respectively where $T = \Sigma k^2$ and $[k^* : T^*] < \infty$, it had been shown in the 70s that this property is equivalent to k being dense in its real closure with respect to T, respectively in each of its real closures (Gondard, Ribenboim, McKenna, Prestel). However the general solution was given only recently by Zeng Guangxin in [Zeng 88], see also [Zeng 91]. He showed that generally the weak Hilbert property is equivalent to a "local density" property of (k, T). He also proved that this property can be axiomatized in the language of preordered fields.

Fix a real closed field R, and let $P_{n,d}$ be the set of all forms f in n variables of degree d over R which are positive semidefinite (*psd*), i.e. which over R take non-negative values only. One may regard $P_{n,d}$ as a subset of R^N, where $N = \binom{n+d-1}{d}$ is the number of degree d monomials in n variables. As such, it is a closed convex semialgebraic cone. Identifying a form f with its tuple of coefficients in R^N, one may ask if there is a solution to Hilbert's question which holds uniformly for all psd forms f and whose entries depend rationally on the parameters, i.e. on (the coefficients of) f. To make the question sensible, one has to allow linear combinations of sums of squares with non-negative coefficients (which should depend rationally on f, too). Also one may want to know if such a uniform solution exists which depends at least continuously on f. These questions were posed by G. Kreisel in 1962 (in a more general form), after Daykin had shown that a rational solution exists which is almost uniform in that it needs only finitely many "case distinctions". The answers to Kreisel's questions were given by Ch. Delzell in his 1980 Stanford thesis: On the one hand, a uniform rational solution is impossible whenever $d > 2$. This is proved by showing that the closed set $P_{n,d}$ is not "basic", i.e. cannot be presented by a conjunction of inequalities (see §3). On the other, there is always a uniform solution which depends semialgebraically and continuously on the parameters. For these results, and for more precise formulations, see [Delzell 82a+b] and [Delzell 84]. The existence proof in [Delzell 84] is a nice application of results from real algebra and geometry, like real Nullstellensatz (in a homogeneous version) and semialgebraic triangulation.

Hilbert himself had already shown that a psd form is not in general a sum of squares of forms. Thus one is led to study $\Sigma_{n,d}$, the closed convex subcone of $P_{n,d}$ consisting of the sums of squares of forms, and the relations between the two cones. This has been done in several articles by M.D. Choi, T.Y. Lam and B. Reznick, starting already in the 70s. It is in general a non-trivial task to exhibit psd forms which are not sums of squares (*sos*), and to verify this. Historically, the first (published) example was given only in 1967, by Motzkin. Of particular interest is it to find the extremal psd forms, since they generate $P_{n,d}$ by convex combinations, and since such a form is not sos unless it is a perfect square. Reznick observed that many of the known examples of forms in $P_{n,d} - \Sigma_{n,d}$ could be obtained by substituting monomials into the arithmetic-geometric inequality (which compares arithmetic and geometric mean). In [Reznick 89] he studies these "agiforms" in great detail, thereby obtaining conditions for agiforms to be sos or to be extremal. These conditions can be used to produce many new examples of psd forms which are not sos. They are of geometric nature and involve the counting of integral points inside certain polytopes associated with such forms. The techniques partially extend and generalize methods used earlier by Motzkin, Choi and Lam to study isolated examples. A very explicit case study is made for $d = 6$ in [Choi Lam Reznick 87], where the intersection of the two cones with the space of even symmetric sextics is determined completely. As a by-product, a family of largely new symmetric polynomial inequalities is obtained. See also [Choi Lam Reznick 91], where new psd ternary sextics are discovered which are extremal and which generalize classical inequalities of Lehmus and Schur. The constructions are inspired by considerations in the euclidean geometry of triangles. I also want to mention Reznick's study of the "Hurwitz form" $H_n = x_1^n + \cdots + x_n^n - nx_1 \cdots x_n$ which is closely related to the arithmetic-geometric inequality [Reznick 87]. While Hurwitz had already shown that H_n is sos (for n even), his construction required a great number of squares; Reznick reduces this length considerably by his "parsimonious construction".

I now pass to quantitative problems. A celebrated theorem of Pfister says that the level of a field (also called *Stufe*), if finite, is a power of 2, and that every power of 2 occurs. Recall that the level $s(A)$ of a (commutative) ring A is the minimal n such that there are $a_1, \ldots, a_n \in A$ with $-1 = a_1^2 + \cdots + a_n^2$ (and is ∞ if -1 is not a sum of squares). It is natural to ask what happens over rings which are not fields; in particular, what type of integer can be the level of a ring? This question was raised by R. Baeza [Knebusch 77, Problem 13]. It was answered — and, in fact, much more than that — by Z.D. Dai, T.Y. Lam and C.K. Peng at the beginning of the 80s [Dai Lam Peng 80]. Later, a more elaborated and comprehensive theory was set up by the first two of them [Dai Lam 84]. Plainly, the answer is that any positive integer is the level of a suitable ring A. In fact, A can be chosen to be a domain whose field of fractions has as its level any preassigned power of 2 not exceeding n. But at least as interesting as the answer itself are the methods, which extend way beyond this specific application. The basic idea is that there is a natural relationship between the arithmetic of sums of squares in rings and the topology of spheres. At quite an elementary level, this is present already in the short and very elegant proof of the fact that $A = \mathbb{R}[X_1, \ldots, X_n]/(1 + X_1^2 + \cdots + X_n^2)$ has level n: Assuming $s(A) < n$, one takes a representation by less than n squares and plugs in purely imaginary arguments. But then a direct application of the Borsuk-Ulam theorem to the odd degree part gives the desired contradiction!

The Borsuk-Ulam theorem states that for any map $f : \mathbb{R}^n \to \mathbb{R}^{n-1}$ which satisfies $f(-x) = -f(x)$ for all x there is $x \in S^{n-1}$ with $f(x) = 0$. Shortly after the above application to the level of rings appeared, algebraic proofs were found for the polynomial version of this theorem [Knebusch 82] [Arason Pfister 82]. Arason and Pfister actually obtained a strong generalization of one of the results of [Dai Lam Peng 80], which in particular gives the formula

$$s\Big(k[X_1,\dots,X_n]/(1+X_1^2+\dots+X_n^2)\Big) = \min\{n, s(k)\}$$

for any field k.

But I want to return to the work of Dai and Lam. Given a topological space X with an involution ϵ (i.e. an action of the group $\mathbb{Z}_2$ on X), they form the $\mathbb{R}$-algebra A_X of all equivariant (continuous) maps $X \to \mathbb{C}$ (the involution on $\mathbb{C}$ being complex conjugation). Generalizing the above argument, they show the

Level Theorem [Dai Lam 84, Theorem 3.1]:

$$s(A_X) = \inf\Big\{n \geq 1 : \textit{ there is an equivariant map } X \to S^{n-1}\Big\} =: s(X).$$

The involutions on the spheres are the antipodal maps $x \mapsto -x$. The invariant $s(X)$ of a space with involution, and its dual (for obvious reasons called the level and the colevel of X), had been studied by topologists before, under the names coindex and index; but the relation to algebra had not been observed.

But Dai and Lam do not stick to spheres and to consequences of Borsuk-Ulam. More generally, they consider Stiefel manifolds together with various natural involutions on them. Equivariant maps into these spaces are related to quadratic forms over rings, more specifically, to the study of subforms of the type $r\langle 1\rangle \perp s\langle -1\rangle$ of forms $n\langle 1\rangle$. Applying deep theorems from topology, this yields striking results. For example:

Theorem. *If A is any (commutative) ring, and if $\langle -1\rangle$ is a subform of $n\langle 1\rangle$, then also $\rho(n)\langle -1\rangle$ is a subform of $n\langle 1\rangle$. But in general, $(1+\rho(n))\langle -1\rangle$ is not a subform of $n\langle 1\rangle$.*

Here $\rho(n)$ is the Hurwitz-Radon function, defined by $\rho(2^{4a+b}u) = 8a+2^b$ for integers a, b, u with u odd and $0 \leq b \leq 3$. The first assertion is more or less well known, but the second is quite deep and rests on Adams' results about vector fields on spheres.

Theorem. *Given $q < n$, there are (honest!) rings A over which $q\langle 1\rangle \perp \langle -1\rangle$ is a subform of $n\langle 1\rangle$, but not so $(q+1)\langle 1\rangle \perp \langle -1\rangle$.*

In particular, a Pfister form can be isotropic without being hyperbolic. Many other interesting results are obtained by Dai and Lam, which unfortunately cannot be mentioned here.

The level of the odd-dimensional real projective spaces $\mathbb{RP}^{2m-1}$ (on which, by identification of $\mathbb{R}^{2m}$ with $\mathbb{C}^m$, multiplication with i induces a natural fixed point free involution) has later been computed in [Stolz 89], see also [Pfister Stolz 87]. It is related to systems of complex quadratic forms whose imaginary part is anisotropic.

Another beautiful application of topology to sums of squares is given in [Dai Lam Milgram 81]. Again it was a question of Baeza [Knebusch 77, Problem 12] who asked whether in a ring A the units which are sums of 2^n squares do always form a group (i.e.

are closed under products). Of course, this is known to be true for fields (Pfister), and more generally for semilocal rings (Knebusch). By classical identities it is also true for $2^n = 1, 2, 4, 8$ in any ring. In [Dai Lam Milgram 81] it is shown that there are no other integers m for which $US_m(A)$, the units which are sums of m squares, are a group in *every* ring A. So Pfister's multiplicativity result does not in general extend beyond the classical cases. Not surprisingly, the counterexample is the "generic" $\mathbb{R}$-algebra for this question, namely $A = \mathbb{R}[X_1, \ldots, X_m, Y_1, \ldots, Y_m, (\Sigma X_i^2)^{-1}, (\Sigma Y_i^2)^{-1}]$. If $(\Sigma X_i^2)(\Sigma Y_i^2)$ is a sum of m squares, the authors produce a map $S^{2m-1} \to S^m$ with odd Hopf invariant, which, according to Adams, can exist only for $m = 1, 2, 4, 8$. In fact, significantly stronger results are obtained. Using results of Adams on the J-homomorphism in K-theory, the authors show that a bi-skew (= bi-equivariant) mapping $S^{r-1} \times S^{n-1} \to S^{n-1}$ can exist only if $r \leq \rho(n)$. They conclude that $US_r(A) \cdot US_n(A) \subset US_n(A)$ is true in every ring A (if and) only if $r \leq \rho(n)$. The special case $r = n$ gives back the first result.

With this, one is close to the composition problem for sums of squares, a classical and fascinating subject which dates back to Hurwitz. It asks for formulas

$$(x_1^2 + \cdots + x_r^2)(y_1^2 + \cdots + y_s^2) = z_1^2 + \cdots + z_n^2 \qquad (*)$$

where the x_i, y_j are variables and the $z_\nu = z_\nu(x, y)$ are supposed to be bilinear in x and y. Originally, this was asked over $\mathbb{C}$ or $\mathbb{R}$, but the problem may be considered over any commutative ring k. Hurwitz and Radon completely settled the "classical" case $s = n$, proving that $(*)$ exists (over a field of characteristic not 2) if and only if $r \leq \rho(n)$. In general the smallest n for which $(*)$ exists is written $r *_k s$. Closely related is the question of the existence of a nonsingular bilinear map $\mathbb{R}^r \times \mathbb{R}^s \to \mathbb{R}^n$. Here the minimal n is written $r \# s$. In general, these numbers are very hard to compute. For upper bounds one needs more or less explicit constructions, while lower bounds have mostly been obtained by finding topological obstructions (this goes back to the work of Stiefel and Hopf in the 1940's). I want to mention briefly some remarkable progress which has been made in the last few years. An excellent survey of these matters, as of 1984, is [Shapiro 84]; see also [Lam 84].

Concerning the existence problem, S. Yuzvinsky claimed to have constructed three new infinite series of composition formulas $(*)$, starting respectively with $(r, s, n) = (10, 10, 16)$, $(12, 20, 32)$ and $(14, 40, 64)$ [Yuzvinsky 81], [Yuzvinsky 84]. However, T. Smith and T.Y. Lam discovered serious flaws in his reasoning. Fortunately, they managed to fix the proofs in the first two of the three cases, the third remains open. The construction, which also yields further new pairings, uses ideas from group representations {Lam Smith 90}[1].

On the non-existence side, K.Y. Lam presented new and beautiful geometric ideas which have since then been used to improve many of the previously known lower bounds ([Lam 85], see also [Lam 84]). Associated with a composition formula $(*)$ (over $\mathbb{R}$) is its Hopf map $H : S^{r+s-1} \to S^n$. Lam shows that for any point $Q \in S^n$, the

[1] References given in curly braces had, to the best of my knowledge, not (yet) appeared in print when this article was typed (July 1991). In *these* references, the numbers indicate the year when the corresponding preprint was (probably) written. References in square brackets have been published before July 1991, and here the year of publication is given.

fibre $H^{-1}(Q)$ is a "linear" subsphere, i.e. $H^{-1}(Q) = W \cap S^{r+s-1}$ where W is a linear subspace of $\mathbb{R}^{r+s}$, and that the tangent mapping to H induces a ***nonsingular*** bilinear map $W \times W^{\perp} \to T_Q(S^n)$. One should think of these bilinear maps as "hidden" behind $(*)$. Moreover, he proves that there are points where H has rank $q \geq r \# s$. So, from the existence of a formula $(*)$ one can conclude that there is a nonsingular bilinear map of type $(r+s-q, q, n)$ for some q in the range $r \# s \leq q \leq n$. This gives potential new obstructions to the existence of $(*)$. Many applications show the power of this discovery, see also [Yiu 86], [Yiu 87]. For example, Lam uses it to prove $10 *_{\mathbb{R}} 11 \geq 20$, thus showing for the first time that $r *_{\mathbb{R}} s \neq r \# s$ is possible (since $10 \# 11 = 17$). By going more deeply into the geometry of composition formulas, and by using advanced topological tools like secondary cohomology operations, Lam and Yiu were later able to show $16 *_{\mathbb{R}} 16 \geq 29$ [Lam Yiu 89]. Until [Lam 85] appeared, 23 had been the best lower bound! Since a composition formula (16, 16, 32) is known classically, one seems to be quite close to the solution of the "16-square problem". Over the integers, the corresponding problem has been solved by Yiu, who showed that indeed 32 squares are necessary [Yiu 90]. The arguments are purely combinatorial and highly involved. Reportedly, Yiu has recently managed to determine $r * s$ over $\mathbb{Z}$ for *all* pairs $r, s \leq 16$!

I return to purely algebraic methods. The Pythagoras number $P(A)$ of a ring A is the smallest number q (if one exists) such that every sum of squares in A is actually a sum of q squares, and ∞ otherwise. It is a very delicate invariant which is usually hard to get a hold on, and is most interesting if $s(A) = \infty$ (i.e. if A has some formally real residue field), since otherwise it lies in the range $s(A) \leq P(A) \leq s(A) + 2$ (and even $P(A) \leq s(A) + 1$ if 2 is a unit). Which integers can be the Pythagoras number of a field is still an open problem. The precise value of $P(K)$ is unknown even for the fields $K = \mathbb{R}(X_1, \ldots, X_n)$ of rational functions in more than 2 variables, although one has the bounds $n + 2 \leq P(K) \leq 2^n$ from work of Pfister and Cassels-Ellison-Pfister. This already illustrates how little accessible this number is in general. This being said, [Choi Dai Lam Reznick 82] make a thorough study of this invariant and obtain a variety of previously unknown results. Particular attention is given to affine algebras over a field, and to local rings. One of the basic results is that the polynomial ring $R[X, Y]$ has infinite Pythagoras number, for any real closed field R; the same is true for $\mathbb{Z}[X]$. Before, it was not even known if there was any affine R-algebra with $P(A) = \infty$. Among the consequences drawn from this theorem is that there exists a principal ideal domain A (containing 1/2) and a unit of A which is a sum of squares in the quotient field, but not in A itself. In contrast, $P(A)$ is finite for one-dimensional affine R-algebras; but $P(A)$ can be arbitrarily large, even for the coordinate rings of plane rational (singular) curves. It seems to be unknown if the situation is different for regular curves. In the case of local rings, the threshold value lies one higher. For regular local rings A of dimension ≥ 3 with real residue field it is proved that $P(A) = \infty$, by using the associated graded ring and the result for polynomial rings. This is used to show that, for any affine R-variety V of dimension ≥ 3 with a smooth R-rational point, $P(R[V]) = \infty$. On the other hand, local rings of dimension ≤ 2 tend to have $P(A) < \infty$. This is proved, for example, for certain such rings of geometric type, where the theorems of Cassels and Pfister even allow to give upper bounds for $P(A)$. However, in the dimension two *global* case, it seems to be unknown for general surfaces V/R whether $P(R[V])$ is finite or not. Beyond its main results, the paper contains a great amount of detailed complementary information, and numerous examples are used to discuss and illustrate the general theorems.

Detailed studies of Pythagoras numbers have been made in the case of real AP-curves, that is, one-dimensional noetherian local real domains with real closed residue field which have Artin's approximation property (i.e. are henselian and excellent). For example, analytic and Nash curve germs give such rings. The Pythagoras number of a real AP-curve is always finite. A key invariant of such a ring is its value semigroup. [Ruiz 85a] characterizes those semigroups for which all, respectively no, corresponding curves are pythagorean. [Campillo Ruiz 90] show that a real AP-curve is pythagorean if and only if it is Arf; this is also equivalent to the Pythagoras condition for sums of $2n$-th powers, for any $n \geq 1$. [Ortega 91] studies the algebroid (= complete) case. Fixing an additive semigroup Γ of positive integers, the corresponding curves are parametrized by an algebraic set. This set is partitioned into finitely many semialgebraic subsets, according to the Pythagoras numbers of the curves. Both upper and lower bounds for $P(A)$ are given. In particular, it is shown that $P(A)$ can be arbitrarily large.

I would like to add a remark on recent deep results about Pythagoras numbers, although they do not belong to real algebra. They are concerned with function fields F over a number field k. Colliot-Thélène observed that Kato's Hasse principle for (Galois cohomology of) such function fields in one variable has immediate consequences for their Pythagoras numbers. Indeed, it follows that $P(F) \leq 7$; hence also $s(F) \leq 4$ if F is non-real, and $P\big(F(X)\big) \leq 8$ [Colliot-Thélène 86]. Recently, Jannsen has proved Kato's conjecture for function fields in two variables. With the help of results of Merkur'ev-Suslin and Rost, this gives similar applications: If F/k is a 2-dimensional function field, then $P(F) \leq 8$ and $P\big(F(X)\big) \leq 16$ [Colliot-Thélène Jannsen 91]. The authors remark that Kato's general conjecture, together with well-known conjectures on Milnor K-theory and Galois cohomology, would imply $P(F) \leq 2^{d+1}$ for any function field F/k of dimension $d \geq 2$.

I now turn to L. Mahé's quantitative results for rings of geometric nature. In [Mahé 86] he considered affine varieties over a real closed field R and succeeded in bounding the cokernel of the global signature map (compare Parimala's talk). For this he needed quantative results on the level of affine R-algebras which are of independent interest. In [Mahé 90] the bounds are improved, and new results are added. If V is an R-variety, the ring $\mathcal{R}[V]$ of real regular functions on V consists of those rational functions which are regular in all real points. A function $f \in \mathcal{R}[V]$ is totally positive if it takes strictly positive values on $V(R)$, the real points of V. In general rings, the notion of totally positive elements is defined via the real spectrum. One of the main results in [Mahé 90] is

Theorem. *Let V an affine R-variety of dimension d. If A is either $\mathcal{R}[V]$ or a semilocalization of $R[V]$, then every totally positive unit of A is a sum of 2^d squares in A.*

So, under this aspect, these rings behave like the corresponding function fields. In the semilocal case the remarkable thing is that no regularity condition is needed. A corollary is that in Pfister's theorem, saying that a psd polynomial function on V is a sum of 2^d squares of rational functions, one can in fact find such an expression where the denominators have no more zeros in $V(R)$ than the given function.

The other main result gives quantitative versions of the Positivstellensatz, of the following type. If $f \in R[V]$ is totally positive then there is an equation

$$f \cdot \boxed{m} = 1 + \boxed{n} \quad \text{in } R[V],$$

with $m = 2^d + 7$ and $n = 2^{d+1} + d - 4$, if $d \geq 4$. (Here $\boxed{s}$ stands for a sum of s squares.) If f is a unit, then even the significantly better bounds $m = 2^d$ and $n = 2^d + d - 5$ hold ($d \geq 4$). For $d \leq 3$ one has $(m, n) = (2^{d+1} - 1,\ 2^{d+1} - 1)$, and $(2^d,\ 2^d - 1)$ in the unit case.[2] In particular, this gives immediate bounds for the level of $R[V]$ in case $V(R) = \emptyset$. Whether or not such bounds exist in terms of d had been an open problem in [Dai Lam 84]. Similar, but weaker estimates appeared before in [Mahé 86]. The proofs proceed inductively over the dimension; Mahé starts with Pfister's theorem to get representations "with denominators", and then by clever arguments gets rid of the denominators, thereby slightly increasing the number of squares. Throughout, weak versions of multiplicativity properties are used for Pfister forms over such rings.

One can derive quantitative Stellensätze of more general type from the above ones. See e.g. [Simon 89] where this was done using the weaker estimates of [Mahé 86].

2. Sums of higher powers, real valuations and real holomorphy ring

Recall that E. Becker, in the late 70s, found the notion of an ordering of higher level of a field, and showed that it bears a close relationship to sums of higher (even) powers. This appears most suggestively in the fact that the sums of $2n$-th powers in K are precisely the elements contained in every ordering of level dividing n (assume char $K = 0$). This is completely analogous to Artin's characterization of the sums of squares in a field. From the beginning, the idea was that this similarity should only be part of a more comprehensive extension, to higher levels, of classical Artin-Schreier theory with all of its features. There has been considerable work towards this program, and I will try to give a very brief sketch of some of the main lines. A very clear exposition of this program, together with a survey of what had been done by 1984, can be found in [Becker 84].

An ordering of higher level is defined to be an additively closed subgroup P of K^* for which K^*/P is cyclic (necessarily of finite even order); the (exact) level of P is $\frac{1}{2}[K^* : P]$. Quite soon a shift in emphasis was made from orderings of higher level to signatures of K, i.e. characters $\chi : K^* \to \mu = \mathbb{Q}/\mathbb{Z}$ whose kernel is closed under addition. This point of view has many advantages, as is already demonstrated by the study of extension theory in [Becker Harman Rosenberg 82]. It leads naturally to the definition of the higher level reduced Witt rings [Becker Rosenberg 85a], which completely parallels the description of the reduced Witt ring of quadratic forms as a ring of $\mathbb{Z}$-valued functions on the space of orderings. In the level n case, one takes the space of all signatures which vanish on ΣK^{*2n}, and the ring of (continuous) $\mathbb{C}$-valued functions on it generated by the evaluation maps $\chi \mapsto \chi(a)$, $a \in K^*$. As in the level one case, it is technically important to extend the definition to preorderings more general than ΣK^{2n}. It is amazing to see how the analogy to the quadratic forms case remains

[2] According to Mahé, the figures in the printed version are incorrect in the non-unit cases; here I have given the corrected ones, communicated by Mahé.

valid for almost all aspects of the theory; e.g. prime ideals of the reduced Witt ring and their relation to orderings {Becker Rosenberg 85b}, the characterization of fans, and the representation theorem, which characterizes the reduced Witt ring as a subring of the ring of continuous functions. Further work on higher level Witt rings can be found in [Schwartz 87] and {Becker Schwartz 89}. However, it is still not clear how much these reduced Witt rings have to do with actual higher degree forms, and if there is a good definition of non-reduced Witt rings of higher level.

Marshall's spaces of orderings, the abstract frame for reduced quadratic form theory, have been generalized as to also incorporate the higher level reduced theory. Much of Marshall's original theory has been transferred to these so-called spaces of signatures by C. Mulcahy, V. Powers and M. Marshall. See [Mulcahy 88a+b], [Marshall Mulcahy 90], [Powers 87], [Powers 89].

Signatures also proved helpful in clarifying the relations between real closures at higher level orderings. Such real closures exist, but are in general not unique up to isomorphism, contrary to the case of usual (level one) orderings. By introducing chains of (higher level) orderings, J. Harman was able to give a satisfying treatment for 2-power levels [Harman 82]. After further investigation of this concept by R. Brown [Brown 87a+b], N. Schwartz formulated the notion of chain signatures [Schwartz 84] which seems to be quite an effective tool. See also [Becker 84, § 4] for a summary of some of its features.

Fields which are real closed with respect to some higher level ordering have been called generalized real closed fields. A lot of work has been devoted to the study of their properties, and in particular, their model theory. It was first investigated by B. Jacob, who showed that if R is a field which is real closed at an ordering of level n, the theory $Th(R)$ of R is decidable and model-complete, if the language of fields is enriched by suitable predicates which only depend on n [Jacob 81]. The crucial and remarkable point is his discovery of a certain elementarily definable henselian valuation ring which has an archimedean real closed residue field. His results were used in [Becker Jacob 85] to derive a Nullstellensatz and a solution of a generalization of Hilbert's 17th problem for function fields over generalized real closed fields. The model theory of chain closed fields and of chainable fields (these notions refer to Harman's concept) was much studied by F. Delon, M. Dickmann and D. Gondard; see e.g. [Delon 86], [Dickmann 88], [Gondard 87], [Gondard 90] and literature cited there.

Another branch of recent research tries to understand the role of higher level orderings for real algebraic geometry. See [Barton 88], [Becker Gondard 89], [Berr 89], {Berr 91} for steps in this direction.

I would like to point out two recent papers in which the general theory is applied to give concrete and attractive results. Generalizations of Hilbert's 17th problem to sums of $2n$-th powers are studied in {Becker Berr Delon Gondard 90}, both in a weak and a strong version. The question is whether a regular function on an irreducible k-variety is a sum of $2n$-th powers of rational functions if the values it takes (in closed resp. in k-rational smooth points) are sums of $2n$-th powers. The integers n for which either version holds form a multiplicative semigroup generated by primes, and conversely every such semigroup occurs (for the weak version, at least). The case where k is a generalized real closed field is studied in detail; also classes of base fields k are given (namely function fields over totally archimedean fields) for which one or both versions hold for all exponents. I also want to hint at Berr's recent work {Berr 90} about sums of mixed powers in fields. A nice application is the previously unknown existence of "mixed" Hilbert identities.

The structure of higher level orderings and signatures of a field F is crucially related to the real valuations of F. In this way, also the real holomorphy ring $H(F)$ — the intersection of all residually real valuation rings of F — plays an important role in the understanding of sums of higher powers. $H(F)$ is a Prüfer domain, and has many other remarkable properties. In [Becker 82b] it is amply demonstrated how a good understanding of $H(F)$ and of its ideal structure will have direct consequences for sums of higher powers in F. Not surprisingly, the most interesting results on the holomorphy ring and on real valuations have been obtained in the function field case. If the base field is $\mathbb{R}$, then $H(F)$ consists of all functions which are regular on the real points of *some* projective model (which one can take smooth by Hironaka). A similar description holds for the relative real holomorphy rings $H(F/R)$ over arbitrary real closed fields R. In this way, birational geometry and resolution of singularities become important tools. See [Schülting 82a+b] for this and for other foundational results on the holomorphy rings.

An important point is to reduce the vast class of all real valuations to a more manageable subclass. Since valuations of a function field chosen at random are ugly in many respects, one would like to dispose of a family of valuations with good arithmetic and geometric properties (value group, residue field, center). On the other hand, since various important criteria refer to the entire class of all real valuations, such a family should be large enough to be sufficient for those criteria. For instance, of central importance is Becker's characterization of ΣF^{2n} as those sums of squares f for which the value $v(f)$ is divisible by $2n$ for every real valuation v of F.

Results of this kind have been obtained by H.-W. Schülting. Working with function fields over the real numbers, he showed [Schülting 86a] that the class of real prime divisors shares all these good properties: It suffices for the Becker criterion just described, for representing the real holomorphy ring as an intersection of valuation rings, and also for the local global principle for weakly isotropic quadratic forms (due to Bröcker and Prestel). As an application, a geometric and very handy description of ΣF^{2n} is given, for $F = \mathbb{R}(X)$ the function field of a surface, which involves the multiplicities of infinitely near points of X on (the nonreal part of) the divisor of f. Schülting also shows that signatures of a function field $F/\mathbb{R}$ can be identified with characters of local divisor groups in smooth projective models. From a more general point of view, valuations of function fields F/k are studied in [Bröcker Schülting 86]. Most results apply also if the theory of formally real fields is replaced by a suitable other one, like that of formally p-adic fields, and the base field k is existentially closed in the respective theory. One of the main results is an existence theorem for chains of valuations of F/k with prescribed centers (in a fixed model), value groups and, to some extent, residue fields. Other parts of the paper generalize earlier work of Schülting on holomorphy rings. The essential and non-trivial tool, used throughout in both papers, is Hironaka's resolution of singularities.

It is interesting to compare this to the article [Kuhlmann Prestel 84], written at about the same time. Its main theorem is in its spirit quite close to Schülting's results just mentioned, and in particular, to the existence theorem in [Bröcker Schülting 86].

It also says that in a function field extension F/k of characteristic 0 the "well-behaved" valuations are dense, in a very strong sense, among all valuations of F/k. Moreover, one is still given considerable freedom of how to specify well-behavedness, as demonstrated by various modifications. This has several applications in common with the Bröcker-Schülting theorem. However, Kuhlmann and Prestel proceed model-theoretically: Instead of using Hironaka, their main tool is the Ax-Kochen-Ershov theorem on the model theory of henselian valued fields.

Another theorem on chains of valuations in function fields has been proved by C. Andradas in [Andradas 89] (see also [Andradas 85]). Working in the real spectrum, he also prescribes centers, dimensions, and rank and rational rank of the value groups. It is remarkable that he does not need desingularization theorems nor model completeness results. Instead, he uses elementary methods like curve selection lemma and Noether normalization.

The results from [Kuhlmann Prestel 84] are applied in [Prestel 84] to show the following: In order to check if a polynomial $f \in \mathbb{R}[X_1, \ldots, X_d]$ has value divisible by $2n$ under all real valuations (the essential part of the Becker criterion), it is sufficient to check only those valuations which come from plugging Laurent polynomials (in one variable) into f. But then Prestel gives an instructive example to show that this becomes false if R is non-archimedean and $n > 1$; namely the polynomial $f(X) = X^{2n} + \omega X^2 + 1$, where ω is a very large element. Thus for $n > 1$, to be a sum of $2n$-th powers is not an elementary property in the coefficients of f, contrary to the case $n = 1$! Another way to put this is that in representations

$$f_m(X) = X^{2n} + mX^2 + 1 = \sum_{i=1}^{N_m} \left(\frac{g_{m,i}}{h_m} \right)^{2n},$$

for positive integers m, one cannot keep the N_m and the degrees of the h_m bounded at the same time for $m \to \infty$, although this is possible for the N_m alone. The substitution criterion was later refined in [Prestel 87], where the degrees of the Laurent polynomials were bounded by computable functions.

At this point I want to digress for a moment, to mention recent work on $R(X)$, the field of rational functions in one variable. Becker had already shown $P_4(R(X)) \leq 36$ [Becker 82b] where P_4 is the Pythagoras (or Waring) number for sums of fourth powers. This was improved to 24 in [Schmid 88]. But now in {Choi Lam Prestel Reznick 91}, $P_4(R(X)) \leq 6$ is shown! Also for higher exponents the upper bounds from [Becker 82b] are slightly improved. This uses a new characterization of $\Sigma R(X)^{2n}$. In addition to the obvious necessary conditions for a polynomial f (initial coefficient positive, multiplicities of real roots and degree divisible by $2n$) — which by Becker's criterion suffice if R is archimedean —, a new condition is needed. If one partitions the upper half plane of $R(\sqrt{-1})$ into equivalence classes, by defining two points to be equivalent if their hyperbolic cross-ratio and its inverse are bounded in absolute value by some integer, the condition says that the number of nonreal roots of f lying in each equivalence class must be a multiple of n. Obviously, this condition is implied by the others if R is archimedean. If one keeps these cross-ratios in each class bounded (and fixes the degree of the polynomial), then also the complexity of representations as sums of $2n$-th powers can be bounded. A related result had been proved before over $\mathbb{R}$ in [Prestel Bradley 89].

I conclude this part with a series of beautiful results about ideals in real holomorphy rings. Here once more geometrical methods are used (coming from algebraic geometry and topology) to obtain purely algebraic theorems. There was a long-standing question by Gilmer, who had asked if any finitely generated ideal in a Prüfer domain has actually two generators. Schülting disproved this by taking the real holomorphy ring $H(F)$ of $F = \mathbb{R}(X, Y)$ and showing that the fractional ideal $(1, X, Y)$ is not generated by two elements [Schülting 79]. Let $\mu(H)$ be the minimal number of generators sufficient for all finitely generated ideals of H. Since $H(F)$ has Krull dimension d, for $F/\mathbb{R}$ a d-dimensional real function field, a general result of Heitmann gives $\mu(H(F)) \leq d+1$. After Schülting had given his counterexample in 1979, people began to wonder whether equality might hold in general. In the end, this turned out to be true, but the general result was established only quite recently: After R. Swan had shown equality for $F = \mathbb{R}(X_1, \ldots, X_d)$ [Swan 84] and Schülting had proved $\mu(H(F)) = 3$ for all function fields of real surfaces {Schülting 86b}, W. Kucharz settled the general case in [Kucharz 89a], showing $\mu(H(F)) = d+1$ for $F/\mathbb{R}$ real of dimension d. As Becker had observed much earlier, this implies that there are $f_1, \ldots, f_{d+1} \in F$ such that $f_1^{2n} + \cdots + f_{d+1}^{2n}$ is not a sum of d $2n$-th powers for any $n \geq 1$. If $F/\mathbb{R}$ is purely transcendental one can take $1, X_1, \ldots, X_d$ (this came out already from Swan's proof), and thereby obtain a generalization of a classical theorem of Cassels!

Kucharz's proof uses the description of $H(F)$ as those rational functions which are defined on the real points of some smooth projective model. Since his techniques build upon the ideas of Schülting and Swan, I will skip over the details of their work. Kucharz proceeds roughly like this: A fractional ideal of $H(F)$ corresponds to a line bundle on some smooth projective model X of $F/\mathbb{R}$. Blowing up X in a real point x_0 one gets $\pi : X' \to X$. For the class α of the exceptional fibre $E = \pi^{-1}(x_0)$ one shows that $\alpha^d \neq 0$ in $H^d(X'(\mathbb{R}), \mathbb{Z}/2)$. If the fractional ideal corresponding to E were generated by d elements, a calculation with Stiefel-Whitney classes shows that there would be a multi-blow-up $\sigma : X'' \to X'$ for which α^d maps to zero in $H^d(X''(\mathbb{R}), \mathbb{Z}/2)$. But one can show that the map in cohomology induced by such a σ is always injective.

Actually, Kucharz's paper provides much more information than just $\mu(H) = d+1$. For example, Schülting had studied the binary class group B_F of F, i.e. the quotient of the class group of $H(F)$ by the classes of those ideals which admit two generators. This group plays also a role in Becker-Rosenberg's theory of reduced Witt rings of higher level {Becker Rosenberg 85b}. Schülting had found that $B_F \cong (\mathbb{Z}/2)^s$ if $d = 2$, and $|B_F| = \infty$ for $d \geq 3$, where s is the number of connected components of a smooth projective model {Schülting 86b}. These results are widely generalized by Kucharz to other quotients of the class group.

The result $\mu(H) = d+1$ was later generalized to relative real holomorphy rings $H = H(F/R)$ over arbitrary real closed fields R {Kucharz 89c}. [Buchner Kucharz 89] also studied relative real holomorphy rings $H(F/V)$, where V is an affine irreducible smooth $\mathbb{R}$-variety and F its function field. Here one has $\mu(H(F/V)) \leq \dim V$ if and only if $V(\mathbb{R})$ has no compact connected components. Also this is a generalization of the above main result of Kucharz.

3. Selected applications of real algebra to geometry

I want to start with a few words on the work of J.M. Ruiz on global semianalytic geometry. He not only made available real spectrum methods and applied them with great success; rather, on his way to this goal, he established remarkable results of algebraic nature which can certainly claim independent interest.

The key result, which makes the real spectrum a highly useful tool in semianalytic geometry, is an Artin-Lang property for global analytic functions on a *compact* analytic set X. It means that the globally semianalytic subsets of X correspond bijectively, via an operator "tilda" which preserves set-theoretic operations, to the constructible sets in the real spectrum of $\mathcal{O}(X)$. The Artin-Lang property in turn rests on the study of the real spectrum formal fibres of local analytic rings. The basic result, whose use pervades the whole theory, is the following extension theorem of Ruiz. Suppose A is an excellent local domain and α is an ordering of its field of fractions. If α makes the maximal ideal of A convex, then it extends to an ordering of the total ring of quotients of $\widehat{A}$, the completion of A. Other forms of this result are a formal curve selection lemma and a real dimension theorem, both for excellent rings. See [Ruiz 85b], [Ruiz 86a+b].

As an immediate application, his theorem allowed Ruiz to solve the 17th Hilbert problem for meromorphic functions in the affirmative, on a compact irreducible analytic set [loc.cit.]. Before, this had been known only in very special cases. Another application is the real Nullstellensatz for global analytic functions.

It must be pointed out that the same results for compact analytic sets (Artin-Lang property, Hilbert 17, real Nullstellensatz) have been proved independently by P. Jaworski. In his approach, part of the algebraic reasoning of Ruiz is replaced by the use of Hironaka's theorems [Jaworski 86].

But more effort was needed to establish basic properties of well-behavedness for the real spectra of (local or global) rings of analytic functions. Ruiz showed that the class of excellent rings is a suitable environment: It contains the rings arising in analytic geometry, and on the other hand has "excellent" properties also for the real spectrum. Apart from the above mentioned theorem (which is central for many arguments), this includes dimension theory, the going-down property for regular homomorphisms, constructibility of closures of constructible sets and (under suitable conditions) of connected components. See [Ruiz 86b], [Ruiz 89a+b], [Alonso Andradas 87], [Andradas Bröcker Ruiz 88]. By Artin-Lang, this gives corresponding geometric theorems: If X is a compact analytic set and Z is a globally semianalytic subset of X, then closure and connected components of Z are again globally semianalytic; and there is a "Finiteness Theorem" saying that the operator tilda preserves closures [Ruiz 86b], [Ruiz 88]. (The compactness hypothesis on X is weakened in [Ruiz 90a].) Another application will be mentioned later in this talk. A recent very nice result is the characterization of sums of $2n$-th powers of meromorphic functions, on a compact analytic set X, by their values on curve germs [Ruiz 90b] (in the case of analytic surfaces this was also proved in [Kucharz 89b]): A necessary and sufficient condition for f to lie in $\Sigma\mathcal{M}(X)^{2n}$ is that for every analytic curve germ $\sigma :]-\epsilon,\epsilon[\to X$ one has $f(\sigma(t)) = at^{2nk}+$ higher powers, with $a > 0$, as long as $\sigma \not\subset \operatorname{Sing} X$ and $f \circ \sigma$ is neither undefined nor identically zero. This is a remarkable result, and is of course very much reminiscent of Prestel's substitution criterion mentioned before. The similarity is "explained" and embedded into a more general context in {Ruiz 91}.

In the remainder of this talk I want to record some progress in questions on the complexity of systems of inequalities. A common feature of the following is that it rests on the — apparently completely abstract — theory of preorderings and fans in fields, and on its even more abstract generalization, the spaces of orderings. I will not go here into the purely algebraic aspects of these matters; it seems that the majority of important results in this theory had already been obtained by the beginning of the 80s, and that since then the applications have been more striking. A noteworthy exception are the important papers [Schwartz 83] and [Marshall 84] on saturation and local stability.

The problem is how to describe semialgebraic sets as economically (or "cheaply") as possible, i.e. how to minimize the number of functions in a description, regardless of their complexity. The first major step was done by L. Bröcker [Bröcker 84a]. He considered affine algebraic varieties V over a real closed field R, and what was later called basic open (semialgebraic) sets in $V(R)$; that is, sets which can be written

$$S = \{x \in V(R) : \; f_1(x) > 0, \ldots, f_r(x) > 0\}$$

with polynomials $f_i \in R[V]$. The question is, how many inequalities are really needed to present S in this form? Bröcker gave the surprising answer that there is a common upper bound for all basic open sets in $V(R)$, which depends only on $d = \dim V$. In fact, writing $s(d)$ for the least upper bound valid for all d-dimensional varieties, he showed that

$$d \;\le\; s(d) \;\le\; d!! := \begin{cases} d(d-2)(d-4)\cdots 2 & d \text{ even,} \\ d(d-2)(d-4)\cdots 1 & d \text{ odd.} \end{cases}$$

In particular, $s(d) = d$ for $d = 1, 2, 3$. It is worthwile to recall the main steps of his proof. Given a basic open set S, one starts by looking only at generic presentations of S, that is, at basic open sets T for which the symmetric difference of S and T has dimension $< d$. This is a problem on the orderings of the function field $R(V)$ (suppose V irreducible, for simplicity), which had been solved by Bröcker long before, actually in much greater generality [Bröcker 74]. Namely, $d = \dim V$ inequalities suffice for a generic presentation. And — assuming that $R(V)$ is formally real — this number is the best possible, even if one considers only generic presentations. This becomes the starting point for an induction argument, since one can now pass to a subvariety of smaller dimension. However, in the end one has to glue together these generic presentations on subvarieties, to obtain a true one. For this, Bröcker proved special pasting lemmas, which are consequences of the Hörmander-Łojasiewicz inequality for polynomial functions. The estimate $d!!$ comes from an extra argument which allowed to approximate S even up to codimension two by a generic presentation with d inequalities.

A somewhat simplified version of the proof was later presented in [Mahé 86], where Bröcker's results were used in an essential way to bound the exponent of the cokernel of the global signature map.

By improving the pasting techniques, Bröcker later gave better bounds for $s(d)$, roughly equal to $(\sqrt{2})^{d+1}$ {Bröcker 85}. He also proposed similar questions for basic closed sets (replacing "> 0" by "≥ 0"), and also for general open (resp. closed) semialgebraic sets. Here one tries to write an open or closed set as a union of few basic (open or closed) sets. (The Finiteness Theorem asserts that this is always possible.) In this way one defines corresponding invariants $\bar{s}$, t and $\bar{t}$. In [Bröcker 84b], finiteness of $t(d)$ and $\bar{t}(d)$ was shown (actually, bounds were given), and {Bröcker 85}, {Bröcker 87} gave the estimate $\bar{s}(d) \leq \frac{1}{2}d(d+1)$, with equality for $d = 1, 2$. As in the basic open case, the upper bounds for $\bar{s}$, t, $\bar{t}$ are obtained by pasting from the corresponding generic bounds. However, in contrast to the s-invariant, the t-invariant is even generically little understood, and even the generic bounds are probably much too high.

The Hörmander-Łojasiewicz inequality has proved useful also for different kinds of problems. For example, it is needed for other characterizations of the property of being basic: An open semialgebraic set is basic open iff it does not meet the Zariski closure of its boundary and is generically basic in each irreducible subvariety. For closed sets, the same holds mutatis mutandis, with the condition on the Zariski closure dropped {Bröcker 87}, {Bröcker 88b}. The advantage of such criteria is that properties like being basic or being representable by a conjunction of r inequalities are *generically* well understood, in terms of fan theory of the function field; and under favorable conditions, this fan theory can be reformulated in geometric terms. Other applications of the Łojasiewicz inequality are to separation questions, where also spaces of orderings are very instrumental. I want to digress into this for some moments before returning to the description of semialgebraic sets.

The separation problem starts with two closed disjoint semialgebraic sets A, B on an affine R-variety V, and asks if there is a polynomial f with $f|A > 0$ and $f|B < 0$. One quickly realizes that this is not in general possible. But it becomes so if one allows square roots of everywhere positive polynomials (theorem of Mostowski and Bochnak-Efroymson). Bröcker showed that separation by polynomials is possible if A, B are *basic* closed and one of them has dimension ≤ 2; the dimension hypothesis cannot be dropped [Bröcker 88a]. This uses both pasting and spaces of orderings; in fact, the results of Schwartz and Marshall on saturation cited above are an essential tool. Pasting techniques are also applied in [Fortuna Galbiati 89], where conditions are studied under which separation up to a closed subvariety implies actual separation by some polynomial. Quantitative aspects of the separation problem were studied by Stengle. He asked for the minimal number of square roots that have to be adjoined to make separation possible, and defined the Mostowski number $m(d)$ to be the smallest m such that on all d-dimensional varieties, separation always is possible after adjoining m (suitable) square roots as above. From the standard proof of Mostowski's theorem one gets the upper bound $m(d) \leq \bar{s}(d)\bar{t}(d)$ [Mahé 86]. Stengle was interested in lower bounds, and related them to a notion of complexity for semialgebraic sets. In his notion, one can use the relations $=$ and $\neq$ for free, but one has to pay for every use of $<$, $\leq$ etc. In this way a lower bound for $m(d)$ was obtained which is roughly logarithmic in d [Stengle 84], [Stengle 89]. It was improved in [Bröcker Stengle 90], where $m(d) \geq d - 1$ was shown. Here fans were used again, together with trace forms.

Another remarkable separation result was proved in [Ruiz 84], namely for closed semianalytic germs, in case one of them has dimension ≤ 1: The (punctured) germs can be separated by a polynomial. Again, the Łojasiewicz inequality is used. Ruiz gave nice applications to semialgebraic geometry, thereby partially answering some questions of T. Recio in [Brumfiel 82].

Returning to the description of basic sets, the next important step was done in [Andradas Bröcker Ruiz 88]. Here it is shown that Bröcker's ideas apply also to compact analytic sets X, namely to the description of basic open sets defined by global analytic functions. The main result is that the number of inequalities needed is bounded in terms of $d = \dim X$. Actually the same bounds are given as in the semialgebraic case. Similar to the semialgebraic case, the problem can be translated into the real spectrum of the ring of global analytic functions $\mathcal{O}(X)$, thanks to Artin-Lang property and finiteness theorem. This fact lies at the base of the proof. Generally, one can consider basic open subsets in the real spectrum of any commutative ring A, and ask for the minimum number of inequalities (of the form $a > 0$, with $a \in A$) needed to present them. The supremum of these numbers is an invariant of A which, generalizing terminology from the field case, is called the (geometric) stability index of A. As a consequence of the Positivstellensatz, an "abstract" Łojasiewicz inequality is available on the real spectrum of any ring. So the pasting lemmas hold in general, and the main problem becomes to find the *generic* minimal bound — which again turns out to be the dimension. Using a refined version of Weierstraß preparation theorem, the stability index of the fraction field of a complete local domain is estimated. The transition to henselian local (excellent) rings is accomplished by the use of Artin's conjecture, as proved by C. Rotthaus.

Then I succeeded in showing that indeed $s(d) = d$ holds in general [Scheiderer 89], i.e. the stability index of a real algebraic R-variety is just its dimension. Also Bröcker found a proof which is slightly different {Bröcker 88b}. The proof makes essential use of the fact that each semilocal ring gives rise to a space of orderings, and of a structure theorem about fans in these spaces of orderings. These results are due to [Kleinstein Rosenberg 78, 80] and [Knebusch 81]. Thus the proof is less geometric than Bröcker's former approach. The main difference is that the pasting techniques are completely left aside, and instead the generic presentations are improved inductively. Actually, the proof generalized without trouble to noetherian rings whose (real) singularities behave reasonably well (e.g. excellent rings). Surprisingly, in such a ring the (minimal) length of a generic presentation is already the (minimal) length of an actual one. (If the ring is not regular, one has to modify the meaning of "generic", according to the singularities.) In particular, also for basic open semianalytic sets as considered in [Andradas Bröcker Ruiz 88], the dimension of X is the *exact* upper bound for the number of inequalities. On the other hand, it was soon realized that Bröcker's proof had the advantage not to need any conditions on the singularities [Bröcker 90], {Marshall 89}. There is no more a reduction to a generic situation, but a local global principle for the presentation of basic open sets still holds with respect to the class of *all* residue fields. In particular, one has the "stability formula": The stability index of a noetherian ring is the supremum of the stability indices of its residue fields.

It should be mentioned that also the $\bar{s}$-invariant was determined in [Scheiderer 89] for all R-varieties. Here Bröcker's previous upper bound $\frac{1}{2}d(d+1)$ turned out to be best possible, for any real d-dimensional R-variety. Again, fan theory was used for this.

But that's not yet the end of the story! Soon after the above proofs were circulated, Mahé came up with a new proof of $s(d) = d$ [Mahé 89a], {Mahé 89b}. It is more elementary in that it doesn't use spaces of orderings, nor the Łojasiewicz inequality. If a basic open set S is given, Mahé first inverts all polynomials which are strictly positive on S. This doesn't affect the problem, and has the advantage of producing sufficiently many units, so that one can do some of the things with quadratic forms which one is used to from fields. For example, a weak form of transversality holds. Mahé then proceeds by a careful analysis of Pfister forms over such rings. In this way, the proof uses only explicit calculations with forms! The only exception to this is the Tsen-Lang theorem, which is used to get the stability index of the function fields.

The very latest progress is a recent preprint of M. Marshall {Marshall 91}. He proves the stability formula in perhaps the greatest generality one can think of, namely for arbitrary commutative rings A: If a basic open subset S of the real spectrum can be presented by $r \geq 1$ inequalities in each residue field of A, then S itself has a presentation by r inequalities. (The real spectrum of A can even be replaced by any intersection of basic subsets.) Marshall essentially follows Mahé's proof; but where Mahé argues by induction over the dimension to get a weak representation, Marshall uses a more abstract tool, namely a local global principle (due originally to [Bröcker 82]) for modules over preorderings in rings. So after all, the former "problem of the stability index" has been settled in a truely elementary and satisfying way, and in the greatest generality, with the help of the machinery of abstract real algebra.

References

In the text, references to articles are by name(s) of author(s) and year. For the difference between curly braces and square brackets see footnote[1] in the text.

[GARFQ] Géométrie Algébrique Réelle et Formes Quadratiques, Proc. Conf. Rennes 1981, ed. J.-L. Colliot-Thélène, M. Coste, L. Mahé, M.-F. Roy, Lect. Notes Math. **959**, Springer, 1982.

[OFRAG] Proc. Special Session on Ordered Fields and Real Algebraic Geometry, ed. D.W. Dubois, T. Recio, Contemp. Math. **8**, 1982.

[RAAG] Real Analytic and Algebraic Geometry. Proc. Conf. Trento 1988, ed. M. Galbiati, A. Tognoli, Lect. Notes Math. **1420**, Springer, 1990.

M.E. Alonso, C. Andradas:
[87] Real spectra of complete local rings. Manuscripta math. **58**, 155-177 (1987).

C. Andradas:
[85] Real places in function fields. Comm. Algebra **13**, 1151-1169 (1985).
[89] Specialization chains of real valuation rings. J. Algebra **124**, 437-446 (1989).

C. Andradas, L. Bröcker, J.M. Ruiz:
[88] Minimal generation of basic open semianalytic sets. Invent. math. **92**, 409-430 (1988).

J.Kr. Arason, A. Pfister:
[82] Quadratische Formen über affinen Algebren und ein algebraischer Beweis des Satzes von Borsuk-Ulam. J. reine angew. Math. **331**, 181-184 (1982).

S.M. Barton:
[88] The real spectrum of higher level of a commutative ring. Ph. D. thesis, Cornell Univ., Ithaca, 1988.

E. Becker:
[82a] Valuations and real places in the theory of formally real fields. In [GARFQ], pp. 1-40.

[82b] The real holomorphy ring and sums of $2n$-th powers. In [GARFQ], pp. 139-181.

[84] Extended Artin-Schreier theory of fields. Rocky Mountain J. Math. **14**, 881-897 (1984).

E. Becker, R. Berr, F. Delon, D. Gondard:

{90} Hilbert's 17-th problem for sums of $2n$-th powers. Preprint 1990.

E. Becker, D. Gondard:

[89] On rings admitting orderings and 2-primary chains of orderings of higher level. Manuscripta math. **65**, 63-82 (1989).

E. Becker, J. Harman, A. Rosenberg:

[82] Signatures of fields and extension theory. J. reine angew. Math. **330**, 53-75 (1982).

E. Becker, B. Jacob:

[85] Rational points on algebraic varieties over a generalized real closed field: A model theoretic approach. J. reine angew. Math. **357**, 77-95 (1985).

E. Becker, A. Rosenberg:

[85a] Reduced forms and reduced Witt rings of higher level. J. Algebra **92**, 477-503 (1985).

{85b} On the structure of reduced Witt rings of higher level. Preprint 1985 (unpublished).

E. Becker, N. Schwartz:

{89} Reduced Witt rings of infinite level. Preprint 1989.

R. Berr:

[89] Reelle algebraische Geometrie höherer Stufe. Dissertation, Ludwig-Maximilians-Universität München, 1989.

{90} Sums of mixed powers in fields and orderings of prescribed level. Preprint 1990.

{91} The p-real spectrum of a commutative ring. Preprint 1991.

L. Bröcker:

[74] Zur Theorie der quadratischen Formen über formal reellen Körpern. Math. Ann. **210**, 233-256 (1974).

[82] Positivbereiche in kommutativen Ringen. Abh. Math. Sem. Univ. Hamburg **52**, 170-178 (1982).

[84a] Minimale Erzeugung von Positivbereichen. Geom. Dedicata **16**, 335-350 (1984).

[84b] Spaces of orderings and semialgebraic sets. In: Quadratic and Hermitian Forms, Conf. Hamilton 1983, ed. I. Hambleton, C.R. Riehm, Canad. Math. Soc. Conf. Proc. **4**, 1984, pp. 231-248.

{85} Description of semialgebraic sets by few polynomials. Lecture CIMPA, Nice 1985.

{87} Characterization of basic semialgebraic sets. Preprint 1987.

[88a] On the separation of basic semialgebraic sets by polynomials. Manuscripta math. **60**, 497-508 (1988).

{88b} On basic semi-algebraic sets. Preprint 1988 (to appear Expos. Math.).

[90] On the stability index of noetherian rings. In [RAAG], pp. 72-80.

L. Bröcker, H.-W. Schülting:

[86] Valuations of function fields from the geometrical point of view. J. reine angew. Math. **365**, 12-32 (1986).

L. Bröcker, G. Stengle:

[90] On the Mostowski number. Math. Z. **203**, 629-633 (1990).

R. Brown:

[87a] Real closures of fields at orderings of higher level. Pacific J. Math. **127**, 261-279 (1987).

[87b] The behavior of chains of orderings under field extensions and places. Pacific J. Math. **127**, 281-297 (1987).

G. Brumfiel:

[82] Some open problems. In [OFRAG], pp. 19-25.

M.A. Buchner, W. Kucharz:

[89] On relative real holomorphy rings. Manuscripta math. **63**, 303-316 (1989).

A. Campillo, J.M. Ruiz:

[90] Some remarks on pythagorean real curve germs. J. Algebra **128**, 271-275 (1990).

M.D. Choi, Z.D. Dai, T.Y. Lam, B. Reznick:

[82] The Pythagoras number of some affine algebras and local algebras. J. reine angew. Math. **336**, 45-82 (1982).

M.D. Choi, T.Y. Lam, A. Prestel, B. Reznick:

{91} Sums of $2m$-th powers of rational functions in one variable over real closed fields. Preprint 1991.

M.D. Choi, T.Y. Lam, B. Reznick:

[87] Even symmetric sextics. Math. Z. **195**, 559-580 (1987).

[91] Positive sextics and Schur's inequalities. J. Algebra **141**, 36-77 (1991).

J.-L. Colliot-Thélène:

[86] Appendix to: K. Kato, A Hasse principle for two dimensional global fields. J. reine angew. Math. **366**, 181-183 (1986).

J.-L. Colliot-Thélène, U. Jannsen:

[91] Sommes de carrés dans les corps de fonctions. C. R. Acad. Sci. Paris **312**, sér. I, 759-762 (1991).

Z.D. Dai, T.Y. Lam:

[84] Levels in algebra and topology. Comment. Math. Helvetici **59**, 376-424 (1984).

Z.D. Dai, T.Y. Lam, R.J. Milgram:

[81] Applications of topology to problems on sums of squares. Ens. Math. (2) **27**, 277-283 (1981).

Z.D. Dai, T.Y. Lam, C.K. Peng:

[80] Levels in algebra and topology. Bull. Am. Math. Soc. (N.S.) **3**, 845-848 (1980).

F. Delon:

[86] Corps et anneaux de Rolle. Proc. Am. Math. Soc. **97**, 315-319 (1986).

C.N. Delzell:

[82a] A finiteness theorem for open semi-algebraic sets, with applications to Hilbert's 17th problem. In [OFRAG], pp. 79-97.

[82b] Case distinctions are necessary for representing polynomials as sums of squares. In: Proc. Herbrand Symp., Logic Coll. 1981, ed. J. Stern, North Holland, 1982, pp. 87-103.

[84] A continuous, constructive solution to Hilbert's 17th problem. Invent. math. **76**, 365-384 (1984).

M. Dickmann:

[88] The model theory of chain closed fields. J. Symb. Logic **53**, 73-82 (1988).

E. Fortuna, M. Galbiati:

[89] Séparation de semi-algébriques. Geom. Dedicata **32**, 211-227 (1989).

D. Gondard:

[87] Théorie du premier ordre des corps chaînables et des corps chaîne-clos. C. R. Acad. Sci. Paris **304**, sér. I, 463-465 (1987).

[90] Chainable fields and real algebraic geometry. In [RAAG], pp. 128-148.

J. Harman:

[82] Chains of higher level orderings. In [OFRAG], pp. 141-174.

B. Jacob:

[81] The model theory of generalized real closed fields. J. reine angew. Math. **323**, 213-220 (1981).

P. Jaworski:

[86] Extension of orderings on fields of quotients of rings of real analytic functions. Math. Nachr. **125**, 329-339 (1986).

J.L. Kleinstein, A. Rosenberg:

[78] Signatures and semisignatures of abstract Witt rings and Witt rings of semilocal rings. Canad. J. Math. **30**, 872-895 (1978).

[80] Succinct and representational Witt rings. Pacific J. Math. **86**, 99-137 (1980).

M. Knebusch:

[77] Some open problems. In: Conf. Quadratic Forms, Kingston 1976, ed. G. Orzech, Queen's Papers Pure Appl. Math. **46**, 1977, pp. 361-370.

[81] On the local theory of signatures and reduced quadratic forms. Abh. Math. Sem. Univ. Hamburg **51**, 149-195 (1981).

[82] An algebraic proof of the Borsuk-Ulam theorem for polynomial mappings. Proc. Am. Math. Soc. **84**, 29-32 (1982).

W. Kucharz:

[89a] Invertible ideals in real holomorphy rings. J. reine angew. Math. **395**, 171-185 (1989).

[89b] Sums of $2n$-th powers of real meromorphic functions. Monatsh. Math. **107**, 131-136 (1989).

{89c} Generating ideals in real holomorphy rings. Preprint 1989.

F.-V. Kuhlmann, A. Prestel:

[84] On places of algebraic function fields. J. reine angew. Math. **353**, 181-195 (1984).

K.Y. Lam:

[84] Topological methods for studying the composition of quadratic forms. In: Quadratic and Hermitian Forms, Conf. Hamilton 1983, ed. I. Hambleton, C.R. Riehm, Canad. Math. Soc. Conf. Proc. **4**, 1984, pp. 173-192.

[85] Some new results on composition of quadratic forms. Invent. math. **79**, 467-474 (1985).

K.Y. Lam, P.Y.H. Yiu:

[89] Geometry of normed bilinear maps and the 16-square problem. Math. Ann. **284**, 437-447 (1989).

T.Y. Lam, T.L. Smith:

{90} On Yuzvinsky's monomial pairings. Preprint, Berkeley 1990.

L. Mahé:

[86] Théorème de Pfister pour les variétés et anneaux de Witt réduits. Invent. math. **85**, 53-72 (1986).

[89a] Une démonstration élémentaire du théorème de Bröcker-Scheiderer. C. R. Acad. Sci. Paris **309**, sér. I, 613-616 (1989).

{89b} On the geometrical stability index of a ring. Lecture, Quadratic Forms Conf., Oberwolfach 1989.

[90] Level and Pythagoras number of some geometric rings. Math. Z. **204**, 615-629 (1990).

M. Marshall:

[84] Spaces of orderings: Systems of quadratic forms, local structure, and saturation. Comm. Algebra **12**, 723-743 (1984).

{89} Minimal generation of constructible sets in the real spectrum of a Noetherian ring. Preprint 1989.

{91} Minimal generation of basic sets in the real spectrum of a commutative ring. Preprint 1991.

M. Marshall, C. Mulcahy:

[90] The Witt ring of a space of signatures. J. Pure Appl. Algebra **67**, 179-188 (1990).

C. Mulcahy:

[88a] An abstract approach to higher level forms and rigidity. Comm. Algebra **16**, 577-612 (1988).

[88b] A representation theorem for higher level reduced Witt rings. J. Algebra **119**, 105-122 (1988).

J.J. Ortega:

[91] On the Pythagoras number of a real irreducible algebroid curve. Math. Ann. **289**, 111-123 (1991).

A. Pfister, S. Stolz:

[87] On the level of projective spaces. Comment. Math. Helvetici **62**, 286-291 (1987).

V. Powers:

[87] Characterizing reduced Witt rings of higher level. Pacific J. Math. **128**, 333-347 (1987).

[89] Finite spaces of signatures. Canad. J. Math. **41**, 808-829 (1989).

A. Prestel:

[84] Model theory of fields: An application to positive semidefinite polynomials. Mém. Soc. Math. France (N.S.) **16**, 53-65 (1984).

[87] Model theory applied to some questions about polynomials. In: Contributions to General Algebra 5, Proc. Salzburg Conf. 1986, Wien 1987, pp. 31-43.

A. Prestel, M. Bradley:

[89] Representations of a real polynomial $f(X)$ as a sum of $2m$-th powers of rational functions. In: Ordered Algebraic Structures, ed. J. Martinez, Kluwer 1989, pp. 197-207.

B. Reznick:

[87] A quantitative version of Hurwitz' theorem on the arithmetic-geometric inequality. J. reine angew. Math. **377**, 108-112 (1987).

[89] Forms derived from the arithmetic-geometric inequality. Math. Ann. **283**, 431-464 (1989).

J.M. Ruiz:

[84] A note on a separation problem. Arch. Math. **43**, 422-426 (1984).

[85a] Pythagorean real curve germs. J. Algebra **94**, 126-144 (1985).

[85b] On Hilbert's 17th problem and real Nullstellensatz for global analytic functions. Math. Z. **190**, 447-454 (1985).

[86a] Cônes locaux et complétions. C. R. Acad. Sci. Paris **302**, sér. I, 67-69 (1986).

[86b] On the real spectrum of a ring of global analytic functions. Publ. Inst. Rech. Math. Rennes **4**, 84-95 (1986).

[88] On the connected components of a global semianalytic set. J. reine angew. Math. **392**, 137-144 (1988).

[89a] A dimension theorem for real spectra. J. Algebra **124**, 271-277 (1989).

[89b] A going-down theorem for real spectra. J. Algebra **124**, 278-283 (1989).

[90a] On the topology of global semianalytic sets. In [RAAG], pp. 237-246.

[90b] A characterization of sums of $2n$th powers of global meromorphic functions. Proc. Am. Math. Soc. **109**, 915-923 (1990).

{91} The substitution test for sums of even powers. Preprint 1991.

C. Scheiderer:

[89] Stability index of real varieties. Invent. math. **97**, 467-483 (1989).

J. Schmid:

[88] Eine Bemerkung zu den höheren Pythagoraszahlen reeller Körper. Manuscripta math. **61**, 195-202 (1988).

H.-W. Schülting:

[79] Über die Erzeugendenanzahl invertierbarer Ideale in Prüferringen. Comm. Algebra **7**, 1331-1349 (1979).

[82a] On real places of a field and their holomorphy ring. Comm. Algebra **10**, 1239-1284 (1982).

[82b] Real holomorphy rings in real algebraic geometry. In [GARFQ], pp. 433-442.

[86a] Prime divisors on real varieties and valuation theory. J. Algebra **98**, 499-514 (1986).

{86b} The binary class group of the real holomorphy ring. Preprint 1986.

N. Schwartz:

[83] Local stability and saturation in spaces of orderings. Canad. J. Math. **35**, 454-477 (1983).

[84] Chain signatures and real closures. J. reine angew. Math. **347**, 1-20 (1984).

[87] Chain signatures and higher level Witt rings. J. Algebra **110**, 74-107 (1987).

D.B. Shapiro:

[84] Products of sums of squares. Expos. Math. **2**, 235-261 (1984).

O. Simon:

[89] Aspects quantitatifs de Nullstellensätze et de Positivstellensätze. Nombres de Pythagore. Comm. Algebra **17**, 637-667 (1989).

G. Stengle:

[84] A lower bound for the complexity of separating functions. Rocky Mountain J. Math. **14**, 925-927 (1984).

[89] A measure for semialgebraic sets related to boolean complexity. Math. Ann. **283**, 203-209 (1989).

S. Stolz:

[89] The level of real projective spaces. Comment. Math. Helvetici **64**, 661-674 (1989).

R.G. Swan:

[84] n-generator ideals in Prüfer domains. Pacific J. Math. **111**, 433-446 (1984).

P.Y.H. Yiu:

[86] Quadratic forms between spheres and the non-existence of sums of squares formulae. Math. Proc. Camb. Phil. Soc. **100**, 493-504 (1986).

[87] Sums of squares formulae with integer coefficients. Canad. Math. Bull. **30**, 318-324 (1987).

[90] On the product of two sums of 16 squares as a sum of squares of integral bilinear forms. Quart. J. Math. Oxford (2) **41**, 463-500 (1990).

S. Yuzvinsky:

[81] Orthogonal pairings of euclidean spaces. Michigan Math. J. **28**, 131-145 (1981).

[84] A series of monomial pairings. Linear Multilin. Algebra **15**, 109-119 (1984).

G. Zeng:

[88] A characterization of preordered fields with the weak Hilbert property. Proc. Am. Math. Soc. **104**, 335-342 (1988).

[91] On preordered fields related to Hilbert's 17th problem. Math. Z. **206**, 145-151 (1991).

TOPOLOGY OF REAL PLANE ALGEBRAIC CURVES

E.I.Shustin
(Math. Dept., Samara State University, ul. Acad. Pavlova 1, 443011 Samara, USSR)

In this article we'll consider some ideas, methods, results found over the last ten years in topology of real plane algebraic curves. We'll also speak of some unsolved problems.

Topology of real curves is expounded in many known surveys [1, 4, 9, 18, 26, 28, 29]. We'll consider matter not included in them. The text consists of 4 parts: in the first part we speak of methods of getting restrictions, in the second part we speak of real curves constructions, in the third part we speak of connection between algebraic curves and singularities smoothings, in the fourth part we formulate some conjectures. Used below (without refering) classical results in real curves topology are described in [1, 4, 9, 18, 26, 28, 29] quite completely.

Introduction

By plane real algebraic curve we mean a homogeneous real polynomial $F(x_0, x_1, x_2)$ (being considered modulo a constant factor) and the set

$$\mathbb{R}F = \{z \in \mathbb{R}P^2 \mid F(z) = 0\}.$$

Connected non-singular two-sided components of $\mathbb{R}F$ are called ovals.

The classification problem of real curves of a given degree can be considered from three points of view:

(i) The real classification - up to continuous isotopies in $\mathbb{R}P^2$ (i.e. the ovals arrangements classification)

(ii) The equivariant one - up to equivariant isotopies in $\mathbb{C}P^2$ connecting the given curves complexifications

$$\mathbb{C}F = \{z \in \mathbb{C}P^2 \mid F(z) = 0\}.$$

(iii) The rigid one - up to isotopies consisting of algebraic curves of a given degree.

We'll deal with the first problem.

1. Restrictions

All the known restrictions to an ovals arrangement arise from investigation of various structures on objects connected with a real curve. Such objects are

(i) the pairs $(\mathbb{R}P^2, \mathbb{R}F)$, $(\mathbb{C}P^2, \mathbb{C}F)$,

(ii) the pair $(S^4, (\mathbb{C}F/conj) \cup \mathbb{R}F_{\pm})$, where $S^4 = \mathbb{C}P^2/conj$, $\mathbb{R}F_\varepsilon = \{z \in \mathbb{R}P^2 \mid \varepsilon F(z) \geq 0\}$, $\varepsilon = \pm$,

and also the double covering of $\mathbb{C}P^2$ branched at $\mathbb{C}F$ or at $(\mathbb{C}F/conj) \cup \mathbb{R}F_{\pm}$, if a curve degree is even.

The technical means of investigation are algebraic-topological (Smith theory, complex orientations, quadratic form of intersection index in 2-dimensional homology of double covering etc.) or algebraic-geometrical (Bezout's theorem, a position of a curve with respect to special straight lines or conics etc.).

Here we'll give the detailed description of one method of getting restrictions (see [15, 24, 28, sec 4.15]), which is based on some topological and geometrical properties of curves.

Namely, let F be a real non-singular curve of the even degree m, let p (resp. n) denote the number of ovals lying inside even (odd) number of other ovals. Put $j = g + 1 - p - n$, where g is a curve genre. Let $\xi : \mathcal{X} \to \mathbb{C}P^2$ be the double covering with branching at $\mathbb{C}F$. The complex conjugation $conj : \mathbb{C}P^2 \to \mathbb{C}P^2$ is covered by two antiholomorphic involutions $conj_+, conj_-$ on $\mathcal{X}$. Assume that $\xi(Fix\, conj_-)$ is non-orientable. It is well-known [2, 11] that $H_2(\mathcal{X})$ is the unimodular lattice (without torsion) with respect to intersection index, and it contains the maximal sublattice $H_+ \oplus H_- \oplus H$ where

$$H = \{\alpha \in H_2(\mathcal{X}) \mid conj_{\pm,*}\alpha = -\alpha\},$$

$$H_\varepsilon = \{\alpha \in H_2(\mathcal{X}) \mid conj_{\varepsilon,*}\alpha = -conj_{-\varepsilon,*}\alpha = \alpha\}, \ \varepsilon = \pm$$

$$rk\, H = 1, \ rk\, H_+ = (m^2 - 3m)/2 + 1 + p - n,$$

$$rk\, H_- = (m^2 - 3m)/2 + 2 - p + n,$$

$$\sigma^+(H) = 1, \ \sigma^-(H) = 0, \ \sigma^+(H_\pm) = (m^2 - 6m + 8)/8,$$

$$\sigma^-(H_+) = 3(m^2 - 2m)/8 + p - n,$$

$$\sigma^-(H_-) = 3(m^2 - 2m)/8 + 1 - p + n,$$

$$discr\, H_+ = \pm 2^{j+1}, \ discr\, H_- = \pm 2^j$$

Here $(\sigma^+, \sigma^-, \sigma^0)$ is a signature. To get restrictions we'll first construct the sublattice $T \subset H_\varepsilon$, which is connected with ovals arrangement, and second verify relations

$$\sigma^{\pm}(T)+\sigma^{\circ}(T)\le\sigma^{\pm}(H_\varepsilon) \tag{1}$$

and

$$\operatorname{discr} T = r^2\cdot \operatorname{discr} H_\varepsilon \quad \text{if } \operatorname{rk} T=\operatorname{rk} H_\varepsilon . \tag{2}$$

The first approach to construction of cycles realizing classes in $H_2(\mathcal{X})$ is as follows. Let M be an immersed in $\mathbb{C}P^2$ surface with boundary $\partial M\subset \mathbb{C}F$ such that

(i) $(M\setminus\partial M)\cap\mathbb{C}F=\emptyset$, $\operatorname{conj} M=M$,

(ii) for any $\lambda\in H_1(M\setminus\partial M)$

$$<w_1(M\setminus\partial M),\lambda>=0 \Leftrightarrow \lambda\in 2H_1(\mathbb{C}P^2\setminus\mathbb{C}F).$$

Then the surface $\xi^{-1}(M)$ is orientable and represents (being equipped with a fixed orientation) some integral 2-cycle in $\mathcal{X}$. Consider examples of such membranes M giving classes in H_ε:

1. Arnold's construction [2, 29]. As M it is possible to take any connected component of $\mathbb{R}F_\varepsilon$, orientable if $m\equiv 0 \pmod 4$.

2. Rokhlin's construction [17]. Let Φ be any connected component of $\mathbb{R}F_{-\varepsilon}$, let $l_1\subset\mathbb{C}F\setminus\mathbb{R}F$ and $l_2\subset \operatorname{Int}\Phi$ be arcs connecting two points on $\partial\Phi$. There exists an immersed in $\mathbb{C}P^2\setminus\mathbb{C}F$ disc $\mathcal{D}$ with boundary $l_1\cup l_2$. Then put $M=\mathcal{D}\cup\operatorname{conj}\mathcal{D}$.

3. Viro's construction (see [15, 28]). Consider the segment $\Pi=[L_1,L_2]$ of a real straight lines pencil, where L_1, L_2 are tangents to F and don't cross $\mathbb{C}F\setminus\mathbb{R}F$. Assume that $\operatorname{card}(\mathbb{C}L\cap\mathbb{C}F\setminus\mathbb{R}F)\le 4$ for any $L\in\Pi$. The set $\bigcup_{L\in\Pi}(\mathbb{C}L\setminus\mathbb{R}L)$ consists of two connected components transposing by conj. Denote $\mathbb{C}\Pi$ one of them. Put

$$S_1=\mathbb{C}\Pi\cap\mathbb{C}F,\quad \mathbb{R}\Pi=\bigcup_{L\in\Pi}\mathbb{R}L\cap\mathbb{R}F_{-\varepsilon}.$$

Suppose that points of contact of F with lines $L\in\Pi$ are connected pairwise in $\mathbb{R}\Pi$ by smooth arcs $l_1,\dots,l_r$, where $l_i\cap l_k=\emptyset$, $i\ne k$, $\operatorname{card}((l_1\cup\dots\cup l_r)\cap\mathbb{R}L)\le 2$,

$$\operatorname{card}(l_i\cap\mathbb{R}L)\le 1,\ L\in\Pi,\ i=1,\dots,r.$$

Then $S=S_1\cup l_1\cup\dots\cup l_r$ meets $\mathbb{C}L$ at two points if $L\in\Pi\setminus\{L_1,L_2\}$, and at one point if $L=L_1$ or L_2. Connect the points $S\cap\mathbb{C}L$ by an arc in $\mathbb{C}L$, which depends on L continuously. The union of these arcs is a disc $\mathcal{D}$ with boundary S. Put $M=\mathcal{D}\cup\operatorname{conj}\mathcal{D}$. The Viro construction gave maximal sublattices in H_ε for the first time, and allowed to obtain restrictions from (2) [28], what stimulated the search of other constructions.

The fourth construction was suggested by Kharlamov and author (see [24]). It differs from above constructions. Namely, let $Q_1, \dots, Q_r$ be the set of different straight lines and conics crossing $\mathbb{C}F$ transversally and only in real points. Then $\xi^{-1}(\mathbb{C}Q_i) \setminus \mathrm{Fix}\, conj_\varepsilon$ has two connected components $\mathcal{D}_i^{(1)}, \mathcal{D}_i^{(2)}$ transposing by involution $conj_\varepsilon$. We'll orient $\mathcal{D}_i^{(1)}, \mathcal{D}_i^{(2)}$ so that the chain $\mathcal{D}_i^{(1)} + \mathcal{D}_i^{(2)}$ is invariant with respect to $conj_\varepsilon$. Then $\partial \mathcal{D}_i^{(1)} = \partial \mathcal{D}_i^{(2)} = q_i$ is 1-cycle on $\mathrm{Fix}\, conj_\varepsilon$. If $2\sum [q_i] = 0 \in H_1(\mathrm{Fix}\, conj_\varepsilon)$ then there is 2-chain S, consisting of closures of the set $\mathrm{Fix}\, conj_\varepsilon \setminus \bigcup q_i$ connected components such that $\partial S = 2\sum q_i$. Then 2-cycle $\sum(\mathcal{D}_i^{(1)} + \mathcal{D}_i^{(2)}) - S$ represents some class in H_ε. For $r = 0$ it is Arnold's construction.

The cycles appearing in [2] gave the Arnold inequalities, and the second construction allowed to give an elementary proof of strong Arnold-Petrovsky inequalities (cf. [29], where the proof is based on Smith theory):

$$p \le 3(m^2 - 2m)/8 + 1 + n^-, \tag{3a}$$

$$n \le 3(m^2 - 2m)/8 + p^-, \tag{3b}$$

$$p \le (m^2 - 6m + 12 + 4\cdot(-1)^{m/2})/8 + p^+, \tag{3c}$$

$$n \le (m^2 - 6m + 8)/8 + n^+, \tag{3d}$$

where p^+, p^-, p^0 (resp. n^+, n^-, n^0) are numbers of even (odd) ovals bounding the components of $\mathbb{R}P^2 \setminus \mathbb{R}F$ with positive, negative and zero Euler characteristic from the outside. Namely, let $\Phi_1, \dots, \Phi_t$ be all the components of $\mathbb{R}F_+$ homeomorphic to a circle or a ring. It is clear that $t = p^+ + p^0$. Consider the sublattice $T \subset H_+$, generated by classes $\varkappa_i = [\xi^{-1}(\Phi_i)],\ i = 1, \dots, t$. According to [2]

$$\begin{cases} \varkappa_i^2 = \begin{cases} -2, & \text{if } \Phi_i \text{ is a circle} \\ 0, & \text{if } \Phi_i \text{ is a ring} \end{cases} \\ \varkappa_i \circ \varkappa_j = 0, \quad i \ne j \end{cases} \tag{4}$$

For every component Ψ of $\mathbb{R}F_-$, let $\Psi_1, \ldots, \Psi_q$ denote all the adjoining conponents of $\mathbb{R}F_+$ so that $\chi(\Psi_1) \leq \ldots \leq \chi(\Psi_q)$. If $\chi(\Psi_q) \geq 0$ and $q \geq 2$ then we connect Ψ_1 with $\Psi_2, \ldots, \Psi_q$ by some $(q-1)$ disjoint smooth lines in Ψ. The graph Γ of adjacency relation of $\Phi_1, \ldots, \Phi_t$ and these lines consists of disjoint trees. If any tree contains $\ell \geq 2$ terminal points, corresponding to some lines, then we'll remove $(\ell-1)$ of last lines. Let $L_1, \ldots, L_r$ denote all the retained lines. The Rokhlin construction provides classes $\varkappa_{t+1}, \ldots, \varkappa_{t+r} \in H_+$ respectively. It is easy to show (see [24]), that

$$\varkappa_{t+i} \circ \varkappa_{t+j} \equiv 0 \ (\mathrm{mod}\ 2), \quad 1 \leq i \leq j \leq r,$$

$$\varkappa_{t+i} \circ \varkappa_j \equiv \begin{cases} \pm 1 \ (\mathrm{mod}\ 4), & \Phi_j \cap L_i \neq \emptyset \\ 0 \ (\mathrm{mod}\ 4), & \Phi_j \cap L_i = \emptyset \end{cases} \tag{5}$$

In the case $p^- > 0$ every tree in Γ contains exactly one terminal point corresponding to some line. Then (4), (5) imply

$$\det(\varkappa_i \circ \varkappa_j) \equiv 1 \ (\mathrm{mod}\ 2),$$

hence $\mathrm{rk}\, T = t = p^+ + p^0$, and also according to (1)

$$p^+ + p^0 \leq \sigma^-(H_+) = 3(m^2 - 2m)/8 + p - n,$$

what is equivalent to (3b). If $p^- = 0$ then it is easy to see that $n = p^0 \leq p^0 + p^+ = p$. Using the Harnack inequality

$$n + p \leq g + 1$$

we obtain

$$n \leq (g+1)/2 = (m^2 - 3m + 4)/4 \leq 3(m^2 - 2m)/8$$

The inequalities (3a), (3c), (3d) can be got analogously.

By means of all the above constructions, Viro 28 and author 24 gave complete classification of the degree 8 curves of type $\langle a \amalg 1\langle b\rangle \amalg 1\langle c\rangle \amalg 1\langle d\rangle\rangle$. Probably, complete classification of all curves of degree 8 can be obtained by means of those constructions. Also I think that the important problem of real curves topology - the Ragsdale-Petrovsky conjecture (see [26,28])

$$\max\{p, n\} \leq 3(m^2 - 2m)/8 + 1 \tag{6}$$

can be proved in limits of a given method.

It should be mentioned that this approach was used in other cases: (i) Viro and Kharlamov prohibited some topological types of the degree 5 surfaces in $\mathbb{R}P^3$, (ii) Kharlamov proved non-exactness of the Harnack estimate for singularities (see sec. 3.1).

2. Constructions

2.1. Viro's method. The Viro method is a basic method of real algebraic curves construction at present [25, 27]. Its algorithm is simple and consists in the following.

Given the set of real polynomials

$$F_k(x,y) = \sum_{(i,j)\in\Delta_k} A_{ij}x^iy^j, \quad k=1,\ldots,N,$$

with Newton polygons $\Delta_1,\ldots,\Delta_N$. Assume that: (i) $\Delta = \bigcup\Delta_k$ is a convex polygon, (ii) the non-empty intersection $\Delta_k\cap\Delta_l$ is a common vertex or a common edge, (iii) there is the convex continuous function $\nu: \Delta\to\mathbb{R}$ linear on each Δ_k and non-linear on each union $\Delta_k\cup\Delta_l$ (it is said to be a regular subdivision of Δ), (iv) curves $F_1,\ldots,F_N$ are non-singular in $\mathbb{C}^2\setminus\{xy=0\}$, (v) the sections of each polynomial F_k on edges of Δ_k are non-degenerate (i.e. they haven't multiple factors different from x, y).

Any curve F_k should be lifted from plane upon the toric manifold defined by polygon Δ_k. Let $\widetilde{\Delta}_k$ be a union of Δ_k and polygons symmetric to Δ_k with respect to axes and origin. Then $\mathbb{R}F_k$ can be naturally projected from the toric manifold to $\widetilde{\Delta}_k$ so that an intersection of $\mathbb{R}F_k$ with any quadrant is projected to the corresponding polygon. This image is called the chart of polynomial F_k. The basic statement [25, 27] is as follows: the chart of polynomial

$$F(x,y) = \sum_{(i,j)\in\Delta} A_{ij}x^iy^jt^{\nu(i,j)}, \quad t\to+0,$$

with Newton polygon Δ is a topological gluing of charts of $F_1,\ldots,F_N$.

The same method gives a construction of smoothings of singular points determined by their Newton diagram [25].

Almost all the curves types discovered over the last ten years [7, 14, 15, 23, 25, 27] were constructed by Viro method. It should be noted that constructions of Chevallier [5, 6] and Korchagin [13] keep within limits of Viro method.

2.2. Regular triangulations of Newton polygon. Consider the particular case of Viro method. We'll call a real curve a triangular curve if it is defined by threenomial with non-degenerate Newton triangle. The chart of triangular curve is determined by signes of its threenomial coefficients. Let Δ be the Newton triangle of the degree m curve with vertexes $(0;0)$, $(m;0)$, $(0;m)$. Then by means of Viro method we can assign to each couple

(regular triangulation of Δ, function $\lambda: \Delta \cap \mathbb{Z}^2 \to \{\pm 1\}$)

some isotopic type of non-singular curve of degree m. Further on the curve having such an isotopic type we'll call T-curve. Apparently, there are curves different from T-curves. At that time all the Harnack curves are T-curves (see [5, 13]). Moreover, "almost all" non-singular curves are T-curves. We'll show that by the following reasoning of S.Yu.Orevkov. Let us introduce affine coordinates in the space of curves of degree m, and then substitute them for logarithmic coordinates [25, 27] in each octant. Then in each octant a suitable polyhedron $\mathcal{D}$ similar to Newton polyhedron of discriminant cuts out the discriminant hypersurface the part of chart corresponding to a given octant. According to [8] the components of complement to discriminant containing vertexes of polyhedron consist of T-curves. Let Θ mean the union of those components, then

$$vol(\Theta \cap \mathcal{D}) / vol(\mathcal{D}) \to 1$$

if we'll enlarge $\mathcal{D}$ infinitely.

On the other side the class of T-curves is quite visible. I think the following conjectures are accessible combinatorical problems.

Conjecture 1. T-curves satisfy the inequality (6).

Conjecture 2 (on cancellation). Any ovals arrangement obtained by elimination of one empty oval of T-curve can be realized by some curve of a given degree.

That follows from

Conjecture 3. Any T-curve is isotopic to a suitable non-singular curve of a given degree, which is obtained by smoothing nodal curve having in $\mathbb{R}P^2$ one connected component and, possibly, isolated points.

2.3. Two modifications of the Viro method. The first modification enlarges the application range of the Viro method to smoothing singular points. We call a singular point quasiordinary if (i) at this point the sections of Taylor series on the Newton diagram edges are non-degenerate, or (ii) this point will satisfy the above condition after a suitable local analytic diffeomorphism. For any singular point z we put

(i) $b(z)=\mu(z)-1$, if z is zero-modal,

(ii) $b(z)$ is equal to the sum of orders of z and its infinitely near points in full resolution of z except nodes of complete inverse image of a given curve, if the modality of z is positive.

Proposition 1 [22]. If F is irreducible and

$$\sum_{z\in \operatorname{Sing} F} b(z) < 3 \deg F,$$

then it is possible to realize independently the smoothings of arbitrary quasiordinary points $z\in \operatorname{Sing} F$ isotopic to their prescribed Viro smoothings by means of small variation of coefficients of F, while all other singularities are retained.

Application examples of this statement are constructions (a) of the 8th degree curve of type $\langle 4 \amalg 3\langle 5\rangle\rangle$ [22], (b) of certain affine sextic curves [15], (c) of quartic curves with a prescribed position of inflexion points [10].

The second modification allows to eliminate the condition (iv) from sec. 2.1 in certain cases. Take all notations of sec. 2.1 and introduce the oriented graph Γ of adjoining the polygons $\Delta_1,\dots,\Delta_N$ without cycles.

Proposition 2 [20]. If $F_1,\dots,F_N$ are irreducible in $\mathbb{C}^2\setminus\{xy=0\}$, and

$$\sum_{z\in \operatorname{Sing} F_k\setminus\{xy=0\}} b(z) < \sum_{\rho_k} (\operatorname{card}(\rho_k\cap \mathbb{Z}^2)-1), \quad k=1,\dots,N$$

where ρ_k runs through all edges of Δ_k not corresponding to arcs of Γ coming in Δ_k. Then there exists the curve F with Newton polygon Δ, whose chart is the gluing of charts of $F_1,\dots,F_N$, and whose singularities collection in $\mathbb{C}^2\setminus\{xy=0\}$ is the union of sungularities collections of $F_1,\dots,F_N$ in $\mathbb{C}^2\setminus\{xy=0\}$. In particular, it is always possible to glue (even reducible) curves $F_1,\dots,F_N$ with only nodes in $\mathbb{C}^2\setminus\{xy=0\}$.

This statement allows to construct real curves with prescribed sungularities collection. For example, there are curves of degree m with arbitrary number of real cusps from zero to $2m^2/9$.

3. Algebraic curves and smoothings of singular points

Let $z \in \mathbb{R}P^2$ be a singular point of a real curve F, the closed ball $\mathcal{D} \subset \mathbb{C}P^2$ centred at z be small enough, and a real curve G be sufficiently close to F. Then the set $\mathcal{D} \cap G$ is called a perturbation of the point $z \in F$. An isotopy of the point z perturbations means the equivariant isotopy in $\mathcal{D} \setminus \{z'\}$, where $z' \in \partial\mathcal{D} \cap \mathbb{R}P^2 \setminus F$ is a fixed point. An isotopy of the point z smoothings (non-singular perturbations) means the continuous isotopy in $\mathcal{D} \cap \mathbb{R}P^2 \setminus \{z'\}$.

3.1. The classification problem for singular points perturbations is like the classification problem for algebraic curves. In fact, all methods of getting restrictions to topology of real curves can be extended to perturbations of singular points. Namely, we have for the non-singular curve F of degree m and the smoothing Φ of the singular point z with Milnor number μ:

(i) The Harnack inequality for ovals number

$$v \leq (m^2 - 3m + 1 + (-1)^m)/2 \qquad (7)$$

has local analogous one [16]

$$v \leq (\mu - r_0 + 1)/2, \qquad (8)$$

where r_0 is a number of real local branches centred at z, or more precisely

$$\tilde{v} \leq (\mu - r + 3)/2, \qquad (9)$$

where r is a number of all the local branches centred at z, and $\tilde{v}$ is a number of connected components of union of $\mathbb{R}\Phi$ with real local branches links.

(ii) The Petrovsky inequality for even m

$$|2\chi(\mathbb{R}F_{\pm}) - 1| \leq (3m^2 - 6m + 4)/4 \qquad (10)$$

has local analogous one [3]

$$|\chi(\mathbb{R}\Phi_{\pm}) - 1| \leq h^{1,1}_{\lambda=1},$$

where $h^{1,1}_{\lambda=1}$ is a Hodge number on the space of vanishing cohomology corresponding to the eigenvalue $\lambda = 1$ of monodromy operator.

(iii) Congruences for M- and (M-1)-curves of even degree

$$\chi(\mathbb{R}F_+) \equiv m^2/4 \pmod 8$$

$$\chi(\mathbb{R}F_+) \equiv m^2/4 \pm 1 \pmod 8$$

correspond to analogous local congruences [12] for M- and (M-1)-smoothings (that means $\tilde{v} = (\mu - \tau + 3)/2$ or $(\mu - \tau + 1)/2$ in (9))

$$\chi(\mathbb{R}\Phi_+) \equiv B(\tilde{q}_\Delta) \pmod 8$$

$$\chi(\mathbb{R}\Phi_+) \equiv B(\tilde{q}_\Delta) \pm 1 \pmod 8,$$

where $B(\tilde{q}_\Delta)$ is the Brown invariant of some quadratic form $\tilde{q}_\Delta$ with the range $\mathbb{Z}/4$, which depends only on a singular point type.

(iv) For complex orientations of the type I curves (i.e. $\mathbb{C}F \setminus \mathbb{R}F$ is not connected) there are Rokhlin's formulae

$$v + 2(\Pi^- - \Pi^+) = m^2/4, \quad m \equiv 0 \pmod 2)$$

$$v + v^- - v^+ + 2(\Pi^- - \Pi^+) = (m^2 - 1)/4, \quad m \equiv 1 \pmod 2),$$

where Π^- (Π^+) is a number of negative (resp. positive) ovals pairs, and v^- (v^+) is a number of ovals equipped with negative (resp. positive) orientation with respect to one-sided branch. For the type I smoothings there are analogous formulae

$$v + 2(\Pi^- - \Pi^+) + \sum (v^-(l) - v^+(l)) = a$$

$$v^-(\partial\mathcal{D} \cap \mathbb{R}P^2) - v^+(\partial\mathcal{D} \cap \mathbb{R}P^2) = b,$$

where l runs through all the non-closed components of $\mathbb{R}\Phi$, and $v^\pm(\Lambda)$ means a number of ovals equipped with positive or negative orientation with respect to line Λ. Here the numbers a, b depend on the singular point z type, a non-closed branches position and an orientation of $\partial\mathcal{D} \cap \mathbb{R}P^2$.

However it should be noted that there is the considerable difference: the inequality (7) is exact, but (8), (9) are not exact. That was showed by Kharlamov for singularity $x^4 - y^6 = 0$.

3.2. The Viro method establishes an immediate correspondence between the set $I_s(z)$ of isotopic types of the semi-quasihomogeneous singular point z smoothings and the set $I_c(z)$ of isotopic types of curves with Newton polygon lying under the point z Newton diagram, and whose number of local real branches centred at non-proper straight line is equal to the number of real branches centred at z. We'll say that z holds the property A, if $I_s(z) \simeq I_c(z)$, and holds the property B, if $I_s(z)$ is included into $I_c(z)$.

Zero-modal singular points, an ordinary 4th order point, point of quadratic contact of 3 smooth branches [25, 27], an ordinary 5th order point [19], a point of quadratic contact of 4 smooth branches [19, 21] hold the property A. An ordinary 6th order point holds the property B [15]. In known cases $I_S(Z)$ is included into $I_C(Z)$, although, probably, it is not in general.

4. Certain problems

Besides Ragsdale-Petrovsky conjecture (6) and the conjecture on cancellation (see sec. 2.2) the problem of discovering new approaches to investigation of the odd degree curves is important and interesting, because here the results list is rather less then one for curves of even degree. For example, Korchagin [14] suggested conjectures

$$n - p \leq 3(m-1)^2/8$$

$$n \leq 3(m-1)^2/8$$

analogous to inequalities (10), (6). We conclude with the following Arnold problem: whether the discriminant $\mathcal{D}$ in the space of curves of a given degree is M-variety (i.e. $b_*(\mathbb{C}\mathcal{D}, \mathbb{Z}/2) = b_*(\mathbb{R}\mathcal{D}, \mathbb{Z}/2)$) ?

References

1. A'Campo N.: Sur la 1e're partie du 16e probléme de Hilbert. Seminaire Bourbaki, 31é année, 1978-79, no. 537.

2. Arnold V.I.: The arrangement of the ovals of real plane algebraic curves, involutions of four-dimensional manifolds and the arithmetic of integral quadratic forms. Function. Anal. Appl. 5, 169-175 (1972)

3. Arnold V.I.: The index of vector field singular point, Petrovsky-Olejnik inequalities and mixed Hodge structures. Function. Anal. Appl. 12, 1-14 (1979)

4. Arnold V.I., Olejnik O.A.: Topology of real algebraic varieties. Vestnik MGU, ser. 1, no. 6, 7-17 (1979) (Russian)

5. Chevallier B.: Sur le courbes maximales de Harnack. C.R. Acad. Sci. Paris, ser. 1, 300, no. 4, 109-114 (1985)

6. Chevallier B.: Courbes maximales et courbes reduites. C.R. Acad. Sci. Paris, ser. 1, 301, no. 2, 35-40 (1985)

7. Chislenko Yu.S.: M-curves of tenth degree. In: Investigations on topology IV (Zap. nauch. semin. LOMI, vol. 122), Nauka, Leningrad, 1982, pp. 147-162 (Russian)

8. Gel'fand I.M., Zelevinsky A.V., Kapranov M.M.: On discriminant of polynomials in several variables. Function. Anal. Pril. 24, no. 1, 1-4 (1990) (Russian)

9. Gudkov D.A.: Topology of real projective algebraic varieties. Uspekhi math. nauk 29, no. 4, 3-79 (1974) (Russian)

10. Gudkov D.A.: Plane real projective quartic curves. In: Lect. Notes in Math., Springer, 1988, vol. 1346, pp. 341-347.

11. Kharlamov V.M.: Topologikal types of non-singular surfaces of degree 4 in $\mathbb{R}P^3$. Function. Anal. Appl. 10 (1977)

12. Kharlamov V.M., Viro O.Ya.: Extensions of the Gudkov-Rokhlin congruence. In: Lect. Notes in Math., Springer, 1988, vol. 1346, pp.687-717.

13. Korchagin A.B.: On the singularities reduction and the classification of non-singular affine sextic curves. Deposit. in VINITI 14.02.86, no. 1107-B, 16p. (Russian)

14. Korchagin A.B.: New M-curves of degrees 8 and 9. Sov. Math. Doklady 39, no. 3, 569-572 (1989)

15. Korchagin A.B., Shustin E.I.: Affine curves of degree 6 and smoothings of non-degenerate 6th order singular point. Math. USSR Izvestiya 33, no. 3, 501-520 (1989)

16. Risler J.-J.: Un analogue local du théoréme de Harnack. Inv. Math. 89, no. 1, 119-137 (1987)

17. Rokhlin V.A.: Proof of Gudkov's conjecture. Function. Anal. Appl 6, 136-138 (1972)

18. Rokhlin V.A.: Complex topological characteristics of real algebraic curves. Uspekhi math. nauk 33, no. 5, 77-89 (1978) (Russian)

19. Shustin E.I.: The Hilbert-Rohn method and smoothings of real algebraic curves singular points. Sov. Math. Doklady 31, no. 2, 282-286 (1985)

20. Shustin E.I.: Gluing of singular algebraic curves. In: Methods of Qualitative Theory, Gorky Univ. Press, Gorky, 1985, pp. 116-128 (Russian)

21. Shustin E.I.: The independence of singular points smoothings and new M-curves of degree 8. Uspekhi math. nauk 40, no. 5, 212 (1985) (Russian)

22. Shustin E.I.: New M-curve of degree 8. Math. Zametki 42, no. 2, 180-186 (1987) (Russian)

23. Shustin E.I.: New M- and (M-1)-curves of degree 8. In: Lect. Notes in Math., Springer, 1988, vol. 1346, pp. 487-493.

24. Shustin E.I.: New restrictions to topology of real curves of degree dividing by 8. Izvestiya Acad. Nauk SSSR, ser. math.

54, no. 5, 1069-1089 (1990) (Russian)

25. Viro O.Ya.: Gluing of algebraic hypersurfaces, smoothing singularities and construction of curves. In: Proc. Leningrad Int. Topol. Conf., Nauka, Leningrad, 1983, pp. 149-197 (Russian)

26. Viro O.Ya.: Progress in topology of real algebraic varieties over the last 5 years. In: Proc. Intern. Congress of Math., Warszawa 1983, aug. 16-24, pp. 603-619 (Russian)

27. Viro O.Ya.: Gluing of plane real algebraic curves and constructions of curves of degrees 6 and 7. In: Proc. Leningrad Intern. Topol. Conf., Leningrad 1982/ Lect. Notes in Math., Springer, 1984, vol. 1060, pp. 187-200.

28. Viro O.Ya.: Progress in topology of real algebraic varieties over the last 6 years. Uspekhi math. nauk 41, no. 3, 45-67 (1986) (Russian)

29. Wilson G.: Hilbert's sixteenth problem. Topology 17, no. 1, 53-73 (1978)

Moduli Problems in Real Algebraic Geometry

R. SILHOL

Université Montpellier II
URA CNRS 1407

0. Introduction.

To quote Hartshorne, one can say that the guiding problem in algebraic geometry is the classification problem and typically such a problem falls into two parts,

(i) a discrete part, that is, finding numerical invariants;
(ii) a continuous part, this is the moduli problem.

To rephrase this, a moduli problem, is the problem of finding a space that classifies up to isomorphism, all objects with a given set of numerical invariants (actually the distinction between (i) and (ii) is not always clear, even in classical algebraic geometry — see for example Catanese [Ca] —, and this is even more so in real algebraic geometry since a real algebraic space, even irreducible, may have many connected components).

In the above informal definition some words are voluntarly inprecise, in the classical case "space" means algebraic space (but not necessarily algebraic variety), "isomorphism" means isomorphism over an algebraically closed field and "numerical invariants" depend on the problem considered.

For example one can consider the problem of constructing,

$$\mathcal{M}^g = \{\text{iso. classes of curves of genus } g\}$$

Stated this way the problem is too general and does not have a solution, but one can build,

$$\mathcal{M}^g_{\mathbf{C}} = \{\text{complex iso. classes of complex curves of genus } g\}$$

and this space is a normal complex space and an irreducible quasi-projective variety.

In real algebraic geometry, the corresponding problem is to build the space,

$$\mathcal{M}^g_{\mathbf{R}} = \{\text{real iso. classes of real curves of genus } g\}.$$

Typeset by $\mathcal{A}\mathcal{M}\mathcal{S}$-TEX

This will be our guideline problem, but one can consider many other situations, vector bundles with given invariants over a given variety, varieties with given Hodge numbers, Chern numbers ... (see section **5** for comments and references).

This change has many consequences. For example, as is noted in [Si₂], one can not ask for the space to be algebraic but rather semi-algebraic (see [Se&Si]) and in general only real analytic.

1. Complex moduli versus real moduli.

Historically moduli problems in real algebraic geometry are quite old, and one can say that they originate with the famous 1876 paper of Harnack, not only because his result can be interpreted as giving the lower bound

$$g + 2 \leq \#(\mathcal{M}^g_{\mathbf{R}})$$

for the number of connected components of $\mathcal{M}^g_{\mathbf{R}}$, but also because the paper of Harnack ends with explicit references to the moduli problem in terms of periods of abelian integrals on a real curve. But the story gets really started with F. Klein and G. Weichhold and, slightly later, A. Comessatti. Implicitly this first approach is to consider what one can formulate in modern terms as $\mathcal{M}^g_{\mathbf{C}}(\mathbf{R})$, (see below for a definition) the real part of the complex moduli space, rather than $\mathcal{M}^g_{\mathbf{R}}$ (but for obvious reasons this is not always completely clear).

This approach has beared many fruits, but before discussing one of these results, I would like to point out here the incovinences of this approach.

To introduce this let me define $\mathcal{M}^g_{\mathbf{C}}(\mathbf{R})$ and introduce a notation. Let C be a complex curve of genus g, then the correspondence $C \mapsto C^\sigma$, that to a curve assigns its complex conjugate, induces an anti-holomorphic involution on $\mathcal{M}^g_{\mathbf{C}}$ and hence a real structure. Thus it makes sense to consider the set of real points of $\mathcal{M}^g_{\mathbf{C}}$, which we denote by $\mathcal{M}^g_{\mathbf{C}}(\mathbf{R})$. Now an obvious relation between $\mathcal{M}^g_{\mathbf{C}}(\mathbf{R})$ and $\mathcal{M}^g_{\mathbf{R}}$ is given by

$$\varphi : \mathcal{M}^g_{\mathbf{R}} \longrightarrow \mathcal{M}^g_{\mathbf{C}}$$

the map that maps the real isomorphy class of a real curve to its complex isomorphy class. Obviously $Im(\varphi) \subset \mathcal{M}^g_{\mathbf{C}}(\mathbf{R})$ and we will consider in the sequel φ as a map from $\mathcal{M}^g_{\mathbf{R}}$ to $\mathcal{M}^g_{\mathbf{C}}(\mathbf{R})$. The problem is that,
(1) *φ is not injective.* For an explicit example one can look at the two curves defined by,

$$y^2 = P(x)$$
$$y^2 = -P(x)$$

where, $P(x)$ is a real polynomial with no real roots. These two curves are complex isomorphic, under the isomorphism $(x, y) \mapsto (x, iy)$ but one has real points while the other has empty real part, and hence the curves can not be real isomorphic.
(2) φ *is not surjective.* This came somewhat as a surprise, and I strongly believe that people at the time of Klein were not conscious of this problem. The first examples of this fact are due to C. Earle (1971) [Ea] and, independently, to G. Shimura (1972) [Sh]. The examples given by both authors are based on the same idea, the existence of curves C, with an antiholomorphic automorphism ψ of order 4 but with no antiholomorphic involution. Such a curve will be isomorphic to its complex conjugate, by the antiholomorphic automorphism (and hence will define a point in $\mathcal{M}^g_{\mathbf{C}}(\mathbf{R})$), but will not be defined over $\mathbf{R}$, since a necessary condition for this, is the existence of an antiholomorphic *involution*, and hence will not define a point in $\mathcal{M}^g_{\mathbf{R}}$. The explicit example of this type given by G. Shimura is the following. Consider the curve defined by,

$$y^2 = a_0 x^m + (\sum_{k=1}^{m} a_k x^{m+k} + (-1)^k \bar{a}_k x^{m-k})$$

where m is odd, $a_0 \in \mathbf{R}$ and $a_k \in \mathbf{C}\backslash\mathbf{R}$ for $k \neq 0$. This curve is complex isomorphic to its complex conjugate via the isomorphism $(x, y) \mapsto (-x^{-1}, ix^{-m}y)$. But if a_0, the a_k's and the $\bar{a}_k$'s are chosen algebraically independent over $\mathbf{Q}$, C will have Id and $-Id$: $(x, y) \mapsto (x, -y))$ (the hyperelliptic involution) as only holomorphic automorphism and there will be no antiholomorphic involution.
(3) In my mind this is the worst objection, $\mathcal{M}^g_{\mathbf{C}}$ is not of any use to study deformations. As an example for this problem one can consider the case of curves of genus 1. $\mathcal{M}^1_{\mathbf{C}} = \mathbf{C}$, via the j invariant, and $\mathcal{M}^1_{\mathbf{C}}(\mathbf{R}) = \mathbf{R}$. On the other hand real curves of genus 1 may have a real part consisting of 0, 1 or 2 components. A real elliptic curve E has one real component if $j(E) < 1728$ two (or zero) if $j(E) > 1728$. The case $j(E) = 1728$ corresponds to the curves,

$$y^2 = x^3 + x$$
$$\text{and} \quad y^2 = x^3 - x$$

which have respectively one and two real components. But obviously it is impossible to deforme, while staying in the category of smooth real curves, a curve with one component into one with two.

These remarks are quite general, the same type of objections appear in many other situations, and motivate specific constructions for real moduli spaces.

People at the time of F. Klein were certainly conscious of (1) and (3), but not, as said above, of (2). F. Klein [Kl] and G. Weichhold [We] were nevertheless able to give the exact number of components of $\mathcal{M}^g_{\mathbf{R}}$,

$$\#(\mathcal{M}^g_{\mathbf{R}}) = \begin{cases} \frac{1}{2}(3g+3) \text{ if } g \text{ is odd} \\ \frac{1}{2}(3g+4) \text{ if } g \text{ is even} \end{cases}$$

A. Comessatti [Co] found the same type of results for $\mathcal{A}^n_{\mathbf{R}}$, the moduli space of principally polarized real abelian varieties of dimension n.

2. Construction of $\mathcal{M}^g_{\mathbf{R}}$.

Various constructions of $\mathcal{M}^g_{\mathbf{R}}$ have been proposed. The first, up to my knowledge is due to C. Earle (1971) [Ea]. Other constructions are due to S. Natanzon (1980) [Na], B.H. Gross and J. Harris (1981) [Gro&Ha], M. Seppälä and — (1986) [Se&Si] and possibly still others I do not know of. I do not have the time here to go into the details of these different constructions, but I will outline the basic ideas of the construction of Earle. First fix an orientation reversing involution σ on a topological, compact, connected, orientable surface Σ_g of genus g. It has been known since Klein, that such an involution is characterised, up to conjugation by an homeomorphism, by two invariants: k the number of connected components of the fixed part, Σ_g^σ, of Σ_g under σ; μ (equal to 1 or 2) the number of components of $\Sigma_g \setminus \Sigma_g^\sigma$ (these invariants are the ones used by Klein and Weichhold to prove the result alluded to in section 1). Let $\mathfrak{X}_g(\sigma)$ be the set of complex structures on Σ_g that make σ antiholomorphic.

Now let $Homeo_g^0$ and $Homeo_g^+$ be the group of homeomorphisms of Σ_g homotopic to zero (resp. orientation preserving), and let $Homeo_g^0(\sigma)$ (resp. $Homeo_g^+(\sigma)$) be the subgroup of elements that commute with σ (the normalizer). Then, the group $Homeo_g^+(\sigma)$ acts on $\mathfrak{X}_g(\sigma)$ by pull-back and

$$\mathcal{M}_{\mathbf{R}}^{(g,k,\mu)} = \mathfrak{X}_g(\sigma)/Homeo_g^+(\sigma)$$

is the set of real isomorphy classes of curves of genus g such that, $\#(C(\mathbf{R})) = k$, $\#(C(\mathbf{C}) \setminus C(\mathbf{R}) = \mu$. Moreover this space is connected, and has a natural structure of stratified analytic space (this is obtained by considering the space

$$\mathfrak{T}_g^\sigma = \mathfrak{X}_g(\sigma)/Homeo_g^0(\sigma),$$

which is an equivalent of the Teichmüller space in this case, where connectedness and the existence of a real analytic structure are easier to prove — $\mathfrak{T}_g$, the ordinary Teichmüller space, is a ball and $\mathfrak{T}_g^\sigma$ is just the fix part of $\mathfrak{T}_g$ under the action of σ).

From this it is easy to build the moduli space of real genus g curves by letting,

$$\mathcal{M}^g_{\mathbf{R}} = \coprod_{(g,k,\mu)} \mathcal{M}_{\mathbf{R}}^{(g,k,\mu)} \ .$$

REMARK: Here one can discuss what the correct numerical invariants are. If one is only interested in smooth structures, then, of course, (g, k, μ) is the correct set. On the other hand if one considers the problem of compactifying the moduli space, then g appears to be the natural invariant one must consider.

3. Applications of Real Moduli Spaces.

Many problems in real algebraic geometry can be formulated in terms of moduli. This does not always lead to a solution of the problems though, and often the reverse happens. But let me formulate a few.

One important question in the theory of real algebraic surfaces is to determine the number of distinct possible topological types of a surface of degree n in $\mathbf{P}^3_{\mathbf{R}}$. A partial answer to this problem could be given in the following way: complex surfaces of degree n have specific numerical invariants (essentially its Hodge numbers and Chern numbers). Call $\underline{\mu}$ this set of invariants (think of this as the equivalent of the genus of a curve). Let $\mathcal{M}^{\underline{\mu}}_{\mathbf{R}}$ be the moduli space of real surfaces such that a minimal complexification has invariants $\underline{\mu}$ (this makes sense if the surfaces we are looking at are not ruled, which is the case if we start with a surface of degree ≥ 4). Then knowing the number of connected components of $\mathcal{M}^{\underline{\mu}}_{\mathbf{R}}$ will give bounds, possibly stronger than the ones we know today, for the number of distinct topological types. This would give in particular an answer to the Ragsdale-Petrovski-Viro conjecture for surfaces (see Viro [Vi]). In its simplest instance this conjecture can be formulated in the following way, let X be a smooth and projective real algebraic surface such that, $X(\mathrm{C})$ (the complex part) is simply connected (for example X is a surface in $\mathbf{P}^3$ or more generally a complete intersection), then the conjecture states that

$$\dim \mathrm{H}_1(X(\mathbf{R}), \mathbf{Z}/2) \leq \dim \mathrm{H}^{1,1}(X) .$$

By analogy with the situation of double coverings ramified along a curve, one can also formulate the following conjecture,

$$\# X(\mathbf{R}) = \dim \mathrm{H}_0(X(\mathbf{R})) \leq \frac{1}{2}(\dim \mathrm{H}^{1,1}(X)) + 1 .$$

In a completely different setting let me give an *à priori* not obvious application of the knowledge of moduli spaces. From the results of S. Novikov, T. Shiota and E. Arbarello and C. De Concini one knows that an abelian variety is the Jacobian of an algebraic curve if, and only if, one can construct from the Theta function associated to a realisation of the abelian variety a solution of the famous K.-P. equations (K.d.V. in the case of hyperelliptic curves). Some authors belive that if you start with a real curve then you obtain in this way a real solution to the K.-P. equations. As noted by B. Dubrovin and S. Natanzon [Du&Na] this turns out to be false, such a solution is real if and only if $\mu = 2$, i.e. the curve is a dividing curve.

Let me indicate how this can be shown using real moduli spaces. Let C be a complex curve of genus g, let $\gamma_1, \ldots, \gamma_{2g}$ be a basis of $\mathrm{H}_1(C, \mathbf{Z})$ and let $\omega_1, \ldots, \omega_g$ be a basis of $\mathrm{H}^0(C, \Omega^1)$ the space of holomorphic differentials. The $g \times 2g$ matrix $\left(\int_{\gamma_j} \omega_i\right)_{i,j}$ is called a period matrix of C. One can normalize (globally) the choices made, in such a way that the period matrix is of the form $(\Omega, 1_g)$ with Ω symmetric and $\Im m(\Omega)$ positive definite.

In the real case one can normalize still further. One can normalize in such a way that $2\Re e(\Omega)$ has interger coefficients (in fact a curve is real if and only if one can normalize in this way — see for example [Se&Si]). Now let M be the reduction mod.2 of $2\Re e(\Omega)$. Then under the above normalization (to simplify we assume here that $\#(C(\mathbf{R})) \neq \emptyset$),

$$k = \#(C(\mathbf{R})) = g + 1 - \operatorname{rank} M$$

and $\mu = \#(C(\mathbf{C}) \setminus C(\mathbf{R})) = 2$ if the bilinear form defined by M is of type II (or even, or orthosymmetric, or ...) that is, such that

$${}^t\!nMn \equiv 0(\text{mod}.2), \text{ for all } n \in \mathbf{Z}^g.$$

Conversely $\mu = \#(C(\mathbf{C}) \setminus C(\mathbf{R})) = 1$ if the bilinear form is of type I (or odd, or diasymmetric, or ...).

The important point here, is that, by the characterisation of the connected components of the space $\mathcal{M}^g_{\mathbf{R}}$ one can reformulate the above result by saying that $M = 2\Re e(\Omega)\text{mod}.2$ is constant, of given rank and type, on the connected components of $\mathcal{M}^g_{\mathbf{R}}$.

Now the Theta function expressed in terms of a given period matrix Ω is defined by,

$$\vartheta(z, \Omega) = \sum_{n \in \mathbf{Z}^g} exp(\pi i\, {}^t\!n\Omega n + 2\pi i\, {}^t\!nz).$$

From this expression it is easy to see that $\vartheta(z, \Omega)$ is a real function in z if and only if $2\Re e(\Omega)$ is the matrix of a form of type II (one way is obvious, the other comes from the fact that the above expression is a developement in Fourier series of the Theta function). This proves the result of Dubrovin and Natanzon.

4. Compactifications of Moduli Spaces.

Real moduli spaces are, as indicated in section 1, the correct setting for studying deformations. But since the study of deformations cannot, of course, be limited to smooth structures, one needs to enlarge somewhat the spaces and this leads to the compactification problem.

There are many ways to do this but there are a certain number of standard procedures. In the case of $\mathcal{M}^g_{\mathbf{C}}$, for example, the standard procedure is to compactify by,

$$\overline{\mathcal{M}^g_{\mathbf{C}}} = \{\text{complex iso. classes of complex stable genus } g \text{ curves}\},$$

where an, eventually singular, curve C is said to be stable if,

(i) the only singularities of C are ordinary quadratic or nodes (i.e. analytically isomorphic — over $\mathbf{C}$ — to $xy = 0$ at 0);

(ii) $Aut(C)$ is finite.

In the real case one can try to do the same and compactify $\mathcal{M}^g_{\mathbf{R}}$ by,

$$\overline{\mathcal{M}^g_{\mathbf{R}}} = \{\text{real iso. classes of real stable genus } g \text{ curves}\}.$$

This has been done by M. Seppälä in [Se]. The main difficulty with this approach is that, whereas in the complex case there is only one type of ordinary quadaratic singularity, in the real case there are 3,

(i) singularities of type $xy = 0$;
(ii) singularities of type $x^2 + y^2 = (x + iy)(x - iy) = 0$;
(iii) a pair of complex conjugate nodes.

The construction of $\overline{\mathcal{M}^g_{\mathbf{R}}}$, given by M. Seppälä, takes into account all three types of nodes. The same type of problem appear in compactifying other moduli spaces, in particular for the moduli space of principally polarized abelian varieties (see [Si$_2$] for an account).

As an application of this construction, or rather of these constructions, I can give the following,
(1) (this small application can actually be proved, in this case, in a more elementary way but it can be generalized to situations where no such elementary proof exists — i.e. not using the compactification of moduli spaces). Let,

$$y^2 = 4x^3 - p(t)x - q(t)$$

be a family of elliptic curves, non singular if $t \neq 0$, with a singular, but semi-stable, fibre at $t = 0$. Then the fibre at $t = 0$ is of the form,

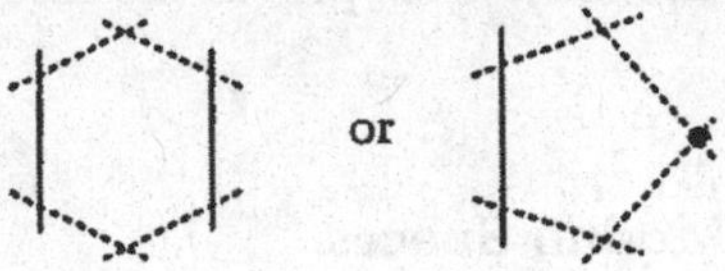

(where — is a real line and - - - a complex line and there are m, even or odd, components) if and only if,

$$\lim_{t \to 0} \frac{\sqrt{p^3(t)}}{q(t)} > 0$$

(in such a situation $\lim p(t)$ must be positive). On the other hand the fibre is of the form

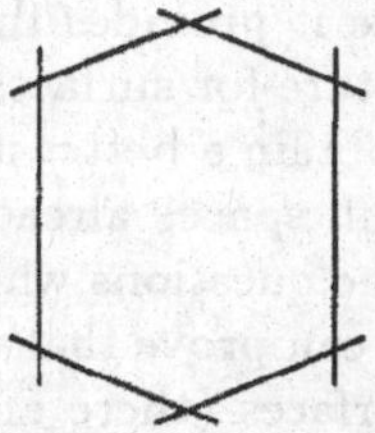

if and only if,

$$\lim_{t\to 0}\frac{\sqrt{p^3(t)}}{q(t)} < 0 \ .$$

(2) Let $\Omega = \frac{1}{2}M + iY$ be the period matrix of a real curve normalized as in 3. Then a real genus 2 curve C has empty real part if and only if $M \equiv \begin{pmatrix} 0 & 1 \\ 1 & 0 \end{pmatrix}$ mod.2, and $\det Y \geq \frac{1}{4}$. This can be proved using the compactification of the moduli space because a curve with empty real part that degenerates to a semi-stable singular curve can only degenerate to one with isolated real points (see [Si$_2$]).

5. Other Moduli, and Some Open Questions.

Apart from $\mathcal{M}^g_{\mathbf{R}}$ other moduli spaces in real algebraic geometry have been constructed, here are some,
(1) Real abelian varieties (Comessatti [Co] and — [Si$_1$]);
(2) Real K3 surfaces (Nikulin [Ni], see also [Si$_1$]);
(3) Real elliptic surfaces (Seiler [Sei]);
(4) Some surfaces of general type (Ballico [Ba]);
(5) Stable rank r vector bundles over $\mathbf{P}^2(\mathbf{R})$ (Ballico [Ba]).

The first open problem is to build still more moduli spaces, especially in the cases when the corresponding complex moduli space has been constructed. Among these the most reachable seems to be the moduli space for so-called numerical quintics (i.e. surfaces such that a minimal smooth complexification has same numerical invariants as surfaces of degree 5 in $\mathbf{P}^3$ — the complex moduli space for these surfaces has been constructed by Horikawa [Ho]). This space is of special importance since it provides the first non-trivial case for the Ragsdale-Petrovski-Viro conjecture for surfaces (see section **3**).

The second open problem is to obtain a better understanding of the topology and the geometry of the moduli spaces already built and of their compactifications. In this direction some questions which do not have a complex equivalent appear, for example one can prove that the connected components of the moduli space of real K3 surfaces (there are 66 of these) are not all homeomorphic.

The third problem is, for the moment, just a reformulation of a well known and well studied problem, the construction of the real moduli space for degree d hypersurfaces in $\mathbf{P}^n$. Of course such a problem will probably be solved in some special cases without a moduli problematic, but on the other hand a close analysis of the solutions found, up to now, for curves of degree 7 and degree 8 and 9 in $\mathbf{P}^2$ (partial solutions in the last two cases) shows that these problems would benefit from a more global approach that can eventually be given by a formulation in terms of moduli, or in terms of deformations.

The fourth problem, concerns compactifications. There is of course the problem of compactifying the moduli spaces already constructed but not yet compactified (e.g. K3 surfaces) but there is also the problem of studying other types of compactifications in the case when one has already been constructed. For example one can consider torodial compactifications of $\mathcal{A}^n_{\mathbf{R}}$ (the real moduli space for principally polarized abelian varieties of dimension n) and study how such constructions can be generalized to other situations (K3 surfaces ? ....?). Closely linked with this problem is the problem of studying real algebraic deformations which is also a theme of great interest.

References

[Ba]. E. Ballico, *Real moduli of complex objects: surfaces and bundles*, preprint (1990).

[Ca]. F. Catanese, *Moduli of algebraic surfaces; in*, "Theory of Moduli (E. Sernesi Ed.)," Springer Lecture Notes in Maths, Berlin Heidelberg New York, 1988, pp. 1–83.

[Co]. A. Comessatti, *Sulle varietà abeliane reale I*, Ann. Mat. Pura Appl. **2** (1924), 67–106; *e II*, 4 (1927–1928), 299–317.

[Du&Na]. B.A. Dubrovin and S. Natanzon, *Real Theta function solutions of the Kadomtsev Petviashvili equation*, Math. USSR Izvestiya **32** (1989), 269–288.

[Ea]. C. Earle, *On moduli of closed Riemann surfaces with symmetries; in*, "Advances in the Theory of Riemann Surfaces," Annals of Mathematics Studies 66, Priceton Univ. Press, Princeton New Jersey, 1971, pp. 119–130.

[Gro&Ha]. B.H. Gross and J. Harris, *Real algebraic curves;*, Ann. scient. Ec. Norm. Sup. **14** (1981), 157–182.

[Ha]. A. Harnack, *Über die Vieltheiligkeit algebraischen Kurven*, Math. Ann. **10** (1876), 189–198.

[Ho]. E. Horikawa, *On deformations of quintic surfaces*, Inventiones Math. **31** (1975), 43–85.

[Kl]. F. Klein, "Über Riemanns Theorie der algbraischen Funktionen und ihrer Integrale. —Eine Ergänzung der gewöhnlichen Darstellungen.," B.G. Teubner, Leipzig, 1882.

[Na]. S. Natanzon, *Moduli spaces of real curves*, Trans. Moscow Math. Soc. (1980), 233–272.

[Ni]. V.V. Nikulin, *Involutions of integral quadratic forms and their applications to real algebraic geometry*, Math; USSR Izvestiya **22** (1984), 99–172.

[Sei]. W. Seiler, *Global moduli for polarized elliptic surfaces*, Compositio Math. **62** (1987), 187–213.

[Se]. M. Seppälä, *Moduli space of stable real algebraic curves; preprint 1990*, to appear in Ann. Sci. E.N.S..

[Se&Si]. M. Seppälä and R. Silhol, *Moduli spaces for real algebraic curves and real abelian varieties*, Math. Z. **201** (1989), 151–165.

[Sh]. G. Shimura, *On the field of rationality for an abelian variety*, Nagoya Math. J. **45** (1972), 167–178.

[Si$_1$]. R. Silhol, "Real Algebraic Surfaces," Springer Lecture Notes in Maths. 1392, Berlin Heidelberg New York, 1989.

[Si$_2$]. R. Silhol, *Compactifications of Moduli spaces in real algebraic geometry*, to appear in Inventiones Math..

[Vi]. O. Viro, *Curves of degree 7, curves of degree 8, and the Ragsdale conjecture*, Soviet Math. Dokl. **22** (1980), 566–570.

[We]. G. Weichhold, "Über symmetrische Riemannsche Flächen und die Periodizitätsmodulen der zugehörigen Abelschen Normalintegrale erster Gattung," Leipziger Dissertation, 1883.

Université Montpellier II, Dépt. de Mathématiques, Place E. Bataillon 34095 Montpellier Cedex 5 France

Constructing Strange Real Algebraic Sets

by S. Akbulut* and H. King**

Recall that algebraic sets are characterized by resolution tower structures **[AK1]**, **[AK2]**. A topological space with such a structure is called a **TR** space for short. In low dimensions existence of resolution tower structures can often be detected by combinatorial invariants of the underlying topological spaces. For example, let X be a locally cone-like Euler stratified space (i.e., a locally conelike stratified set so that the Euler characteristics of the links of all strata are even). Then for every component of the codimension one stratum U of X we define $\kappa_X(U)$ to be the number of points in the link of U. For every component of the codimension two stratum W of X we define $\alpha_i(W)$ to be the number of vertices v of the link L of W so that $\kappa_L(v) = i \bmod 8$, where $i = 0, \ldots, 7$. Then define

$$\begin{aligned}
\epsilon_0(W) &= \alpha_0(W) \bmod 2\\
\epsilon_1(W) &= \alpha_6(W) \bmod 2\\
\epsilon_2(W) &= (\alpha_0(W) + \alpha_4(W))/2 \bmod 2\\
\epsilon_3(W) &= (\alpha_2(W) + \alpha_6(W))/2 \bmod 2
\end{aligned}$$

and define $\epsilon_X(W) = (\epsilon_0(W), \epsilon_1(W), \epsilon_2(W), \epsilon_3(W)) \in \mathbf{Z}_2^4$. It is an easy exercise to show that $\epsilon_X(W)$ is well defined, i.e., $\alpha_0 + \alpha_4$, and $\alpha_2 + \alpha_6$ are even. For every component V of the codimension three stratum of X and every $\epsilon \in \mathbf{Z}_2^4$ we define $\beta_\epsilon(V) \in \mathbf{Z}_2$ to be the number of vertices v of the link L of V with $\epsilon_L(v) = \epsilon$.

Theorem. (**[AK2]**) *For any compact topological space X, the following are equivalent:*

a) *X is homeomorphic to a real algebraic set of dimension ≤ 3.*

b) *X is a **TR** space of dimension ≤ 3.*

c) *X is a locally cone-like Euler stratified space of dimension ≤ 3 and for all codimension three strata x of X we have:*

$$\begin{aligned}
\beta_{1110}(x) + \beta_{1111}(x) &= 0\\
\beta_{0100}(x) + \beta_{0101}(x) &= 0\\
\beta_{1000}(x) + \beta_{1001}(x) &= 0\\
\beta_{1100}(x) + \beta_{1101}(x) &= 0
\end{aligned}$$

Hence whether or not a stratified space of dimension less then four is homeomorphic to an algebraic set can be decided by purely combinatorial data. In **[AK1]**, **[AK2]** a similar characterization, with homeomorphism replaced by a stratified set isomorphism, is also given. (One needs all β_ϵ's to be 0 and if all strata have trivial normal bundle then this suffices.)

This gives some surprising examples of real algebraic sets. For example, in **[CK]** this was used to give an example of a real algebraic set which is topologically imbedded in

---

* Supported in part by the N.S.F.

** Supported in part by the Graduate Research Board, University of Maryland

$\mathbf{R}^4$ but cannot be an algebraic subset of $\mathbf{R}^4$. The purpose of this paper is to indicate a construction of such algebraic sets which, for these special cases, is simpler than the general construction given in [AK2]. It falls short of giving explicit equations, but comes much closer to an explicit description.

The construction of the Example

Let us now proceed with the construction, which follows the spirit, but differs quite a bit from the letter, of that in [AK2]. We shall construct below a compact three dimensional manifold M with boundary and closed subsets A, B_0 and B_1 so that:

1) A is a spine of M, in fact there is a diffeomorphism $h: \partial M \times (0,1] \to M - A$ so that $h(x,1) = x$ for all $x \in \partial M$ and $h^{-1}(B_i) = (\partial M \cap B_i) \times (0,1]$.
2) B_0 and B_1 are disjoint.
3) If you take ∂M, crush $\partial M \cap B_0$ to a point and then crush $\partial M \cap B_1$ to a point, you obtain a two dimensional sphere.
4) Everything can be made algebraic. In particular, there is a compact nonsingular three dimensional real algebraic set $V \subset \mathbf{R}^n$, algebraic subsets C, D_0 and D_1 and an imbedding $g: M \to V$ so that $C = g(A)$, $B_i = g^{-1}(D_i)$ and $D_0 \cap D_1 = \emptyset$.

By Corollary 2.5.14 of [AK2] we may as well assume that the V in 4) is projectively closed (i.e., it is Zariski closed in projective space), since any compact real algebraic set is isomorphic to a projectively closed algebraic set. Alternatively, using the fact we see below that A and B_i are unions of codimension one submanifolds in general position, this is a consequence of Theorem 2.10 of [AK4]. So there is an overt polynomial $r: \mathbf{R}^n \to \mathbf{R}$ with $V = r^{-1}(0)$. Overtness means that if r has degree d and r_d is the sum of the monomials of r of degree d, then $r_d^{-1}(0) = \{0\}$. Pick polynomials $q_i: \mathbf{R}^n \to \mathbf{R}$ so that $q_i(x) \geq 0$ for all x and so $q_i^{-1}(0) = C \cup D_i$. For example, q_i could be the sum of squares of generators of the ideal of polynomials vanishing on $C \cup D_i$. Let $q = q_0 + q_1$ and pick an integer e bigger than the degrees of q_0 and q_1. Note that $C = q^{-1}(0)$.

Let $\alpha: V \to \mathbf{R}^{n+2}$ be the map

$$\alpha(x) = (\, q_0(x) q_1(x) x \,,\, q_0(x) \,,\, q_1(x) \,)$$

and let $Z \subset \mathbf{R}^{n+2}$ be the algebraic set

$$Z = \{\, (y, t_0, t_1) \mid y = 0 \text{ and } t_0 t_1 = 0 \,\}.$$

We claim that $\alpha(V) \cup Z$ is a real algebraic set and near 0 it has strange behavior which was remarked upon in [CK], see their question (¿d). In particular, the line segment $\{\, y = 0,\ t_0 = 0,\ t_1 < 0 \,\}$ is open in $\alpha(V) \cup Z$ but $\{y = 0,\ t_0 = 0,\ t_1 > 0\}$ is contained in a part of $\alpha(V) \cup Z$ which is a topological manifold. See Figure 1, which is a fairly truthful representation. In particular, $\alpha(V)$ lies over the first quadrant and $Z - \alpha(V)$ consists of two half lines.

To see our claim that $\alpha(V) \cup Z$ is a real algebraic set, let $r': \mathbf{R}^{n+2} \to \mathbf{R}$ and $q_i': \mathbf{R}^{n+2} \to \mathbf{R}$ be the polynomials

$$r'(y, t_0, t_1) = (t_0 t_1)^d \, r(y / t_0 t_1)$$

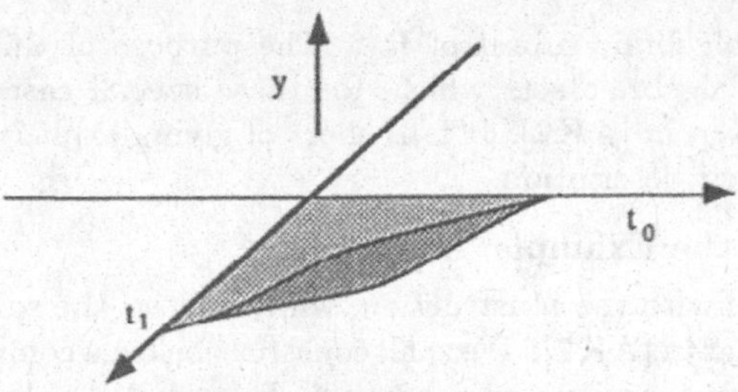

Figure 1: $\alpha(V) \cup Z$

$$q_i'(y, t_0, t_1) = (t_0 t_1)^e (t_i - q_i(y/t_0 t_1))$$

Let W be the algebraic set $W = r'^{-1}(0) \cap q_0'^{-1}(0) \cap q_1'^{-1}(0)$. Then certainly $\alpha(V) \subset W$ and $Z \subset W$. But in fact $\alpha(V) \cup Z = W$. To see this, pick any $(y, t_0, t_1) \in W$. If $t_0 t_1 = 0$ then $0 = r'(y, t_0, t_1) = r_d(y)$ so $y = 0$ by overtness of r and hence $(y, t_0, t_1) \in Z$. But if $t_0 t_1 \neq 0$ then setting $x = y/(t_0 t_1)$ we have $r(x) = 0$ so $x \in V$ and $(y, t_0, t_1) = \alpha(x)$.

Now let us investigate the topology of W near 0. We think of W as the quotient space of $V \cup Z$ via the map α. Suppose $\alpha(x) = \alpha(x')$ for some $x, x' \in V$. If $x \notin C \cup D_0 \cup D_1$ then $q_0(x) q_1(x) \neq 0$. But $q_i(x') = q_i(x)$ and $q_0(x) q_1(x) x = q_0(x') q_1(x') x' = q_0(x) q_1(x) x'$ so $x = x'$. If $x \in D_i - C$ then $q_i(x) = 0$ and $q_{1-i}(x) \neq 0$ so $q_i(x') = 0$ and $q_{1-i}(x') \neq 0$ so $x' \in D_i - C$ also and $q(x) = q(x')$. If $x \in C$ then $q_0(x) = q_1(x) = 0$ so $q_0(x') = q_1(x') = 0$ and $x' \in C$ also. So the map α crushes C to a point and maps D_i to the half line $y = 0$, $t_i = 0$, $t_{1-i} \geq 0$.

For small enough $\epsilon > 0$ we have a homeomorphism $f: q^{-1}(\epsilon) \times (0, \epsilon] \to q^{-1}((0, \epsilon])$ so that $qf(x, t) = t$ and $f^{-1}(D_i) = (D_i \cap q^{-1}(\epsilon)) \times (0, \epsilon]$. This follows from Thom's first isotopy lemma, although in fact C and D_i are unions of codimension one submanifolds in general position so if one wanted, one could take f to be a diffeomorphism. Anyway we have an induced homeomorphism $f': \alpha(q^{-1}(\epsilon)) \times (0, \epsilon] \to \alpha(q^{-1}((0, \epsilon]))$ where $f'(\alpha(x), t) = \alpha(f(x, t))$. Thus near 0, $\alpha(V)$ is the cone on $\alpha(q^{-1}(\epsilon))$. Consequently near 0, W is the cone on the disjoint union of $\alpha(q^{-1}(\epsilon))$ and two points. But $\alpha(q^{-1}(\epsilon))$ is obtained from $q^{-1}(\epsilon)$ by crushing $q^{-1}(\epsilon) \cap D_0$ to one point and $q^{-1}(\epsilon) \cap D_1$ to another. But this space must be the two sphere. The reason is that the two product structures f and gh give an invertible cobordism from the triple $(q^{-1}(\epsilon), q^{-1}(\epsilon) \cap D_0, q^{-1}(\epsilon) \cap D_1)$ to $(\partial M, \partial M \cap B_0, \partial M \cap B_1)$. But because the dimensions are so small the cobordism is trivial. So there is a homeomorphism between the two triples.

This algebraic set W has the following curious property. Consider one of the lines in Z, say $y = 0$, $t_0 = 0$. Then if $t_1 < 0$ it is locally open in W. But if $t_1 > 0$, then W is a three dimensional topological manifold nearby. Thus it has the curious behavior remarked upon in **[CK]**.

One could modify this example a bit to obtain the specific space mentioned in **[CK]**. Let $W' \subset \mathbf{R}^{n+3}$ be the union of $W \times 0$ and the plane $\{(y, t_0, t_1, t_2) \mid y = 0 \text{ and } t_0 = 0\}$. Then near 0, W' is the cone on the disjoint union of a point and the wedge of a 2-sphere and a circle. To obtain a real algebraic set homeomorphic to the suspension of this, one can use the Lemma below.

Lemma. *Let X be a real algebraic set and pick $x_0 \in X$. It is well known that x_0 has a neighborhood in X homeomorphic to the cone on some compactum L. Then there is a real algebraic set Y homeomorphic to the suspension of L.*

Proof: We may suppose $X \subset \mathbf{R}^n$ and $x_0 = 0$. For small enough $\epsilon > 0$ we know that $X \cap B_\epsilon$ is homeomorphic to the cone on $X \cap \partial B_\epsilon$ where B_ϵ is the ball around 0 of radius ϵ and ∂B_ϵ is its boundary sphere. We then take

$$Y = \{ (x,t) \in X \times \mathbf{R} \mid t^2 + |x|^2 = \epsilon^2 \}.$$

Clearly Y is the suspension of $X \cap \partial B_\epsilon$. But L and $X \cap \partial B_\epsilon$ are invertibly cobordant (see [K]) so their products with $\mathbf{R}$ are homeomorphic and hence their suspensions are homeomorphic also. ∎

The construction of M, A and the B_i's

We promised above the construction of the manifold M and its subsets A and B_i. We start out with part of the boundary of M. Let E be a two dimensional disc with three handles attached as in Figure 2. For convenience we do not smooth out corners in the figures although in reality they are smooth. Let E_1, E_2 and E_3 be the curves shown in Figure 2. Note that the quotient space $E/(E_1 \cup E_2 \cup E_3)$ is a disc. For another description, note that E is obtained from a disc by blowing up three times and the E_i's are the strict transforms of the exceptional divisors.

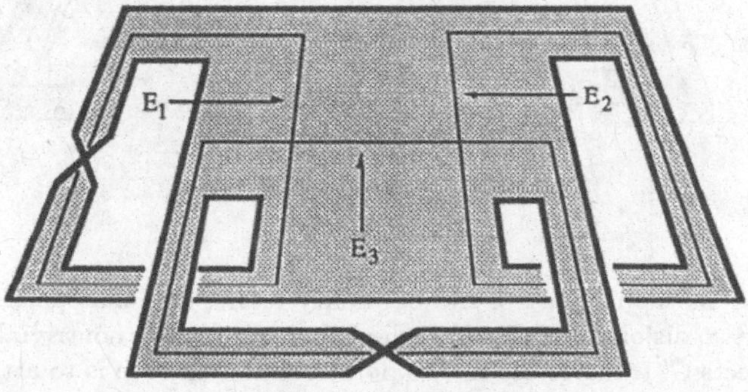

Figure 2: E

Next we consider the manifold $F = E \times [0,1]$. We show it in Figure 3, although it does not quite imbed in $\mathbf{R}^3$. In F we have an immersed surface $H = (E_1 \cup E_2 \cup E_3) \times [0,1]$. The manifold F will be part of M and H will be part of B_0.

Our next step is to add a handle to F as in Figure 4. In particular, letting $J = [-1,1]$, we pick disjoint imbeddings $\varphi_i \colon J \times J \to E \times 1$ for $i = -1, 1$ so that $\varphi_{-1}(0,0)$ is the point $(E_1 \cap E_3) \times 1$ and $\varphi_1(0,0)$ is the point $(E_2 \cap E_3) \times 1$, so that $\varphi_i^{-1}(E_3 \times 1) = 0 \times J$, $\varphi_1^{-1}(E_2 \times 1) = J \times 0$ and $\varphi_{-1}^{-1}(E_1 \times 1) = J \times 0$. Furthermore the imbeddings are oriented

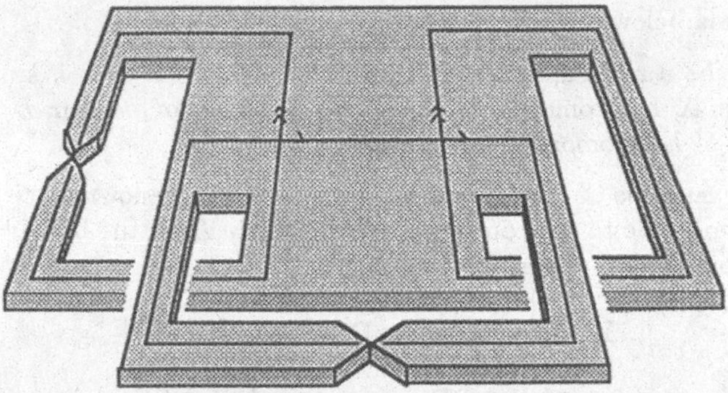

Figure 3: F

as shown by the arrows in Figure 3. This gives the handle a twist if we try to imbed it in $\mathbf{R}^3$ as we do in Figure 4. We let $F' = F \cup J \times J \times J$ where we identify (i, s, t) with $\varphi_i(s,t)$ for $i = -1, 1$ and $(s,t) \in J \times J$. We let $H' = H \cup J \times 0 \times J \cup J \times J \times 0$ and $G' = 0 \times J \times J$. This G' will be part of A.

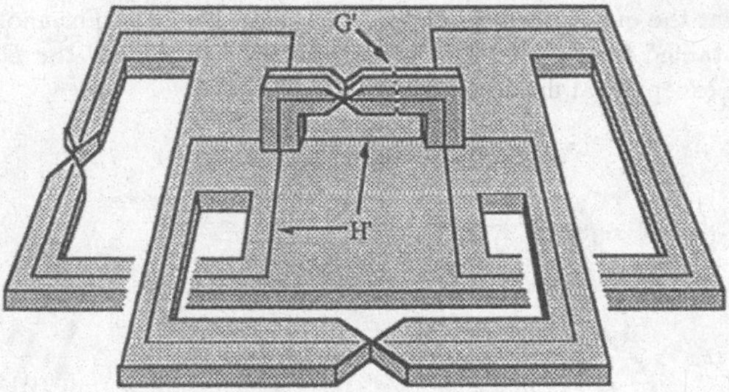

Figure 4: F

Note that the the upper part of the boundary of H', i.e., that part which is not in $E \times 0$, consists of two disjoint circles and each of these circles has nontrivial normal bundle in $\partial F'$ and intersects G' transversely in two points. Our next step is to attach a thickened cylinder to these two circles.

In particular, let K be a Moebius band with central circle K' and let K'' be two transverse arcs as in Figure 5. Then we can pick disjoint imbeddings $\theta_i : K \to \partial F'$ for $i = -1, 1$ so that $\theta_i^{-1}(H') = K'$ and $\theta_i^{-1}(G') = K''$. We now let $F'' = F' \cup J \times K$ where we identify (i, x) with $\theta_i(x)$ for $i = -1, 1$ and $x \in K$. We let $H'' = H' \cup J \times K'$ and $G'' = G' \cup J \times K'' \cup 0 \times K$.

Our construction has the following property. There is a diffeomorphism $f: E \times (0,1] \to F'' - G''$ so that $f(E \times 1) = E \times 0$ and $f((E_1 \cup E_2 \cup E_3) \times (0,1]) = H'' - G''$. For example,

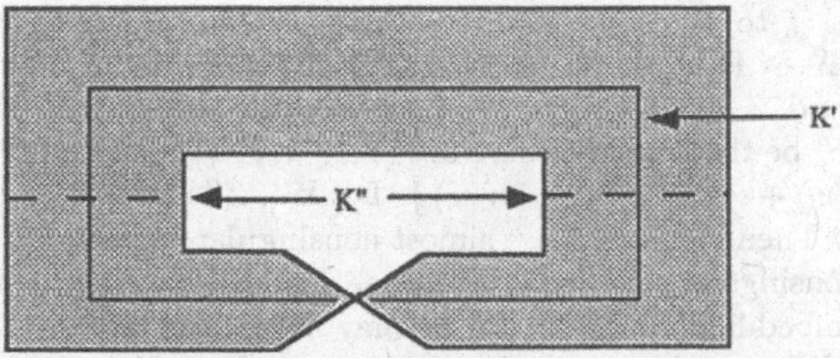

Figure 5: K

f can be obtained by integrating a vector field tangent to H'' and to $\partial F'' - E \times 0$ and pointing inward on $E \times 0$.

We now let M be obtained by doubling F'' along $\partial F''_+ = \text{Cl}(\partial F'' - E \times 0)$. In other words, $M = F'' \times \{0,1\} \cup \partial F''_+ \times [0,1]$. We let $A = G'' \times \{0,1\} \cup \partial G'' \times [0,1]$ and $B_i = H'' \times i$.

Making M, A and the B_i's algebraic

In order to facilitate making M, A and the B_i's algebraic, we modify this example slightly by blowing up F'' with center the double point arc of H''. After doing so, G'' and H'' become unions of proper codimension one submanifolds of F'' in general position. (Before doing this, H'' was only immersed, not imbedded.) After doubling F'' along $\partial F''_+$ to obtain M, we see that A and the B_i's are still unions of proper codimension one submanifolds in general position. We may now double M along its boundary to obtain a closed smooth manifold M'. Inside M' are B'_i, the doubles of B_i and A' which is a copy of A in one of the copies of M in M'. Then A' and the B'_i's are unions of codimension one submanifolds in general position. Hence by Corollary 2.8.10 of **[AK2]**, or more directly by Theorem 2.10 of **[AK4]**, we may assume that M' is a nonsingular real algebraic set and A' and the B'_i's are algebraic subsets of M'. Consequently we have made M, A and the B_i's algebraic as in 4) above.

However, if one tried to construct explicit polynomial equations for M by following the construction in **[AK2]** or **[AK4]**, many steps would be done which are not necessary for this particular example. Thus we indicate here some simplifications.

First, it is only necessary to make F'', G'' and H'' algebraic and then an explicit construction allows us to make M, A and the B_i's algebraic. In particular, suppose we have a three dimensional real algebraic set X and algebraic subsets W, Y and Z of X and an imbedding $f: F'' \to X$ so that $f^{-1}(Y) = G''$, $f^{-1}(Z) = H''$ and $f(H'' \cap G'') = W \subset Y \cap Z$. It makes the construction easier if we also assume that Y and Z are finite unions of nonsingular real algebraic sets, or more generally that they are almost nonsingular in the sense of **[AK5]**. We may suppose by Theorem 2.5.13 of **[AK2]** that W is projectively closed. Pick an overt polynomial $p: X \to \mathbf{R}$ so that $W = p^{-1}(0)$ and suppose that p has degree d. Let $X' = \{(x,t) \in X \times \mathbf{R} \mid p(x)^2 + t^{2d} = \epsilon^{2d}\}$ for a small $\epsilon > 0$. Note that X' is projectively closed. Topologically, X' is the double of a small neighborhood of W in X. Let $Y' = X' \cap Y \times \mathbf{R}$, $Z' = X' \cap Z \times \mathbf{R}$ and $W' = W \times \epsilon$. Since $G'' \cap H''$ is a spine of F'', i.e., $F'' - G'' \cap H'' = \partial F'' \times (0,1]$ we may as well assume that the image of the imbedding f lies in a small neighborhood of W, in particular that $f(F'') \subset p^{-1}([-\epsilon^d, \epsilon^d])$. We may then lift the imbedding f to $f': F'' \to X'$ by setting $f'(x) = (\, f(x)\,,\, + \sqrt[2d]{\epsilon^{2d} - pf(x)^2}\,)$. Let $r: X \to \mathbf{R}$ and $s: X \to \mathbf{R}$ be the sums of squares of generators of the ideal of polynomials vanishing on Y and Z respectively. We now blow up X' via the ideal $\langle r, s, t+\epsilon \rangle$. In other words let X'' be the Zariski closure of $\{\, (x, t, [u, v, w]) \in X' \times \mathbf{RP}^2 \mid t \neq -\epsilon,\ wr(x) = u(t+\epsilon),\ ws(x) = v(t+\epsilon) \text{ and } us(x) = vr(x)\,\}$. Let Y'', Z'' and W'' be the strict transforms of Y', Z' and W'. Then if Y and Z are almost nonsingular we have $W'' = Y'' \cap Z''$. If they are not almost nonsingular, one must do a more careful blowing up to have this. Again we may lift the imbedding f' to X''. So we may as well replace X, Y, Z and W by X'', Y'', Z'' and W''. What we have gained is that we now have X projectively closed and also $Y \cap Z = W$.

So $Z = p^{-1}(0)$ for some overt polynomial p. Say p has degree d. (Of course p is different from the p in the preceeding paragraph.) Then for small $\epsilon > 0$ let $V = \{\, (x,t) \in X \times \mathbf{R} \mid p(x)^2 + t^{2d} = \epsilon^{2d}\,\}$ and $C = V \cap Y \times \mathbf{R}$, $D_0 = Z \times (-\epsilon)$ and $D_1 = Z \times \epsilon$. Then V is projectively closed and satisfies 4) above. The point is that $F'' - H''$ is diffeomorphic to the product $\partial F''_+ \times (0,1]$ via a diffeomorphism preserving G''. But for small $\epsilon > 0$, the set $p^{-1}([-\epsilon^d, 0) \cup (0, \epsilon^d])$ is also a product with $(0,1]$. Consequently because the dimensions are so low that invertible cobordisms are trivial, we know $\partial F''_+$ imbeds in $p^{-1}(\pm\epsilon^d)$ in a way which takes $\partial G''$ to $p^{-1}(\pm\epsilon^d) \cap Y$. Consequently we see F'' doubled along $\partial F''_+$ in V. But this is just M.

A certain amount of staring at Figure 4 convinces one that F'' can be imbedded in $\mathbf{R}^4$. This is also seen from the fact that $G'' \cap H''$ is a spine of F'', i.e., $F'' - G'' \cap H'' = \partial F'' \times (0,1]$. So F'' is just a thickening of $G'' \cap H''$ which is the curve in Figure 6.

Figure 6: $G'' \cap H''$

If we had modified F'' by blowing up the double points of H'', then $G'' \cap H''$ would be a little different, the double point of the figure eight would be replaced by a circle, but F'' would still imbed in $\mathbf{R}^4$. One could then make F'', G'' and H'' algebraic in the following explicit way. Note $G'' \cap H''$ is three circles intersecting in a certain way, so it is easy to construct a projectively closed real algebraic set W in $\mathbf{R}^4$ homeomorphic to $G'' \cap H''$. In fact we may let W be the union of the three circles $x = w = 0, y^2 + z^2 = 1$ and $y = w = 0, x^2 + z^2 = 1$ and $z = w = 0, x^2 + 2y^2 = 1$. Imbed F'' in $\mathbf{R}^4$ so that W is $G'' \cap H''$. Let U be a small neighborhood of F'' in $\mathbf{R}^4$ and let $\beta: U \to \mathbf{RP}^4$ be a map which is transverse to $\mathbf{RP}^3$ so that $\beta^{-1}(\mathbf{RP}^3) = F''$. For example, the restriction of β to F'' could be the map which takes a point to the line perpendicular to F'' at that point. Extend this map to U using the fact that the normal bundle of $\mathbf{RP}^3$ in $\mathbf{RP}^4$ is the canonical line bundle over $\mathbf{RP}^3$. Suppose H'' and G'' are made up of k codimension one submanifolds $C_1, \ldots, C_k$, say $H'' = \bigcup_{i=1}^{m} C_i$ and $G'' = \bigcup_{i=m+1}^{k} C_i$. Let $\gamma_i: U \to \mathbf{RP}^4$ for $i = 1, \ldots, k$ be maps transverse to $\mathbf{RP}^3$ so that $F'' \cap \gamma_i^{-1}(\mathbf{RP}^3) = C_i$. For example γ_i could send a point in C_i to the line tangent to F'' and perpendicular to C_i. By doing the constructions carefully one may make sure that the restriction of β and each γ_i to W is a rational function. Now approximate β and the γ_i's by rational functions using Lemma 2.4 of [AK4] or Theorem 2.8.3 of [AK2]. One then obtains a real algebraic set V (in some higher dimensional affine space), an open set $O \subset V$, a diffeomorphism $\varphi: O \to U$ and rational functions $r: W \to \mathbf{R}$ and $q_i: W \to \mathbf{R}$ whose restrictions to O approximate $\beta \circ \varphi$ and $\gamma_i \circ \varphi$. Furthermore V contains W and r and q_i coincide with β and γ_i on W. We may then set $X = \beta^{-1}(\mathbf{RP}^3)$, $Y = X \cap \bigcup_{i=m+1}^{k} \gamma_i^{-1}(\mathbf{RP}^3)$ and $Z = X \cap \bigcup_{i=1}^{m} \gamma_i^{-1}(\mathbf{RP}^3)$. By transversality we know that near W; X, Y and Z look like F'', G'' and H''. Then by the construction we did above we are done.

Note that a variation on this example provides a real algebraic set homeomorphic to the suspension of the disjoint union of a point and the connected sum of three $\mathbf{RP}^2$'s. The only difference is that we replace α with the map

$$\alpha(x) = (\, q_0(x)x \,,\, q_0(x) ,\, q_1(x)\,)$$

and replace Z by the algebraic set

$$Z = \{\, (y, t_0, t_1) \mid y = 0 \text{ and } t_0 = 0 \,\}.$$

[AK1] S. Akbulut and H. King, The topological classification of 3-dimensional algebraic sets, M.S.R.I. preprint 12912-85 (1985).

[AK2] S. Akbulut and H. King, Topology of real algebraic sets, M.S.R.I. Publications Vol. 25, Springer-Verlag (1992).

[AK3] S. Akbulut and H. King, The topology of real algebraic sets, L'Enseignement Math., 29 (1983), 221-261.

[AK4] S. Akbulut and H. King, The topology of real algebraic sets with isolated singularities, Annals of math., Vol. 113 (1981), 425-446.

[AK5] S. Akbulut and H. King, On approximating submanifolds by algebraic sets and a solution to the Nash conjecture, Inventiones Math., Vol. 107 (1992), 87-98.

[CK] M. Coste and K. Kurdyka, On the link of a stratum in a real algebraic set, Topology, Vol. (1991).

[K] H. King, Real analytic germs and their varieties at isolated singularities, Inv. Math., Vol. 37 (1976), 193-199.

[S] D. Sullivan, Invariants of analytic spaces, Proc. Liverpool Singularities I, Lecture Notes 192, Springer-Verlag (1971), 165-168.

MORE ON BASIC SEMIALGEBRAIC SETS

Carlos Andradas*
Departamento de Algebra
Jesús M. Ruiz*
Departamento de Geometría y Topología
Facultad de Ciencias Matemáticas, Universidad Complutense de Madrid
28040 Madrid, Spain

Introduction

Let $X \subset \mathbf{R}^n$ be a real algebraic variety, and let $\mathcal{P}(X)$ denote the ring of polynomial functions on X. Recall that a subset $S \subset X$ is called *semialgebraic* if there exist polynomials $f_{ij}, g_i \in \mathcal{P}(X)$ such that

$$S = \bigcup_{i=1} : p\{x \in X : f_{i1}(x) > 0, ..., f_{ir_i}(x) > 0, g_i(x) = 0\}.$$

As is well known, if S is open the g_i's in this expression can be omitted. Recall also that an open semialgebraic set is called *basic open* if furthermore $p = 1$. These basic open sets have attracted a lot of interest in recent times, see :Br2,3!, [Sch], [Mh], [AnBrRz1], culminating in the beautiful theorem that a basic open set S always has a description of the form

$$S = \{x \in X : f_1(x) > 0, ..., f_s(x) > 0\}$$

where $s \leq \dim(X)$. On the other hand the problem of characterizing basic open sets in geometric terms seems to be wide open. First of all, there is an immediate necessary condition for a semialgebraic open set S to be basic; namely, that S does not meet the Zariski closure of its usual boundary. If this is the case, a theorem of Bröcker says that S is basic if and only if its intersection $S \cap Y$ with any irreducible subvariety Y of X is *generically basic*, which means that $S \cap Y \setminus Z$ is basic for some proper subvariety $Z \subset Y$. This brings the problem to the birational setting, or in other words to the fields $\mathcal{K}(Y)$ of rational functions of the Y's. Then using the so-called *tilda operation*, which relates semialgebraic subsets of Y and constructible subsets of the space of orderings of the field $\mathcal{K}(Y)$, the theory of fans can be used to decide whether $S \cap Y$ is or not generically basic. For the details of all of this we refer to [Br3] or the forthcoming book [AnBrRz2]. The purpose of this paper is to improve the theorem of Bröcker quoted above in the following way:

* *Partially supported by a grant Del Amo, Universidad Complutense de Madrid, 1990, and DGICYT, PB 89-0379-C02-02*

Theorem 1. *Let $S \subset X$ be an open semialgebraic set which does not meet the Zariski closure of its boundary. Then S is basic if and only if $S \cap Y$ is generically basic for any irreducible algebraic surface $Y \subset X$.*

This theorem shows that the obstructions to basicness appear in the smallest possible dimension, because as is well known everything is basic in dimension 1. The result will follow from a more precise statement characterizing generically basic sets through their intersections with irreducible surfaces. Of course the main tool in the proofs is the theory of fans mentioned above, and more specifically, the Theorem 1.2 below: an approximation theorem which shows that for the matter we are discussing it is enough to consider fans built up from real prime divisors.

Since we have not even said what a fan is, let us at least describe a fan built up from a real prime divisor. First we start with a prime divisor V of the field $\mathcal{K}(X)$, that is, V is a discrete rank one valuation ring of $\mathcal{K}(X)$ whose residue field k is a finitely generated extension of $\mathbf{R}$ of transcendence degree $d-1$ where $d = \dim(X)$. Let t be a uniformizer of V. Then every element $f \in V$ can be written in the form $f = ut^n$, where n is the value of f and u is a unit of V, so that its residue class $\overline{u}$ in k is not zero. Now, we pick two orderings τ_1 and τ_2 in k. (In particular, we are assuming that k is formally real, and that is why V is called a *real* prime divisor.) It is easy to lift τ_1 and τ_2 to get four orderings σ_1, σ_3 and σ_2, σ_4 in V, and so in the field $\mathcal{K}(X)$. Namely, for an $f = ut^n$ as above, put

$$\sigma_1(f) = \tau_1(\overline{u}), \quad \sigma_3(f) = (-1)^n \tau_1(\overline{u})$$

$$\sigma_2(f) = \tau_2(\overline{u}), \quad \sigma_4(f) = (-1)^n \tau_2(\overline{u})$$

(for an ordering α and an element x we let $\alpha(x)$ denote the sign of x in α). In other words, the units of V have the sign of their residue classes, and the uniformizer is declared either positive or negative. It follows from the construction that $\sigma_4 = \sigma_1 \cdot \sigma_2 \cdot \sigma_3$, which is in fact the definition of a 4-element fan $F = \{\sigma_1, \sigma_2, \sigma_3, \sigma_4\}$ (orderings are multiplied as signatures).

Now we can formulate rigorously our vague assertion that the tilda operation makes the connection between our initial problem and fans. To start with, given a semialgebraic set $S \subset X$ we consider the set $\tilde{S} \subset Spec_r(\mathcal{K}(X))$ defined by any formula that also defines S. Then we will prove:

Theorem 2. *Let $X \subset \mathbf{R}^n$ be an irreducible algebraic variety, and $S \subset X$ a semialgebraic set. Then S is generically basic if and only if $\#(\tilde{S} \cap F) \neq 3$ for every 4-element fan F built up from a real prime divisor.*

This improves the following theorem of Bröcker:

Theorem 3. *Let $X \subset \mathbf{R}^n$ be an irreducible algebraic variety, and $S \subset X$ a semialgebraic set. Then S is generically basic if and only if $\#(\tilde{S} \cap F) \neq 3$ for every 4-element fan F.*

What is new in our statement is the very special type of fans which are enough to check generic basicness. To give some hint of how this helps in our special problem, suppose S is not generically basic. Then, by the if part of Theorem 2, there are a real prime divisor V of $\mathcal{K}(X)$, two orderings τ_1 and τ_2 in its residue field k, and four liftings σ_1, σ_3 and σ_2, σ_4, such that $\sigma_1, \sigma_3, \sigma_2 \in \tilde{S}$ and $\sigma_4 \notin \tilde{S}$.Here is the clue for the proof of Theorem 1: by a suitable aplication of Bertini's theorem we can decrease the transcendence degree of the residue field k, till in the end we get transcendence degree 1. Hence k is the function field of a real algebraic curve, and we get a homomorphism $V \to k[[t]]$ (where t is a uniformizer of V). After some work with M. Artin's approximation theorem we may suppose the homomorphism is actually into the ring of algebraic power series $V \to k[[t]]_{\mathrm{alg}}$. Furthermore, we can suppose that V contains the ring $\mathcal{P}(X)$ of polynomial functions on X, and so get a homomorphism $\mathcal{P}(X) \to k[[t]]_{\mathrm{alg}}$. This already gives the surface we were looking for: it is the zero set of the kernel of the homomorphism above. Finally, the four orderings σ_i we had in V extend easily to $k[[t]]$, then restrict to $\kappa(\mathfrak{p})$. Thus we obtain a 4-element fan in $\kappa(\mathfrak{p}) = \mathcal{K}(Y)$ which shows that the intersection $S \cap Y$ is not generically basic (only if part of Theorem 2).

The paper is organized as follows. Section 1 contains all results on fans in function fields needed in Section 2 to prove Theorems 1 and 2. We finish in Section 3 with a counterexample showing why our result fails in both the Nash and the analytic categories.

This work started while the authors where participating in the Special Year 1990-91 on Quadratic Forms and Real Algebraic Geometry organized by Profs. T.Y. Lam and R. Robson in the University of California, Berkeley. We are glad to thank the RAGSQUADers, who listened to our first presentation of these ideas. We also want to thank Profs. J. Bochnak, M. Coste and P. Milman, who helped us to understand the Nash and analytic cases; in particular, we completed the construction of the counterexample which ends the paper during the conference in La Turballe, using a surface found by E. Bierstone and P. Milman.

1. The approximation theorem for fans

Let K be a field and V a valuation ring of K with residue field k. Consider an ordering σ of K. We say that σ and V are *compatible* if the maximal ideal $\mathfrak{m}_V$ of V is convex with respect to σ. Then σ induces in the obvious way an ordering τ in k, and we write $\sigma \to \tau$. Now consider a fan $F = \{\sigma_1, \sigma_2, \sigma_3, \sigma_4\}$ of K, that is, a set of four orderings of K such that the product of three of them gives always the fourth. We say that F and V are compatible if every σ_i is compatible with V; we say that F *trivializes along* V if the σ_i's induce in k no more than two distinct orderings. Then the situation looks like

$$\begin{array}{cccc} \sigma_1 & \sigma_3 & \sigma_2 & \sigma_4 \\ \searrow & \swarrow & \searrow & \swarrow \\ \tau_1 & & \tau_2 & \end{array}$$

The key result concerning the relation between fans and valuations is the following ([Br1]):

Theorem 1.1. *Let F be a fan of the field K. Then F trivializes along a valuation ring of K.*

Henceforth we will think of a 4-element fan as a tuple $(\sigma_1, \sigma_2, \sigma_3, \sigma_4)$ rather than a set $\{\sigma_1, \sigma_2, \sigma_3, \sigma_4\}$. This way fans become points in the product Σ of four copies of the space of orderings $Spec_r(K)$. Since Σ carries naturally the product topology of the Harrison topology, we can make statements in terms of that topology. Formally we should identify points in Σ up to permutations, but such a formalism is not needed for our purposes.

From now on we suppose that K is a function field over $\mathbf{R}$, and denote by n its *dimension*, that is, its transcendence degree over $\mathbf{R}$. We recall that V is a *prime divisor* if it is a discrete rank one valuation ring of K and its residue field k is a function field over $\mathbf{R}$ of dimension $n-1$. With this terminology, we have:

Theorem 1.2. *The set of fans of K which trivialize along prime divisors is dense in the set of all fans.*

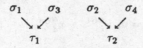

Proof: Let X be a compact model of K, that is, a compact irreducible algebraic variety $X \subset \mathbf{R}^n$ whose field of rational functions is K. Then every polynomial is bounded on X, from which it follows that every real valuation ring of K contains the ring $\mathcal{P}(X)$ of polynomial functions of X. Let $F = (\sigma_1, \sigma_2, \sigma_3, \sigma_4) \in \Sigma$ be a fan of K and $U = U_1 \times U_2 \times U_3 \times U_4$ an open neighborhood of F, with $U_i = \{f_{i1} > 0, \ldots, f_{ir_i} > 0\}$, $f_{ij} \in \mathcal{P}(X)$; after shrinking the U_i's, we may assume they are pairwise disjoint. By Theorem 1.1 F trivializes along a valuation ring V of K: the σ_i's are compatible with V and induce two orderings in the residue field k of V according to the picture

$$\begin{array}{ccccccc} \sigma_1 & & \sigma_3 & & \sigma_2 & & \sigma_4 \\ & \searrow & \swarrow & & & \searrow \swarrow & \\ & & \tau_1 & & & \tau_2 & \end{array}$$

(possibly $\tau_1 = \tau_2$).

Now we apply resolution of singularities so that after finitely many blowings-up X is non-singular and all the f_{ij}'s are normal crossings. Let $\mathfrak{p} \subset \mathcal{P}(X)$ be the center of V in $\mathcal{P}(X)$. Then $A = \mathcal{P}(X)_\mathfrak{p}$ is a local regular ring of dimension say d, and there exists a regular system of parameters $x_1, \ldots, x_d$ such that for all i, j

$$f_{ij} = u_{ij} x_i^{\alpha_{ij1}} \cdots x_d^{\alpha_{ijd}},$$

where the u_{ij} are units of A and the α_{ijk} are non-negative integers.

In this situation the residue field $\kappa(\mathfrak{p})$ of A is a subfield of the residue field k of V, and τ_1, τ_2 restrict to two orderings in $\kappa(\mathfrak{p})$ that we still denote τ_1, τ_2. Then notice that the signs of the elements f_{ij} in the ordering σ_p are completely determined by the signs of the parameters x_l in σ_p and the signs of the units (or more properly of their residue classes) in τ_q, where $p \equiv q \bmod 2$.

Reordering the parameters we may suppose that $\sigma_1(x_l) = \sigma_3(x_l)$ for $l \leq r$ and $\sigma_1(x_l) \neq \sigma_3(x_l)$ for $l > r$. Note that $r < d$ since otherwise σ_1 and σ_3 would coincide over the f_{1j}'s, which is imposible because these elements define a neighborhood of σ_1 which does not contain σ_3. On the other hand, it might very well happen that $r = 0$. Recalling that $\sigma_1 \cdot \sigma_2 \cdot \sigma_3 \cdot \sigma_4 = 1$, we see that also $\sigma_2(x_l) = \sigma_4(x_l)$ for all $l \leq r$ and $\sigma_2(x_l) \neq \sigma_4(x_l)$ for all $l > r$.

Consider the diagram

$$\begin{array}{ccccc}
A = \mathcal{P}(X)_\mathfrak{p} \to B = A[x_1, \ldots, x_r, x_{r+1}/x_d, \ldots, x_{d-1}/x_d, x_d]_{(x_1,\ldots,x_r,x_d)} & \to & W = B_{(x_d)} \subset K \\
\downarrow \qquad\qquad \downarrow & & \downarrow \\
\qquad\qquad B/(x_d) & \subset & k_W \\
\qquad\qquad\qquad \downarrow & & \\
\kappa(\mathfrak{p}) \quad \to \quad k' = \kappa(\mathfrak{p})[\overline{x_{r+1}/x_d}, \ldots, \overline{x_{d-1}/x_d}] & &
\end{array}$$

Let us analyse the ingredients of this figure. First, the local ring B is regular of dimension $r+1$ with regular system of parameters $\{x_1, \ldots, x_r, x_d\}$, and dominates A. Furthermore, its residue field is generated over the residue field of A, which is $\kappa(\mathfrak{p})$, by the residue classes $\overline{x_{r+1}/x_d}, \ldots, \overline{x_{d-1}/x_d}$, which are algebraically independent over $\kappa(\mathfrak{p})$.

Second, the ring W is a discrete rank one valuation ring of K, and it is in fact a prime divisor. Indeed, its residue field k_W is by construction a finitely generated extension of $\mathbf{R}$. Moreover, k_W is the quotient field of $B/(x_d)$, which is a regular local ring of dimension r and whose residue field k' is that of B, and has transcendence degree $d-1-r$ over $\kappa(\mathfrak{p})$. As the latter has transcendence degree $e = \dim(\mathcal{P}(X)/\mathfrak{p})$ over $\mathbf{R}$, we see that the transcendence degree of k_W over $\mathbf{R}$ is $\geq r + (d-1-r) + e = d + e - 1$. But

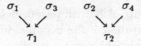

$d + e = \mathrm{ht}(\mathfrak{p}) + \dim(\mathcal{P}(X)/\mathfrak{p}) = \dim(\mathcal{P}(X)) = n$, and we find that k_W is a function field over $\mathbf{R}$ of dimension $n-1$.

Third, we have the following coincidences of signs:

$$\sigma_1(x_l) = \sigma_3(x_l), l \leq r; \quad \sigma_1(x_l/x_d) = \sigma_3(x_l/x_d), r < l < d; \quad \sigma_1(x_d) \neq \sigma_3(x_d)$$

and since $\sigma_1 \cdot \sigma_2 \cdot \sigma_3 \cdot \sigma_4 = 1$,

$$\sigma_2(x_l) = \sigma_4(x_l), l \leq r; \quad \sigma_2(x_l/x_d) = \sigma_4(x_l/x_d), r < l < d; \quad \sigma_2(x_d) \neq \sigma_4(x_d)$$

Finally, taking all of this into account, we can extend and lift orderings through the preceding diagram, starting from $\kappa(\mathfrak{p})$ with $\tau_i, i = 1, 2$, as follows:

• Extend τ_i to an ordering τ_i' in k' such that $\tau_i'(\overline{x_l/x_d}) = \sigma_i(x_l/x_d)$ for $r < l < d$. This extension is possible since k' is a purely transcendental extension of $\kappa(\mathfrak{p})$ with $\{\overline{x_{r+1}/x_d}, \ldots, \overline{x_{d-1}/x_d}\}$ as a transcendence basis.

• Lift τ_i' to an ordering γ_i in k_W such that $\gamma_i(x_l) = \sigma_i(x_l)$ for $1 \leq l \leq r$. This lifting follows easily after remarking that the completion of $B/(x_d)$ is a ring of formal power series $k'[[\tilde{x}_1, \ldots, \tilde{x}_r]]$, where $\{\tilde{x}_1, \ldots, \tilde{x}_r\}$ is the regular system of parameters of $B/(x_d)$ induced by $\{x_1, \ldots, x_r\}$.

• Lift γ_i to two orderings σ_i' and σ_{i+2}' of K such that $\sigma_j'(x_d) = \sigma_j(x_d)$. This can be done as before using the completion $k_W[[x_d]]$ of $W = B_{(x_d)}$, or as was described in the introduction for a discrete rank one valuation ring like W.

We claim that the four orderings of K just constructed form a 4-element fan F' which trivializes along the prime divisor W, and that F' is in the neighborhood U of F fixed at the beginning. Indeed, they form a fan because they are built up from the real prime divisor W, as explained in the introduction, which implies also that F' trivializes along W. Furthermore,

$$\sigma_p'(u_i) = \sigma_p(u_i), \quad \sigma_p'(x_l) = \sigma_p(x_l)$$

for all p, i, l, and so

$$\sigma_p'(f_{ij}) = \sigma_p(f_{ij})$$

for all i, j. Hence $\sigma_p' \in U_p$ for all p, and $F' = (\sigma_1', \sigma_2', \sigma_3', \sigma_4') \in U$. ∎

2. Characterizations of basic sets

We start by deducing Theorem 2 of the introduction as an immediate consequence of Theorem 1.2:

Proof of Theorem 2: It is immediate that in case S is generically basic, $\#(\tilde{S} \cap F) \neq 3$ for any fan F, built up or not from a real prime divisor. Conversely, by Theorem 3, if S is not generically basic, there is a fan F such that $\#(\tilde{S} \cap F) = 3$, and the problem is to replace F by another fan F', built up from a real prime divisor. To do it, let h_j be the polynomial functions involved in a fixed description of S, and $\sigma_1, \sigma_2, \sigma_3, \sigma_4$ the orderings of F. Then consider the neighborhood $U_i = \{f_{i1} > 0, \ldots, f_{ir} > 0\}$ of σ_i where $f_{ij} = h_j$ if $h_j(\sigma_i) > 0$, $f_{ij} = -h_j$ if $h_j(\sigma_i) < 0$. Then $U = U_1 \times U_2 \times U_3 \times U_4$ is a neighborhood of F in the sense of Theorem 1.2, and there is a fan F' in U which trivializes along a prime divisor, which means that $F' = (\sigma'_1, \sigma'_2, \sigma'_3, \sigma'_4)$ is built up from that real prime divisor. By construction the signs of σ'_i and σ_i at h_j coincide, and so $\sigma'_i \in \tilde{S}$ if and only if $\sigma_i \in \tilde{S}$. Consequently, $\#(\tilde{S} \cap F') = \#(\tilde{S} \cap F) = 3$, and we are done. ∎

Now we are ready to state and prove the characterization of generically basic sets through surfaces:

Theorem 2.1. *Let $X \subset \mathbf{R}^n$ be an irreducible algebraic variety and $S \subset X$ a semialgebraic set. Let $Z \subset X$ be an algebraic set $\neq X$ containing both the singular locus of X and the boundary of S, $\partial S = \overline{S} \setminus S^\circ$. Then S is generically basic if and only if $S \cap Y$ is generically basic for any irreducible surface $Y \subset X$ not contained in Z.*

Proof: Assume first that S is generically basic, and let $Y \subset X$ be an irreducible surface not contained in Z; denote by $\mathfrak{p}$ the ideal of Y. Since Y is not contained in the singular locus of X the localization $\mathcal{P}(X)_\mathfrak{p}$ is a regular local ring, and there is a discrete valuation ring V of the quotient field $\mathcal{K}(X)$ of $\mathcal{P}(X)$ dominating $\mathcal{P}(X)_\mathfrak{p}$ whose residue field is $k = qf(\mathcal{P}(X)/\mathfrak{p}) = \mathcal{K}(Y)$. Now suppose that $S \cap Y$ is not generically basic. Then there is a fan $(\tau_1, \tau_2, \tau_3, \tau_4)$ of k such that $\tau_1, \tau_2, \tau_3 \in S$ and $\tau_4 \notin S$. Using V one easily lifts this fan to another $(\sigma_1, \sigma_2, \sigma_3, \sigma_4)$ of $\mathcal{K}(X)$, with $\sigma_i \to \tau_i$, and we claim that $\sigma_1, \sigma_2, \sigma_3 \in S$ and $\sigma_4 \notin S$. First, since Z does not contain Y and Z contains ∂S, $\tau_i \in S \setminus Z \subset S^\circ$ and consequently $\sigma_i \in S$, for $i = 1, 2, 3$. Second, since $\tau_4 \in Y \setminus Z \cup S$, we see that $\tau_4 \notin \overline{S}$ and $\sigma_4 \notin S$. Whence, S is not generically basic, as claimed. Note that for this implication we do not need fans trivializing along prime divisors.

Now assume S is not generically basic. By Theorem 1.2 there is a 4-element fan F of the field $K = \mathcal{K}(X)$ with $\#(S \cap F) = 3$, and F trivializes along a prime divisor V of K; say $F = \{\sigma_1, \sigma_2, \sigma_3, \sigma_4\}$, $\sigma_1, \sigma_3 \to \tau_1$ and $\sigma_2, \sigma_4 \to \tau_2$, where τ_1 and τ_2 are two orderings in the residue field k of V. Substituting X by its one-point compactification we can assume that

X is compact, and so V contains the ring $\mathcal{P}(X)$ of polynomial functions of X. The residue field k of V is a function field over $\mathbf{R}$: there is an irreducible algebraic variety $W \subset \mathbf{R}^m$ whose field of rational functions $\mathcal{K}(W)$ is k, or in other words, k is the quotient field of the ring $\mathcal{P}(W)$ of polynomial functions on W.

Now let H stand for a generic irreducible hyperplane section of W, $\mathfrak{p}$ for the ideal of H in $\mathcal{P}(W)$ and k' for the field of rational functions on H; note that k' is the residue field of $\mathfrak{p}$, that is, the quotient field of $\mathcal{P}(W)/\mathfrak{p}$. With all of this we have the following diagram

$$\begin{array}{ccc} \mathcal{P}(X) \subset V \xrightarrow{\phi} \hat{V} = & \mathcal{K}(W)[[t]] \\ & \cup \\ \mathcal{P}(W)_{\mathfrak{p}}[[t]] \xrightarrow{\varphi} & \mathcal{K}(H)[[t]] \end{array}$$

where t is a uniformizer of V and the homomorphism φ is the obvious extension of the canonical mapping $\mathcal{P}(W)_{\mathfrak{p}} \to \mathcal{K}(H)$.

Since the ring $\mathcal{P}(X)$ is an algebra finitely generated over $\mathbf{R}$ we can pick finitely many generators $f_i \in \mathcal{P}(X)$ and add to them the equations involved in a description of the semialgebraic set S and an equation of Z. All these functions f_i are in V, and so they have, in its completion $\hat{V} = \mathcal{K}(W)[[t]]$, power expansions $f_i = f_i(t) = \sum_l (g_{il}/h_{il})t^l$, where $g_{il}, h_{il} \in \mathcal{P}(W)$. As our hyperplane section H is generic, we can suppose that no g_{il}, h_{il} vanishes on H (although there are infinitely many g_{il}, h_{il}'s, their number is countable, and working over the reals we can use Baire's theorem). In particular $h_{il} \notin \mathfrak{p}$ implies that the $f_i(t)$'s are well defined elements of $\mathcal{P}(W)_{\mathfrak{p}}[[t]]$. Finally, since the f_i's generate $\mathcal{P}(X)$ we get $\phi(\mathcal{P}(X)) \subset \mathcal{P}(W)_{\mathfrak{p}}[[t]]$ and consequently we have the map

$$\psi = \varphi\phi : \mathcal{P}(X) \to \mathcal{P}(H)[[t]].$$

Moreover $g_{il} \notin \mathfrak{p}$ implies that the coefficients of the $f_i(t)$'s are units in $\mathcal{P}(W)_{\mathfrak{p}}$ and so $\psi(f_i) = \varphi(f_i(t)) = \sum_l \overline{(g_{il}/h_{il})}t^l$ is a non-zero element of $\mathcal{K}(H)[[t]]$ (here $\overline{}$ stands for residue class mod $\mathfrak{p}$).

Now we complete the choice of the generic hyperplane section H. We have the two orderings σ_1, σ_3 in $\hat{V} = \mathcal{K}(W)[[t]]$, liftings of the same ordering τ_1 in $\mathcal{K}(W)$. It is easy to find an open neighborhood G_1 of τ_1 such that for any ordering $\tau_1' \in G_1$ its two liftings σ_1', σ_3' have at the f_i's the same signs as σ_1, σ_3, where the liftings are chosen so that $\sigma_i'(t) = \sigma_i(t)$. Analogously, we find a neigborhood G_2 of τ_2. This implies that for any two orderings $\tau_1' \in G_1$ and $\tau_2' \in G_2$ the fan F' built up from them and the prime divisor V verifies $\#(S \cap F') = 3$.

Now, G_1, G_2 will also denote two open semialgebraic subsets of W corresponding to the neigborhoods just constructed. These semialgebraic sets are Zariski dense in W, which guarantees that we can choose the generic hyperplane section H to meet both of them. This implies that there are $\tau_1' \in G_1$ and $\tau_2' \in G_2$ which make the ideal $\mathfrak{p}$ of H convex. In other words, τ_1' and τ_2' induce two orderings γ_1 and γ_2 in the residue field of $\mathfrak{p}$, which is $\mathcal{K}(H)$. Then we lift γ_1 and γ_2 to four orderings δ_1, δ_3 and δ_2, δ_4 which form a fan F'' of $\mathcal{K}(H)[[t]]$ with the conditions $\delta_i(t) = \sigma_i(t)$.

After this preparation, we have the following diagram

$$
\begin{array}{ccccc}
 & \phi \swarrow & \mathcal{P}(X) & \searrow \psi & \\
\mathcal{P}(W)_{\mathfrak{p}}[[t]] & & \longrightarrow & & \kappa(\mathfrak{p})[[t]] \\
\downarrow & & & & \downarrow \\
\mathcal{P}(W)_{\mathfrak{p}} & & \longrightarrow & & \kappa(\mathfrak{p})
\end{array}
\qquad
\begin{array}{ccccc}
\sigma_1' \quad \sigma_3' & \sigma_2' \quad \sigma_4' & \longrightarrow & \delta_1 \quad \delta_3 & \delta_2 \quad \delta_4 \\
\searrow \swarrow & \searrow \swarrow & & \searrow \swarrow & \searrow \swarrow \\
\tau_1' & \tau_2' & \longrightarrow & \gamma_1 & \gamma_2
\end{array}
$$

Now consider the kernel $\mathfrak{q}$ of the homomorphism ψ. Its zero set is an algebraic set $Y \subset X$ with $\mathcal{P}(Y) = \mathcal{P}(X)/\mathfrak{q}$, and $\dim(Y) = \dim(X) - \mathrm{ht}(\mathfrak{q})$. Furthermore, the fan $F'' = \{\delta_1, \delta_2, \delta_3, \delta_4\}$ induces a fan F^* in $\mathcal{K}(Y)$ such that $\#(S \cap F^*) = 3$, because by construction the signs of the δ_i's at the $\psi(f_j)$'s coincide with those of the σ_i''s. Consequently, the semialgebraic set $S \cap Y$ is not generically basic. Furthermore, since among the f_i's there is an equation of Z, and no f_i is in $\mathfrak{p}$, Y is not contained in Z. Hence it only remains to show that we can impose the further condition $\dim(Y) < \dim(X)$ and from that the proof will end by induction.

Now consider any other homomorphism $\psi' : \mathcal{P}(X) \to \kappa(\mathfrak{q})[[t]]$ which approximates ψ to an order ν, that is, such that $\psi'(f) \equiv \psi(f) \bmod t^\nu$ for every $f \in \mathcal{P}(X)$. If we look at the requirements that $\psi : \mathcal{P}(X) \to \kappa(\mathfrak{q})[[t]]$ fulfils, it is clear that for ν large enough ψ' will fulfil them too. Then consider the ideal $\mathfrak{n} = \psi^{-1}(t)$ and the localization $A = \mathcal{P}(X)_{\mathfrak{n}}$. The homomorphism ψ extends to the henselization A^h, which is a quotient of a ring $\kappa(\mathfrak{n})[[x_1, \dots, x_r]]_{\mathrm{alg}}$ of algebraic power series. Using this remark and M. Artin's approximation theorem we can arbitrarily approximate ψ by $\psi' : A^h \to \kappa(\mathfrak{q})[[t]]_{\mathrm{alg}}$ (see [Tg] III §5). Hence, substituting ψ by ψ' we may merely suppose $\psi(\mathcal{P}(X)) \subset \kappa(\mathfrak{q})[[t]]_{\mathrm{alg}}$. It follows that ψ induces an embedding $\mathcal{P}(Y) \hookrightarrow \kappa(\mathfrak{q})[[t]]_{\mathrm{alg}}$, which extends to the quotient fields $\mathcal{K}(Y) \hookrightarrow \kappa(\mathfrak{q})((t))_{\mathrm{alg}}$. Counting transcendence degrees over $\mathbf{R}$ we find

$$
\begin{aligned}
\dim(Y) = \mathrm{tr.deg.}[\mathcal{K}(Y) : \mathbf{R}] \leq \mathrm{tr.deg.}[\kappa(\mathfrak{q})((t))_{\mathrm{alg}} : \mathbf{R}] = \\
1 + \mathrm{tr.deg.}[\kappa(\mathfrak{q}) : \mathbf{R}] = 1 + \dim(H) < 1 + \dim(W) = \dim(X)
\end{aligned}
$$

as wanted. ∎

We finish this section with the following:

Proof of Theorem 1: It is clear that if S is basic open any intersection $S \cap Y$ with an irreducible subvariety $Y \subset X$ is also basic, and so generically basic. Conversely, suppose S is not basic open. Then by Bröcker's theorem there is an irreducible subvariety $X' \subset X$ such that $S \cap X'$ is not generically basic. Then, by Theorem 2.1 there is an irreducible surface $Y \subset X'$ such that $S \cap Y$ is not generically basic, and we are done. ∎

3. The Nash and the analytic cases

The approach followed here to study semialgebraic sets works the same in the Nash and the analytic categories (with some compactness assumption in the latter) [Rz],[AnBrRz1]. Even the approximation theorem for fans proved in Section 1 could be proved in both the Nash and the analytic cases. However any attempt to prove Theorem 2.1 or Theorem 1 would fail. The obstruction is that no Bertini type theorem is available in those cases. An easy counterexample is given by the standard torus T in $\mathbf{R}^3$: a lot of hyperplane sections of T consist of two circles, and so cannot be irreducible in either the Nash or the analytic category. More difficult is to produce a counterexample to Theorem 1 itself. Here we present one based on a surface used by Bierstone and Milman in dealing with arc-analytic functions, [BM] Ex.1.2(3).

Example. Let $\mathbf{R}P_3$ be the real projective 3-space with homogeneous coordinates $(x_0 : x_1 : x_2 : x_3)$, and $Y_0 \subset \mathbf{R}P_3$ the surface $x_0^2x_1^2 = x_2^4 + x_3^4$. This surface has two singular points, namely $a = (1:0:0:0)$ and $b = (0:1:0:0)$, and $Y_0 \setminus \{a, b\}$ consists of two connected components whose closures Y_1 and Y_2 meet exactly at a and b. Moreover, Y_0 is irreducible from the analytic viewpoint, but has the property that any analytic arc through one singular point has to be completely contained in either Y_1 or Y_2. This implies that any irreducible analytic curve $Z \subset Y_0$ has to be completely contained in either Y_1 or Y_2.

After these preliminaries consider the semialgebraic set $S \subset \mathbf{R}P_3$ defined as follows. First S will be a subset of the affine space $\mathbf{R}^3 = \{x_0 \neq 0\} \subset \mathbf{R}P_3$, in which we take coordinates $x = x_1/x_0, y = x_2/x_0, z = x_3/x_0$. Then $Y_0 \cap \mathbf{R}^3$ is the surface of equation $x^2 = y^4 + z^4$, and we have the sets:

- B_1 = ball centered at $(1,0,1) \in Y_1$ with a very small radius.
- $B_1^+ = B_1 \cap \{x^2 \geq y^4 + z^4\}$, $\quad B_1^- = B_1 \cap \{x^2 \leq y^4 + z^4\}$.
- B_2 = ball centered at $(-1,0,1) \in Y_2$ with a very small radius.
- $B_2^+ = B_2 \cap \{x^2 \geq y^4 + z^4\}$, $\quad B_2^- = B_2 \cap \{x^2 \leq y^4 + z^4\}$.

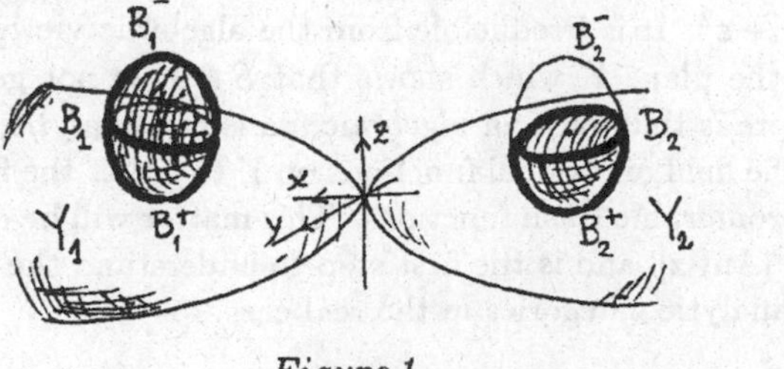

Figure 1

Now put $S = B_1 \cup B_2^+$ (see *Figure 1*). We claim that this set is not generically basic, but its intersection with any surface is generically basic, which gives the announced counterexample.

The assertion that S is not generically basic follows easily by constructing a 4-element fan $F = \{\sigma_1, \sigma_2, \sigma_3, \sigma_4\}$ where $\sigma_1 \in B_1^+, \sigma_2 \in B_2^+, \sigma_3 \in B_1^-, \sigma_4 \in B_2^-$. Now suppose that there is a surface $Y \subset \mathbf{R}P_3$ such that $S \cap Y$ is not generically basic. Then there is a fan $F = \{\sigma_1, \sigma_2, \sigma_3, \sigma_4\}$ in the field K of meromorphic functions on Y such that $\sigma_1, \sigma_2, \sigma_3 \in S$ and $\sigma_4 \notin S$. As $S = B_1 \cup B_2^+$ and $B_1, \mathbf{R}P_3 \setminus B_1, B_2, B_2^+$ are basic, we necessarily have, up to reordering, $\sigma_1, \sigma_3 \in B_1, \sigma_2 \in B_2^+, \sigma_4 \in B_2^-$. Moreover, there is a valuation ring V of K along which F trivializes, so that $\sigma_1, \sigma_3 \to \tau_1$ and $\sigma_2, \sigma_4 \to \tau_2$ where τ_1, τ_2 are orderings in the residue field k of V (it cannot be $\sigma_1, \sigma_2 \to \tau_1$, because this would imply τ_1 is adherent to both B_1 and B_2, which is impossible if our choice of the radii was small enough). Now, since Y is compact, V contains the ring $\mathcal{O}(Y)$ of analytic functions on Y; let $\mathfrak{p}$ stand for the center of V in $\mathcal{O}(Y)$ and Z for the zero set of $\mathfrak{p}$. The field L of meromorphic functions on Z is the residue field of $\mathfrak{p}$ and is contained in k. There are two possibilities:

- $\mathrm{ht}(\mathfrak{p}) = 2$. Then Z is a point z and the four orderings in the fan specialize to it. Since $\sigma_1 \in B_1$ and $\sigma_2 \in B_2$ the point z is adherent to both B_i's, which is impossible.

- $\mathrm{ht}(\mathfrak{p}) = 1$. Then Z is an irreducible curve and τ_2 restricts to an ordering in the field of meromorphic functions of Z, ordering still denoted by τ_2. Since $\sigma_2 \in B_2^+, \sigma_4 \in B_2^-$ and $\sigma_2, \sigma_4 \to \tau_2$ we conclude $\tau_2 \in B_2^+ \cap B_2^- \subset Y_0$. It follows that $Z \subset Y_0$, and, Z being an irreducible curve, the special choice of Y_0 guarantees that Z is completely contained in either Y_1 or Y_2, say Y_1. Since $\sigma_2 \in B_2$ and $\sigma_2 \to \tau_2$, this τ_2 is adherent to B_2, and for a small enough choice of the radius of B_2 this implies that τ_2 is not adherent to Y_1. However, $\tau_2 \in Z \subset Y_1$, a contradiction. The argument is analogous if $Z \subset Y_2$.

Whence, $S \cap Y$ is generically basic for any surface $Y \subset \mathbf{R}P_3$. ∎

This example shows that Theorems 1 and 2.1 fail in the Nash and analytic cases. However it is very easy to find an irreducible algebraic surface Y such that $S \cap Y$ is not generically basic from the algebraic viewpoint. Namely, take any plane $Y : y = \varepsilon$, with ε very small. The intersection of such a plane with our distinguished surface Y_0 is the curve $y = \varepsilon, \quad x^2 = z^4 + \varepsilon^2$: It is irreducible from the algebraic viewpoint, and gives rise to a 4-element fan in the plane Y which shows that $S \cap Y$ is not generically basic. Of course the key point here is that such an *algebraic* fan is not *analytic*, in the sense that it does not extend from the field of rational functions on Y to either the field of meromorphic functions or that of meromorphic Nash functions. This matter will be discussed in detail in the forthcoming paper [AnRz], and is the first step to understand the connection between the algebraic and the analytic categories in the real case.

References

[AnBrRz1] C. Andradas, L. Bröcker, J.M. Ruiz: Minimal generation of basic open semianalytic sets, *Invent. math.* **92** (1988) 409-430.

[AnBrRz2] C. Andradas, L. Bröcker, J.M. Ruiz: Real algebra and analytic geometry, *to appear.*

[AnRz] C. Andradas, J.M. Ruiz: Analytic fans versus algebraic fans, *in preparation.*

[BM] E. Bierstone, P. Milman: Arc-analytic functions, *Invent. math.* **101** (1990) 411-424.

[BCR] J. Bochnak, M. Coste, M.-F. Roy: Géométrie algébrique réelle, *Ergeb. Math. 12*, Springer-Verlag, Berlin-Heidelberg-New York, 1987.

[Br1] L. Bröcker: Characterization of fans and hereditarily pythagorean fields, *Math. Z.* **151** (1976) 149-163.

[Br2] L. Bröcker: Minimale Erzeugung von Positivbereichen, *Geom. Dedicata* **16** (1984) 335-350.

[Br3] L. Bröcker: On basic semialgebraic sets, *to appear in Exposit. Math.*

[Mh] L. Mahé: Une démostration élémentaire du théorème de Bröcker-Scheiderer, *C. R. Acad. Sci. Paris* **309**, Serie I (1989) 613-616.

[Rz] J.M. Ruiz: On the real spectrum of a ring of global analytic functions, *Publ. Inst. Recherche Math. Rennes* **4** (1986) 84-95.

[Sch] C. Scheiderer: Stability index of real varieties, *Invent. math.* **97** (1989) 467-483.

[Tg] J.-C. Tougeron: Idéaux de fonctions différentiables, *Ergeb. Math. 71*, Springer-Verlag, Berlin-Heidelberg-New York, 1972.

MIRROR PROPERTY FOR NONSINGULAR MIXED CONFIGURATIONS OF ONE LINE AND K POINTS IN $\mathbb{R}^3$

Alberto Borobia*
Dpto. Matemáticas
Facultad de Ciencias, U.N.E.D.
28040 Madrid, SPAIN

ABSTRAC: *Given* $h,k\geq 0$, $M(h,k)$ *denotes the set of nonsingular mixed configurations of* h *lines and* k *points in* $\mathbb{R}^3$. *We will say that* $f\in M(h,k)$ *is mirror if it is isotopic to its mirror image in any plane. The following problem has been proposed by Viro and Drobotukhina* [3]: *given* $h,k\geq 0$, *does there exist some mirror configuration on* $M(h,k)$? *A satisfactory answer is given when* $h=1$: $f\in M(1,k)$ *is mirror if and only if* $k\leq 3$.

0. Introduction

By a *nonsingular mixed configuration of lines and points in* $\mathbb{R}^3$ we understand a collection of lines and points in general position in $\mathbb{R}^3$, this means that the points not lie on the lines, that no line of $\mathbb{R}^3$ contains 3 of the points, and that no plane of $\mathbb{R}^3$ contains 4 of the points or one of the lines and 2 of the points or 2 of the lines.

Given $h,k\geq 0$, $M(h,k)$ will denote the topological space whose points are the nonsingular mixed configurations of h lines and k points in $\mathbb{R}^3$. The topology induced by $\mathbb{R}^3$ in $M(h,k)$ is the obvious one. We will call each connected component of $M(h,k)$ a *camera*.

An *isotopy* will be a continuous function $\gamma:[0,1]\longrightarrow M(h,k)$, that is, γ is a path joining two points that lie in the same camera. Therefore, if $f,g\in M(h,k)$ lie in the same camera we will say that f and g are *isotopic*.

$f\in M(h,k)$ is said to be *mirror* or to have the *mirror property* if f is isotopic to its mirror image in any plane. Otherwise f is said to be *nonmirror*.

Let $f\in M(h,k)$ be a mirror configuration. Given any reflection σ in a plane of $\mathbb{R}^3$ consider an isotopy $\gamma:[0,1]\longrightarrow M(h,k)$ such that $\gamma(0)=\sigma(f)$ and $\gamma(1)=f$. The composition of σ with γ acts over f permuting its h lines and its k points. Any permutation of the lines and the points of f obtained in this way will be called a *mirror permutation of* f.

Given $f\in M(h,k)$, if $f'\in M(h',k')$ is composed of $h'\leq h$ lines and $k'\leq k$ points of f we will say that f' is a *subconfiguration of* f.

1980 Mathematics Subject Classification: Primary 51A20
* Partially suported by DGICYT PB89-0201

Let $f\in M(h,k)$ be a mirror configuration, φ a mirror permutation of f and $f'\in M(h',k')$ a subconfiguration of f. If $\varphi(f')=f'$ then it is easy to see that f' is mirror and that $\varphi'=\varphi|_{f'}$ is a mirror permutation of f'. In particular the configurations of $M(h,0)$ and $M(0,k)$ composed respectively of the lines and the points of f are mirror.

Let $h,k\geq 0$ and $s:M(h,k)\longrightarrow\mathbb{R}$. We will say that s *is preserved under isotopies* if it is constant in each camera. And that s *is antisymmetric* if $s(\sigma(f))=-s(f)$ when $\sigma(f)$ is a mirror image of f. Therefore, if $f\in M(h,k)$ is mirror and $s:M(h,k)\longrightarrow\mathbb{R}$ is preserved under isotopies and antisymmetric then $s(f)=0$.

In this work I am principally concerned about the following problem: given $k\geq 0$, does there exist some mirror configuration on $M(1,k)$? I will prove that $f\in M(1,k)$ is mirror if and only if $k\leq 3$.

A big part of this paper is contained in Viro's paper [2]. Namely: linking coefficients were introduced into the subject; it contains the construction that I will employ in order to define adjacent points in $f\in M(1,k)$; part of the results of lemmas 2 and 5 are inmediate consecuence of following Viro's lemma: $M(0,k)$ does not contain mirror configurations if $k\equiv 6 (\mathrm{mod}\, 8)$ or if $k\equiv 3 (\mathrm{mod}\, 4)$ and $k\geq 7$; moreover, the arguments that I use here on the proof of lemma 2 and on part (iii) of the proof of lemma 5 are slight modifications of Viro's arguments in order to show that lemma.

The study of configurations in general position was introduced by O. Ya. Viro in [2] under the influence of V. M. Kharlamov's account of their conection with the problem of classifying nonsingular real algebraic surfaces of degree 4 (see [1]). The mirror property can be used to prove in an elementary way that certain of these surfaces are not amphicheiral. Other questions related to configurations in general position can be found in [3], and in their references. I would like to thank Professor O. Ya. Viro for introducing me to this problem, and Professor H. R. Morton and I. Itemberg for his helpful comments. This work was done while the author was visiting the University of Liverpool during July and August of 1991. I thank very much the Department of Mathemathics for its hospitality.

1. Mirror configurations in M(1,k)

Lemma 1.- *$f\in M(1,k)$ is mirror if $k=0,1,2$ or 3.*

Proof: (i) It is easy to see that if $k\leq 2$ then $M(1,k)$ has only one camera, hence any $f\in M(1,k)$ with $k\leq 2$ is mirror.

(ii) Given $f=\{l;p_1,p_2,p_3\}\in M(1,3)$, consider the plane π_f that contains to p_1, p_2 and p_3. In this plane draw the lines l_{12}, l_{13} and l_{23}, where l_{ij} denotes the line containing the points p_i and p_j. The plane π_f is divided in three different kinds of regions (denoted by different letters in figure 1).

It is easy to see that $f,g\in M(1,3)$ are isotopic if and only if the line of g intersects the same kind of region on π_g that the line of f on π_f. From this we deduce that all the configurations of $M(1,3)$ are mirror ∎

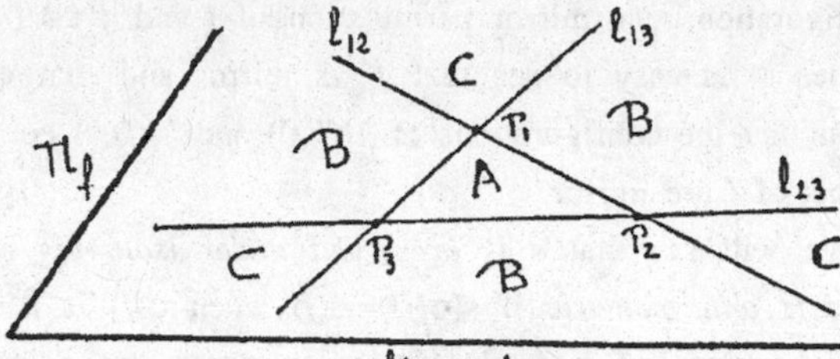

figure 1

Let $\{l_1,l_2\}\in M(2,0)$ and $\{l^*_1,l^*_2\}$ the same lines endowed with some orientations; $\{l^*_1,l^*_2\}$ has *linking coefficient* $\mathrm{lk}(l^*_1,l^*_2)$ equal to +1 or -1 depending if it is isotopic (by means of an isotopy consisting of configurations of $M(2,0)$ with oriented lines) to those lines in figure 2 or figure 3 respectively. The sign of the linking coefficient changes if we reverse the orientation of one of the lines, or if we consider a mirror image of $\{l^*_1,l^*_2\}$.

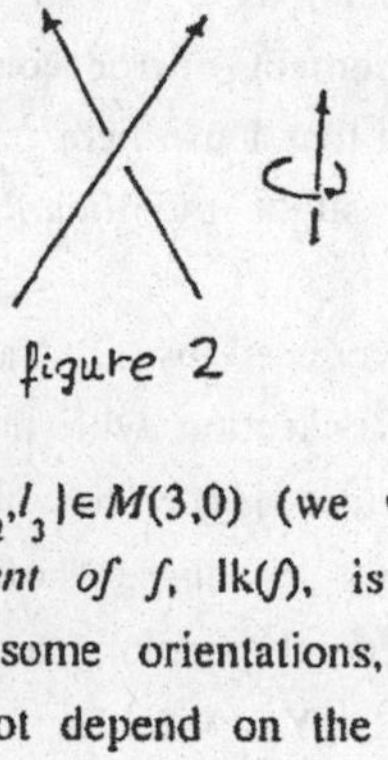

figure 2

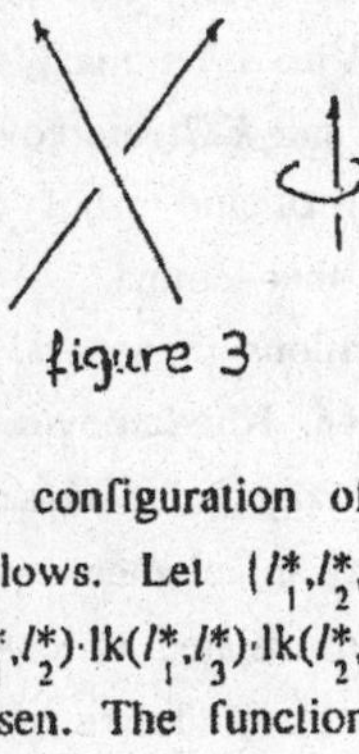

figure 3

Let $f=\{l_1,l_2,l_3\}\in M(3,0)$ (we will call to any configuration of $M(3,0)$ a *triple*). The *linking coefficient of* f, $\mathrm{lk}(f)$, is defined as follows. Let $\{l^*_1,l^*_2,l^*_3\}$ be the same lines endowed with some orientations, then $\mathrm{lk}(f)\equiv\mathrm{lk}(l^*_1,l^*_2)\cdot\mathrm{lk}(l^*_1,l^*_3)\cdot\mathrm{lk}(l^*_2,l^*_3)$. It is well defined since it does not depend on the orientations chosen. The function $\mathrm{lk}:M(3,0)\longrightarrow\{1,-1\}$ is preserved under isotopies and antisymmetric in $M(3,0)$.

Lemma 2.- *$f\in M(1,k)$ is nonmirror if $k\equiv 4$, 5, 6 or 7(mod8).*

Proof: Let $f\in M(1,k)$ with $k\geq 4$. Define $s(f)\in\mathbb{Z}$ to be the sum of the linking coefficients of the $\binom{k}{2}\cdot\binom{k-2}{2}/2$ triples that consist of the line of f and 2 lines determined by two disjoint pairs of points of f. Clearly, s is preserved under isotopies and antisymmetric in $M(1,k)$. If k=4, 5, 6 or 7(mod8) then $s(f)$ is an odd number (since it is a sum of an odd number of ±1's) and therefore f is nonmirror ∎

Let $f=\{l;p_1,p_2,...,p_k\}\in M(1,k)$ with $k\geq 3$. Consider the bundle of halfplanes S_l with boundary l. As f is in general position, then any halfplane of S_l can contain at most one point of f. Let S_i denote the halfplane of S_l that contains p_i. We will say that p_i and p_j are *adjacent in* f if some component of $\mathbb{R}^3\setminus\{S_i\cup S_j\}$ does not contain points of f. Construct a bijection ψ between the k points of f and the vertices of a regular polygon P_k of k edges such that adjacent points in f correspond to adjacent vertices of P_k (an example with k=6 is given in figure 4).

On the other hand, let $\gamma:[0,1] \longrightarrow M(1,k)$ be any isotopy with $\gamma(0)=f$. Then p_i and p_j are adjacent in f if and only if their images are adjacents in $\gamma(1)$. Suppose the opposite. Then, as γ is a continuous transformation, for some $t \in [0,1]$ $\gamma(t)$ would be a configuration that is not in general position. In the same way, if σ is any reflection in a plane of $\mathbb{R}^3$ then p and q are adjacents in f if and only if $\sigma(p)$ and $\sigma(q)$ are adjacents in $\sigma(f)$.

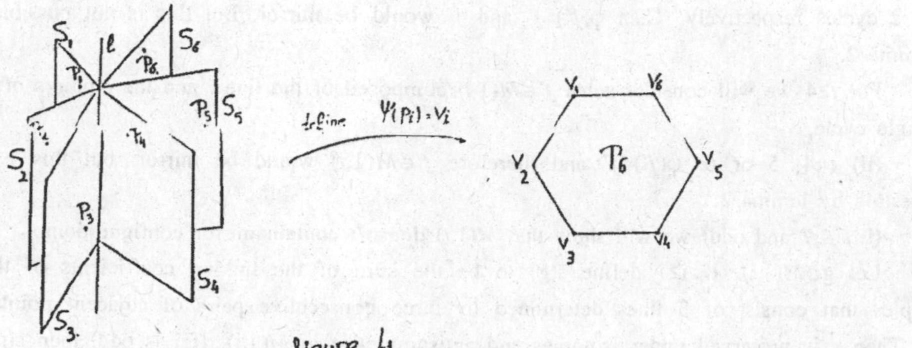

figure 4

In what follows we will suppose that $f=[l;p_1,..,p_k]\in M(1,k)$ with $k\geq 8$ is mirror and φ is the mirror permutation of f obtained from the composition of one reflection σ with one isotopy $\gamma:[0,1] \longrightarrow M(h,k)$ such that $\gamma(0)=\sigma(f)$ and $\gamma(1)=f$. From the preceding paragraph we deduce that φ transforms adjacent points into adjacent points. Then φ induces (by means of ψ) the action over P_k of some $\tau \in D_{2k}$, where D_{2k} denote the dihedral group of $2k$ elements. Let C_k denote the cyclic group of k elements, then

Lemma 3.- $\tau \in C_k \subset D_{2k}$.

Proof: Suppose $\tau \in D_{2k} \setminus C_k$. Any non-orientable element of D_{2k} acts on P_k as a reflection; then τ acts on the set of vertices of P_k as a permutation with 0 or 2 vertices fixed if k is even, or 1 vertex fixed if k is odd. The remaining vertices of P_k are divided into disjoint transpositions. As $k\geq 8$ we have more than 2 transpositions. Let f' be composed of l and 4 points of f whose images by ψ are the vertices corresponding to two transpositions. Then $\varphi(f')=f'$ and $f'\in M(1,4)$ would be mirror. But this is not possible by lemma 2 ■

Corollary 4.- *If the line l is endowed with some orientation then φ reverses it.*

Proof: Construct one loop ϕ passing through all the points of f and such that any halfplane of S_l intersects ϕ in one single point. We can extend the action of σ and γ over $l\cup\phi$ in such a way that this properties will be preserved in the images of the loop during the isotopy, and that if we denote by η to the composition of σ with γ then $\eta(\phi)=\phi$. Consider l and ϕ endowed with some orientations. As η is composed of one reflection and l and ϕ are linked, η must reverse the orientation on l or on ϕ. By lemma 3, η doesn't reverse the orientation of ϕ ■

Now, we can prove the following

Lemma 5.- *$f\in M(1,k)$ is nonmirror if $k\geq 8$.*

Proof: Suppose $f=\{l;p_1,p_2,...,p_k\}\in M(1,k)$ with $k\geq 8$ is mirror. Let φ be a mirror permutation of f. Then, from lemma 3 we deduce that φ permutes the set of points of f in such a way that $\{p_1,p_2,...,p_k\}$ will be divided in disjoint cycles, all of the same length t. We will see that no value for t is possible.

(i) $t=1$, 2 or 3: consider some $f'\in M(1,6)$ composed of the line l and the points of 6, 3 or 2 cycles respectively. Then $\varphi(f')=f'$ and f' would be mirror. But this is not possible by lemma 2.

For $t\geq 4$ we will consider some $f'\in M(1,t)$ composed of the line l and the t points of one single cycle.

(ii) $t=4$, 5 or 6: $\varphi(f')=f'$ and therefore $f'\in M(1,t)$ would be mirror, but this is not possible by lemma 2.

(iii) $t\geq 7$ and odd: we will show that $M(1,t)$ doesn't contain mirror configurations.

Let $g\in M(1,i)$. If $i\geq 7$ define $s(g)$ to be the sum of the linking coefficients of the i triples that consist of 3 lines determined by three consecutive pairs of adjacent points in g. Then s is preserved under isotopies and antisymmetric in $M(1,i)$. If i is odd, then s(g) is odd and g is nonmirror.

(iv) $t\geq 8$ and even: let p be a point of f', then $f'=\{l;p,\varphi(p),...,\varphi^{t-1}(p)\}$ and $\varphi^{t}(p)=p$. We will denote by l' to the line determined by p and $\varphi^{t/2}(p)$ endowed with some orientation, and by $\varphi^j(l')$ to the line determined by $\varphi^j(p)$ and $\varphi^j(\varphi^{t/2}(p))=\varphi^{j+(t/2)}(p)$ endowed with the orientation induced by φ^j and l'. Then, l' and $\varphi^{t/2}(l')$ denote the same line with opposite orientations. We know that φ is composed of one reflection, therefore if the line l of f is endowed with some orientation we will have

$$\mathrm{lk}(l',l) \overset{\varphi}{=} -\mathrm{lk}(\varphi(l'),\varphi(l)) \overset{\mathrm{Cor.4}}{=} \mathrm{lk}(\varphi(l'),l).$$

Hence, applying $t/2$ times φ we obtain $\mathrm{lk}(l',l)=\mathrm{lk}(\varphi^{t/2}(l'),l)$ which is a contradiction ■

By lemmas 1, 2 and 5 we have the following:

Theorem 6.- *$f\in M(1,k)$ is mirror if and only if $k\leq 3$* ■

Note: Since this paper was completed, the problem addressed in it (the existence of mirror configurations on $M(h,k)$) has been soved by this author for an extense quantity of cases by using differents methods (see "Mirror property for nonsingular mixed configurations of lines and points in $\mathbb{R}^3$", in preparation).

Bibliography

[1] V. M. Kharlamov, *Non-amphicheiral surfaces of degree 4 in* $\mathbb{R}P^3$, Topology and Geometry (Rokhlin Seminar), Lecture Notes in Math., vol.1346, Springer, 1988, pp. 349-356.

[2] O. Ya. Viro, *Topological problems concerning lines and points of three-dimensional space*, Soviet Math. Dokl. 32 (1985), no.2, pp. 528-531.

[3] O. Ya. Viro and Yu. V. Drobotukhina, *Configurations of skew lines*, Leningrad Math. J. 1 (1990), no.4, pp. 1027-1050.

Families of Semialgebraic Sets and Limits

by

Ludwig Bröcker

1 Introduction and notations

The motivation for this article is the question, to what extent a semialgebraic set $S \subset \mathbb{R}^n$ can be approximated by algebraic sets W. More generally, let $U \subset \mathbb{R}^n$ be semialgebraic open and let $S \subset U$ be semialgebraic and closed in U. Do there exist algebraic sets $W \subset \mathbb{R}^n$ such that $W \cap U$ is arbitrarily close to S?

Some comments are needed:

1. The question makes no sense if S itself is of the form $S = W \cap U$ where $W \subset \mathbb{R}^n$ is algebraic.

2. The question makes no sense unless the algebraic sets W are required to belong to a fixed finite - parameter family.

Before we get more precise, let us consider an example:
Let $n = 2$ and $U = Q^2 = \{x \in \mathbb{R}^2 \mid x_1^2 < 1, x_2^2 < 1\}$ the open unit cube. We shall also use the short notation

$$U = \{X_1^2 < 1, X_2^2 < 1\}\ .$$

Let $S \subset Q^2$ be the right angle

$$S = Q^2 \cap (\{X_1 = 0,\ X_2 \geq 0\} \cup \{X_2 = 0,\ X_1 \geq 0\})$$

So we want to approximate S in U by algebraic curves W which belong to a fixed family. For instance, we may require that $\deg(W) \leq d$ where d is fixed. Clearly, S cannot be approximated in U by curves of degree $d \leq 2$ and one can also see that for $d = 3$ this is not possible. But an approximation of S

in U by algebraic curves of degree 4 is possible. For instance, for W we can take a couple of very flat ellipses which are very close to S.

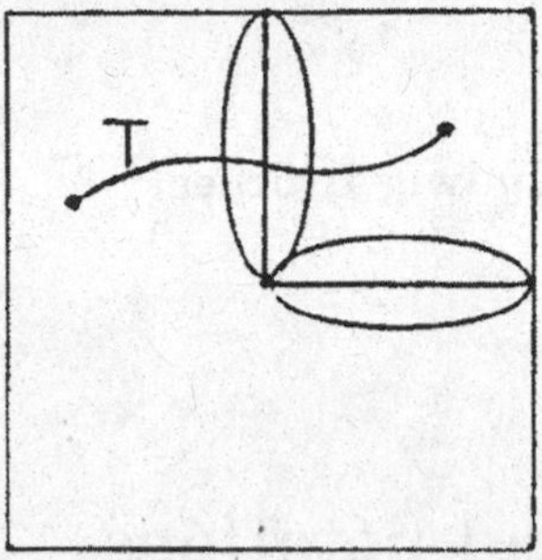

By a small perturbation W can also be made irreducible. However, this kind of approximation might be considered to be not good, since it is two fold or of multiplicity 2. Of course, it has to be defined precisely, what multiplicity means, and a major part of this paper and [Br] is concerned with this. We are mainly interested in multiplicities mod. 2 which have a simple topological meaning. Let us stay in our example:
Let $\deg(W) \leq 4$, $W \subset U_\delta(S) = \{x \in \mathbb{R}^2 \mid d(x, S) < \delta\}$ and δ sufficiently small. We consider W as a semialgebraic chain (after omitting isolated points and intersecting with a sufficiently large ball if W is not bounded). Consider another semialgebraic 1-chain $T \subset Q^2$ such that $\partial T \cap U_\delta(S) = \emptyset$. Then the intersection number $T \circ W$ is defined [Do,VII § 4] and the claim is that it is even. This does not only hold for the family of curves of degree ≤ 4 but for any finite parameter family of curves, if δ is (depending on the family) sufficiently small.
The main goal of this article is to show this kind of non-approximability of semialgebraic sets by algebraic sets in any dimension. For the precise statement see Th. 4.8. I am grateful to M. Coste, who gave me a hint how to simplify and clarify my earlier proof of that theorem.
In § 1 we shall introduce families of semialgebraic sets and we consider Hausdorff limits for sequences of semialgebraic sets in a family. The main result is that these Hausdorff limits are again semialgebraic. This is done by interpreting these limits as reductions of semialgebraic sets defined over a real closed real valued field R. Now this kind of reduction is well understood [Br]. Let me quote the main results:
Let R be a real closed field with real valuation ring V and residue field $\overline{R}$. For $x \in V$ let $\overline{x} \in \overline{R}$ the residue class. Also for $x = (x_1, \dots, x_n)$ let $\overline{x} =$

$(\overline{x}_1, \ldots, \overline{x}_n)$. This defines the reduction map

$$V^n \to \overline{R}^n \; ; \; x \mapsto \overline{x} \quad .$$

For $S \subset V^n$ we denote by $\overline{S}$ the image of S under the reduction map.

1. If $S \subset V^n$ is definable (in the language of valued real closed fields [Ch-Di]), then $\overline{S}$ is semialgebraic.

2. If S is semialgebraic, then $\dim(\overline{S \cap V^n}) \leq \dim(S)$.

3. The reduction map extends to the real spectrum. If $S \subset V^n$ is semialgebraic and $\alpha \in \tilde{\overline{S}}$ with $\dim(\alpha) = \dim(S)$, then the number $\mu_S(\alpha)$ of elements $\beta \in \tilde{S}$ which are mapped to α, is finite. Moreover, the map $\alpha \mapsto \mu_S(\alpha)$ is locally constant on the subspace of elements $\alpha \in \tilde{\overline{S}}$ of dimension equal to $\dim(S)$.

Here " $\sim$ " denotes the usual correspondence between semialgebraic sets and constructible sets in real spectra [B-C-R, 7.2][Kn-Schei, III § 5]. The numbers $\mu_\alpha(S)$ are algebraic analogues of the visual multiplicities we observed (but did not define) in the example.

4. The multiplicities allow to define the reduction of semialgebraic p-chains which leads to a reduction of homology groups. Also intersection numbers of p- and q-chains, $p + q = n$, are well behaved under the reduction (here we use coefficients in $\mathbb{Z}/2\mathbb{Z}$).

The latter point leads to the direct topological interpretation of multiplicities we considered in our example.

For all this work, we shall use the following notations. Let $S \subset R^n$ be semialgebraic where R is a real closed field. Then $Cl(S)$ = closure of S, $\mathrm{Int}(S)$ = interior of S, $Bd(S)$ = boundary of S, $Cl_Z(S)$ = Zariski-closure of S. $\dim(S)$ = dimension of S. $\dim_x(S)$ = local dimension of S at x. $S^* = \{x \in S \mid \dim_x(S) = \dim(S)\}$.

We provide R^n with the euclidean distance d. Then we also set, for $\varepsilon > 0$:

$$\begin{aligned} U_\varepsilon(S) &= \{x \in R^n \mid d(x, S) < \varepsilon\} \\ B_\varepsilon(S) &= \{x \in R^n \mid d(x, S) \leq \varepsilon\} \\ S_\varepsilon(S) &= \{x \in R^n \mid d(x, S) = \varepsilon\} \end{aligned}$$

The real spectrum of a ring A [B-C-R, chap.7], [Kn-Schei, Kap. III], will be denoted by $Sper(A)$ and the maximal real spectrum by $Mar(A)$.

2 Limits of semialgebraic sets

Let R be a real closed field.
Consider $R^{n+k} = R^n \times R^k$. Let $\pi : R^{n+k} \to R^k$ be the canonical projection. Let $S \subset R^n \times R^k$ be semialgebraic. So $S = \{(x,y) \in R^n \times R^k \mid \Phi(x,y)\}$. S is considered as a family of semialgebraic sets, $S = \cup S_y$ where $y \in \pi(S)$ and $S_y = \pi^{-1}(y) \cap S$ for $y \in R^k$. R^k (or $\pi(S)$ if we want S_y to be non empty) is called the *parameter space*. We write S_T for $\pi^{-1}(T) \cap S$ if $T \subset R^k$ is semialgebraic.
We shall also consider the underlying real spectra of $R[Y]$ and $R[X,Y]$ for $X = (X_1, \ldots, X_n)$ and $Y = (Y_1, \ldots, Y_k)$. We use the tilda operator, which associates to every semialgebraic set S the constructible set $\tilde{S}$ in the corresponding real spectrum.
We are looking for limits of semialgebraic sets S_y, depending on the parameter y. There are various ways, how they occur. Let us consider some of them and the relationships between them.

Semialgebraic limits 2.1

Let $T \subset \pi(S)$ be semialgebraic, and let $y_0 \in Cl(T) =$ closure of T. Then we put

$$\lim_{T \to y_0} S_T := Cl(\pi^{-1}(T \setminus \{y_0\}) \cap S) \cap \pi^{-1}(y_0) \quad .$$

So this limit is obviously semialgebraic in the linear space $\pi^{-1}(y_0)$.

Abstract limits 2.2

Let $\alpha \in \widetilde{\pi(S)}$ and let R_α be the corresponding real closure. Then the fiber $S_\alpha \subset R_\alpha^n$ is defined as follows. Take a formula $\Phi(X,Y)$ for S and set

$$S_\alpha := \{x \in R_\alpha \mid \Phi(x, \alpha(Y)\} \quad .$$

By model completeness this definition does not depend on the particular description Φ for S. Now consider a specialisation $\alpha \to \beta$ in $\mathrm{Sper}(A)$ for $A = R[Y]$. This yields a real valuation ring V of R_α where we assume R_α to be algebraic over $k(\mathrm{supp}(\alpha))$, namely

$$V = \text{ convex hull of } (A/\mathrm{supp}(\alpha))_{\mathrm{supp}(\beta)}$$

in R_α See [B-C-R, 10.2]. We refer to this valuation of R_α as the *natural valuation* of $\alpha \to \beta$. We set

$$\lim_{\alpha\to\beta} S_\alpha := \overline{S_\alpha \cap V^n}$$

where "bar" denotes the reduction map of section 0 with respect to V.

Hausdorff limits 2.3

For simplicity we assume that $R = \mathbb{R}$. For any subset $M \subset \mathbb{R}^n$ and $\varepsilon > 0$ let $U(\varepsilon, M)$ be the open ε-neighbourhood of M. Moreover, for any two subsets $M, N \subset \mathbb{R}^n$ we set

$$d(M,N) := \inf\{\varepsilon \mid M \subset U(\varepsilon, N), N \subset U(\varepsilon, M)\}$$

(Hausdorff distance). Now let $\mathcal{C}$ be the collection of all closed sets $C \subset \mathbb{R}^n$. For ε, $\rho \in \mathbb{R}$, ε, $\rho > 0$, we set

$$U_{\varepsilon,\rho}(C) = \{C' \in \mathcal{C} \mid d(C' \cap B_\rho(0), C \cap B_\rho(0)) < \varepsilon\} \quad .$$

For each $C \in \mathcal{C}$ we take the sets $U_{\varepsilon,\rho}(C)$, ε, $\rho > 0$, as a basis of neighbourhoods, thus defining a topology on $\mathcal{C}$. Now let $\{y_n\}$ be a sequence in the parameter space, tending to y_0. If the corresponding sequence $Cl(S_{y_n})$ converges in $\mathcal{C}$, we write $\lim_{\{y_n\}\to y_0} S_{y_n}$ for the limit and call it *Hausdorff limit.* (We hope that, in the sequel, the running index n will not be confused with the dimension of $\mathbb{R}^n$.)

Let us consider the relations between these limits. From the geometrical point of view we are interested in Hausdorff limits. By [Br, § 1] we get information on abstract limits. The semialgebraic limits are somehow between them.

First let us visualize the map $V \to \overline{R}$ where (R_α, V) is given as in 2.2 and $\overline{R}$ is the residue field. For this we feel free to consider the easy case that $R = \mathbb{R}$ and $\beta = y_0$ is an ordinary point in $\mathbb{R}^k$, that is, supp(β) is a maximal ideal. Let φ be the prime filter corresponding to α. Recall that R_α is the field of all Nash functions, living on some $T \in \varphi$, T open in supp(φ), modulo the obvious equivalence relation defined by restrictions. Then $V \subset R_\alpha$ is the ring of all elements which have a bounded representant. Now let Γ be the graph of the Nash function f on T, so $T \subset \mathbb{R}^k$ and $\Gamma \subset \mathbb{R} \times \mathbb{R}^k$. So Γ is a family of semialgebraic sets. However, for $y \in T$ the fiber Γ_y consists of the single element $f(y)$. Nevertheless $\lim_{T\to y_0} \Gamma_T$ is not a single point but a closed semialgebraic subset of $\mathbb{R}$ in general. For instance, let $k = 2$,

$T = \{X_1 > 0, X_2^2 < X_1^2\}$, $y_0 = (0,0)$ and $f = Y/X$. Then $\lim_{T \to y_0} \Gamma_T = [-1,1] = \{X^2 \leq 1\}$.
Note that $\lim_{T \to y_0} \Gamma_T$ is bounded and non empty, if f is bounded. So it is compact.
Now having chosen one representation T, f for $a \in R_\alpha$, $a \in V$ we look at all smaller $T' \in \varphi$ (ultrafilter of α), T' open in supp(φ). Then

$$\bigcap_{T'} \lim_{T' \to y_0} \Gamma_{T'} =: F$$

is compact and non empty. We claim

Lemma 2.4 *F consists of a single point and this coincides with the residue class $\overline{a}$ of a.*

Proof: Assume, there are c_1, $c_2 \in F$. Take neighbourhoods $U_\varepsilon(c_i)$ such that $2\varepsilon \leq |c_1 - c_2|$. Now let

$$Y_i := \Big\{(x_1, x_2) \in \Gamma_T \,|\, x_1 \in U_\varepsilon(c_i)\Big\}, \quad i = 1,2 \quad .$$

and $Z_i = \pi(Y_i)$.
Then Z_i is open in T and $Z_1 \cap Z_2 = \emptyset$. Thus at most one of the Z_i belongs to φ. Contradiction. So $F = \{c\}$ for $c \in \mathbb{R}$ and we have to show that $c = \overline{a}$. For this it is enough that there are representations f, T of a for which $f - c$ is arbitrarily small, which is clear by construction. □

Remark 2.5 *If we work over an arbitrary real closed field R we can make a similar argument, but then $\overline{a}$ defines only an ultrafilter ψ on the residue field R_β, which corresponds to a point in $Mar(R_\beta[X])$. This agrees with the fact that $\overline{R}_\alpha$ does not coincide with R_β in general, but $\overline{R}_\alpha$ is an archimedean extension of R_β.*

Now let us look at the relation between semialgebraic limits and abstract limits. We consider the same situation as before: $R = \mathbb{R}$, $\alpha \to \beta = y_0$. Let φ be the ultrafilter corresponding to α. Then we get

Proposition 2.6

$$\lim_{\alpha \to y_0} S_\alpha = \bigcap_{T \in \varphi} \lim_{T \to y_0} S_T \quad .$$

Note that a fortiori it is not clear that the right hand side is semialgebraic, and in fact, if we do not work over the ordinary reals, it is only defined over an extension of R_{y_0}.

Proof: Let $a = (a_1, \ldots, a_n) \in S_\alpha$ and let $T \in \varphi$. Then the a_i can be represented by Nash function f_i, living on an open set $T' \in \varphi$, $T' \subset T$ and T' is contained in the variety of supp(α). Moreover $(f_1(y), \ldots, f_n(y)) \in S_y$ for all $y \in T'$. Hence from 2.4 we get

$$\overline{a} = (\overline{a}_1, \ldots, \overline{a}_n) \subset \lim_{T' \to y_0} S_{T'} \subset \lim_{T \to y_0} S_T \quad .$$

Conversely, assume that

$$c \in \bigcap_{T \in \varphi} \lim_{T \to y_0} S_T, \quad c \notin \lim_{\alpha \to y_0} S_\alpha \quad .$$

Note that both sets are closed. Take a neighbourhood $U_\varepsilon(c) \subset \mathbb{R}^n_{y_0}$ such that $U_{2\varepsilon}(c)$ does not intersect $\lim_{\alpha \to y_0} S_\alpha$. Similary as before consider

$$Y := \left\{(x, y) \in S \mid x \in U_\varepsilon(c), \; y \neq y_0\right\}$$

and $Z = \pi(Y)$. Then $y_0 \in Cl(Z)$ by construction. If $Z \cap T = \emptyset$ for some $T \in \varphi$ we get immediately a contradiction. So let's assume $Z \in \varphi$. By Hardt's theorem [B-C-R, 9.3], [Ha] we have a decomposition: $Z = Z_1 \cup \ldots \cup Z_r$ where the Z_i are open in their Zariski closure and Nash sections $g_i : Z_i \to Y$ for $i = 1, \ldots, r$. So replacing Z by one of the Z_i we may even assume that there is a Nash section $f = (f_1, \ldots, f_n) : Z \to Y$. this defines an element $a \in R^n_\alpha$ with $a \in S_\alpha$ for which, by construction and 2.4, we get $\overline{a} \in Cl(U_\varepsilon(c))$. Contradiction. □

Now let us see, how Hausdorff limits fit into this. Again assume that $R = \mathbb{R}$. Let $\{y_n\}$ be a sequence tending to y_0 where y_n, $y_0 \in \mathbb{R}^k$, and assume that $\lim_{\{y_n\} \to y_0} S_{y_0}$ exists. Let φ' be the filter of all semialgebraic sets $T \subset \mathbb{R}^k$ which contain almost all y_n. Let $\varphi \supset \varphi'$ be any ultrafilter and α the corresponding element in Sper(A). Then $\alpha \to \beta = y_0$. With these notations we have

Proposition 2.7

$$\lim_{\{y_n\} \to y_0} S_{y_n} = \lim_{\alpha \to y_0} S_\alpha \quad .$$

Proof: First observe, that any $T \in \varphi$ contains a subsequence of $\{y_n\}$. Thus, if $c \in \lim_{\{y_n\}\to y_0} S_{y_n}$, so $c \in \lim_{T\to y_0} S_T$, from which by 2.6 we get $c \in \lim_{\alpha\to y_0} S_\alpha$. Conversely, assume

$$c \in \lim_{\alpha\to y_0} S_\alpha, \quad c \notin \lim_{\{y_n\}\to y_0} S_{y_n} \quad .$$

Again choose a neighbourhood $U_\varepsilon(c)$ of c in $\pi^{-1}(y_n)$ such that $U_{2\varepsilon}(c) \cap \lim_{\{y_n\}\to y_0} S_{y_n} = \emptyset$, and consider $Y = \{(x,y) \in S \mid x \in U(\varepsilon, c)\}$ and $Z = \pi(Y)$. Apparently, $Z \notin \varphi$. Hence we find $T \in \varphi$ with $Z \cap T = \emptyset$. Then $c \notin \lim_{T\to y_0} S_T$ and, by 2.6, $c \notin \lim_{\alpha\to y_0} S_\alpha$. Contradiction. □

Corollary 2.8 *Every Hausdorff limit* $S_0 = \lim_{\{y_n\}\to y_0} S_{y_n}$ *is semialgebraic. Moreover,* $\dim S_0 \leq \liminf(\dim(S_{y_n}))$.

Proof: Apply [Br, Th. 1.7] and [Br, Prop. 2.5].
One should compare this with the limits considered in [v.d.D.].

3 Continuity

For all this section we assume that $R = \mathbb{R}$. Let $S \subset \mathbb{R}^n \times \mathbb{R}^k$ be a semialgebraic family. First we consider 1 - parameter families, that is, $k = 1$.

Proposition 3.1 *Let S be a 1 - parameter family. Then for all sequences* $\{y_n\} \to y_0$, $\forall n\ y_n < y_0$, *the Hausdorff limit* $\lim_{\{y_n\}\to y_0} S_{y_n}$ *exists and the following limits coincide.*

1. $\lim_{T\to y_0} S_T$ *for all T such that $T =]y_0 - \varepsilon, y_0[$ in a neighbourhood of y_0.*

2. $\lim_{\alpha\to y_0} S_\alpha$ *for the left hand generization of y_0.*

3. $\lim_{\{y_n\}\to y_0} S_{y_n}$ *for all sequences* $\{y_n\} \to y_0$, $\forall n\ y_n < y_0$.

Proof: First we observe that the limit in 1) is independent of T subject to the condition on T. So the coincidence of the limits 1) and 2) follows from 2.6. For 3) we fix an interval $T =]z, y_0[$, $z < y_0$ and a sequence $\{y_n\} \to y_0$, $y_n \in T$. Let $\varepsilon, \rho > 0$. It remains to show that $d\big(S_{y_n} \cap B_\rho(0), S_0 \cap B_\rho(0)\big) < \varepsilon$ for sufficiently large n where d is the Hausdorff distance and $S_0 = \lim_{T\to y_0} S_T$. Let $U = U_\varepsilon(S_0 \cap B_\rho(0))$ be the ε-neighbourhood of $S_0 \cap B_\rho(0)$ in $\mathbb{R}^n$. Since $B_\rho(0)$ is bounded, it is clear that $S_{y_n} \cap B_\rho(0) \subset U$ for $n > n_0$. Now assume that there are infinitely many $m \in \mathbb{N}$ such that $(S_0 \cap B_\rho(0)) \not\subset U(\varepsilon, S_{y_m} \cap B_\rho(0))$. In particular, there is a point $z_m \in S_0 \cap B_\rho(0)$ such that $z_m \notin U(\varepsilon, S_{y_m} \cap B_\rho(0))$. Since $S_0 \cap B_\rho(0)$ is compact, we may assume that $\{z_m\} \to z_0$. Let $U = U_{\varepsilon/2}(z_0)$ in $\mathbb{R}^n \times \mathbb{R}$. By construction, $U \cap S_{y_m} = \emptyset$ for all m with $z_m - z_0 < \varepsilon/2$. So now $T_1 := \{y \in T \mid S_y \cap U = \emptyset\}$ is infinite and $y_0 \in Bd(T_1)$. On the other hand, T_1 is semialgebraic. So we may assume $T_1 = T$. But then $z_0 \notin \lim_{T \to y_0} S_T$. Contradiction. □

Let us consider further properties of a family S.

Proposition and Definition 3.2 *For a semialgebraic family $S \subset \mathbb{R}^n \times \mathbb{R}^k$ the following properties are equivalent:*

a) For any sequence $\{(x_n, y_n)\}$ in S with $\{(x_n, y_n)\} \to (x_0, y_0)$ and $y_n \neq y_0$ one has $(x_0, y_0) \in Cl(S_{y_0})$. (The vertical closure of S is closed.)

b) For any semialgebraic set $T \subset \mathbb{R}^k$, $y_0 \in Cl(T)$, one has $\lim_{T \to y_0} S_T \subset Cl(S_{y_0})$.

b') Same as b), but with the further condition that $\dim(T) = 1$.

c) $\lim_{\mathbb{R}^k \to y_0} S_{\mathbb{R}^k} \subset Cl(S_{y_0})$.

d) For all $\alpha \in \operatorname{Sper}(\mathbb{R}[y])$, $\alpha \to y_0$ one has $\lim_{\alpha \to y_0} S_\alpha \subset Cl(S_{y_0})$.

d') Same as d), but with the further condition that $\dim(\operatorname{supp}(\alpha)) = 1$.

e) $\lim_{\{y_n\} \to y_0} S_{y_n} \subset Cl(S_{y_0})$ whenever the left hand side exists.

If these conditions hold, the family S is called semicontinuous.

Proof: Obviously, $a) \to b) \to c) \to b)$, $b) \to b')$ and $d) \to d')$. Moreover, by 2.6 : $b) \to d)$, by 3.1 and the curve selection lemma [B-C-R, 2.5]: $b') \leftrightarrow d')$ and $e) \to b')$.
Finally, by 2.7 : $d) \to e)$. So it remains to show $b') \to a)$: Let $\{(x_n, y_n)\} \to (x_0, y_0)$ and assume that $(x_0, y_0) \notin S_{y_0}$. By the curve selection lemma (loc. cit) we find a semialgebraic arc $\gamma : [0, \varepsilon[\to S$ with $\gamma(0) = (x_0, y_0)$ and $\gamma(t) \notin \pi^{-1}(y_0)$ for $t > 0$. Let $T := \pi \circ \gamma$. Then $\dim(T) = 1$ and by construction

$$(x_0, y_0) \in \lim_{T \to y_0} S_T ,$$

which contradicts $b')$. □

Remark 3.3 *In fact, the semicontinuity is a condition at the points $y_0 \in \mathbb{R}^k$. "Natural" families will be semicontinuous, but by brute force one can replace fibres by bad ones at certain points to make it non-semicontinuous. From property c) above we learn, that for an arbitrary semialgebraic family S the subset $T \subset \mathbb{R}^k$ of points at which S is not semicontinuous, is semialgebraic. If S is semicontinuous, then $\pi(S)$ is closed.*

It should be clear, what we do now and how we do it.

Proposition and Definition 3.4 *For a semialgebraic family $S \subset \mathbb{R}^n \times \mathbb{R}^k$ the following properties are equivalent:*

a) For any semialgebraic set $T \subset \pi(S)$, $y_0 \in Cl(T)$ one has $\lim_{T \to y_0} S_T = Cl(S_{y_0})$.

a') Same as a), but with the further condition that $\dim(T) = 1$.

b) For all $\alpha \in \mathrm{Sper}(\mathbb{R}[Y])$, $\alpha \to y_0$ one has $\lim_{\alpha \to y_0} S_\alpha = Cl(S_{y_0})$.

b') Same as b), but with the further condition that $\dim(\mathrm{supp}(\alpha)) = 1$.

c) For any sequence $\{y_n\}$ with $y_n \in \pi(S)$ and $\{y_n\} \to y_0 \in S$ one has $\{S_{y_n}\} \to Cl(S_{y_0})$ (in the Hausdorff topology).

If these conditions hold, the family S is called continuous.

Proof: Obviously $a) \to a')$ and $b) \to b')$. Moreover, by 2.6: $a) \to b)$ and by 3.1 and the curve selection lemma (loc. cit): $a') \to b')$.
$b') \to c)$: Let $\{y_n\} \to y_0$. Assume also that $y_n \neq y_0$. Suppose $\{S_{y_n}\}$ does not tend to $Cl(S_{y_0})$. Then there exists $\varepsilon, \rho > 0$ and infinitely many $m \in \mathbb{N}$ such that $d(S_{y_m} \cap B_\rho(0), S_{y_0} \cap B_\rho(0)) \geq \varepsilon$. If $(S_{y_m} \cap B_\rho(0)) \not\subset U_\varepsilon(S_{y_0} \cap B_\rho(0))$ for infinitely many m, we find a curve $\gamma : [0, \rho[\to \mathbb{R}^n \times \mathbb{R}^k, \gamma(]0, \rho[) \subset S$, $\gamma(0) = z \in \pi^{-1}(y_0) \setminus (U_\varepsilon(S_{y_0}) \cap B_\rho(0))$ and $\gamma(t) \notin \pi^{-1}(y_0)$ for $t > 0$. Then $\pi(\gamma)$ defines an element $\alpha \in \mathrm{Sper}(\mathbb{R}[y])$ such that $\alpha \to y_0$. For this, by 3.1, we get $z \in \lim_{\alpha \to y_0} S_\alpha$, which contradicts $b')$. So assume that $(S_{y_0} \cap B_\rho(0)) \not\subset U_\varepsilon(S_{y_m} \cap B_\rho(0))$ for infinitely many m. As in the proof of 3.1 we find $z_0 \in Cl(S_{y_0}) \cap B_\rho(0)$ and a ball U around z_0 in $\mathbb{R}^n \times \mathbb{R}^k$ such that $U \cap S_{y_m} = \emptyset$ for infinitely many m. Therefore we find a curve $\delta : [0, \rho[\to \pi(S), \delta(0) = y_0$ and $\delta(t) \neq y_0$ for $t > 0$, such that $S_{\delta(t)} \cap U = \emptyset$ for all $t \in]0, \rho[$. Let α be the element in $\mathrm{Sper}(\mathbb{R}[y])$ defined by δ. Then $\alpha \to y_0$ but, by 3.1, $z_0 \notin \lim_{\alpha \to y_0} S_\alpha$,

which again contradicts b').
$c) \to a)$: Let $T \subset \pi(S)$ be semialgebraic with $y_0 \in Cl(T)$. Then we find a sequence $\{y_n\} \to y_0$, $y_n \in T \setminus \{y_0\}$. Clearly $Cl(S_{y_0}) = \lim_{y_n \to y_0} S_{y_n} \subset \lim_{T \to y_0} S_T$. The converse inclusion follows from 3.2. □

Remark 3.5 *Again, continuity is defined at points. One might ask for the size of the set $D \subset \pi(S)$ of all points, where S is not continuous. A first information is given by the following* partition into continuous parts. *There is a decomposition $\pi(S) = T_1, \dots, T_r$ into disjoint semialgebraic sets such that the following hold:*

a) T_i is open in its Zariski closure.

b) $S|T_i$ is continuous.

This is an immediate consequence of Hardt's theorem [Ha]. In particular, D is nowhere dense in $\pi(S)$. May be that D is also semialgebraic, but I didn't see anything which makes it evident.

4 Limits of chains

Let R be a real closed field.

chains 4.1

A semialgebraic q–chain S in R^n is a bounded closed semialgebraic set $S \subset R^n$ of dimension q such that $S = S^*$, that means, also the local dimension $\dim_x(S) = q$ for all $x \in S$. The sum of two q–chains S and T is defined to be 0 if $S = T$ or $Cl(S \Delta T)$ (symmetric difference of S and T) if $S \neq T$. The boundary ∂S of a q–chain S is defined via any triangulation of S but it does not depend on the particular triangulation.
For semialgebraic sets $A \subset X \subset R^n$ the semialgebraic q–chains give rise to a definition of homology groups $H_q(X/A, \mathbb{Z}/2\mathbb{Z})$. For $R = \mathbb{R}$ these coincide with the usual ones and in general they coincide with those defined in [D-Kn] or [B-C-R, 11.7] (see [Br, § 4]).

Example 4.2 *Let $S \subset R^n$ be algebraic of dimension q such that S^* is bounded and closed. Then S^* is a semialgebraic q-cycle. Such a cycle is called* algebraic.

Reduction of chains 4.3

Assume that the real closed field R admits a real valuation v with residue field $\overline{R}$ and valuation ring V. As before, for a semialgebraic set $S \subset V^n$ let $\overline{S} := \{x' \in \overline{R}^n \mid \exists x \in S : x' = \overline{x}\}$. If S is a q-chain, we define $r(S)$, the reduction of S, as follows: If $\dim(\overline{S}) < q$ then $r(S) := 0$. If $\dim(\overline{S}) = q$, there exist semialgebraic q-chains $T_1, \dots, T_m \subset \overline{R}^n$ such that $\overline{S}^* = T_1 + \dots + T_m$ and the multiplicity $\mu_S(\alpha)$ is constant $= r_i$ for all $\alpha \in \tilde{T}_i$ such that $\dim(\alpha) = q$ (see § 1 or [Br, § 4]). Now

$$r(S) := r_1 T_1 + \dots + r_m T_m \quad .$$

Then r commutes with ∂ [Br, Th. 4.4] and thus for semialgebraic sets $A \subset X \subset V^n$, the reduction r defines a homomorphism

$$r : H_*(X/A, \mathbb{Z}/2\mathbb{Z}) \to H_*(\overline{X}/\overline{A}, \mathbb{Z}/2\mathbb{Z}) \quad .$$

We want to understand these multiplicities in the context of limits in a purely geometric way, at least modulo 2. For this we need another tool.

Intersection numbers 4.4

Let η be a p-chain and ζ a q-chain in $\mathbb{R}^n$ with $p + q = n$. Then η and ζ are called admissible if $|\eta| \cap |\partial\zeta| = \emptyset$ and $|\zeta| \cap |\partial\eta| = \emptyset$. In this case, an intersection number $\eta \circ \zeta \in \mathbb{Z}$ is defined [Do, VII § 4], [S-T, § 73]. We denote by $\eta \circ_2 T$ the intersection number mod.2. If $\mathbb{R}$ is replaced by any real closed field, $S \circ_2 T$ is still defined for admissible semialgebraic p-chains S and q-chains T [Br, § 5]. For an isolated point $x \in S \cap T$ we denote by $i_2(x, S, T)$ the local intersection number mod. 2 at x. If $x \notin S \cap T$ we set $i_2(x, S, T) = 0$. Now let us return to families $S \subset \mathbb{R}^n \times \mathbb{R}^k$ of semialgebraic sets. Such a family S is called a *family of q-chains* if S_y is a q-chain for all $y \in \pi(S)$. It is easily seen, using Hardt's theorem [Ha], [B-C-R, 9.3] or similar arguments, that

$$\partial S := \bigcup_{y \in \pi(S)} \partial S_y$$

is semialgebraic, so it is a subfamily of S.
Now assume that $S \subset \mathbb{R}^n \times \mathbb{R}^k$ is a family of p-chains, $T \subset \mathbb{R}^n \times \mathbb{R}^k$ a family of q-chains, $p + q = n$ and $\pi(S) = \pi(T)$. Let $y_n \in \pi(S) = \pi(T)$ for $n \in \mathbb{N}$ and $\{y_n\} \to y_0$. We assume furthermore, that all Hausdorff limits

$$\lim_{\{y_n\} \to y_0} S_{y_n} =: S_0, \ \lim_{\{y_n\} \to y_0} \partial S_{y_n} =: (\partial S)_0$$

$$\lim_{\{y_n\} \to y_0} T_{y_n} =: T_0, \ \lim_{\{y_n\} \to y_0} \partial T_{y_n} =: (\partial T)_0$$

exist and are bounded. By 2.7 there exists an element $\alpha \in (\pi(S))^\sim$ with $\alpha \to y_0$ such that (with respect to the natural valuation v of the specialization $\alpha \to y_0$),

$$S_0 = \lim_{\alpha \to y_0} S_\alpha = \overline{S}_\alpha, \quad (\partial S)_0 = \lim_{\alpha \to y_0} (\partial S)_\alpha = \overline{(\partial S)}_\alpha$$
$$T_0 = \lim_{\alpha \to y_0} T_\alpha = \overline{T}_\alpha, \quad (\partial T)_0 = \lim_{\alpha \to y_0} (\partial T)_\alpha = \overline{(\partial T)}_\alpha .$$

In this situation we have

Proposition 4.5 *Assume that $S_0 \cap (\partial T)_0 = \emptyset$ and $T_0 \cap (\partial S)_0 = \emptyset$. Then*

a) S_α and T_α are admissible.

b) $r(S_\alpha)$ and $r(T_\alpha)$ are admissible.

c) There exists $n_0 \in \mathbb{N}$ such that S_{y_n} and T_{y_n} are admissible and $S_{y_n} \circ_2 T_{y_n} = S_\alpha \circ_2 T_\alpha$ for $n \geq n_0$.

d) $S_\alpha \circ_2 T_\alpha = r(S_\alpha) \circ_2 r(T_\alpha)$.

Proof: a): Obvious.
b): This follows from the fact that ∂ commutes with v [Br, Th. 4.4].
c): Obviously, S_{y_n} and T_{y_n} are admissible for sufficiently large n, and from the general theory of intersection numbers [Do, VII § 4] it is also clear that $S_{y_n} \circ_2 T_{y_n}$ becomes constant. This follows also from b) and the construction below. First of all, by the compactness of $(\pi(S))^\sim$ with respect to the constructible topology, there is a semialgebraic set $D \subset \pi(S) = \pi(T)$ such that $\alpha \in \tilde{D}$ and S_β, T_β are admissible for all $\beta \in \tilde{D}$. Now for $y \in D$ the statement, that $S_y \circ_2 T_y$ takes a certain value, is elementary. Therefore, there is a semialgebraic partition: $D = T_1 \cup \ldots \cup T_m$ such that $S_y \circ_2 T_y$ is constant for $y \in T_i$. But $\alpha \in \tilde{T}_i$ for exactly one $i \in \{1, \ldots, m\}$, say $\alpha \in \tilde{T}_1$. By construction, $y_n \in T_1$ for infinitely many n and for these we have

$$S_{y_n} \circ_2 T_{y_n} = S_\alpha \circ_2 T_\alpha$$

d): This is just stated by the intersection – reduction formula [Br, Th. 5.3].

□

As a consequence, we get a geometric description of multiplicities mod. 2. for this, let the family S of p–chains be given as before, and also the sequence

$\{y_n\} \to y_0$, the limits S_0 and $(\partial S)_0$ and the valuation α which leads to the reduction of chains:

$$S_\alpha \mapsto r(S_\alpha) \subset \overline{S}_\alpha \quad .$$

Corollary 4.6 *Let $x \in \overline{S}_\alpha$ and let $T \subset \mathbb{R}^n$ be a q-chain such that $\partial T \cap \overline{S}_\alpha = \emptyset$, $T \cap (\overline{\partial S})_\alpha = \emptyset$ and $T \cap \overline{S_\alpha} = \{x\}$. Then T and S_{y_n} are admissible for almost all n and $T \circ_2 S_{y_n}$ is stationary. Moreover*

a) If $x \notin r(S_\alpha)$, then $T \circ_2 S_{y_n} = 0$ for almost all n.

b) If $x \in r(S_\alpha)$, then $T \circ_2 S_{y_n} = T \circ_2 \overline{S}^*_\alpha$ for almost all n.

We conclude this work by a result on the non approximability of proper semialgebraic sets by algebraic ones. We consider a relative situation. So let us fix an open semialgebraic set $U \subset \mathbb{R}^n$.

Definition 4.7 *A semialgebraic set $B \subset U$ is essentially algebraic in U, if there is an algebraic subset $W \subset \mathbb{R}^n$ such that $B^* = (W \cap U)^*$.*

If S is a p-chain and $\emptyset \neq S \cap U$ is essentially algebraic in U, then $\partial S \cap U = \emptyset$. Any p-dimensional semialgebraic set $B \subset U$ admits a unique maximal essentially algebraic subset of pure dimension p. This subset will be denoted by B^a (but note that B^a does not only depend on B but also on U).

Theorem 4.8 *Let $U \subset \mathbb{R}^n$ bc semialgebraic and open. Let $S \subset \mathbb{R}^n \times \mathbb{R}^k$ be a semialgebraic family for which the following conditions hold:*

- *$\pi(S)$ is bounded.*
- *S_y is a p-chain for all $y \in \pi(S)$.*
- *$S_y \cap U$ is essentially algebraic in U for all $y \in \pi(S)$.*

Let $B \subset U$ be semialgebraic such that

- *B is closed in U.*
- $\dim(B) \leq p$.

Then there exists $\varepsilon > 0$ such that

- *for all δ with $0 < \delta \leq \varepsilon$,*
- *for all q-chains $T \subset U \setminus B^a$ where $p + q = n$ and $\partial T \cap U_\delta(B) = \emptyset$.*

– *for all* $y \in \pi(S)$ *with* $(S_y \cap U) \subset U_\delta(B)$

one has $T \circ_2 S_y = 0$.

This looks a little complicated. So let us first consider an example. Let $n = 2$ and $Q = Q^2$ the unit cube:
$Q = \{x \in \mathbb{R}^2 \,|\, |x_i| \leq 1\}$. We take $U = \mathrm{Int}(Q)$. Let $S' \subset \mathbb{R}^2 \times \mathbb{R}^k$ be the family of all curves of degree ≤ 3. So $\mathbb{R}^k$ is ths space of the coefficients. By normalizing we can achieve that $\pi(S')$ is bounded. Let $S \subset S'$ such that $S_y = (S'_y \cap Q)^*$. Then S is a family of 1–chains such that $S_y \cap U$ is essentially algebraic in U for all y. We set

$$B := U \cap (\{X_2 = 0\} \cup \{X_1 = 0, X_2 \geq 0\}) \quad .$$

Here $B^a = U \cap \{X_2 = 0\}$. Now, of course, for an arbitrarily small $\delta > 0$ we find $y \in \pi(S)$ such that $(S_y \cap U) \subset U_\delta(B)$, but then the Theorem says, that for any 1–chain T in U with $T \cap B^a = \emptyset$ and $\partial T \cap U_\delta(B) = \emptyset$ one has $T \circ_2 S_y = 0$.

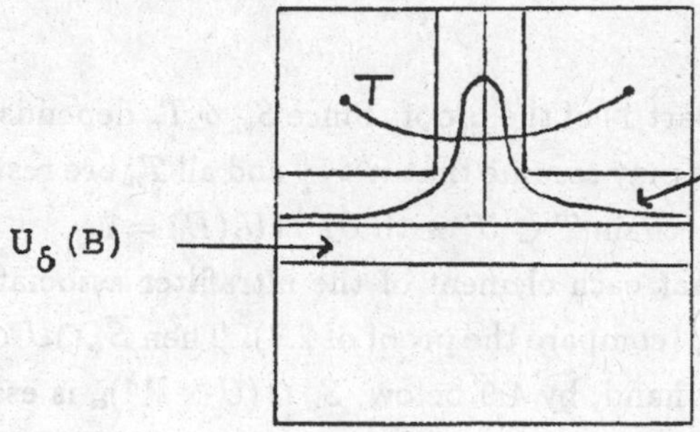

It might be very difficult to find the smallest possible ε for which the Theorem holds, even for the above example.

proof of the theorem:

1) Let $U' := U \setminus B^a$. We set for $\delta > 0$

$$H_q(\delta) := H_q(U'/(U' \setminus U_\delta(B)), \mathbb{Z}/2\mathbb{Z})$$

and

$$H_q(0) := H_q(U'/(U' \setminus B), \mathbb{Z}/2\mathbb{Z}) \quad .$$

Then

$$H_q(0) = \lim_{\delta > 0} H_q(\delta) \quad .$$

We need a little more: First of all, $H_q(0)$ and all $H_q(\delta)$ are finite $\mathbb{Z}/2\mathbb{Z}$–vectorspaces. Moreover, by Hardt's theorem [Ha], [Br-C-R, 9.34] there exists $\varepsilon > 0$ such that $\pi : U' \supset (U_\delta(B) \cap U') \mapsto \delta$ is a trivial fibration for $\delta \in]0, \varepsilon]$. In fact, for $0 < \delta_1 < \delta_2 \leq \varepsilon$ the inclusion

$$(U', U' \setminus U_{\delta_2}(B)) \subset (U', U' \setminus U_{\delta_1}(B))$$

is a retraction, hence the canonical maps

$$H_q(\varepsilon) \to H_q(\delta_2) \to H_q(\delta_1) \to H_q(0)$$

are isomorphisms.

2) Assume that the statement of the Theorem is wrong. Then there exists a convergent sequence $\{y_n\} \to y_0$, $y_n \in \pi(S)$ and a sequence $\{T_n\}$ of semialgebraic q–chains with the following properties:

i) $(S_{y_n} \cap U) \subset U_{\frac{1}{n}}(B)$.

ii) $T_n \subset U' = U \setminus B^a$ and $\partial T_n \cap U_{\frac{1}{n}}(B) = \emptyset$.

iii) $S_{y_n} \circ_2 T_n = 1$.

We choose ε according to part 1) of the proof. Since $S_{y_n} \circ_2 T_n$ depends only on the class of T_n in $H_q\left(\frac{1}{n}\right)$ we may assume that $n > \frac{1}{\varepsilon}$ and all T_n are restrictions of a single semialgebraic q–chain $T \subset U'$ with $\partial T \cap U_\varepsilon(B) = \emptyset$.
Now pick $\alpha \in \widetilde{\pi(S)}$ such that each element of the ultrafilter associated to α contains infinitely many y_n (compare the proof of 2.7). Then $\overline{S}_\alpha \cap U \subset B$ and $S_\alpha \circ_2 T = 1$. On the other hand, by 4.9 below, $S_\alpha \cap (U \times \mathbb{R}^k)_\alpha$ is essentially algebraic in $(U \times \mathbb{R}^k)_\alpha$. Hence by [Br,Cor. 3.8] we get $r(S_\alpha) \cap U \subset B^a$. Since $T \cap B^a = \emptyset$ and $T = rT$ we get $rT \circ_2 rS_\alpha = 0$. But this is a contradiction to the intersection–reduction formula [Br, Th. 5.3] saying $S_\alpha \circ_2 T = rS_\alpha \circ_2 rT$. □

For the Lemma below we use the following notation: Let $S \subset \mathbb{R}^n$ be semialgebraic and let $x_1, \ldots, x_n \in \mathbb{R}^n$. Then S is called *essentially algebraic at* $x_1, \ldots, x_n$, if there is an open set U containing $x_1, \ldots, x_n$ such that S is essentially algebraic in U.

Lemma 4.9 *Let $U \subset \mathbb{R}^n$ be semialgebraic and open and let $B \subset U$ be semialgebraic. Then the statement "B is essentially algebraic in U" is elementary.*

Proof: This is an immediate consequence of the following facts.

a) B is essentially algebraic in U if and only if B is essentially algebraic at any two points $x_1, x_2 \in U$.

b) Let $\dim(B) < n-1$. Then B is essentially algebraic at $x_1, \dots, x_n$ if and only if for all $y \in S^{n-1}$ and all $\varepsilon > 0$ there exists $z \in U_\varepsilon(y)$ such that $\pi_z(B)$ is essentially algebraic at $\pi(x_1), \dots, \pi(x_n)$, where π_z is the orthogonal projection: $\mathbb{R}^n \to \langle z \rangle^\perp$.

c) The statement is true if $\dim(B) = n-1$.

Here a) is clear and b) holds, since the image of an algebraic set under an injective polynomial map is essentially algebraic. For c) first one finds among the polynomials, which describe B, equations for an algebraic set $W \supset B$ with $\dim(W) = \dim(B)$. Since $\mathbb{R}[X]$, $X = X_1, \dots, X_n$ is factorial, the degree of equations for the Zariski closure of B can be bounded.

□

References

[B-C-R] Bochnak, J., Coste, M., Roy, M.F.: Géométrie Algébrique Réelle. Springer, Berlin Heidelberg New York, 1987

[Br] Bröcker, L.: On the reduction of semialgebraic sets by real valuations. To appear.

[Ch-Di] Cherlin, G., Dickmann, M.: Real closed rings II. Model theory. Annals of pure and applied Logic 25, 213-231 (1983)

[D-Kn] Delfs, H., Knebusch, M.: On the homology of algebraic varieties over real closed fields. J. reine angew. Math. 335, 122-163 (1982)

[Do] Dold, A.: Lectures on Algebraic Topology. Springer, Berlin Heidelberg New York 1972

[Ha] Hardt, R.: Semi-algebraic local triviality in semi-algebraic mappings. Amer. J. Math. 102, 291-302 (1980)

[Kn-Schei] Knebusch, M., Scheiderer, C.: Einführung in die reelle Algebra. Vieweg, Braunschweig Wiesbaden 1989 [v.d.D.] van den Dries, L.: Tarskis Problem and Pfaffian functions. Logic Colloquium 1984, ed. I.B. Paris, A.J. Wilkie, G.M. Wilmers, North Holland 59-90 (1986)

Ludwig Bröcker
Mathematisches Institut
Einsteinstr. 62
4400 Münster

Germany

A Hopf Fixed Point Theorem for Semi-Algebraic Maps

G. W. BRUMFIEL*

Introduction.

This note is a sequel to [2], in which we proved the Brouwer fixed point theorem for the real spectrum compactification of $\mathbb{R}^n$, that is, for $\mathsf{A}^n = \mathrm{Spec}_R(\mathbb{R}[x_1 \dots x_n])$, the real spectrum of the polynomial ring in n variables over the real numbers. Here, we fix a real closed field K and a semi-algebraic subset $X \subseteq K^n$. Let $\tilde{X} \subseteq \mathsf{A}^n_K = \mathrm{Spec}_R(K[x_1 \dots x_n])$ be the constructible subset corresponding to $X \subseteq K^n$. Any continuous semi-algebraic map $\varphi: \ X \to X$ extends to a continuous map $\tilde{\varphi}: \tilde{X} \to \tilde{X}$. Define the Lefschetz number of φ to be

$$\mathrm{tr}(\varphi_*) = \Sigma(-1)^i \ \mathrm{trace}(\varphi_*: \ H_i(X, \mathbb{Q}) \to H_i(X, \mathbb{Q})).$$

Homology of semi-algebraic sets $X \subseteq K^n$ is defined in [3]. The Hopf fixed point theorem basically states that if $\mathrm{tr}(\varphi_*) \neq 0$ then $\tilde{\varphi}: \tilde{X} \to \tilde{X}$ has a fixed point.

Before giving a precise statement, we make a few remarks. It seems reasonable to investigate topological properties of $\mathrm{Fix}(\tilde{\varphi})$, the fixed point set of $\tilde{\varphi}$. For example, we can show that $\mathrm{Fix}(\tilde{\varphi}) \subseteq \tilde{X}$ is always closed. I am indebted to Claus Scheiderer for the proof of this fact given below. His proof replaces a more cumbersome argument in an earlier version of this note, which only worked under the added hypothesis that $X \subset K^n$ was locally closed. Note that $\mathrm{Fix}(\tilde{\varphi})$ is obviously closed in the Tychonoff, or constructible topology on $\tilde{X}$, because the Tychonoff topology is compact and Hausdorff and $\tilde{\varphi}$ is Tychonoff continuous. A Tychonoff closed subset is closed in the real spectrum topology, precisely when it is closed under specialization $\alpha \to \beta$ [1, Proposition 2.11]. We will show below that if $\alpha, \beta \in \tilde{X}$, $\alpha \to \beta$, and $\tilde{\varphi}(\alpha) = \alpha$, then $\tilde{\varphi}(\beta) = \beta$. Thus we obtain

Proposition 1. *If $X \subseteq K^n$ is any semi-algebraic set and $\varphi: X \to X$ is any continuous semi-algebraic map, then $\mathrm{Fix}(\tilde{\varphi})$ is closed in $\tilde{X}$.*

The classical Hopf fixed point theorem concerns maps $\varphi: \ Y \to Y$, where $Y \subset \mathbb{R}^n$ is a finite closed simplicial complex. If $\mathrm{tr}(\varphi_*) \neq 0$ then φ has a fixed point in Y. A key step in our general Hopf fixed point theorem is to extend this result to arbitrary real closed ground fields K.

Proposition 2. *Suppose $Y \subset K^n$ is a complete semi-algebraic set, that is, Y is closed and bounded in K^n. Suppose $\varphi: \ Y \to Y$ is a continuous semi-algebraic map with $\mathrm{tr}(\varphi_*) \neq 0$. Then φ has a fixed point in Y.*

---

* Research supported by NSF grant DMS 85-06816 at Stanford University

The assumption that Y is complete is equivalent to assuming that Y is a closed finite simplicial complex over K. In fact, we use triangulations in a crucial way in our proofs. Specifically, we exploit the fact that any semi-algebraic set $X \subseteq K^n$ can be triangulated and admits strong deformation retractions onto complete subspaces $Y \subset X$. Define Fringe($\tilde{X}$) $\subset \tilde{X}$ to be all those points which do not belong to any constructible $\tilde{C} \subset \tilde{X}$, where $C \subset X$ is a complete subspace. Here is a precise statement of our main theorem.

Theorem. *If $\varphi : X \to X$ is any continuous semi-algebraic map with $tr(\varphi_*) \neq 0$ then either $\varphi : X \to X$ has a fixed point or $\tilde{\varphi} : \tilde{X} \to \tilde{X}$ has a fixed point in Fringe($\tilde{X}$). Moreover, $\tilde{\varphi} : \tilde{X} \to \tilde{X}$ always has a closed fixed point, and, if either $X \subseteq K^n$ is locally closed or if the ground field K is $\mathbf{R}$, then $\tilde{\varphi} : \tilde{X} \to \tilde{X}$ always has a closed fixed point in Fringe($\tilde{X}$) $\cup X$.*

The proof of the Brouwer fixed point theorem for $\mathbf{A}^n$ in [2] was somewhat unsatisfactory because at one stage a transcendental argument was used to construct certain retractions from $\mathbf{R}^n$ onto closed disks in $\mathbf{R}^n$. In fact, the details of that transcendental argument, which amounted to constructing a suitable gradient vector field and integrating, were not included. The argument in this paper is purely algebraic. Of course, if $X \subseteq K^n$ is any contractible space, then $\mathrm{tr}(\varphi_*) = 1$ for any $\varphi : X \to X$, so the Hopf fixed point theorem certainly implies the Brouwer fixed point theorem.

Finally, we point out that it would be desirable to have an even more general Hopf-Lefschetz fixed point formula for $\tilde{\varphi} : \tilde{X} \to \tilde{X}$, which expresses $\mathrm{tr}(\varphi_*)$ in terms of local data near $\mathrm{Fix}(\tilde{\varphi}) \subset \tilde{X}$. However, this looks rather messy. For example, if K contains an infinitesimal ϵ relative to $\mathbf{Q} \subset K$, consider the map $\varphi : I \to I$ given by $\varphi(x) = x + \epsilon x(1-x)$, where $I = [0,1] \subset K$. Then $\tilde{\varphi} : \tilde{I} \to \tilde{I}$ has a very complicated fixed point set. Somehow, only the two boundary fixed points 0 and 1 should be relevant, but for each $r \in K$, $0 \leq r \leq 1$, the Dedekind cuts of $I \subset K$ defined by the sequences $r + n\epsilon$ or $r - n\epsilon$, $n = 1, 2, 3 \ldots$, determine points of $\tilde{I}$ which are fixed by $\tilde{\varphi}$. There are also many other fixed points. Even when $K = \mathbf{R}$, if $X \subset \mathbf{R}^n$ is not compact, the point-set nature of the fixed points of $\tilde{\varphi} : \tilde{X} \to \tilde{X}$ in Fringe($\tilde{X}$) will often resemble the bad fixed point set of the map $\tilde{\varphi} : \tilde{I} \to \tilde{I}$ above. Finding a good generalization of the Hopf-Lefschetz fixed point formula looks challenging.

Proof of the theorem. We first point out that if $\tilde{\varphi} : \tilde{X} \to \tilde{X}$ has any fixed point then it has a closed fixed point. Namely, if $\tilde{\varphi}(\alpha) = \alpha$ and $\alpha \to \beta$ with β closed in $\tilde{X}$. Then $\tilde{\varphi}(\alpha) = \alpha \to \tilde{\varphi}(\beta)$. Since the specializations of α form a chain, either $\beta \to \tilde{\varphi}(\beta)$ or $\tilde{\varphi}(\beta) \to \beta$. But, since β is closed, $\beta \to \tilde{\varphi}(\beta)$ implies $\beta = \tilde{\varphi}(\beta)$, and, since $\dim(\tilde{\varphi}(\beta)) \leq \dim(\beta)$, $\tilde{\varphi}(\beta) \to \beta$ also implies $\beta = \tilde{\varphi}(\beta)$.

Next, consider the fringe of $\tilde{X}$, defined as

$$\mathrm{Fringe}(\tilde{X}) = \tilde{X} - \bigcup_{\substack{C \subseteq X \\ C \text{ complete}}} \tilde{C}.$$

Since each constructible $\tilde{C}$ is Tychonoff open, Fringe($\tilde{X}$) is Tychonoff closed, hence compact. If $\alpha \in$ Fringe($\tilde{X}$) and $\tilde{\varphi}(\alpha) \neq \alpha$ then $\tilde{\varphi}(\alpha) \not\to \alpha$ since $\dim(\tilde{\varphi}(\alpha)) \leq \dim(\alpha)$. Thus, there exists a polynomial $f \in K[x_1, \ldots x_n]$ with $f(\alpha) < 0$ and $f(\tilde{\varphi}(\alpha)) \geq 0$. The set $\tilde{V}_\alpha = \left\{\beta \in \tilde{X} \;\middle|\; f(\beta) < 0 \text{ and } f(\tilde{\varphi}(\beta)) \geq 0\right\}$ is a constructible set containing α, with the property that $\beta \in \tilde{V}_\alpha$ implies $\tilde{\varphi}(\beta) \notin \tilde{V}_\alpha$. If $\tilde{\varphi} : \tilde{X} \to \tilde{X}$ has no fixed points in Fringe($\tilde{X}$), we can find finitely many such constructible sets $\tilde{V}_i$ with Fringe($\tilde{X}$) $\subset \cup \tilde{V}_i$.

Let $V_i = \tilde{V}_i \cap X$, a semi-algebraic set in X. Choose a semi-algebraic triangulation of X such that all the V_i are subcomplexes. Recall that a semi-algebraic triangulation is simply a semi-algebraic homeomorphism between X and some union of open simplices of a closed finite simplicial complex $Z \subset K^n$. Boundary faces of simplexes of X need not belong to X and also, of course, the V_i are not necessarily closed subcomplexes of X.

Next, we take two barycentric subdivisions Z'' of Z. As in [3], there is now a canonical strong deformation retraction, $r : X \to Y$, of X onto a closed subcomplex Y of Z''. The picture below illustrates this construction when X is a single closed 2-simplex with one vertex and two edges removed.

$X \subset Z = \Delta^2$

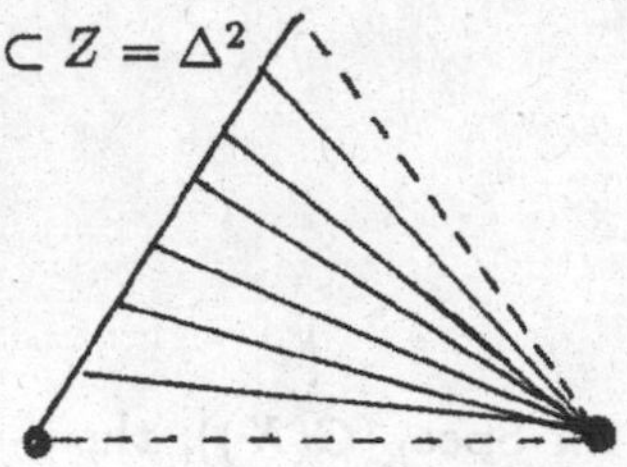

$Y \subset X \subset Z''$

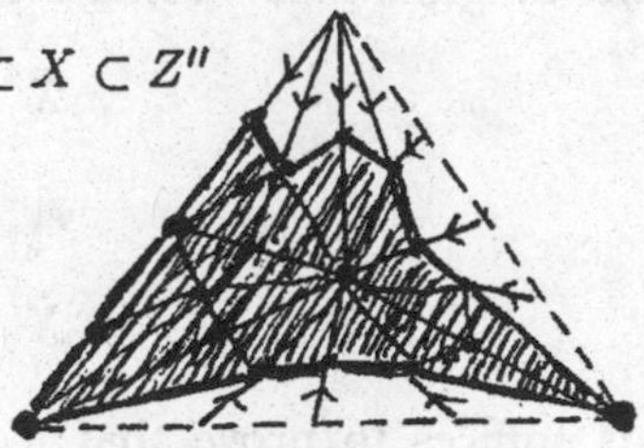

The general formulation can be found in [3]. For our purposes, the crucial property of $r : X \to Y$ is that if σ is any open simplex of the original triangulation of X then $r(\sigma) \subset \sigma$. Thus, if $\sigma \subset V_i$ then $r(\sigma) \subset V_i$ so $r(V_i) \subset V_i$. Also, if $r|_\sigma \neq Id$ then σ must contain fringe points of $\tilde{X}$ so $\sigma \subset V_j$, some j. Thus $X - Y \subset \cup V_i$.

Consider the composition $r\varphi : X \to X \to Y$. By Proposition 2, $r\varphi|_Y : Y \to Y$ has a fixed point $y \in Y$ since $\mathrm{tr}((r\varphi|_Y)_*) = \mathrm{tr}(\varphi_*) \neq 0$. If $\varphi(y) \notin Y$, say $\varphi(y) \in V_i$. Then $r\varphi(y) \in V_i$, and it is impossible that $y = r\varphi(y)$ because $y \in V_i$ implies $\varphi(y) \notin V_i$. Therefore $\varphi(y) \in Y$ and we have thus proved that either $\varphi : X \to X$ has a fixed point or $\tilde{\varphi} : \tilde{X} \to \tilde{X}$ has a fixed point in Fringe($\tilde{X}$).

If $X \subset K^n$ is locally closed then Fringe($\tilde{X}$) is closed in $\tilde{X}$. This is most easily seen by identifying X with a closed semi-algebraic set $X' \subset K^{n+1}$. Then Fringe($\tilde{X}$) corresponds to the set of points of $\tilde{X}'$ at infinity relative to K, which is closed in $\tilde{X}'$. Therefore, in our theorem, we can conclude that for locally closed X, $\tilde{\varphi} : \tilde{X} \to \tilde{X}$ either has a fixed point in X or a closed fixed point in Fringe($\tilde{X}$).

We can also make this same conclusion for arbitrary $X \subset \mathbf{R}^n$. Namely, suppose $\alpha \in \text{Fringe}(\tilde{X})$, $\tilde{\varphi}(\alpha) = \alpha$, and $\alpha \to \beta$ with β closed in $\tilde{X}$. We know $\tilde{\varphi}(\beta) = \beta$. If $\beta \notin \text{Fringe}(\tilde{X})$, then $\beta \in \tilde{C}$ for some complete $C \subset X$. But when the groundfield is $\mathbf{R}$, the only closed points in a complete $\tilde{C}$ are actually points of C, hence $\beta \in C$ and $\varphi(\beta) = \beta$ is a fixed point of φ in X.

It seems plausible that this stronger conclusion, that $\tilde{\varphi} : \tilde{X} \to \tilde{X}$ either has a fixed point in X or a closed fixed point in $\text{Fringe}(\tilde{X})$, holds for arbitrary $X \subset K^n$. However, if $X \subset K^n$ is not locally closed and $K \neq \mathbf{R}$, there will often exist $\alpha \in \text{Fringe}(\tilde{X})$ with $\alpha \to \beta$, β closed, and $\beta \notin \text{Fringe}(\tilde{X}) \cup X$. Conceivably, the only closed fixed points β of $\tilde{\varphi} : \tilde{X} \to \tilde{X}$ could be of this type.

Proof of Proposition 1. We introduce the ring $C(X)$ of continuous K-valued semi-algebraic functions on $X \subseteq K^n$. If $\alpha \in \tilde{X}$ corresponds to a homomorphism $ev(\alpha)\colon K[x_1 \dots x_n] \to k(\alpha)$, where $k(\alpha)$ is a real closed field, algebraic over the image of $K[x_1 \dots x_n]$, then every $f \in C(X)$ can also be evaluated at α. That is, $ev(\alpha)$ extends to $ev(\alpha)\colon C(X) \to k(\alpha)$. In this way, we obtain a continuous inclusion $\tilde{X} \subset \text{Spec}_R(C(X))$, which, in fact, is a homeomorphism if X is locally closed in K^n, [4]. But, in any case, the map $\varphi\colon X \to X$ induces a ring homomorphism $\varphi^*\colon C(X) \to C(X)$, which, in turn, induces $\varphi_*\colon \text{Spec}_R(C(X)) \to \text{Spec}_R(C(X))$ such that the diagram below commutes.

$$\begin{array}{ccc} \tilde{X} & \subset & \text{Spec}_R(C(X)) \\ {\scriptstyle\tilde{\varphi}}\downarrow & & \downarrow{\scriptstyle\varphi_*} \\ \tilde{X} & \subset & \text{Spec}_R(C(X)) \end{array}$$

It thus suffices to prove that if $\varphi_*(\alpha) = \alpha$ and $\alpha \to \beta$ in $\text{Spec}_R(C(X))$, then $\varphi_*(\beta) = \beta$.

The assertion $\varphi_*(\alpha) = \alpha$ is equivalent to the existence of an order preserving K-endomorphism $i\colon k(\alpha) \to k(\alpha)$ such that the following diagram commutes

$$\begin{array}{ccc} C(X) & \xrightarrow{ev(\alpha)} & k(\alpha) \\ {\scriptstyle\varphi^*}\downarrow & & \downarrow{\scriptstyle i} \\ C(X) & \xrightarrow{ev(\alpha)} & k(\alpha) \end{array}$$

Since $k(\alpha)$ is real closed and has finite transcendence degre over K, the inclusion i is an isomorphism. Thus, i induces a permutation of the convex valuation rings in $k(\alpha)$ which contain K, and since these valuation rings form a finite chain under inclusion, each such valuation ring maps isomorphically onto itself under i. On the other hand, the specializations of α in $\text{Spec}_R(C(X))$ correspond, by the real place existence theorem, to the prime ideals of $C(X)$ which are the inverse images under $ev(\alpha)$ of the maximal ideals of those convex valuation rings in $k(\alpha)$ which contain the image of $C(X)$. From the commutative diagram above, we see that for each such prime ideal p, $(\varphi^*)^{-1}(p) = p$, hence $\alpha_*\colon \mathrm{Spec}_R(C(X)) \to \mathrm{Spec}_R(C(X))$ is the identity map on the specializations of α, as desired.

Proof of Proposition 2. We have a continuous semi-algebraic map $\varphi : Y \to Y$, where $Y \subset K^n$ is a closed finite simplicial complex. The graph $\Gamma(\varphi) \subset Y \times Y \subset K^n \times K^n$ is defined in terms of finitely many polynomial inequalities with coefficients in K. We know that if $K \subseteq \mathbf{R}$ and $\mathrm{tr}(\varphi_*) \neq 0$ then φ has a fixed point in Y, because the classical proof of the Hopf fixed point theorem works for any Archimedean ground field K. We will deduce the proposition for arbitrary K by transfer from the real algebraic numbers, using Tarski's principle.

We may assume that Y itself is defined over $\mathbf{Z}$, since we may as well assume the vertices of Y are vertices of some standard simplex $\Delta^N \subset K^{N+1}$. We regard the coefficients of the polynomials used to define $\Gamma(\varphi) \subset Y \times Y$ as parameters and we regard the statement that $\Gamma(\varphi)$ is the graph of a continuous function $\varphi : Y \to Y$ as an elementary statement about these parameters. Also, the desired conclusion that φ has a fixed point is an elementary statement about these parameters. For example, a fixed point of φ is just a point of $\Gamma(\varphi) \cap \Delta(Y)$, where $\Delta(Y) \subset Y \times Y$ is the diagonal.

The crucial point is that the hypothesis $\mathrm{tr}(\varphi_*) \neq 0$ is also an elementary statement about the parameters defining φ. Assuming this, the truth of the proposition is clear, because if the propositon were false, there would exist in K some values of the coefficient parameters for which (a) $\Gamma(\varphi) \subset Y \times Y$ is the graph of a continuous semi-algebraic function, (b) $\mathrm{tr}(\varphi_*) \neq 0$, and (c) $\Gamma(\varphi) \cap \Delta(Y) = \emptyset$. By Tarski's principle, there would also exist in the real algebraic numbers some values of the parameters for which (a), (b), (c) hold, which is a contradiction.

The claim that the statement $\mathrm{tr}(\varphi_*) \neq 0$ is an elementary statement amounts to understanding how the homology maps $\varphi_* : H_i(Y, \mathbf{Q}) \to H_i(Y, \mathbf{Q})$ can be computed in terms of the polynomial inequalities which define $\Gamma(\varphi) \subset Y \times Y$. We have the cover of Y by (open) semi-algebraic simplices, σ_i. Since the $\varphi^{-1}(\sigma_i) \subset Y$ are also semi-algebraic, we can retriangulate Y so that all the sets $\varphi^{-1}(\sigma_i)$, as well as the original σ_i, are unions of (new) simplices. The important point is that this retriangulation can be done constructively.

We now have a second closed, finite, simplicial complex Y', semi-algebraically identified with Y. We claim that the original map $\varphi\colon Y' \to Y$ is homotopic to a simplicial map $\varphi'\colon Y' \to Y$. Specifically, for each vertex $v'_j \in Y'$, $\varphi(v'_j)$ is in some unique open simplex σ_j of Y. Choose $\varphi'(v'_j)$ to be any vertex of this σ_j. Suppose τ' is an open simplex of Y' with vertices $\{v'_0, \ldots, v'_m\}$ and suppose σ is the simplex of Y with $\tau' \subset \varphi^{-1}(\sigma)$. Then $\varphi(\bar{\tau}') \subset \bar{\sigma}$. Thus, the $\{\varphi'(v'_j)\}$ are vertices of σ, hence span a face of σ, which shows that our vertex assignment extends to a simplicial map $\varphi'\colon Y' \to Y$. From the construction it is clear that φ' is canonically homotopic to φ, since for any $y' \in Y'$, both $\varphi(y')$ and $\varphi'(y')$ belong to the same closed simplex of Y.

The vertex choices made above are constructive because of the way Y' was obtained from $\varphi: Y \to Y$. Thus, φ' induces a constructively obtained chain map $t_*: C_*(Y') \to C_*(Y)$. There is also an algebraic subdivision map of chain complexes $s_*: C_*(Y) \to C_*(Y')$, since each simplex of Y is a union of simplices of Y'. The chain map s_* induces a homology isomorphism. Identifying $H_*(Y, \mathbb{Q}) = H_*(Y', \mathbb{Q})$ via s_*, we have $\mathrm{tr}(\varphi_*) = \mathrm{tr}(s_*t_*)$ and the Lefschetz number of $s_*t_*: C_*(Y') \to C_*(Y) \to C_*(Y')$ can certainly be constructively computed.

The above proof of Proposition 2 might seem to just be an adaptation of the classical proof of the Hopf fixed point theorem, using the existence of simultaneous triangulations of a finite collection of semi-algebraic sets, rather than barycentric subdivisions. If we just wanted to prove that $\tilde{\varphi}: \tilde{Y} \to \tilde{Y}$ had a fixed point whenever $\mathrm{tr}(\varphi_*) \neq 0$ then we could adapt the classical argument in this way, and it would not be necessary to appeal to the Tarski principle. (In fact, this is essentially what we did in the proof of the main theorem.) However, the example of $\varphi: I \to I$ discussed in the last paragraph of the introduction shows that there is not such a close connection between fixed points of $\tilde{\varphi}: \tilde{Y} \to \tilde{Y}$ and fixed points of $\varphi: Y \to Y$. Thus, something additional is needed to derive the strong conclusion of Proposition 2 that $\varphi: Y \to Y$ has a fixed point.

In the chain map $s_*t_*: C_*(Y') \to C_*(Y) \to C_*(Y')$ which we constructed in the course of the proof, we do not claim that all cells of Y' are "moved far away from themselves" if $\varphi: Y \to Y$ has no fixed point. (This would obviously imply $\mathrm{tr}(s_*t_*) = 0$.) Instead, we simply exploit the facts that $\mathrm{tr}(s_*t_*) = \mathrm{tr}(\varphi_*)$ and that computation of $\mathrm{tr}(s_*t_*)$ is constructive to validate the use of Tarski's principle. There is something a little mysterious about this.

The classical Brouwer fixed point theorem over K is much easier to prove than the Hopf theorem. That is, if $B^n \subset K^{n+1}$ is the standard unit disk and $\varphi: B^n \to B^n$ is any continuous semi-algebraic map, then φ has a fixed point in B^n. Namely, if not, one constructs a continuous semi-algebraic retraction $r: B^n \to S^{n-1} = \partial B^n$ in the usual way, by following rays from $\varphi(x) \in B^n$ through $x \in B^n$ until the rays hit ∂B^n. Thus, $r\Big|_{S^{n-1}} = Id$, and the existence of the homology functor is enough to get a contradiction, since $H_{n-1}(S^{n-1}) \neq 0$ and $H_{n-1}(B^n) = 0$, [3].

Curiously, it isn't clear whether this proof of the Brouwer theorem for B^n is adequate to deduce the Hopf theorem for K^n, or, more precisely, for $\mathbb{A}^n_K$. The proof of the Hopf theorem depends on a strong deformation retraction $K^n \to Y$ for some finite simplicial complex $Y \subset K^n$, which is, of course, contractible. What is not clear is whether Y can always be assumed to be semi-algebraically a closed disk. This seems to be closely related to the problem of semi-algebraic uniqueness of cone neighborhoods of points in semi-algebraic sets. But, in any case, Proposition 2 applies to Y.

References

1. E. Becker, "On the real spectrum of a ring and its application to semi-algebraic geometry", *Bull. Amer. Math. Soc.* 15 (1986) 19–60.

2. G. W. Brumfiel, "A semi-algebraic Brouwer fixed point theorem for real affine space", *Contemporary Mathematics* 74 (1988), Geometry of Group Representations, 77–82.

3. H. Delfs and M. Knebusch, "Homology of algebraic varieties over real closed fields", *Journal für die reine und angewandte Mathematik*, 335 (1982) 122–163.

4. N. Schwartz, "Real closed spaces", *Habilitationsschrift*, München, 1984.

On regular open semi-algebraic sets

G. W. Brumfiel*

Introduction. Let k be an ordered field with real closure K. If $f \in k[x_1 \cdots x_n]$, let $U(f) = \{\bar{x} \in K^n \mid f(\bar{x}) > 0\} \subset K^n$ Given a finite collection of polynomials $\{f_i\} \subset k[x_1 \cdots x_n]$, let $U\{f_i\} = \bigcap_i U(f_i) \subset K^n$. The Finiteness Theorem asserts that any semi-algebraic subset of K^n which is open (in the local sense that it contains a ball around each of its points) is, in fact, a *finite union* of sets of the form $U\{f_i\}$. [2], [5], [6], [7], [10].

A regular open set V in a topological space is an open set which satisfies $V = \overline{V}^\circ$, where bar indicates closure and circle indicates interior. For any open set V, $\overline{V}^\circ$ is regular. Also, finite intersections of regular open sets are regular.

In the semi-algebraic context of the first paragraph above, let $V(f) = \overline{U(f)}^\circ \subset K^n$ and $V\{f_i\} = \bigcap_i V(f_i) \subset K^n$. By the Tarski principle, the $V(f_i)$ are semi-algebraic sets. From the second paragraph above, they are also regular open sets. The goal of this note is to prove a regular version of the Finiteness Theorem. Specifically, we show that any open semi-algebraic set $U \subset K^n$ is a finite union of sets of the form $V\{f_i\}$.

In general, $U(f)$ need not be regular. For example, $U = U(\Sigma x_i^2) = K^n - (0)$ is not regular, since $\overline{U}^\circ = K^n$. Of course, $U(\Sigma x_i^2) = \bigcup_i U(\pm x_i)$ and the $U(\pm x_i) = V(\pm x_i)$ are regular. It is conceivable that some algorithmic proof of the standard Finiteness Theorem always produces regular $U(f_i)$ in the process of decomposing a given open U. The argument we give in this note, producing $V(f_i)$, is an existence argument only and does not clarify this fine point of whether one can get by with regular $U(f_i)$. We pose this as an open question.

Actually, we prove a relative form of the regular Finiteness Theorem. That is, we fix a semi-algebraic subset $X \subset K^n$. The sets $U(f)$ and $V(f)$ above, as well as closures, interiors, open sets, and regular open sets, now refer to the subspace topology on X. It is not true that a regular open set in K^n necessarily intersects X in a regular open set in X. For example, if $X = K^1 \subset K^2$ then $U = U(x^2 - y^2)$ is regular in K^2 but $U \cap X = K^1 - (0)$ is not regular in X.

The relative regular Finiteness Theorem requires one assumption on X. For example, if $X = [0, \infty) \subset K^1$ and if V is any set containing an open interval $(0, \epsilon) \subset X$, then $0 \in \overline{V}^\circ$ (relative to X). Thus, $U = (0, \infty) \subset X$ is not a finite union of regular open sets.

We will say that X has *boundary of codimension greater than 1* if no point of X has a neighborhood semi-algebraically homeomorphic to a neighborhood of a boundary point in a half-space (of any dimension). Note that algebraic sets have this property, for example, by Sullivan's even local Euler characteristic condition, [11]. Also, if X has boundary of codimension greater than 1 and if X' is obtained from X by deleting a semi-algebraic subset W with the property that for all $p \in \overline{W} - W$, $\dim(\overline{W}, p) < \dim(X, p)$, then $X' = X - W$

---

* Research supported by NSF grant DMS85-06816 at Stanford University

has boundary of codimension greater than 1. Here $\dim(\overline{W}, p)$ and $\dim(X, p)$ mean the local dimensions of $\overline{W}$ and X at p.

It is rather easy to see what the boundary condition above says if one makes use of the existence of "triangulations" of semi-algebraic sets X, [8]. This means that X is semi-algebraically homeomorphic to some finite union, Y, of open simplices (of any dimensions) of some closed, linear, finite simplicial complex in affine space over K. Boundary faces of a simplex of Y need not belong to Y. The condition "boundary codimension greater than 1" on X just says that the picture below *cannot* occur as an open subset of Y. The simplex σ must adhere to at least one other simplex $\mu \neq \tau$ with $\dim(\mu) > \dim(\sigma)$.

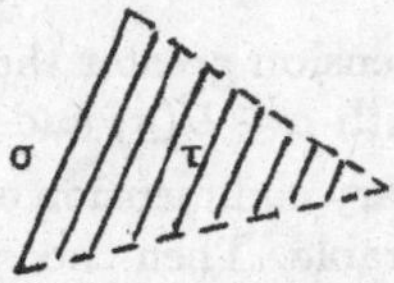

$$\dim(\tau) = \dim(\sigma) + 1$$

Triangulations Y of X can be chosen so that any preassigned finite family of semi-algebraic subsets of X correspond to subcomplexes of Y. We will use triangulations in the proof of the rational, relative, regular Finiteness Theorem, which is stated as follows.

Theorem: If the semi-algebraic set $X \subset K^n$ has boundary of codimension greater than 1, then every open semi-algebraic subset of X is a finite union of regular open sets of the form $V\{g_i\}$, $\quad g_i \in k[x_1 \cdots x_n]$.

Remark: This result was "conjectured", at least in a special case, in [3, page 233], and also at the end of [4]. I would like to thank J. M. Ruiz for some correspondence concerning the proof which follows.

Proof: The proof will use properties of the real spectrum, $\text{Spec}_R(K[x_1 \cdots x_n])$, which can be found in [1] and [9]. As far as the "rationality" aspect of the theorem is concerned, that is, finding polynomials with coefficients in k rather than K, the point is that $\text{Spec}_R(K[x_1 \cdots x_n])$ is homeomorphic, in both the constructible and Harrison topologies, to the subset of $\text{Spec}_R(k[x_1 \cdots x_n])$ which extends the given order on k.

Let $U \subset X$ be open and let $W = X - U$. By the Artin-Lang Theorem, X, U, and W correspond to constructible sets $\widetilde{X}, \widetilde{U}$, and $\widetilde{W} \subset \text{Spec}_R(K[x_1 \cdots x_n])$. By the Finiteness Theorem, $\widetilde{U}$ is open in $\widetilde{X}$. Now, $\widetilde{U}$ and $\widetilde{W}$ are compact in the constructible topology. Fix $\alpha \in \widetilde{U}$. We will first construct a certain constructible cover of $\widetilde{W}$ and extract a finite subcover. This will lead to a regular open neighborhood $V\{g_i\}$ of α, disjoint from $\widetilde{W}$, that is, contained in $\widetilde{U}$. Since these neighborhoods cover $\widetilde{U}$, which is compact, the Theorem follows. Here, if $g \in k[x_1 \cdots x_n]$, we abuse notation somewhat and also write $U(g)$ and $V(g)$ for the constructible subsets of $\text{Spec}_R(K[x_1 \cdots x_n])$ corresponding to $U(g)$ and $V(g) \subset K^n$. This is so that bars, circles, and tildes don't accumulate too much. The point is, the Artin-Lang correspondence preserves closures and interiors, by the Finiteness Theorem, so $V(g) = \overline{U(g)}^{\circ}$ holds in $\text{Spec}_R(K[x_1 \cdots x_n])$, as well as in K^n.

Consider $\beta \in \widetilde{W}$. Since $\widetilde{W}$ is closed, $\beta \not\to \alpha$, where $\to$ indicates specialization of points in the real spectrum. Thus, either $\alpha \to \beta$ or α and β are not comparable, that is, neither specializes the other. In the second case, there exists $g_\beta \in k[X_1 \dots X_n]$ such that

$g_\beta(\alpha) > 0$ and $g_\beta(\beta) < 0$. In particular, $\beta \in U(-g_\beta) \subset \overline{U(-g_\beta)}$. In the first case, that is, if $\alpha \to \beta$, then we claim that there exists $g_\beta \in k[x_1 \cdots x_n]$ such that again $g_\beta(\alpha) > 0$ and $\beta \in \overline{U(-g_\beta)}$. (See Lemma 4.7 of [4].)

Assume this claim for the moment. Then we have a cover of $\widetilde{W}$ by constructible sets $\overline{U(-g_\beta)}$ with $\alpha \in U(g_\beta) \subset V(g_\beta)$. Extract a finite subcover, say $\widetilde{W} \subset \cup \overline{U(-g_i)}$. Now, obviously, $\overline{U(-g_i)} \cap V(g_i) = \emptyset$, since $V(g_i) = \overline{U(g_i)}^{\circ}$ and $g_i \geq 0$ on $\overline{U(g_i)}$. Therefore, $\alpha \in \cap V(g_i) = V\{g_i\}$ is disjoint from $\widetilde{W}$, as desired.

It remains to prove the following separation lemma, which is an improvement of Lemma 4.7 of [4].

Lemma: If $X \subset K^n$ has boundary of codimension greater than 1 and if $\alpha \to \beta$ in $\widetilde{X}$ with $\alpha \neq \beta$, then there exists $g \in k[X_1 \dots X_n]$ with $\alpha \in U(g)$ and $\beta \in \overline{U(-g)}$.

Proof: We claim that because of the boundary codimension assumption, there exists $\gamma \in \widetilde{X}$ with $\gamma \to \beta$, but with α and γ *not* comparable. Then choose $g \in k[x_1 \dots x_n]$ with $g(\alpha) > 0$ and $g(\gamma) < 0$. Since $\gamma \in U(-g)$ and $\gamma \to \beta$, we have $\beta \in \bar{\gamma} \subset \overline{U(-g)}$.

To find γ, triangulate X so that $X \cap \text{Supp}\ (\alpha)$ and $X \cap \text{Supp}\ (\beta)$ are subcomplexes. Here, Supp indicates the irreducible varieties in K^n corresponding to points of the real spectrum. The triangulation decomposes $\widetilde{X}$ into disjoint "constructible simplices". Say $\beta \in \tilde{\sigma}$ and $\alpha \in \tilde{\tau}$. Certainly σ is a proper face of τ, since $\alpha \to \beta$ and $\dim(\tau) = \dim(\text{Supp}\ (\alpha)) > \dim(\text{Supp}\ (\beta)) = \dim(\sigma)$. Now, find a simplex $\mu \neq \tau$, with σ also a face of μ and with $\dim(\mu) = \dim(\tau)$. This may require retriangulating X slightly if the local picture is like one of the following.

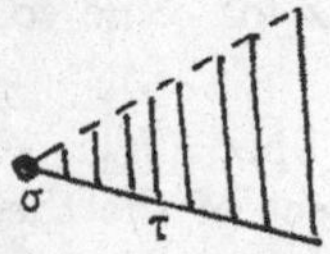

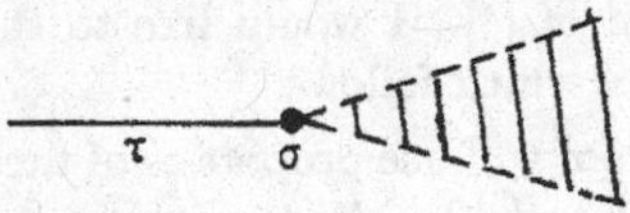

The first barycentric subdivision of a closed simplicial complex which contains X as a union of simplices will induce a suitable new triangulation of X. In any case, here is where we use the boundary codimension greater than 1 hypothesis.

It is now possible to find $\gamma \in \tilde{\mu}$ with $\gamma \to \beta$ and $\dim(\text{Supp}\ (\gamma)) = \dim(\mu) = \dim(\tau) = \dim(\text{Supp}\ (\alpha))$. In particular, α and γ are not comparable. The point is, semi-algebraically homeomorphic sets correspond to homeomorphic constructibles, so the structure of $\tilde{\sigma} \subset \tilde{\sigma} \cup \tilde{\mu}$ in $\widetilde{X}$ is exactly like the corresponding structure for a pair of linear affine simplices. The linear simplex case is the same locally as a pair of positive coordinate orthants, as below.

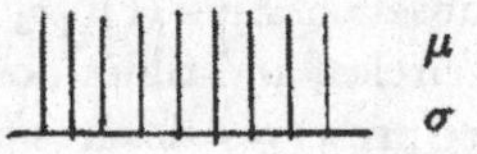

Then the existence of γ is just the observation that any order on a field L extends to $L(x)$, with x infinitesimally small and positive, relative to L.

References

1. **E. Becker**, On the real spectrum of a ring and its application to semi-algebraic geometry, *Bull. A.M.S.*, Vol. 15, No. 1, (1986), 19-60.
2. **J. Bochnak, G. Efroymson**, Real algebraic geometry and the 17th Hilbert problem, *Math. Annalen* 251 (1980), 213-242.
3. **G. Brumfiel**, *Partially ordered rings and semi-algebraic geometry*, Cambridge University Press, 1979.
4. **G. Brumfiel**, The ultrafilter theorem in real algebraic geometry, *Rocky Mountain J. Math.* 19 (1989), 611–628.
5. **M. Coste**, Ensembles semi-algébriques, Proc. Conf. on "Géométrie Algébrique Réele et Formes Quadratiqués", Rennes 1981, *Lecture Notes in Mathematics*, vol. 959, Springen (1982), 109-138.
6. **C. Delzell**, A finiteness Theorem for open semi-algebraic sets with applications to Hilbert's 17th problem, *Contemporary Math* 8 (1982), 79-97.
7. **L. v.d. Dries**, Some applications of a model theoretic fact to (semi-)algebraic geometry, *Indag. Math.* 44 (1982), 397-401.
8. **H. Hironaka**, Triangulation of algebraic sets, *Proc. Symp. in Pure Math.* 29 (A.M.S. 1975), 165-185.
9. **T.Y. Lam**, An introduction to real algebra, *Rocky Mountain J. Math* 14 (1984), 767-814.
10. **T. Recio**, Una decomposición de un conjunto semi-algebráico, Actas del V Congresso de la Agrupacion de Matemáticos de Expresión Latina, Madrid (1978).
11. **D. Sullivan**, Combinatorial invariants of analytic spaces, Proc. of Liverpool Singularities Symposium I, *Lecture Notes in Mathematics*, vol. 192, Springer 1871, 165-168.

SUMS OF 2n-th POWERS OF MEROMORPHIC FUNCTIONS WITH COMPACT ZERO SET

Ana Castilla

1. Introduction

Geometric characterizations of sums of $2n$−th powers by means of analytic arcs have been thoroughly studied in different contexts: for polynomials, [Pr], for meromorphic functions on a compact analytic surface, [Kz], and finally for meromorphic germs and meromorphic functions on compact analytic manifolds of arbitrary dimension, [Rz2]. In all the situations they are based on the so called Becker's valuative criterion, [Be]:

Theorem. *Let K be a field and $f \in K$. Then f is a sum of $2n$−th powers of elements of K if and only if f is a sum of squares in K and for any real valuation v of K, we have $2n|v(f)$.*

The aim of this short note is to give a geometric condition for certain meromorphic functions on an arbitrary paracompact analytic manifold to be sum of $2n$−th powers, and which generalizes the resull of [Rz]. To be more precise, let M be a connected paracompact real analytic manifold, let $\mathcal{O}(M)$ be the ring of analytic functions on M and let K be the field of meromorphic functions on M, that is, the field of fractions of $\mathcal{O}(M)$. Then we show

Theorem 1. *Let $f \in \mathcal{O}(M)$ be such that its zero set is compact. Then f is a sum of $2n$−th powers in K if and only if for every analytic curve $c : [-\varepsilon, \varepsilon] \to M$ such that $f(c(t)) = at^m + \cdots$, with $a \neq 0$, it holds $a > 0$ and $2n|m$.*

Our proof is based on Becker's criterion, Ruiz's result for the local case, as well as the attachment to each ordering $\beta \in Spec_r(K)$ of an ultrafilter $\mathcal{U}_\beta$ of closed semianalytic subsets of M, which allows to give some geometric criteria for a function f to be positive in β. This method was used in [An-Be] and is similar to the one exposed by P. Jaworski in his contribution to this volume. In fact the results he presents were also independently obtained by us, as he kindly points out. Therefore we omit the background concerning the attachment of $\mathcal{U}_\beta$, which can be seen in either of these references, and will appear in my dissertation, [Ca].

2. Proof of the result.

Let $\mathcal{O}_b(M)$ denote the ring of bounded analytic functions on M. Let β be an order of K, and let W_β be the convex hull of $\mathbf{R}$ in K. Then, W_β is a valuation ring containing $\mathcal{O}_b(M)$ and with residue field $\mathbf{R}$. Let $\mathfrak{m}_\beta$ denote its maximal ideal, and let $\mathcal{S}$ be the family of closed global semianalytic subsets of M. Thus the set

$$\mathcal{U}_\beta = \{X \in \mathcal{S} | X \cap f^{-1}([-1,1]) \neq \emptyset \text{ for all } f \in \mathfrak{m}_\beta \cap \mathcal{O}_b(M)\}.$$

is an ultrafilter among the filters of sets of the family $\mathcal{S}$, cf. [Jw, Prop 1]. The main feature we are going to use is the following

Lemma 1. *Let $f \in \mathcal{O}_b(M)$ and assume that there is $X \in \mathcal{U}_\beta$ such that $|f|$ is bounded from below over X. Then f is a unit in W_β.*

Proof: [Jw].

Before entering in the proof of Theorem 1, we need a technical lemma.

Lemma 2. *Let $X \subset M$ be a closed global semianalytic set, and let $f \in \mathcal{O}_b(M)$ be such that $f_{|X} > 0$. Then for each $n \in \mathbf{N}$ there are $u, g \in \mathcal{O}_b(M)$ such that $u_{|X} > 0$, u is bounded from below over X and $uf = g^n$.*

Proof: Let $v^* : M \to \mathbf{R}$ be a continuous function such that $v^*(x) > 0$ for all $x \in M$, and $v^*_{|X} = f^{n-1}$, which exists by Tietze's Theorem. Now let $v : M \to \mathbf{R}$ be an analytic approximation of v^* such that $|v(x) - v^*(x)| < \frac{1}{4}v^*(x)$. We set $u_1 = f^{n-1}/v$ and $u = u_1/(1+u_1^2) \in \mathcal{O}_b(M)$. Since for $x \in X$ we have that $u_1(x) > 4f^{n-1}(x)/5v^*(x) = 4/5$, we get that u is bounded from below on X. Now, $uf = f^n\,(1/v(1+u_1^2))$, and since $v(1+u_1^2)$ is strictly positive, we have that $v(1+u_1^2) = h^n$ for some $h \in \mathcal{O}_b(M)$. Altogether we get $uf = g^n$, where $g = f/h$.

Proof of Theorem 1:

Necessary condition. If f is a sum of $2n$-powers, obviously f is nonnegative on M, so that the only thing to be shown is the condition on the divisibility of the exponent m by $2n$. Take any representation of f as sum of $2n$-th powers:

$$f = \sum g_i^{2n}/h^{2n}, \text{ with } g_i, h \in \mathcal{O}(M), h \neq 0,$$

and let $c : [0, \varepsilon] \to M$ be an analytic curve. We set $x = c(0)$. If $h(c(t)) \neq 0$, since $h^{2n}(c(t))f(c(t)) = \sum g_i(c(t))^{2n}$ then

$$2n \operatorname{ord}(h(c(t))) + \operatorname{ord}(f(c(t))) = 2n \operatorname{ord} \sum g_i(c(t))$$

and easily we get the result. In general, we need to use suitable valuations of the field K associated to the point x.

Let $\mathfrak{m}$ be the maximal ideal corresponding to x. The curve c defines a sequence

$$\mathcal{O}(M)_{\mathfrak{m}} \hookrightarrow \mathcal{O}_x(M) \xrightarrow{\phi} \mathbf{R}\{t\},$$

where $\phi(f)$ is the expansion at 0 of $f \circ c$ and $\mathcal{O}_x(M)$ denotes the ring of analytic germs at the point x.

Let $\mathfrak{q} = (ker\phi) \cap \mathcal{O}(M)_{\mathfrak{m}} = ker(\phi \circ i)$. Thus, the homomorphism ϕ induces an embedding

$$k(\mathfrak{q}) \hookrightarrow \mathbf{R}(\{t\}),$$

where $k(\mathfrak{q})$ is the quotient field of $\mathcal{O}(M)_{\mathfrak{m}}/\mathfrak{q}$ and $\mathbf{R}(\{t\})$ denotes the quotient field of $\mathbf{R}\{t\}$.We restrict the ordinary valuation of $\mathbf{R}(\{t\})$ to a valuation $\bar{v}$ of $k(\mathfrak{q})$. In particular the class $\bar{f}$ in $k(\mathfrak{q})$ verifies $\bar{v}(\bar{f}) = \text{ord}(f \circ c) = m$.

Since $\mathcal{O}(M)_{\mathfrak{q}}$ is regular, there is a valuation w of K with residue field $k(\mathfrak{q})$. Finally let v be the composite valuation of $\bar{v}$ and w. We have $\Gamma_w = \Gamma_v/\Gamma_{\bar{v}}$. Since f is a sum of $2n$-th powers in K, by Becker's criterion $2n$ divides both $v(f)$ and $w(f)$, and therefore $2n$ divides $\bar{v}(\bar{f}) = m = \text{ord}(f(c(t)))$.

Sufficient condition:

Assume that $f \notin \sum K^{2n}$. If $f(x) < 0$ for some $x \in M$, then for any curve $c : [0, \varepsilon] \to M$ with $c(0) = x$, we have $f(c(t)) = at^m + \cdots$ with $a < 0$. Therefore we may assume that f is nonnegative over M. Then, by Artin-Lang, see [Jw], [Ca], f is a sum of squares. In particular, it follows from Becker's criterion, that it exists a real valuation w of K such that $2n$ does not divide $w(f)$. Let β be an ordering of K compatible with w, and let $V = W_\beta(\mathbf{R}, K)$ be the convex hull of $\mathbf{R}$ in K with respect to β. Thus, V is contained in the valuation ring of w. In particular $\Gamma_w = \Gamma_v/\Delta$ where v is the valuation defined in V and Δ is an isolated subgroup of Γ_v. Therefore, $2n$ can not divide $v(f)$. Now, let $\mathfrak{m}_\beta$ be the maximal ideal of the ring V and let $\mathcal{U}_\beta$ be the ultrafilter of global closed semianalytic sets defined by

$$\mathcal{U}_\beta = \{X \in \mathcal{S} | X \cap f^{-1}([-1,1]) \neq \emptyset \text{ for all } f \in \mathfrak{m}_\beta \cap \mathcal{O}_b(M)\}.$$

If $f^{-1}(0) \notin \mathcal{U}_\beta$, there exists $X \in \mathcal{U}_\beta$ such that $f_{|X} > 0$. Then we can apply Lemma 2 to the function f: there exist a unit u in V, and $g \in \mathcal{O}_b(M)$ such that $uf = g^{2n}$. In particular $2n$ divides $v(f)$, contradiction.

Next, assume that $f^{-1}(0) \in \mathcal{U}_\beta$. Since $f^{-1}(0)$ is compact, necessarily $\mathcal{U}_\beta$ is the principal ultrafilter defined by a point, say $p \in M$. In particular β is centered at p, that is, the maximal ideal $\mathfrak{m}_p$ of p is convex with respect to β.

Now, we have

$$(\mathcal{O}_b(M))_{\mathfrak{m}_p} = \mathcal{O}(M)_{\mathfrak{m}_p} \hookrightarrow \mathcal{O}_p(M) \simeq \mathbf{R}\{x_1, \ldots, x_n\}$$

Since, V dominates $\mathcal{O}(M)_{\mathfrak{m}_p}$ we can argue as in [Rz] to find and analytic curve germ $c:[0,\varepsilon]\to M$ with $c(0)=p$ and such that $f(c(t))=at^m+\cdots$, where $2n$ does not divide m.

3. References.

[An-Be] C. Andradas, E. Becker: "A note on the Real Spectrum of Analytic functions on an Analytic Manifold of dimension one". Proceedings of the Conference on Real Analytic and Algebraic Geometry (Trento 1988), Lect. Notes in Math. no. 1420, 1-21, Berlin-Heidelberg-New York, Springer, 1990.

[Be] E. Becker: "The real holomorphic ring and sums of $2n$−th powers", in Lect. Notes Math. 959, Berlin-Heidelberg-New York, Springer, 1982.

[Ca] A. Castilla: Dissertation, in preparation.

[Jw] P. Jaworski: "The 17−th Hilbert problem for noncompact real analytic manifolds", this volume.

[Kz] W. Kucharz: "Sums of $2n$−th powers of real meromorphic functions", Monatshefte für Mathematik. 107 (1989) 131-336.

[Pr] A. Prestel: "Model theory applied to some questions about polynomials", Proceedings of the Salzburg Conference, May 29 - June 1, 1986.

[Rz1] J. Ruiz: "On Hilbert's 17−th problem and real Nullstellensatz for global analytic functions", Math. Z. 190, 499-514 (1985).

[Rz2] J. Ruiz: "A characterizatization of sums of $2n$−th powers of global meromorphic functions", Proceedings of the A.M.S. 109, 915-923 (1990).

PSEUDOORTHOGONALITY OF POWERS OF THE COORDINATES OF A HOLOMORPHIC MAPPING IN TWO VARIABLES WITH THE CONSTANT JACOBIAN

Zygmunt Charzyński and Przemysław Skibiński

Summary. In the paper the equivalence of the constancy of Jacobian of holomorphic mapping in two variables with the pseudoorthogonality of powers of its coordinates is showed. There is also obtained a characterization of holomorphic mapping with the constant Jacobian by one coordinate and the restriction of the other coordinate of the mapping to one of the axes.

Introduction. This note is devoted to holomorphic mappings with the constant Jacobian. In section 1 there is given an auxiliary information. In section 2 it is showed that the constancy of the Jacobian is equivalent to the integral pseudoorthogonality of powers of the coordinates of the mapping (theorem 1). Also the formulae for the coefficients of the inverse mapping are derived (theorem 2). In section 3 more general expressions than those representing the mentioned coefficients are investigated (lemmas 1, 2 and theorem 3). In section 4 a simplification of the formula from the previous section is obtained, namely, some integrals of functions of two variables are reduced to some integrals of functions of one variable, in particular, those representing the coefficients of the inverse mapping (theorem 4). These facts are applicated in the next section. Here a characterization of the mappings under consideration by one of its coordinates and by the values of the second one on one of the axes is obtained (theorem 5).

1. Terms, notations and auxiliary information. 1. Let $0 < \alpha < 1$ be a number fixed for the sequel and $r > 0$ a number which will be later precised. Let further

$$(1.1) \quad E_r = \{(x,y) \in \mathbb{C}^2 : 0 < |x| < r, \quad 0 < |y| < r, \quad \alpha < |\frac{x}{y}| < \frac{1}{\alpha}\}.$$

The set (1.1), as one can easily see, is a Reinhardt domain (comp. [2], p. 299).

Let be given the pair of plain circles

$$(1.2)\qquad \begin{aligned} C &: x(s) = \xi e^{is}, \quad 0 \le s \le 2\pi, \quad \xi > 0, \\ D &: y(t) = \eta e^{it}, \quad 0 \le t \le 2\pi, \quad \eta > 0. \end{aligned}$$

This pair, treated as a mapping in two variables s,t, and its range is called the bicircle with the radii ξ,η. We shall assume that $(\xi,\eta) \in E_r$ i.e. the bicircle (1.2) is lying in E_r.

2. Let (P,Q) be a holomorphic mapping in a neighbourhood of the point $(0,0)$ in the variables x,y with the coordinates of the form

$$(1.3)\qquad P(x,y) = x + P_2(x,y) + \ldots, \qquad Q(x,y) = y + Q_2(x,y) + \ldots$$

where P_j and Q_j denote homogeneous polynomials of the degree j.

It is easily seen that from form (1.3) of the mapping (P,Q) it follows that the Jacobian of this mapping at the point $(0,0)$ is equal to 1. Thus, in a neighbourhood of the point $(0,0)$ of the variables x,y, the mapping is invertible. Denote by (U,V) the holomorphic inverse mapping. Then in a neighbourhood of the point $(0,0)$ of the variables z,w the coordinates of this mapping expand in the series

$$(1.4)\qquad U(z,w) = \sum_{k,l=0}^{\infty} c_{kl} z^k w^l, \qquad V(z,w) = \sum_{k,l=0}^{\infty} d_{kl} z^k w^l.$$

Simultaneously, in a certain neighbourhood of the point $(0,0)$ in the variables x,y, the values of $(P(x,y),Q(x,y))$ are contained in the lately mentioned neighbourhood of the variables z,w.

Of course, in this situation, there is

$$(1.5)\qquad \begin{aligned} x &= \sum_{k,l=0}^{\infty} c_{kl}(P(x,y))^k(Q(x,y))^l, \\ y &= \sum_{k,l=0}^{\infty} d_{kl}(P(x,y))^k(Q(x,y))^l. \end{aligned}$$

Now, let us notice that for r sufficiently small the set (1.1) is contained in the neighbourhood of the defintion of (P,Q).

Similarly, for sufficiently small r the following inequalities hold

$$(1.6)\qquad \left|\sum_{j=2}^{\infty} \frac{P_j(x,y)}{x}\right| < 1, \qquad \left|\sum_{j=2}^{\infty} \frac{Q_j(x,y)}{y}\right| < 1;$$

at the same time we have

$$\frac{1}{P(x,y)} = \frac{1}{x(1+\sum_{j=2}^{\infty}\frac{P_j(x,y)}{x})} = \frac{1}{x}\sum_{g=0}^{\infty}(-\sum_{j=2}^{\infty}\frac{P_j(x,y)}{x})^g$$

(1.7)

$$\frac{1}{Q(x,y)} = \frac{1}{y(1+\sum_{j=2}^{\infty}\frac{Q_j(x,y)}{y})} = \frac{1}{y}\sum_{g=0}^{\infty}(-\sum_{j=2}^{\infty}\frac{Q_j(x,y)}{y})^g$$

and, in consequence, the functions $1/P(x,y)$ and $1/Q(x,y)$ are holomorphic in E_r and expand there in Laurent series centred at $(0,0)$ (comp. [2], p. 301). Moreover, it is easily seen from (1.7) that in the first of the series in (1.7) the variable y appears in non-negative powers only and in the second series in (1.7) the variable x appears in non-negative powers only; in both series there appear the terms of degree not smaller than -1 only.

3. Simultaneously in (1.1) there exist some single-valued branches of logarithms

$$\log\frac{P(x,y)}{x} = \sum_{g=1}^{\infty}\frac{(-1)^{g-1}}{g}(\sum_{j=2}^{\infty}\frac{P_j(x,y)}{x})^g,$$

(1.8)

$$\log\frac{Q(x,y)}{y} = \sum_{g=1}^{\infty}\frac{(-1)^{g-1}}{g}(\sum_{j=2}^{\infty}\frac{Q_j(x,y)}{y})^g.$$

Similarly, as above, the branches (1.8) expand in Laurent series; in the first one the variable y appears in non-negative powers only and in the second one the variable x appears in non-negative powers only; in both series there appear the terms of degree not smaller than 1 only.

We easily check that the functions

$$(\log\xi + is) + \log\frac{P(\xi e^{is},\eta e^{it})}{\xi e^{it}},$$

(1.9) $\qquad 0 \le s \le 2\pi,\quad 0 \le t \le 2\pi,$

$$(\log\eta + it) + \log\frac{Q(\xi e^{is},\eta e^{it})}{\eta e^{it}},$$

are some branches of the logarithm $P(x,y)$ and $Q(x,y)$ "along" bicircle (1.2). We shall denote them by $\log P(x,y)$ and $\log Q(x,y)$.

We see that if one of the variables x and y is fixed, then $\log P(x,y)$ is a branch of the logarithm of the function $P(x,y)$ along the circle C or D and its increment along this circle is equal to $2\pi i$ or 0. Analogously for the branch of logarithm of the function $Q(x,y)$.

4. By W_T we shall denote the linear Whitney operator (comp. [1]) assigning to any holomorphic function S in a neighbourhood of the point $(0,0)$ the expression $S_x T_y - S_y T_x$. Here T is a holomorphic function in this neighbourhood, fixed for a moment. The iterations of Whitney's operator we shall denote as powers. In particular, for the operator W_Q where Q is the coordinate of the considered mapping (P,Q) we shall apply the following notations

$$W_Q^j(S) = S_j, \qquad j = 0,1,\ldots, \qquad S_o = S. \tag{1.10}$$

In the sequel as r we choose an arbitrary fixed number so small that in the set E_r relation (1.6) and representations (1.5), (1.7), (1.8) and the corresponding Laurent expansions hold as well as the set E_r is contained in all neighbourhoods of the variables x,y mentioned in the points 2, 3 and 4.

2. Pseudoorthogonality. Let us adopt the constructions, notations and agreements introduced in the previous section.

THEOREM 1. For the functions (1.3) the following Jacobian condition

$$P_x Q_y - P_y Q_x = 1 \tag{2.1}$$

holds if and only if for every integers k and l the following pseudoorthogonality relations hold

$$\frac{1}{(2\pi i)^2} \int_C \left(\int_D \frac{1}{(P(x,y))^{k+1}(Q(x,y))^{l+1}}\, dy\right) dx = \begin{cases} 0 & \text{for } (k,l) \neq (0,0), \\ 1 & \text{for } (k,l) = (0,0). \end{cases} \tag{2.2}$$

P r o o f. First, let us notice that the following equalities hold

$$\frac{1}{(2\pi i)^2} \int_C \left(\int_D \frac{P_x(x,y)Q_y(x,y) - P_y(x,y)Q_x(x,y)}{(P(x,y))^{k+1}(Q(x,y))^{l+1}}\, dy\right) dx \tag{2.3}$$

$$= \begin{cases} 0 & \text{for } (k,l) \neq (0,0), \\ 1 & \text{for } (k,l) = (0,0). \end{cases}$$

Indeed, assume that $(k,l) \neq (0,0)$. Consider the case $k \neq 0$. Passing in the first of the integrals in (2.3) to common integrals with respect to the parameters s and t by the substitutions (1.2) and applying the integration by parts we have

$$(2.4) \qquad \int_D \Big(\int_C \frac{P_x(x,y)}{(P(x,y))^{k+1}} \frac{Q_y(x,y)}{(Q(x,y))^{l+1}} dx \Big) dy$$

$$= \frac{1}{k} \int_D \Big(\int_C \frac{1}{(P(x,y))^k} \Big(\frac{Q_{yx}(x,y)}{(Q(x,y))^{l+1}} - \frac{(l+1)Q_x(x,y)Q_y(x,y)}{(Q(x,y))^{l+2}} \Big) dx \Big) dy$$

and

$$(2.5) \qquad \int_C \Big(\int_D \frac{P_y(x,y)}{(P(x,y))^{k+1}} \frac{Q_x(x,y)}{(Q(x,y))^{l+1}} dy \Big) dx$$

$$= \frac{1}{k} \int_C \Big(\int_D \frac{1}{(P(x,y))^k} \Big(\frac{Q_{xy}(x,y)}{(Q(x,y))^{l+1}} - \frac{(l+1)Q_x(x,y)Q_y(x,y)}{(Q(x,y))^{l+2}} \Big) dy \Big) dx.$$

Subtracting (2.4) and (2.5) and changing the order of the integration in (2.4) we obtain the first equality of (2.3).

In the case $l \neq 0$, we proceed analogously.

Assume that $(k,l) = (0,0)$. Applying again the integration by parts and taking into account the remark concerning the existence of a branch of logarithm (comp. section 1.3), we have

$$(2.6) \qquad \int_D \Big(\int_C \frac{P_x(x,y)}{P(x,y)} \frac{Q_y(x,y)}{Q(x,y)} dx \Big) dy = \int_D \Big(\log P(\xi e^{i2\pi},y)$$

$$- \log P(\xi e^{i0},y) \Big) \frac{Q_y(\xi,y)}{Q(\xi,y)} dy$$

$$- \int_D \Big(\int_C \log P(x,y) \frac{Q_{yx}(x,y)Q(x,y) - Q_y(x,y)Q_x(x,y)}{(Q(x,y))^2} dx \Big) dy,$$

$$(2.7) \qquad \int_C \Big(\int_D \frac{P_y(x,y)}{P(x,y)} \frac{Q_x(x,y)}{Q(x,y)} dy \Big) dx = \int_C \Big(\log P(x,\eta e^{i2\pi})$$

$$- \log P(x,\eta e^{i0}) \Big) \frac{Q_x(x,\eta)}{Q(x,\eta)} dx$$

$$-\int_C\left(\int_D \log P(x,y)\,\frac{Q_{xy}(x,y)Q(x,y)-Q_x(x,y)Q_y(x,y)}{(Q(x,y))^2}\,dy\right)dx.$$

At the same time we have

$$\int_D \frac{Q_y(\xi,y)}{Q(\xi,y)}\,dy = \log Q(\xi,\eta e^{i2\pi}) - \log Q(\xi,\eta e^{i0}). \tag{2.8}$$

Subtracting (2.6) and (2.7), changing the order of integration in (2.6) and taking into account (2.8) as well as remarks concerning the increment of the logarithm of the function P or Q (comp. section 1.3), we obtain the second equality from (2.3).

The necessity of relation (2.2) is immediately obtained, substituting, according to (2.1), the numerator of the integrand in (2.3) by 1.

The sufficiency follows from the following observations.

First, let us notice that according to the choice of r (comp. section 1.4) in E_r the representations (1.5) hold. Consequently, the Jacobian of the mapping can also be represented in the form of the following series

$$P_x(x,y)Q_y(x,y) - P_y(x,y)Q_x(x,y) = \sum_{h,j=0}^{\infty} e_{hj}(P(x,y))^h(Q(x,y))^j. \tag{2.9}$$

Next, by the assumption that the bicircle (1.2) is contained in E_r and by (2.9) we get

$$\int_C\left(\int_D \frac{P_x(x,y)Q_y(x,y)-P_y(x,y)Q_x(x,y)}{(P(x,y))^{k+1}(Q(x,y))^{l+1}}\,dy\right)dx \tag{2.10}$$

$$= \sum_{h,j=0}^{\infty} e_{hj}\int_C\left(\int_D \frac{1}{(P(x,y))^{k+1-h}(Q(x,y))^{l+1-j}}\,dy\right)dx.$$

Since, according to (2.2), the integrals on the right-hand side of (2.10) vanish for all $(h,j) \neq (k,l)$, then by (2.10) and (2.3) we have

$$e_{kl} = \begin{cases} 0 & \text{for } (k,l) \neq (0,0), \\ 1 & \text{for } (k,l) = (0,0). \end{cases}$$

The above and (2.9) give (2.1).

THEOREM 2. For the mapping (P,Q) fulfilling condition (2.1), the coefficients c_{kl}, d_{kl} (comp. (1.4)) of the inverse mapping (U,V) are given by the formulae

$$c_{kl} = \frac{1}{(2\pi i)^2} \int_C \Big(\int_D \frac{x}{(P(x,y))^{k+1}(Q(x,y))^{l+1}} dy\Big) dx,$$

(2.11)

$$d_{kl} = \frac{1}{(2\pi i)^2} \int_C \Big(\int_D \frac{y}{(P(x,y))^{k+1}(Q(x,y))^{l+1}} dy\Big) dx.$$

P r o o f. From the definition of (U,V) (comp. section (1.2) and from (1.5) we obtain

$$x = \sum_{k,l=0}^{\infty} c_{kl}(P(x,y))^k (Q(x,y))^l,$$

$$y = \sum_{k,l=0}^{\infty} d_{kl}(P(x,y))^k (Q(x,y))^l,$$

whence, following the analogous pattern as in the proof of the preceding theorem and by this theorem, we obtain (2.11).

3. Generalization and reduction of degree. Let us adopt the constructions, notations and agreements introduced in section 1.

Here we shall assume that the functions (1.3) fulfil condition (2.1).

LEMMA 1. Let k be a positive integer, l - an integer, M - a holomorphic function in a neighbourhood of the point $(0,0)$ containing E_r. Then

$$\text{(3.1)} \qquad \frac{1}{(2\pi i)^2} \int_C \Big(\int_D \frac{M(x,y)}{(P(x,y))^{k+1}(Q(x,y))^{l+1}} dy\Big) dx$$

$$= \frac{1}{(2\pi i)^2} \frac{1}{k!} \int_C \Big(\int_D \frac{1}{P(x,y)} \frac{M_k(x,y)}{(Q(x,y))^{l+1}} dy\Big) dx.$$

P r o o f. Applying the integration by parts we have

$$\text{(3.2)} \qquad \int_D \Big(\int_C \frac{P_x(x,y)}{(P(x,y))^{k+1}} \frac{Q_y(x,y)M(x,y)}{(Q(x,y))^{l+1}} dx\Big) dy$$

$$= \frac{1}{k} \int_D \Big(\int_C \frac{1}{(P(x,y))^k} \Big(\frac{Q_{yx}(x,y)M(x,y) + Q_y(x,y)M_x(x,y)}{(Q(x,y))^{l+1}}$$

$$- \frac{(l+1)Q_y(x,y)Q_x(x,y)M(x,y)}{(Q(x,y))^{l+2}})dx)dy,$$

(3.3)
$$\int_C(\int_D \frac{P_y(x,y)}{(P(x,y))^{k+1}} \frac{Q_x(x,y)M(x,y)}{(Q(x,y))^{l+1}} dy)dx$$

$$= \frac{1}{k} \int_D(\int_C \frac{1}{(P(x,y))^k} (\frac{Q_{xy}(x,y)M(x,y) + Q_x(x,y)M_y(x,y)}{(Q(x,y))^{l+1}}$$

$$- \frac{(l+1)Q_x(x,y)Q_y(x,y)M(x,y)}{(Q(x,y))^{l+2}})dx)dy.$$

Now, let us denote by $I_{kl}(M)$ the integral on the left-hand side of (3.1). Subtracting (3.2) and (3.3) and taking account of (2.1) and (1.10), we get

$$I_{kl}(M_o) = \frac{-1}{k} \int_C(\int_D \frac{1}{(P(x,y))^k} \frac{M_1(x,y)}{(Q(x,y))^{l+1}} dy)dx$$

$$= \frac{1}{k} I_{k-1\ l}(M_1).$$

Hence, by induction, we obtain (3.1).

LEMMA 2. Let l be an integer, N - a holomorphic function in a neighbourhood of the point $(0,0)$ containing E_r. Then for every positive integer ν the following equality holds.

(3.4)
$$\frac{1}{(2\pi i)^2} \int_C(\int_D \frac{1}{P(x,y)} \frac{N(x,y)}{(Q(x,y))^{l+1}} dy)dx$$

$$= \frac{1}{2\pi i} \sum_{h=1}^{\nu} \frac{(-1)^{h-1}}{(h-1)!} \int_D (P(\xi,y))^{h-1}$$

$$\times \frac{Q_y(\xi,y)N_{h-1}(\xi,y)}{(Q(\xi,y))^{l+1}} dy + \frac{1}{(2\pi i)^2} \frac{(-1)^\nu}{(\nu-1)!} \int_C(\int_D (P(x,y))^{\nu-1}$$

$$\times [\log P(x,y) - (1+\frac{1}{2}+\dots+\frac{1}{\nu-1})] \frac{N_\nu(x,y)}{(Q(x,y))^{l+1}} dy)dx$$

where we assume that $0! = 1$ and for $\nu = 1$ the expression in the square brackets on the right-hand side of (3.4) reduces to the first summand (for N_ν compare also sec. 1.4 (1.10)).

P r o o f. We shall do it by induction

1° Integrating by parts and taking into account the remarks concerning the branch of logarithm (comp. section 1.3), we obtain

$$(3.5)\qquad \int_D\Big(\int_C \frac{P_x(x,y)}{P(x,y)}\,\frac{Q_y(x,y)N(x,y)}{(Q(x,y))^{l+1}}\,dx\Big)dy$$

$$= \int_D \big(\log P(\xi e^{i2\pi},y) - \log P(\xi e^{i0},y)\big)\,\frac{Q_y(\xi,y)N(\xi,y)}{(Q(\xi,y))^{l+1}}\,dy$$

$$- \int_D\Big(\int_C (\log P(x,y))\Big(\frac{Q_{yx}(x,y)N(x,y) + Q_y(x,y)N_x(x,y)}{(Q(x,y))^{l+1}}$$

$$- \frac{(l + 1)Q_y(x,y)Q_x(x,y)N(x,y)}{(Q(x,y))^{l+2}}\,dx\Big)dy,$$

$$(3.6)\qquad \int_C\Big(\int_D \frac{P_y(x,y)}{P(x,y)}\,\frac{Q_x(x,y)N(x,y)}{(Q(x,y))^{l+1}}\,dx\Big)dy$$

$$= \int_C \big(\log P(x,\eta e^{i2\pi}) - \log P(x,\eta e^{i0})\big)\,\frac{Q_x(x,\eta)N(x,\eta)}{(Q(x,\eta))^{l+1}}\,dx$$

$$- \int_C\Big(\int_D (\log P(x,y)\Big(\frac{Q_{xy}(x,y)N(x,y) + Q_x(x,y)N_y(x,y)}{(Q(x,y))^{l+2}}$$

$$- \frac{(l + 1)Q_x(x,y)Q_y(x,y)N(x,y)}{(Q(x,y))^{l+2}}\,dy\Big)dx.$$

Subtracting (3.5) and (3.6) and taking into account (2.1) and the remarks concerning the increment of the branch of the logarithm of P (comp. section 1.3) as well as (1.10), we get

$$(3.7)\qquad \int_C\Big(\int_D \frac{1}{P(x,y)}\,\frac{N(x,y)}{(Q(x,y))^{l+1}}\,dy\Big)dx = 2\pi i \int_D \frac{Q_y(\xi,y)N(\xi,y)}{(Q(\xi,y))^{l+1}}\,dy$$

$$- \int_C\Big(\int_D \log P(x,y)\,\frac{N_1(x,y)}{(Q(x,y))^{l+2}}\,dy\Big)dx,$$

which gives (3.4) for $\nu = 1$.

2° Assume that (3.4) hold for ν. Applying the integration by parts to the second integral on the right-hand side of (3.4), we obtain

$$(3.8)\quad \int_D\Big(\int_C (P(x,y))^{\nu-1}[\log P(x,y) - (1 + \tfrac{1}{2} + \ldots + \tfrac{1}{\nu-1})]\, P_x(x,y)$$

$$\times \frac{Q_y(x,y)N_\nu(x,y)}{(Q(x,y))^{l+1}}\,dy\Big)dx = \int_D [\tfrac{1}{\nu}(P(\xi e^{i2\pi},y))^\nu \log P(\xi e^{i2\pi},y)$$

$$- \frac{1}{\nu^2}(P(\xi e^{i2\pi},y))^\nu - (1 + \tfrac{1}{2} + \ldots + \tfrac{1}{\nu-1})\tfrac{1}{\nu}(P(\xi e^{i2\pi},y))^\nu$$

$$- \frac{1}{\nu}(P(\xi e^{i0},y))^\nu \log P(\xi e^{i0},y) + \frac{1}{\nu^2}(P(\xi e^{i0},y))^\nu$$

$$+ (1 + \tfrac{1}{2} + \ldots + \tfrac{1}{\nu-1})\tfrac{1}{\nu}(P(\xi e^{i0},y))^\nu]$$

$$\times \frac{Q_y(\xi,y)N_\nu(\xi,y)}{(Q(\xi,y))^{l+1}}\,dy - \int_D\Big(\int_C [\tfrac{1}{\nu}(P(x,y))^\nu \log P(x,y)$$

$$- \frac{1}{\nu^2}(P(x,y))^\nu - (1 + \tfrac{1}{2} + \ldots + \tfrac{1}{\nu-1})\tfrac{1}{\nu}(P(x,y))^\nu]$$

$$\times \Big(\frac{Q_{yx}(x,y)N_\nu(x,y) + Q_y(x,y)N_{\nu x}(x,y)}{(Q(x,y))^{l+1}}$$

$$- \frac{(l+1)Q_y(x,y)Q_x(x,y)N_\nu(x,y)}{(Q(x,y))^{l+2}}\Big)dx\Big)dy,$$

$$(3.9)\quad \int_C\Big(\int_D (P(x,y))^{\nu-1}[\log P(x,y) - (1 + \tfrac{1}{2} + \ldots + \tfrac{1}{\nu-1}]$$

$$\times P_y(x,y)\,\frac{Q_x(x,y)N_\nu(x,y)}{(Q(x,y))^{l+1}}\,dy\Big)dx$$

$$= \int_C [\tfrac{1}{\nu}(P(x,\eta e^{i2\pi}))^\nu \log P(x,\eta e^{i2\pi}) - \frac{1}{\nu^2}(P(x,\eta e^{i2\pi}))^\nu$$

$$- (1 + \tfrac{1}{2} + \ldots + \tfrac{1}{\nu-1})\tfrac{1}{\nu}(P(x,\eta e^{i2\pi}))^\nu$$

$$- \frac{1}{\nu}(P(x,\eta e^{i0}))^\nu \log P(x,\nu e^{i0})) + \frac{1}{\nu^2}(P(x,\eta e^{i0}))^\nu$$

$$+ (1 + \tfrac{1}{2} + \ldots + \tfrac{1}{\nu-1})\tfrac{1}{\nu}(P(x,\eta e^{i0}))^\nu]\,\frac{Q_y(x,\eta)N_\nu(x,\eta)}{(Q(x,\eta))^{l+1}}\,dx$$

$$- \int_D\Big(\int_C [\tfrac{1}{\nu}(P(x,y))^\nu \log P(x,y) - \frac{1}{\nu^2}(P(x,y))^\nu$$

$$- (1 + \tfrac{1}{2} + \ldots + \tfrac{1}{\nu-1})\tfrac{1}{\nu}(P(x,y))^\nu]$$

$$\times \Big(\frac{Q_{xy}(x,y)N_\nu(x,y) + Q_x(x,y)N_{\nu y}(x,y)}{(Q(x,y))^{l+1}}$$

$$- \frac{(l+1)Q_x(x,y)Q_y(x,y)N_\nu(x,y)}{(Q(x,y))^{l+2}}\Big)dx\Big)dy.$$

Subtracting (3.8) and (3.9), taking into account (2.1) and remarks on the increment of a branch of logarithm (comp. section 1.3) as well as (1.10), we obtain

$$(3.10)\quad \int_C\Big(\int_D (P(x,y))^{\nu-1}\,[\log P(x,y) - (1 + \tfrac{1}{2} + \ldots + \tfrac{1}{\nu - 1})]$$

$$\times \frac{N_\nu(x,y)}{(Q(x,y))^{l+1}}\,dy\Big)dx = \frac{2\pi i}{\nu}\int_D (P(\xi,y))^\nu\,\frac{Q_y(\xi,y)N_\nu(\xi,y)}{(Q(\xi,y))^{l+1}}\,dy$$

$$- \int_C\Big(\int_D \frac{1}{\nu}(P(x,y))^\nu\,(\log P(x,y) - (1 + \tfrac{1}{2} + \ldots + \tfrac{1}{\nu}))$$

$$\times \frac{N_{\nu+1}(x,y)}{(Q(x,y))^{l+1}}\,dy\Big)dx.$$

Putting (3.10) into (3.4) we obtain (3.4) for $\nu + 1$.

The induction ends the proof.

THEOREM 3. Let k be a non-negative integer, l - an integer, ν - a positive integer, M - a holomorphic function in a neighbourhood of the point $(0,0)$ containing E_r. Then

$$(3.11)\quad \frac{1}{(2\pi i)^2}\int_C\Big(\int_D \frac{M(x,y)}{(P(x,y))^{k+1}(Q(x,y))^{l+1}}\,dy\Big)dx$$

$$= \frac{1}{2\pi i}\,\frac{1}{k!}\sum_{h=1}^{\nu}\frac{(-1)^{h-1}}{(h-1)!}\int_D (P(\xi,y))^{h-1}$$

$$\times \frac{Q_y(\xi,y)M_{k+h-1}(\xi,y)}{(Q(\xi,y))^{l+1}}\,dy$$

$$+ \frac{1}{(2\pi i)^2}\,\frac{(-1)^\nu}{k!}\int_C\Big(\int_D \frac{1}{(\nu-1)!}(P(x,y))^{\nu-1}$$

$$\times (\log P(x,y) - (1 + \tfrac{1}{2} + \ldots + \tfrac{1}{\nu - 1}))\,\frac{M_{k+\nu}(x,y)}{(Q(x,y))^{l+1}}\,dy\Big)dx.$$

P r o o f. Putting in (3.4) $N = M$, we obtain (3.11) for $k = 0$, and putting in (3.4) $N = M_k$, by (3.1), we obtain (3.11) for $k = 1, 2, \ldots$.

4. Reduction to single integrals. Let us adopt the constructions, notations and agreements introduced in section 1.

THEOREM 4. Let k be a non-negative integer, l - an integer, ν - a positive integer such that $\nu > l$, M - a holomorphic function in a neighbourhood of the point $(0,0)$ containing E_r. Then

$$(4.1)\qquad \frac{1}{(2\pi i)^2} \int_{C_o} \left(\int_{D_o} \frac{M(x,y)}{(P(x,y))^{k+1}(Q(x,y))^{l+1}} dy\right)dx$$

$$= \frac{1}{2\pi i} \frac{1}{k!} \sum_{h=1}^{\nu} \frac{(-1)^{h-1}}{(h-1)!} \int_{D_o} (P(0,y))^{h-1} \times \frac{Q_y(0,y)M_{k+h-1}(0,y)}{(Q(0,y))^{l+1}} dy$$

where (C_o, D_o) is an arbitrary bicircle of the type (1.2) with the radii ξ_o, η_o, lying in E_r.

P r o o f. Let us consider besides the circles appearing in (4.1) also the circles appearing in (3.11).

We shall base our reasoning on the four mentioned below observations:

(i) The term on the left-hand side of (4.1) is identical with the term on the left-hand side of (3.11). Indeed, according to the facts given in section 1.2 the integrand is the sum of a Laurent series in x,y. In this situation the considered integrals are reduced to the integrals of terms containing x and y in the powers -1 and their values are, as one can see, equal. This gives the announcement.

(ii) The term on the right-hand side of (4.1) is identical with the term

$$(4.1')\qquad \frac{1}{2\pi i} \frac{1}{k!} \sum_{h=1}^{\nu} \frac{(-1)^{h-1}}{(h-1)!} \int_{D} (P(0,y))^{h-1} \times \frac{Q_y(0,y)M_{k+h-1}(0,y)}{(Q(0,y))^{l+1}} dy$$

arising from (4.1) by substitution of the circle D_o by the circle D. Indeed, it follows immediately from the fact that with r sufficiently small the rational integrand in (4.1) and (4.1) does not have any poles in the closed ring bounded by the circles D_o and D.

(iii) The difference between the first term on the right-hand side of (3.11) and term (4.1´) tends to zero as ξ,η tend to 0,0. Indeed, let us represent the investigated difference in the form

$$(4.2)\qquad \frac{1}{2\pi i}\ \frac{1}{k!}\int_D \sum_{h=1}^{\nu} \frac{(-1)^{h-1}}{(h-1)!}\,[(P(\xi,y))^{h-1} \times \frac{Q_y(\xi,y)M_{k+h-1}(\xi,y)}{(Q(\xi,y))^{l+1}} - (P(0,y))^{h-1} \times \frac{Q_y(0,y)M_{k+h-1}(0,y)}{(Q(0,y))^{l+1}}]dy.$$

According to the facts given in section 1.2 we easily see that the expressions in the square brackets in (4.2) are the sums of some Laurent series in ξ,y with the terms divisible by ξ so, containing the positive powers of ξ. In this situation the considered integral reduces to the sum of integrals of the terms of the series which have the degree -1 with respect to y, i.e. to the sum of a power series with respect to ξ with the positive powers of ξ. Of course, this sum tends to 0 as ξ tends to 0. This gives the announcement.

(iv) The second term on the right-hand side of (3.11) tends to 0 as ξ,η tend to 0,0. Indeed, according to the facts given in section 1.2 and 1.3 we see that the integrand can be represented in the form

$$(4.3)\qquad \frac{1}{(\nu-1)!}[(P(x,y))^{\nu-1}\frac{M_{k+\nu}(x,y)}{(Q(x,y))^{l+1}}(\log\frac{P(x,y)}{x} - (1+\frac{1}{2}+\dots+\frac{1}{\nu-1}))] + \frac{1}{(\nu-1)!}\ [(P(x,y))^{\nu-1}\frac{M_{k+\nu}(x,y)}{(Q(x,y))^{l+1}}]\{\log\xi + is\}.$$

One can readily see that by the assumption $\nu > 1$ the expressions in the square brackets in (4.3) are the sums of some Laurent series in x,y with the terms of degree not smaller than -1. In this situation the integral of the first summand from (4.3) reduces, of course, to zero. On the other hand, the integral of the second summand from (4.3) with respect to y reduces to the sum of a power series with respect to x multiplied by the factor in the brace. Hence it follows immediately that the considered integral is of the type $O(\xi \log \xi)$ and tends to 0 as ξ,η tend to $0,0$. This gives the announcement.

Passing to the demonstration of the assertion of the theorem, we see that according to observations (i), (ii), equality (3.11) can be represented in the form

$$(4.4)\quad \frac{1}{(2\pi i)^2}\int_{C_o}\Big(\int_{D_o}\frac{M(x,y)}{(P(x,y))^{k+1}(Q(x,y))^{l+1}}\,dy\Big)dx$$

$$= \frac{1}{2\pi i}\;\frac{1}{k!}\;\sum_{h=1}^{\nu}\frac{(-1)^{h-1}}{(h-1)!}\int_{D_o}(P(0,y))^{h-1}$$

$$\times\,\frac{Q_y(0,y)M_{k+h-1}(0,y)}{(Q(0,y))^{l+1}}\,dy + \frac{1}{2\pi i}\;\frac{1}{k!}\;\Big[\sum_{h=1}^{\nu}\frac{(-1)^{h-1}}{(h-1)!}$$

$$\times\int_{D}(P(\xi,y))^{h-1}\;\frac{Q_y(\xi,y)M_{k+h-1}(\xi,y)}{(Q(\xi,y))^{l+1}}\,dy$$

$$-\sum_{h=1}^{\nu}\frac{(-1)^{h-1}}{(h-1)!}\int_{D}(P(0,y))^{h-1}\,\frac{Q_y(0,y)M_{k+h-1}(0,y)}{(Q(0,y))^{l+1}}\,dy\Big]$$

$$+\,\frac{1}{(2\pi i)^2}\;\frac{(-1)^{\nu}}{k!}\;\int_{C}\Big(\int_{D}\frac{1}{(\nu-1)!}\,(P(x,y))^{\nu-1}$$

$$\times\,\Big(\log P(x,y) - \Big(1+\frac{1}{2}+\ldots+\frac{1}{\nu-1}\Big)\Big)\;\frac{M_{k+\nu}(x,y)}{(Q(x,y))^{l+1}}\,dy\Big)dx.$$

Passing to the limit on both sides of (4.4) with ξ,η tending to $0,0$ we obtain, according to observations (iii) and (iv), the required equality (4.1).

<u>5. Characterization of a mapping by values of a coordinate on axis.</u> Let us adopt constructions, notations and agreements introduced in section 1.

THEOREM 5. The mapping of form (1.3) with the Jacobian equal to 1 is uniquely characterized by the second coordinate Q and the first one P restricted to the axis of ordinates.

P r o o f. Let us notice that with the specification $M(x,y) = x$ or $M(x,y) = y$ the left-hand side of (4.1), according to theorem 2, represents the coefficient c_{kl} or d_{kl} of the inverse mapping of (P,Q).

On the other hand, the right-hand side of (4.1) depends on the values of the first coordinate P on the axis of ordinates and, by the definition of M_{k+h-1} (comp. (1.10)) - depends on the second coordinate Q.

Thus, the inverse mapping of (P,Q) is uniquely characterized by the coordinate Q and by the coordinate P restricted to the axis of ordinates. This gives immediately the assertion of the theorem.

The theorems 1 and 2 can be generalized for the mappings of several variables. It will be published in Bulletin de la Société des Sciences et des Lettres de Łódź, Serie: Recherches sur les déformations (the address of the editor: Narutowicza 56, 90-136 Łódź, Poland).

Bibliography

[1] Charzyński Z., Chądzyński J., Skibiński P., A contribution to Keller´s Jacobian conjecture III, Bull. Soc. Sci. Lettres Łódź, 39 (58) (1989), 1-8.

[2] Шабат Б.В. Бьедение в комплексный анализ, Москва 1969.

Institute of Mathematics
University of Łódź
ul. S. Banacha 22
90-238 Łódź, Poland

Trivialités en famille

Michel Coste et Miloud Reguiat

IRMAR (CNRS, U.R.A. 305), Université de Rennes 1, Campus de Beaulieu,
35042 Rennes Cedex, France

Introduction

On s'intéresse ici à des problèmes de trivialisation de familles d'ensembles ou de fonctions semi-algébriques. On utilise en quelques endroits les techniques du spectre réel, qui sont bien adaptées à ce genre de problèmes : elles permettent de traduire des résultats obtenus de manière élémentaire pour un corps réel clos quelconque en résultats pour les familles. C'est ce qu'on peut appeler "cacher les paramètres dans le corps de base". Une première illustration de ce principe concerne la construction de triangulations lipschitziennes de compacts semi-algébriques. La lipschitzianité ne concerne qu'un sens : du complexe simplicial à l'objet triangulé. On en donne une démonstration dans la première section. On montre dans la deuxième section comment en déduire une version lipschitzienne du théorème de trivialisation semi-algébrique de Hardt [HA], et les classiques bornes métriques uniformes pour les familles. Le même résultat donne dans la troisième section des trivialisations semi-algébriques, au dessus d'un intervalle $]0,\varepsilon]$, de fonctions semi-algébriques propres, qui peuvent se prolonger par continuité en 0. Ceci pourrait aussi s'obtenir en utilisant le théorème de triangulation des fonctions semi-algébriques de Shiota [SH]. Ce dernier théorème, passé à la moulinette du spectre réel dans la quatrième section, fournit une trivialisation semi-algébrique des familles de fonctions (à valeurs dans $\mathbb{R}$) semi-algébriques, copié sur le théorème de Hardt mentionné plus haut. On a ainsi des bornes sur le nombre de types semi-algébriques (et, a fortiori, topologiques) de fonctions semi-algébriques, en fonction récursive du degré des fonctions et de la dimension de l'espace affine de départ, retrouvant des résultats de Benedetti et Shiota [BS] qui généralisent ceux de Fukuda pour les polynômes [FU]. Dans la dernière section, un contre-exemple inspiré de Nakai [NA] montre que, par contre, on ne peut pas étendre les résultats de finitude de types topologiques des applications polynomiales en deux variables de Aoki [AN] aux applications semi-algébriques.

1. Triangulations lipschitziennes

On se propose de montrer comment transformer n'importe quelle triangulation en une triangulation lipschitzienne. On va travailler sur un corps réel clos quelconque R, ce qui nous sera utile par la suite et qui ne complique pas les démonstrations. Par contre, il faut être un peu soigneux à cause du fait que R n'a aucune raison d'être archimédien. On insiste sur le fait que l'on veut une P-lipschitzianité pour un *entier* P. Ceci ne peut bien sûr s'obtenir que si l'ensemble semi-algébrique à trianguler n'est pas infiniment grand par rapport aux entiers, c.-à-d. que s'il est contenu dans un pavé $[-N, N]^n$ où N est un entier. Dans ce qui suit, I désigne l'intervalle $[0, 1]$ de R.

Proposition 1. *Soit $G : I^d \to R^n$ une application semi-algébrique continue. On suppose qu'il existe un entier $N \in \mathbb{N}$ tel que $G(I^d) \subset [-N, N]^n$, et que G est N-lipschitzienne sur les faces de I^d. Alors il existe un homéomorphisme semi-algébrique $H : I^d \to I^d$, qui est l'identité sur les faces de I^d, et un entier P tel que $G \circ H$ soit P-lipschitzienne.*

La démonstration se fait en appliquant plusieurs fois le lemme suivant. On note pour ce lemme (x, y, z) les coordonnées dans R^d, avec $x \in R$, $y \in R^k$ et $z \in R^{d-k-1}$, où $0 \leq k < n$. On reprend les notations et les hypothèses de la proposition, en supposant que $n = 1$ (c.-à-d. que G est une fonction à valeurs dans R).

Lemme 2. *Supposons, en plus des hypothèses de la proposition, que G est une fonction N-lipschitzienne par rapport aux variables y :*

$$\forall x, z \quad |G(x, y', z) - G(x, y, z)| \leq N \, \|y' - y\|.$$

Alors il existe un homéomorphisme semi-algébrique $H : I^d \to I^d$ qui est l'identité sur les faces de I^d, et un entier Q tel que $G \circ H$ et H soient Q-lipschitziens par rapport aux variables (x, y).

Démonstration. On écrit $I^d = I \times I^{d-1}$, où le premier facteur correspond à la variable x. On "démonte" ensuite le cube I^d de la manière suivante : I^{d-1} est divisé en une partition finie en sous ensembles semi-algébriques, et pour chacun des sous-ensembles semi-algébriques A de cette partition, le cylindre $I \times A$ est divisé en tranches au moyen de fonctions semi-algébriques continues :

$$0 = \xi_0 < \xi_1 < \ldots < \xi_p = 1 : A \longrightarrow R$$

telles que, sur chaque tranche $]\xi_{i-1}, \xi_i[$, $i = 1, \ldots, p$ du cylindre $I \times A$, G est une fonction de Nash, et $\partial G / \partial x$ est ou bien constamment majoré par N en valeur absolue, ou bien constamment plus grand que N en valeur absolue. On notera T le nombre maximal de ces tranches, quand A parcourt la partition semi-algébrique de I^{d-1}.

Considérons la fonction semi-algébrique $L : I^d \to [0, 1]$ définie comme suit. On pose toujours $L(0, y, z) = 0$. Soit $(y, z) \in A$, et supposons $L(x, y, z)$ défini pour $x \in [0, \xi_l(y, z)]$. Si, dans la tranche $]\xi_l, \xi_{l+1}[$, $\partial G / \partial x$ est majoré en valeur absolue par N, on pose simplement

$$L(x, y, z) = L(\xi_l(x, y), y, z) + x - \xi_l(y, z) \quad \text{pour } \xi_l(y, z) \leq x \leq \xi_{l+1}(y, z) \, .$$

Si dans la tranche $]\xi_l, \xi_{l+1}[$, $\partial G / \partial x$ est plus grand en valeur absolue que N, on pose

$$L(x, y, z) = L(\xi_l(x, y), y, z) + \frac{1}{N} \, |G(x, y, z) - G(\xi_l(y, z), y, z)| \qquad \text{pour } \xi_l(y, z) \leq x \leq \xi_{l+1}(y, z).$$

On peut remarquer que $\frac{1}{N} \, |G(x, y, z) - G(\xi_l(y, z), y, z)|$ est, dans la tranche, une fonction croissante de x, et minorée par $x - \xi_l(y, z)$. On remarque aussi que, pour (y, z) fixé, la fonction $x \mapsto L(x,y,z)$ est un homéomorphisme semi-algébrique de I sur $[0, L(1,y,z)]$. On notera H la bijection semi-algébrique réciproque de la bijection :

$$(x,y,z) \longmapsto \left(\frac{L(x,y,z)}{L(1,y,z)}, y, z\right)$$
$$I^d \longrightarrow I^d.$$

Notons tout de suite que H est l'identité sur les faces de I^d. En effet, puisque G est N-lipschitzienne sur les faces, on a, pour tout (y,z) appartenant à une face de I^{d-1}, $L(x,y,z) = x$. Il reste à établir les propriétés annoncées de H. La première chose est de montrer que L est continu. Ceci suffira à établir que H^{-1} est continu, et donc que H est un homéomorphisme : une bijection semi-algébrique continue d'un fermé borné sur un fermé borné est bicontinue, même sur un corps réel clos quelconque.

Soit $\varepsilon > 0$, $\varepsilon \in R$. Fixons $(y,z) \in I^{d-1}$. Alors il existe $\delta > 0$, $\delta \in R$ tel que

$$\|(y',z') - (y,z)\| < \delta \text{ entraîne } \forall x \ |G(x,y',z') - G(x,y,z)| < \varepsilon.$$

En effet, même sur un corps réel clos quelconque, une fonction semi-algébrique continue sur un fermé borné est uniformément continue. Soit alors $(y',z') \in I^{d-1}$ vérifiant $\|(y',z') - (y,z)\| < \delta$. L'intervalle I est subdivisé en au plus $2T-1$ intervalles tels que, quand x varie dans l'intérieur d'un de ces intervalles, ni (x,y,z) ni (x,y',z') ne sautent de tranche dans leurs cylindres respectifs. Soient $x_0 = 0 < \ldots < x_m < \ldots < x_q = 1$ les bornes de ces intervalles. Pour $x \in [x_m, x_{m+1}]$, on a suivant les cas :

$$|L(x,y',z') - L(x,y,z)| \le |L(x_m,y',z') - L(x_m,y,z)| + \\ + \begin{cases} 0 \\ |G(x,y',z') - G(x_m,y',z')| / N - x + x_m \\ |G(x,y,z) - G(x_m,y,z)| / N - x + x_m \\ \big| |G(x,y',z') - G(x_m,y',z')| / N - |G(x,y,z) - G(x_m,y,z)| / N \big|. \end{cases}$$

Dans le dernier cas, on a clairement

$$(1) \qquad |L(x,y',z') - L(x,y,z)| < |L(x_m,y',z') - L(x_m,y,z)| + \frac{2\varepsilon}{N}.$$

Le deuxième cas se présente quand $\partial G/\partial x$ est majoré par N en valeur absolue sur $]x_m, x_{m+1}[\times \{(y,z)\}$, et donc

$$|G(x,y',z') - G(x_m,y',z')| < |G(x,y,z) - G(x_m,y,z)| + 2\varepsilon \le N(x - x_m) + 2\varepsilon.$$

On aboutit aussi dans ce cas à la majoration (1). On procède de même dans le troisième cas. Finalement, on obtient

$$\forall x \in I \quad |L(x,y',z') - L(x,y,z)| < (2T-1)\frac{2\varepsilon}{N}.$$

Si on a choisi δ suffisamment petit pour que l'on ait aussi, pour tout $(y,z) \in I^{d-1}$,

$$|x' - x| < \delta \text{ entraîne } |L(x',y,z) - L(x,y,z)| < \varepsilon,$$

alors $\|(x',y',z')-(x,y,z)\| < \delta$ entraîne

$$|L(x',y',z')-L(x,y,z)| < \left(\frac{2(2T-1)}{N}+1\right)\varepsilon.$$

Ceci montre la continuité de L, et donc le fait que H est un homéomorphisme. Il reste à vérifier que H et $G\circ H$ sont Q-lipschitziens, pour un entier Q, par rapport aux variables (x,y). On voit que, si $x'\geq x$, alors $L(x',y,z)-L(x,y,z)\geq x'-x$, et que $L(1,y,z)\leq T\,N$ (on utilise ici l'hypothèse $G(I^d)\subset[-N,N]$). Donc H est $(T\,N)$-lipschitzien par rapport à x, et $G\circ H$ est $(T\,N^2)$-lipschitzien par rapport à x. Voyons maintenant ce qui se passe pour les variables y. Les calculs faits ci-dessus pour la continuité montrent aussi que :

$$|L(x,y',z)-L(x,y,z)|\leq 2(2T-1)\,\|y'-y\|.$$

Posons

$$\xi=\frac{L(x,y,z)}{L(1,y,z)}\qquad \xi'=\frac{L(x,y',z)}{L(1,y',z)}\,.$$

Il vient

$$\begin{aligned}
|\xi'-\xi| &\leq L(1,y,z)\,|L(x,y',z)-L(x,y,z)|+L(x,y,z)\,|L(1,y,z)-L(1,y',z)|\\
&\leq 4\,T\,N\,(2T-1)\,\|y'-y\|\,,\\
\|H(\xi,y',z)-H(\xi,y,z)\| &\leq\\
&\leq \|H(\xi,y',z)-H(\xi',y',z)\|+\|H(\xi',y',z)-H(\xi,y,z)\|\\
&\leq T\,N\,|\xi'-\xi|+\|(x,y',z)-(x,y,z)\|\\
&\leq \left(4\,T^2\,N^2\,(2T-1)+1\right)\|y'-y\|\,,\\
\|G\circ H(\xi,y',z)-G\circ H(\xi,y,z)\| &\leq\\
&\leq \|G\circ H(\xi,y',z)-G\circ H(\xi',y',z)\|+\|G\circ H(\xi',y',z)-G\circ H(\xi,y,z)\|\\
&\leq T\,N^2\,|\xi'-\xi|+\|G(x,y',z)-G(x,y,z)\|\\
&\leq N\left(4\,T^2\,N^2\,(2T-1)+1\right)\|y'-y\|\,.
\end{aligned}$$

En prenant par exemple $Q=16\,T^3\,N^3$, on a que H et $G\circ H$ sont Q-lipschitziens par rapport aux variables (x,y). □

Démonstration de la proposition. On commence par remarquer que l'on peut étendre le lemme au cas où G est à valeurs dans R^n avec n quelconque. Pour cela, on traite successivement chaque coordonnée de G au moyen du lemme. Comme l'homéomorphisme dans le lemme est lipschitzien, on ne perd pas, en rendant une coordonnée lipschitzienne, la lipschitzianité des coordonnées précédentes. En appliquant d fois le lemme ainsi étendu, on transforme G en une application P-lipschitzienne, pour un certain entier P, en toutes les variables. □

Nous passons maintenant du cas d'un pavé fermé, vu dans la proposition, à celui d'un complexe simplicial fini. Nous expliquons d'abord comment "quadranguler" un complexe simplicial fini. Considérons le simplexe standard

$$\Delta^d=\left\{(t_0,\ldots,t_d);\ \sum_{i=0}^{d}t_i=1,\ t_i\geq 0\right\}\subset R^{d+1}.$$

On le divise en $d+1$ morceaux

$$C_i^d = \{(t_0, \ldots, t_d) \in \Delta^n;\ \forall j\ t_j \leq t_i\},$$

chacun projectivement isomorphe au pavé fermé I^d. La formule pour l'isomorphisme de C_0^d sur I^d est :

$$(t_0, \ldots, t_d) \longmapsto \left(\frac{t_1}{t_0}, \ldots, \frac{t_d}{t_0}\right).$$

On remarque que la subdivision induite sur les faces de Δ^d est bien la subdivision des simplexes de dimensions inférieures à d : les subdivisions se recollent bien. Si maintenant K est un complexe simplicial fini dans R^m, on le quadrangule en transportant, pour chaque simplexe σ de dimension d de K, la quadrangulation du simplexe standard Δ^d par un isomorphisme affine de Δ^d sur σ. On a alors K comme réunion de boîtes projectivement isomorphes à des pavés fermés, l'intersection de deux boîtes étant une face (itérée) commune, ou vide. On notera K_{quad} le complexe ainsi quadrangulé ; le k-squelette de K_{quad} est la réunion des boîtes de dimensions inférieures ou égales à k.

Théorème 3. *Soit $K \subset R^m$ un complexe simplicial fini, dont les sommets ont des coordonnées entières. Soit $G : K \to R^n$ une fonction semi-algébrique continue, avec $G(K) \subset [-N, N]^n$ pour un certain entier N. Alors il existe un homéomorphisme semi-algébrique $H : K \to K$ et un entier P tels que $G \circ H$ soit P-lipschitzien.*

Démonstration. On quadrangule K et on construit H par induction sur les k-squelettes de K_{quad}. Les homéomorphismes que l'on construit à chaque étape préservent les boîtes de K_{quad}. Supposons que l'on ait un tel homéomorphisme H_k tel que $G \circ H_k$ soit P_k-lipschitzien sur le k-squelette de K_{quad}, pour un entier P_k. Alors la proposition permet d'obtenir un homéomorphisme semi-algébrique h_{k+1} du $k+1$-squelette sur lui-même, préservant les boîtes, qui est l'identité sur le k-squelette, et tel que $G \circ H_k \circ h_{k+1}$ soit P_{k+1}-lipschitzien, pour un nouvel entier P_{k+1}. On peut ensuite étendre, par une construction de cône, $H_k \circ h_{k+1}$ à un homéomorphisme semi-algébrique H_{k+1} de K_{quad} sur lui-même, préservant les boîtes. □

Remarques. a) Vu la façon dont on l'a construit, on peut ajouter dans le théorème que l'homéomorphisme H est semi-algébriquement homotope à l'identité.

b) Il n'y a pas d'effectivité dans le théorème, mais on peut en rajouter après coup. Rappelons que pour mesurer la complexité d'un objet semi-algébrique, on considère son degré, qui est la somme des degrés des polynômes qui interviennent dans sa description. On obtient alors, pour $N = 1$, une borne pour le degré e de H et pour P, en fonctions récursives du nombre de sommets s du complexe K, du degré d de G et de n. En effet, si l'on fait $N = 1$ et si l'on fixe s, d, n, e et P, alors le théorème 3 s'écrit comme un énoncé $\Phi(s, d, n, e, P)$ de la théorie des corps réels clos. Etant donnés s, d, et n, il existe e et P tels que cet énoncé soit un théorème de la théorie des corps réels clos ; sinon, on obtiendrait par ultraproduit un corps réel clos où le théorème 3 serait en défaut. Enfin, l'algorithme de décision pour la théorie des corps réels clos montre que ces e et P sont des fonctions récursives de s, d, et n. Nous suivrons la même démarche pour le théorème 8.

c) Le résultat est évidemment lié au "Triangulation Lemma" que l'on peut trouver dans [GR]. Ce dernier dit que l'on peut subdiviser une triangulation $G : K \to G(K) \subset \mathbf{R}^n$, de façon que pour tout k-simplexe Δ de la subdivision de K, on a un homéomorphisme semi-algébrique H_Δ du k-simplexe standard sur Δ, avec toutes les dérivées partielles jusqu'à l'ordre r (donné à l'avance) de $G \circ H_\Delta$ bornées par 1. Ici on ne parle que de lipschitzianité, mais on ne subdivise pas la triangulation, et l'homéomorphisme H qui rend lipschitzien est défini globalement, de K sur K.

2. Bornes métriques uniformes

Le fait d'avoir montré l'existence de triangulations lipschitziennes sur un corps réel clos quelconque permet d'obtenir des résultats métriques sur les familles. Le passage se fait en "cachant les paramètres dans le corps de base". On obtient une version lipschitzienne du théorème de trivialité semi-algébrique de Hardt [HA]. La lipshitzianité est uniforme dans les fibres, mais on ne peut bien sûr pas espérer de lipschitzianité par rapport aux variables de la base.

Théorème 4. *Soit B un ensemble semi-algébrique, et soit X un sous-ensemble semi-algébrique de $R^n \times B$. On note p la projection de X sur B. On suppose qu'il existe un entier N tel que, pour tout $b \in B$, la fibre X_b soit contenue dans $[-N, N]^n$. Alors il existe un entier P, une partition de B en un nombre fini de sous-ensembles semi-algébriques B^i, avec pour chaque i un ensemble semi-algébrique F^i et un homéomorphisme semi-algébrique*

$$h^i : F^i \times B^i \longrightarrow X \cap p^{-1}(B^i)$$

tel que $p \circ h^i$ soit la projection sur B^i et tel que pour tout $b \in B^i$, l'application $h^i_b : F^i \to X_b$ soit P-lipschitzienne.

Démonstration. On suit fidèlement la seconde démonstration du théorème de Hardt dans [BCR], 9.3.1. Soit d'abord $\overline{X}$ l'adhérence de X dans $R^n \times B$. on note comme d'habitude $\widetilde{B}$ le constructible du spectre réel correspondant à B, et pour un point α de $\widetilde{B}$, $k(\alpha)$ désigne le corps réel clos associé. La fibre $\overline{X}_\alpha$ est fermée dans $[-N, N]^n_{k(\alpha)}$. On a donc une triangulation P-lipschitzienne, pour un certain entier P, de $\overline{X}_\alpha$, compatible avec X_α, par un complexe simplicial fini contenu dans $k(\alpha)^n$, dont les sommets sont à coordonnées entières. Soit $F_{k(\alpha)}$ la réunion des simplexes ouverts de ce complexe dont l'image par la triangulation est dans X_α. Cet ensemble semi-algébrique $F_{k(\alpha)}$ est l'extension à $k(\alpha)$ d'un sous-ensemble semi-algébrique F de R^n (on a spécifié que les sommets étaient à coordonnées entières pour en être sûrs). On récupère donc un sous-ensemble semi-algébrique B' de B, avec $\alpha \in \widetilde{B'}$, et un homéomorphisme $h : F \times B' \to X \cap p^{-1}(B')$, compatible avec les projections sur B', et qui est P-lipschitzien par rapport aux variables dans R^n. La compacité de $\widetilde{B}$ permet ensuite de trouver la partition finie du théorème. □

Le résultat précédent permet d'obtenir directement les classiques bornes métriques uniformes pour les familles, sans faire appel à la géométrie intégrale. Pour pouvoir parler de volume, il faut ici se limiter au cas des vrais réels. on note $\operatorname{diam}(A)$ le diamètre d'un compact A de $\mathbb{R}^n$ (borne supérieure des distances entre deux points de A).

Corollaire 5. *Soit B un ensemble semi-algébrique réel, et soit X un sous-ensemble semi-algébrique de $\mathbb{R}^n \times B$. On suppose que, pour tout $b \in B$, X_b est non vide de dimension $k > 0$. Alors il existe une constante C telle que, pour tout $b \in B$, le volume k-dimensionnel de X_b soit majoré par $C \times (\operatorname{diam}(X_b))^k$.*

Démonstration. On peut supposer que l'origine de $\mathbb{R}^n$ est dans X_b pour tout b. La fonction

$$b \longmapsto \rho(b) = \big(\operatorname{diam}(X_b)\big)^{-1}$$

est semi-algébrique (pas continue a priori), et en appliquant l'homothétie de rapport $\rho(b)$ à la fibre X_b on peut se ramener au cas où $X_b \subset [-1,1]^n$ pour tout b. Le théorème 4 donne alors le résultat voulu. □

3. Prolongement de trivialisations

Nous avons déjà évoqué le théorème de trivialité semi-algébrique de Hardt [HA]. Formulons en une conséquence dans un cas très particulier, qui nous occupera dans cette section. Soit X un ensemble semi-algébrique, et $f : X \to R$ une fonction semi-algébrique. Alors il existe un $\varepsilon > 0$ et un homéomorphisme semi-algébrique

$$h : f^{-1}(\varepsilon)\times]0,\varepsilon] \longrightarrow f^{-1}(]0,\varepsilon])$$

tel que $f \circ h$ est la projection sur $]0,\varepsilon]$ et que $h(.,\varepsilon)$ est l'identité sur $f^{-1}(\varepsilon)$. Peut-on prolonger continûment cette trivialisation au-dessus de $]0,\varepsilon]$ en 0 ? Il faut déjà supposer que la fonction f est propre, mais ceci ne suffit pas. Prenons pour f la projection de $[0,1]\times[0,1]$ sur le deuxième facteur, et pour h la fonction de $[0,1]\times]0,1]$ dans R^2 définie par

$$(x,t) \longmapsto \begin{cases} (x,t) & \text{si } x+t \geq 1 \\ (t\dfrac{x}{1-x}, t) & \text{si } x+t \leq 1. \end{cases}$$

Ce h ne peut pas se prolonger par continuité à $[0,1]\times[0,1]$. Cependant, il y a toujours moyen de choisir la trivialisation pour que le prolongement soit possible.

Théorème 6. *Soit X un ensemble semi-algébrique, et $f : X \to R$ une fonction semi-algébrique propre. Alors il existe un $\varepsilon > 0$ et une application semi-algébrique*

$$h : f^{-1}(\varepsilon) \times [0,\varepsilon] \longrightarrow f^{-1}([0,\varepsilon])$$

tel que $f \circ h$ est la projection sur $]0,\varepsilon]$, que $h(.,\varepsilon)$ est l'identité sur $f^{-1}(\varepsilon)$ et que h en restriction à $f^{-1}(\varepsilon)\times]0,\varepsilon]$ est un homéomorphisme sur $f^{-1}(]0,\varepsilon])$.

Démonstration. Comme f est propre, on peut se ramener au cas où $f^{-1}([0,1])$ est contenu dans un pavé $[-N,N]^n$, avec N entier. On applique alors le théorème, qui nous donne un $\varepsilon > 0$ et un homéomorphisme semi-algébrique

$$g : K\times]0,\varepsilon] \longrightarrow f^{-1}(]0,\varepsilon])$$

où K est un ensemble semi-algébrique compact, tel que $f \circ g$ est la projection sur $]0,1]$ et que, pour tout $t \in]0,\varepsilon]$, $g(.,t) : K \to f^{-1}(t)$ est P-lipschitzien pour un entier P indépendant de t. Soit $y \in K$. Puisque f est propre, alors l'intersection de $\{(y,0)\} \times X$ avec l'adhérence du graphe de g est non vide. Cette intersection est réduite à un point. En effet, soient $(y,0,x)$ et $(y,0,x')$ deux points de cette intersection. Par le lemme de sélection des courbes, il existe deux chemins semi-algébriques

$$\gamma, \gamma' : [0,\delta[\longrightarrow K$$

tels que $\gamma(0) = \gamma'(0) = y$, que $\lim_{t\to 0}(g(\gamma(t),t)) = x$ et que $\lim_{t\to 0}(g(\gamma'(t),t)) = x'$. Alors la propriété de lipschitzianité implique que $x = x'$. On a ainsi montré que g se prolonge par continuité à $K \times [0,\varepsilon]$, ce qui nous donne le résultat. $\square$

Le théorème précédent montre que $f^{-1}([0,\varepsilon])$ se rétracte par déformation semi-algébrique sur $f^{-1}(0)$, tandis que $f^{-1}(]0,\varepsilon])$ est semi-algébriquement homéomorphe à $f^{-1}(\varepsilon)\times]0,\varepsilon]$. On a donc ainsi des morphismes de l'homologie de $f^{-1}(\varepsilon)$ dans celle de $f^{-1}(0)$ (et ceci sur n'importe quel corps réel clos de base). On peut interpréter le théorème précédent comme une généralisation du théorème de structure conique locale pour les ensembles semi-algébriques : soit Y un sous-ensemble semi-algébrique fermé borné de l'ensemble semi-algébrique X fermé dans R^n, et soit f une fonction semialgébrique continue de X dans R, positive ou nulle, propre, et telle que $y = f^{-1}(0)$; par exemple, on peut prendre pour f la distance à Y. Alors le voisinage compact $f^{-1}([0,\varepsilon])$ de Y, pour ε suffisamment petit, est semi-algébriquement homéomorphe au "mapping cylinder" d'une application semi-algébrique continue de $f^{-1}(\varepsilon)$ dans Y.

A. Durfee [DU] a considéré les voisinages semi-algébriques d'ensembles semi-algébriques : ce sont les $f^{-1}([0,\varepsilon])$ comme ci-dessus. Il obtient l'unicité à homéomorphisme près de ces voisinages (ils ne dépendent essentiellement ni du choix de ε, ce que l'on a vu, ni du choix de f). Mais les homéomorphismes sont obtenus par utilisation du premier lemme d'isotopie de Thom, et ils ne sont pas semi-algébriques.

Le théorème 6 s'obtient comme conséquence facile du résultat de triangulation des fonctions semi-algébriques (à valeurs dans R) de Shiota [SH]. Ce résultat dit que, si $f : X \to R$ est une fonction semi-algébrique continue bornée, alors il existe un complexe simplicial fini K sur R, une fonction $g : K \to R$ affine sur chaque simplexe de K, et un homéomorphisme semi-algébrique h d'une réunion U de simplexes ouverts de K sur X tels que $f \circ h = g_{|U}$.

4. Trivialisation de familles de fonctions semi-algébriques

Dans cette section, nous exploitons le résultat de triangulation des fonctions semi-algébriques ci-dessus, et en particulier le fait que la démonstration de [SH] est élémentaire, c.-à-d. valide sur n'importe quel corps réel clos. On peut alors lui faire subir le même traitement que celui qui donne le théorème de trivialité de Hardt à partir du théorème de triangulation des ensembles semi-algébriques, dans [BCR] 9.3.1. Voici le résultat qu'on obtient :

Théorème 7. *Soient $B \subset R^m$ et $X \subset R^n \times B$ des ensembles semi-algébriques. On note p la projection de X sur B. Soit $f : X \to R$ une fonction semi-algébrique telle que pour tout $b \in B$, la fonction $f_b : X_B \to R$ est continue. Alors il existe une partition semi-algébrique* *finie* $B = \bigcup_{i \in I} B^i$, *et pour chaque* $i \in I$ *un carré commutatif*

$$\begin{array}{ccc} X^i = X \cap p^{-1}(B^i) & \xrightarrow{(f_{|X^i}, p_{|X^i})} & R \times B^i \\ \Big\downarrow h^i & & \Big\downarrow g^i \\ F^i & \xrightarrow{\phi^i} & R \end{array}$$

où F^i *est un ensemble semi-algébrique,* h^i, ϕ^i *et* g^i *sont semi-algébriques continues et vérifient que pour tout* $b \in B^i$, $h^i_b : X_b \to F^i$ *et* $g^i_b : R \to R$ *sont des homéomorphismes.*

Démonstration. Sans perte de généralité, on peut supposer que f est bornée, à valeurs dans $[-1, 1]$. Soit $\alpha \in \widetilde{B}$. La fibre $f_\alpha : X_\alpha \to k(\alpha)$ est une fonction semi-algébrique continue bornée, et on peut lui appliquer le théorème de triangulation des fonctions semi-algébriques continues [SH] pour le corps réel clos $k(\alpha)$: la fonction f_α est semi-algébriquement équivalente à la restriction d'une fonction affine par morceaux sur un complexe simplicial fini. On insiste ici pour que le complexe simplicial et la fonction affine par morceaux sont des extensions à $k(\alpha)$ d'objets définis sur R. De façon précise, il y a un complexe simplicial fini K défini sur R, une fonction $\psi : K \to R$, affine sur chaque simplexe, une réunion de simplexes ouverts $F \subset K$, et un carré commutatif

$$\begin{array}{ccc} X_\alpha & \xrightarrow{f_\alpha} & k(\alpha) \\ \Big\downarrow H_\alpha & & \Big\downarrow G_\alpha \\ F_{k(\alpha)} & \xrightarrow{\phi_{k(\alpha)}} & k(\alpha) \end{array}$$

où ϕ est $\psi_{|F}$, et où H_α et G_α sont des homéomorphismes semi-algébriques. On récupère ainsi un sous-ensemble semi-algébrique $B' \subset B$, tel que $\alpha \in \widetilde{B'}$, et un carré commutatif

$$\begin{array}{ccc} X' = X \cap p^{-1}(B') & \xrightarrow{(f_{|X'}, p_{|X'})} & R \times B' \\ \Big\downarrow H & & \Big\downarrow G \\ F \times B' & \xrightarrow{\phi \times B'} & R \times B' \end{array}$$

où toutes les applications sont compatibles avec les projections sur B', et où H et G sont des homéomorphismes semi-algébriques dont les fibres en α sont H_α et G_α respectivement. Pour obtenir l'énoncé du théorème, il ne reste plus qu'à prendre en compte la compacité de $\widetilde{B}$ pour avoir la finitude de la partition. □

On obtient aussi des résultats de finitude et de calculabilité pour la triangulation des fonctions. Ces résultats ont été montré dans [BS] en réexaminant le preuve de [SH]. Ici, on les obtient par un artifice logique. C'est moins fatigant, mais le défaut est que l'on n'a aucune précision sur les bornes obtenues, à part le fait qu'elles sont récursives.

Théorème 8. *Etant donnés n et d, il existe s fonctions semi-algébriques continues*

$$\phi^i : \mathbb{R}^n \supset X^i \longrightarrow R, \quad i = 1, \ldots, s$$

de degré d, et un entier t, tels que pour toute fonction

$$\phi : \mathbb{R}^n \supset X \longrightarrow R$$

de degré d, il existe un entier i, $1 \leq i \leq s$, et des homéomorphismes semi-algébriques

$$h : X \longrightarrow X^i \quad \text{et} \quad g : R \longrightarrow R$$

de degrés au plus t, et qui vérifient $\phi^i \circ h = g \circ \phi$. De plus s et t sont bornés par des fonctions récursives de n et d.

Démonstration. On peut paramétrer, par un ensemble semi-algébrique $B(n,d)$, la famille de toutes les fonctions semi-algébriques continues, à valeurs dans R, et dont le domaine est un sous-ensemble semi-algébrique de R^n. On obtient ainsi une fonction

$$\begin{array}{ccc} \mathcal{X} & \xrightarrow{\quad f \quad} & R \\ {\scriptstyle p}\big\downarrow & & \\ B(n,d) & & \end{array}$$

à qui l'on applique le théorème précédent. On récupère une partition semi-algébrique finie $B(n,d) = \bigcup_{i=1}^{s} B^i$, et en choisissant $b^i \in B^i$, on a pour chaque $i \in I$ un carré commutatif

$$\begin{array}{ccc} \mathcal{X}^i = \mathcal{X} \cap p^{-1}(B^i) & \xrightarrow{(f_{|\mathcal{X}^i}, p_{|\mathcal{X}^i})} & R \times B^i \\ {\scriptstyle h^i}\big\downarrow & & \big\downarrow{\scriptstyle g^i} \\ X^i = \mathcal{X}_{b^i} & \xrightarrow{\phi^i = f_{b^i}} & R \end{array}$$

où h^i et g^i sont semi-algébriques continues, vérifient que pour tout $b \in B^i$, $h^i_b : X_b \to X^i$ et $g^i_b : R \to R$ sont des homéomorphismes et que $h^i_{b^i} = \mathrm{Id}_{X^i}$ et $g^i_{b^i} = \mathrm{Id}_R$. En prenant pour t le maximum des degrés des h^i et des g^i, on obtient la première partie de l'énoncé. Pour obtenir la récursivité des bornes, on remarque que, une fois que l'on a fixé n, d, s et t, la thèse du théorème 8 "Il existe s fonctions semi-algébriques... qui vérifient $\phi^i \circ h = g \circ \phi$." s'écrit comme un énoncé $\Phi(n, d, s, t)$ du premier ordre de la théorie des corps réels clos, que l'on peut produire algorithmiquement à partir de n, d, s et t. L'algorithme pour calculer une borne pour s et t en fonction de n et d consiste alors à appliquer un algorithme de décision pour la théorie des corps réels clos aux énoncés $\Phi(n, d, 1, 1)$, $\Phi(n, d, 2, 2), \ldots$ et en s'arrêtant quand on trouve une formule vraie. La première partie du théorème dit que l'algorithme s'arrête bien. □

5. Un contre-exemple pour les familles d'applications semi-algébriques définies sur R^2

Le résultat de Benedetti et Shiota formulé dans le théorème 8 généralise un résultat de Fukuda [FU] qui concerne les fonctions polynomiales. On connaît aussi un théorème d'Aoki [AN] qui dit que, le nombre de types topologiques de germes d'applications polynomiales de $\mathbb{R}^2$ dans $\mathbb{R}^2$ de degré donné est fini. Mais dans ce cas, on ne peut pas espérer une généralisation aux familles d'applications semi-algébriques continues, comme le montre le contre-exemple suivant inspiré par Nakai [NA].

Soit $a = (a_1, a_2, a_3, a_4) \in \mathbb{R}^4$, avec $0 < a_1 < a_2 < a_3 < a_4$. Soit $f_a : \mathbb{R}^2 \to \mathbb{R}^2$ la fonction semi-algébrique continue définie par

$$f_a : (x, y) \longmapsto (|x|, a_i |x| \, |y|)$$

où i est le numéro du quadrant dans lequel se trouve (x, y) (les quadrants sont numérotés dans l'ordre direct). On va montrer que pour deux quadruplets différents a et b, les applications f_a et f_b ne sont pas en général semi-algébriquement équivalentes. Avec un petit peu plus de travail, on aurait le même résultat pour l'équivalence topologique.

Examinons d'abord les fibres de la fonction f_a :

- $f_a^{-1}(0,0)$ est l'axe des y,
- $f_a^{-1}(u, 0)$, pour $u > 0$, comprend les deux points $(u, 0)$ et $(-u, 0)$,
- $f_a^{-1}(u, v)$, pour $u, v > 0$, comprend les quatres points $(u, \dfrac{v}{a_1 u})$, $(-u, \dfrac{v}{a_2 u})$, $(-u, \dfrac{-v}{a_3 u})$, $(u, \dfrac{-v}{a_4 u})$,
- dans les autres cas la fibre est vide.

Soit maintenant $\gamma : [0, 1] \to \mathbb{R}^2$ un chemin semi-algébrique avec $\gamma(0) = (0, 0)$ et $\gamma(]0, 1]) \subset \{(u, v) \; ; \; u > 0, \; v > 0\}$. Ce chemin arrive à l'origine avec une tangente de coefficient directeur $t(\gamma)$. Si $0 < t(\gamma) < +\infty$, alors $S(\gamma) = f_a^{-1}(\gamma([0, 1]))$ contient les quatre points remarquables d'abscisses (dans l'ordre croissant) $t(\gamma)/(-a_3)$, $t(\gamma)/(-a_4)$, $t(\gamma)/a_2$ et $t(\gamma)/a_1$ sur l'axe des y. Ces points, que l'on désignera par $P_3(\gamma)$, $P_4(\gamma)$, $P_2(\gamma)$ et $P_1(\gamma)$ se remarquent dans $S(\gamma)$ parce que ce sont ceux dont le complémentaire a trois composantes. Ce phénomène ne se produit pas si $t(\gamma) = 0$ ou $+\infty$. Choisissons alors un chemin semi-algébrique γ comme ci-dessus, avec $0 < t(\gamma) < +\infty$. On peut trouver deux suites de chemins semi-algébriques (δ_n) et (ϕ_n), telles que $\delta_0 = \phi_0 = \gamma$, que $P_3(\delta_{n+1}) = P_4(\delta_n)$ et que $P_1(\phi_{n+1}) = P_2(\phi_n)$. Définissons $s(n)$ comme le plus petit entier s tel que ϕ_s soit en dessous de δ_n près de l'origine. Les remarques faites plus haut montrent que

$$t(\delta_n) = t(\gamma)\left(\frac{a_3}{a_4}\right)^n \quad \text{et} \quad t(\phi_n) = t(\gamma)\left(\frac{a_1}{a_2}\right)^n.$$

On en déduit donc que

$$\lim_{n\to\infty} \frac{s(n)}{n} = \frac{\log(a_3) - \log(a_4)}{\log(a_1) - \log(a_2)}$$

On suppose maintenant que f_a et f_b sont semi-algébriquement équivalentes, c'est-à-dire qu'il existe des homéomorphismes semi-algébriques g et h de $\mathbb{R}^2$ sur lui-même tels que $f_b \circ g = h \circ f_a$. D'après ce que l'on a vu pour les fibres, h envoie l'origine sur l'origine, le demi-axe des $u > 0$ sur lui-même, et le premier quadrant ouvert sur lui-même ; g envoie l'axe des y sur lui-même. La situation avec les suites de chemins (δ_n) et (ϕ_n) se transporte par h en une situation semblable, avec la même fonction $s(n)$, et on doit donc avoir

$$\frac{\log(a_3) - \log(a_4)}{\log(a_1) - \log(a_2)} = \frac{\log(b_3) - \log(b_4)}{\log(b_1) - \log(b_2)}.$$

Cette dernière égalité montre que l'on ne peut pas avoir un nombre fini de classes d'équivalence semi-algébrique dans la famille des fonctions f_a quand a varie.

Références

[AN] Aoki K., Nagachi H. : On topological types of polynomial map germs of plane to plane, Memoirs of the School of Science & Engineering 44 (1980), 133-156, Waseda University

[BS] Benedetti R., Shiota M. : Finiteness of semialgebraic types of polynomial functions, à paraître

[BCR] Bochnak J., Coste M., Roy M-F. : Géométrie algébrique réelle, Springer (1987)

[DU] Durfee A. : Neighborhoods of algebraic sets, Trans. A.M.S. 270 (1983), 517-530

[FU] Fukuda T. : Types topologiques des polynômes, Publ. math. I.H.E.S. 46 (1976), 87-106

[GR] Gromov M. : Entropy, homology and semialgebraic geometry (after Y. Yomdin), Astérisque 145-146 (1987), 225-240

[HA] Hardt R. : Semi-algebraic local-triviality in semi-algebraic mappings, Amer. J. Math. 102 (1980), 291-302

[NA] Nakai I. : On topological types of polynomial mappings, Topology 23 (1984), 45-66

[SH] Shiota M. : Piecewise linearization of subanalytic functions II. Dans : Real analytic and algebraic geometry, Lect. Notes Math. 1420, 247-307, Springer

STIEFEL ORIENTATIONS ON A REAL ALGEBRAIC VARIETY

A. I. DEGTYAREV

ABSTRACT. Some natural Stiefel orientations on the normal bundle of the fixed point set of an involution on a smooth manifold are constructed. This result is applied to non-singular real algebraic varieties in order to generalize Rokhlin's construction of complex orientations on a separating real algebraic curve

INTRODUCTION

0.1. The main purpose of this paper is to generalize Rokhlin's construction of complex orientations on a separating real algebraic curve. The original construction is the following (Rokhlin [6]): Denote, respectively, by $\mathbb{C}A$ and $\mathbb{R}A$ the sets of real and complex points of curve A. Since $\mathbb{R}A$ separates $\mathbb{C}A$, the complement $\mathbb{C}A \setminus \mathbb{R}A$ consists of two components $\mathbb{C}A_{\pm}$. The complex orientation of complex manifolds $\mathbb{C}A_{\pm}$ defines a pair of opposite orientations of their common boundary $\mathbb{R}A$.

A direct generalization of this construction is obvious: Let F be an m-dimensional smooth submanifold of a smooth n-dimensional manifold X. Suppose that the fundamental class $[F]$ of F vanishes in $H_m(X)$. Then the linking coefficient map $\operatorname{lk} : H_{n-m-1}(X \setminus F) \longrightarrow \mathbb{Z}_2$ restricted to the boundary of a tubular neighborhood of F in X is an $(n-m-1)$-dimensional Stiefel orientation on the normal bundle ν_F of F in X. (The notion of Stiefel orientation is a generalization of the ordinary orientation and Spin-structure on a vector bundle; see Definition 1.4.1 for details.) This orientation is well defined over the subgroup Ker[inclusion$_*$: $H_{n-m-1}(F) \to H_{n-m-1}(X)$]. O.Viro [8] proposed a further generalization of this construction (see 4.2) for the case when pair (X, F) is supplied with the following additional structure: X is a manifold with an involution $c : X \to X$, and $F = \operatorname{Fix} c$ is the fixed point set of c. In this paper we give another approach to Viro's construction which shows that the arising Stiefel orientations are, in fact, also the linking coefficient functionals on homology groups of some appropriate spaces $\mathcal{S}^{k-1}X \setminus \mathcal{S}^{k-1}F$ associated with pair (X, c). These orientations exist if for some $k \geqslant 1$ class $[F]$ vanishes under some "higher" inclusion maps $e^k_{m+k-1} : H_*(F) \dashrightarrow H_{m+k-1}(X)$, and it is defined over the subgroup $\operatorname{Ker} e^k_{n-m-1}$ (see Theorem 4.1.1 for details).

Remark. The advantage of the approach of this paper is that the construction is connected with the general theory of $\mathbb{Z}_2$-spaces (i.e. spaces with involution). In particular, if F is an M-space (i.e. $\dim H_*(F) = \dim H_*(X)$), this theory enables to express the

---

Key words and phrases. Stiefel orientation, involution, fixed point set, real algebraic variety.

condition necessary for existence of the orientations in terms of characteristic classes of X (Degtyarev [3]).

0.2. Applying Theorem 4.1.1 to an n-dimensional real algebraic manifold A yields a series of partial i-dimensional Stiefel orientations on the tangent bundle of $\mathbb{R}A$, $n-k \leqslant i \leqslant n-1$, provided that $e^k_{n+k-1}[\mathbb{R}A] = 0$. However, this condition seems to be rather difficult to be verified. If A is a projective complete intersection, it can be substantially simplified and reduced to the usual requirement that $[\mathbb{R}A]$ should be either equal to zero or dual to the plane section class in $H_n(\mathbb{C}A)$. In this case some modified construction provides a series of Stiefel orientations of all dimensions on the tangent bundle of the double cover $\widetilde{\mathbb{R}A}$ of $\mathbb{R}A$ corresponding to the real hyperplane section class $\mathbb{R}h \in H^1(\mathbb{R}A)$. These orientations are well defined over the whole group $\widetilde{H}_*(\mathbb{R}A)$ (Theorem 4.3.4).

Remark. If $[\mathbb{R}A]$ is equal to the plane section class, manifold $\mathbb{R}A$ may not possess any Stiefel orientation, but the modified construction is still applicable to $\widetilde{\mathbb{R}A}$. If A is a surface, there is another approach, also due to Viro. It provides a Pin$^-$-structure on $\mathbb{R}A$ which lifts to the 1-dimensional Stiefel orientation on $\widetilde{\mathbb{R}A}$ arising due to Theorem 4.3.4. It would be interesting to find a higher-dimensional generalization of this approach.

Remark. The main difficulty in generalizing Viro's approach (see the previous remark) seems to consist in the following:

In a way, Pin$^-$-structure is the structure corresponding to characteristic class $w_2+w_1^2$. More precisely, let $f : X \to BO_n$ be a characteristic map of bundle ξ, and $P(\omega) \to BO_n$ be a fixed fibration killing some characteristic class $\omega \in H^k(BO_n)$. Then an ω-structure on ξ can be defined as a class $\varkappa \in H^{k-1}(f^*P(\omega))$ which does not vanish when restricted to any fibre. (For example, Stiefel orientations are defined via the Stiefel bundles $V_{n,k}(\eta_n) \to BO_n$, killing w_{n-k+1}; Pin$^-$-structure is defined via the bundle $B\mathrm{Pin}^- \to BO_n$, which kills $w_2 + w_1^2$.) The problem is that there is no universal way to choose the killing fibration $P(\omega)$, so in fact structures of this kind are functorially defined only modulo all $(k-1)$-dimensional characteristic classes. This reduction being accepted, both Viro's Pin$^-$-structure and Nezvetaev's form (Viro [8]; in fact, this form is a w_1^2-structure) can easily be generalized (Degtyarev [2]). But in order to define the absolute (i.e. not reduced modulo characteristic classes) structures some appropriate killing spaces over BO_n and BU_n are necessary which would be functorial under some standard homomorphisms of groups O_n and U_n.

0.3. Among the other results of the paper is a new approach to defining Stiefel orientations. Instead of the associated Stiefel bundles $V_{n,k}(\xi)$ it involves some other spaces which can be constructed using only the total space of the associated sphere bundle $S^{n-1}(\xi)$ and the standard antipodal involution on it (Definition 3.3.1). It is this approach that, being applied to a tubular neighborhood of F in X, makes the construction of Theorem 4.1.1 possible.

0.4. The paper is organized as follows:

In section 1 we remind some known basic facts and introduce some notations, which are not generally accepted.

Section 2 is devoted to $\mathbb{Z}_2$-spaces. Most of the notions and results of this section seem to be familiar to those who deal with involutions, though I could not find any systematic account of them. In 2.1 we discuss singular Smith homology groups, which are more suitable for this paper. In 2.2 the spectral sequence of space $S^\infty \times_{\mathbb{Z}_2} X$ (cf., for example, Hsiang [4]) and of some natural finite dimensional approximations of this space are introduced. In 2.3 we define the stabilized spectral sequence of an involution. This sequence, first introduced by Kalinin [5], is, in fact, just a compact way to express the stabilization theorem (Hsiang [4]). Kalinin also conjectured the description of the differentials of of these spectral sequences given in 2.4. In 2.5 we define higher inclusion homomorphisms $e^r_* : SH_*(X) \dashrightarrow H_*(X)$ and study their properties.

In section 3 the new approach to constructing Stiefel orientations is introduced and its relation to the ordinary orientations is studied.

Section 4 is devoted to the main construction and its applications to real algebraic varieties.

0.5. Acknowledgements. I take this opportunity to express my deep gratitude to O.Ya.Viro for his help and interest to this work.

1. PRELIMINARIES

1.1. Complexes and homology groups. For a topological space X we denote by $S_*(X)$ and $S^*(X)$ its singular chain and cochain complexes with *coefficients* $\mathbb{Z}_2$. All homology and cohomology groups also have coefficients $\mathbb{Z}_2$. When it cannot confuse, we use the same letters to denote cycles and their homology classes.

1.2. *G*-bundles. Let G be a Lie group. The term "G-bundle" (over space X) means a principal G-bundle $\xi : G(\xi) \to X$, i.e. a free right G-space $G(\xi)$ such that $G(\xi)/G = X$.

For a G-bundle ξ and left G-space F we denote by $F(\xi)$ the *associated F-bundle* $G(\xi) \times_G F \longrightarrow X$, where the twisted product $G(\xi) \times_G F$ is the quotient space

$$G(\xi) \times F / \{(x, f) = (xg^{-1}, gf) \quad \text{for any } g \in G\}.$$

1.3. Some standard spaces. We use the following standard notations:

$\mathbb{R}^n$ is the coordinate Euclidean space with the standard basis $(\mathbf{e}_1, \ldots, \mathbf{e}_n)$. For $k \leqslant n$ we identify $\mathbb{R}^k$ with the subspace of $\mathbb{R}^n$ generated by $(\mathbf{e}_1, \ldots, \mathbf{e}_k)$;

S^n and $D^n_\pm$ are the standard sphere and semispheres

$$S^n = \{\sum t^i \mathbf{e}_i \in \mathbb{R}^{n+1} \mid \sum (t^i)^2 = 1\}$$
$$D^n_+ = \{\sum t^i \mathbf{e}_i \in \mathbb{R}^{n+1} \mid \sum (t^i)^2 = 1,\ t^1 \geqslant 0\}$$
$$D^n_- = \{\sum t^i \mathbf{e}_i \in \mathbb{R}^{n+1} \mid \sum (t^i)^2 = 1,\ t^1 \leqslant 0\};$$

Δ^n is the standard simplex $\{\sum t^i \mathbf{e}_i \in \mathbb{R}^{n+1} \mid \sum t^i = 1,\ t^i \geqslant 0\}$;

O_n is the group of orthogonal transformations of $\mathbb{R}^n$;

$V_{n,k}$ is the *Stiefel manifold,* which can be defined as either the quotient space O_n/O_{n-k}, or the space of all orthogonal k-frames in $\mathbb{R}^n$; it is well known that $\widetilde{H}^i(V_{n,k}) = 0$ for $i < n-k$, and $\widetilde{H}^{n-k}(V_{n,k}) = \mathbb{Z}_2$ (see Steenrod, Epstein [7]);

for an O_n-bundle ξ we denote by $S(\xi)$ the associated S^{n-1}-bundle; $S(\xi)$ is called the *sphere bundle* of ξ. Note that $S(\xi) = V_{n,1}(\xi)$.

Let $R = (\boldsymbol{x}_1, \ldots, \boldsymbol{x}_{k+1})$ be an orthonormal $(k+1)$-frame in $\mathbb{R}^{n+1}$. Denote by $\Delta^k(R)$ the singular simplex $\Delta^k \to S^n$, $\sum t^i \boldsymbol{x}_i \longmapsto \sum t^i \boldsymbol{x}_i / \sum (t^i)^2$. We will use the following singular chains in S^n:

$$\Sigma^k_+ = \Delta^n(+\mathbf{e}_1, \ldots, \mathbf{e}_{k+1}) + \sum_{i=2}^{k+1} \Delta^n(+\mathbf{e}_1, \ldots, -\mathbf{e}_i, \ldots, \mathbf{e}_{k+1}),$$

$$\Sigma^k_- = \Delta^n(-\mathbf{e}_1, \ldots, \mathbf{e}_{k+1}) + \sum_{i=2}^{k+1} \Delta^n(-\mathbf{e}_1, \ldots, -\mathbf{e}_i, \ldots, \mathbf{e}_{k+1}),$$

$$\Sigma^k = \Sigma^k_+ + \Sigma^k_-.$$

$\Sigma^k_\pm$ will also be considered as singular chains in $D^k_\pm$. If c is the antipodal involution on S^n, then obviously $c_\# \Sigma^k_+ = \Sigma^k_-$, and $\partial \Sigma^k_\pm = \Sigma^{k-1}_+ + \Sigma^{k-1}_-$.

1.4. Stiefel orientations.

1.4.1. Definition. A (reduced) k-dimensional Stiefel orientation on an O_n-bundle ξ is a class $\varkappa_k \in H^k(V_{n,n-k}(\xi))$ (resp., $\tilde{\varkappa}_k \in \widetilde{H}^k(V_{n,n-k}(\xi))$), such that for any fibre $V_{n,n-k}$ of fibration $V_{n,n-k}(\xi) \to X$ the restriction of $\varkappa_k$ (resp., $\tilde{\varkappa}_k$) to $\widetilde{H}^k(V_{n,n-k})$ is non-zero. The set of all (reduced) Stiefel orientations on ξ is denoted by $\mathrm{St}_k(\xi)$ (resp., $\widetilde{\mathrm{St}}_k(\xi)$).

Remark. A 0-dimensional Stiefel orientation is an ordinary orientation of vector bundle ξ. A 1-dimensional orientation is a Pin^+-structure on ξ. A pair consisting of a 0- and 1-dimensional orientation is a Spin-structure.

Remark. If $k > 0$, obviously $\widetilde{\mathrm{St}}_k(\xi) = \mathrm{St}_k(\xi)$. If $k = 0$, a reduced orientation is a pair of opposite ordinary orientations on ξ. (So this notion is not trivial only if the underlying space X is not connected.)

1.4.2. Theorem (well known). *Let ξ be an O_n-bundle over X. Then* $\mathrm{St}_k(\xi)$ *and* $\widetilde{\mathrm{St}}_k(\xi)$ *are not empty if and only if $w_{k+1}(\xi) = 0$. In this case $\mathrm{St}_k(\xi)$ (resp., $\widetilde{\mathrm{St}}_k(\xi)$) is an affine space over $H^k(X)$ (resp., $\widetilde{H}^k(X)$). The reduction map $\mathrm{St}_k(\xi) \to \widetilde{\mathrm{St}}_k(\xi)$ is affine.*

2. $\mathbb{Z}_2$-SPACES

2.1. Singular Smith homology. Throughout this section X is a fixed $\mathbb{Z}_2$-*space,* i.e. a topological space supplied with an involution $c : X \to X$. We denote by $\mathrm{Fix}\, c$ the fixed point set of c, and by X/c the quotient space. For simplicity we assume that pair $(X, \mathrm{Fix}\, c)$ allows a c-invariant cell partition. (Note, though, that most results can easily be extended to general $\mathbb{Z}_2$-spaces using an equivariant cell approximation of (X, c).)

2.1.1. Notation. Denote $\Lambda = \mathbb{Z}_2[c]/(c^2 = 1)$. Since $\mathbb{Z}_2$ acts on $S_*(X)$ via the induced involution $c_\#$, this complex can be considered as a complex of Λ-modules.

We will also consider the $\mathbb{Z}_2$-algebra $H^*(\mathbb{R}\mathrm{p}^k) = \mathbb{Z}_2[h]/(h^{k+1} = 0)$ and the $H^*(\mathbb{R}\mathrm{p}^k)$-module $H_*(\mathbb{R}\mathrm{p}^k)$. The generator of $H_i(\mathbb{R}\mathrm{p}^k) = \mathbb{Z}_2$ is denoted by h_i; both h_i and h^i are assumed to be zero if either $i < 0$ or $i > k$.

2.1.2. Definition. The *Smith homology groups* $SH_*(X)$ are the homology groups of the *Smith complex* $Sm_*(X) = \mathrm{Ker}[(1+c_\#) : S_*(X) \to S_*(X)]$. The inclusion homomorphism $SH_*(X) \to H_*(X)$ is denoted by sm_X.

2.1.3. Proposition. *(1) Let $C_*(X)$ be the cell complex corresponding to some c-invariant cell partition of pair $(X, \mathrm{Fix}\, c)$. Then $SH_*(X) = H_*(\mathrm{Ker}[(1+c_\#) : C_*(X) \to C_*(X)])$;*

(2) denote by $p : X \to X/c$ the projection. Then the map $x \mapsto \mathrm{rel}\, p_\# x$ induces isomorphism $H_*(S_*(X)/Sm_*(X)) \cong H_*(X/c, \mathrm{Fix}\, c)$;

(3) $Sm_*(X)$ naturally splits into $S_*(\mathrm{Fix}\, c) \oplus S_*(X)/Sm_*(X)$, due to (2) this induces an isomorphism $SH_*(X) \cong H_*(\mathrm{Fix}\, c) \oplus H_*(X/c, \mathrm{Fix}\, c)$.

Proof. The last two statements are well known for cell Smith homology groups (cf. Bredon [1]), so they follow from the first one. Proof of the first statement is absolutely analogous to the standard proof of similar statement about ordinary homology groups $H_*(X)$. We omit the details since they would require developing from the very beginning the theory of functor SH_*, which is absolutely analogous to the standard singular homology theory. □

2.1.4. Corollary. *Let $(X, \mathrm{Fix}\, c)$ is an n-dimensional cell pair. Then $SH_i(X) = 0$ for $i > n$, and $SH_n(X) = \mathrm{Ker}[(1+c_*) : H_n(X) \to H_n(X)]$.*

2.1.5. Corollary. *Let X be an n-dimensional smooth manifold, and c be a smooth involution. Then there exists one and only one class $[X]_S \in SH_n(X)$ such that $\mathrm{sm}_X[X]_S = [X]$. We call this class the* Smith fundamental class *of X.)*

For $(\mathrm{sm}_X)_n : SH_n(X) \to H_n(X)$ is a monomorphism, and $[X]$ is a c_*-invariant class. □

2.2. Spectral sequence of an involution.

2.2.1. Definition. Denote by $\mathcal{S}^k X$, $0 \leqslant k \leqslant \infty$, the twisted product $S^k \times_{\mathbb{Z}_2} X$ supplied with the induced involution and filtration by subspaces $\emptyset \subset \mathcal{S}^0 X \subset \mathcal{S}^1 X \subset \dots \subset \mathcal{S}^k X$. (Here inclusions are induced by the standard inclusions $\emptyset \subset S^0 \subset S^1 \subset \dots \subset S^k$.) The homology and cohomology spectral sequences of filtered space $\mathcal{S}^k X$ are denoted by ${}^kE^r_{pq}(X)$ and ${}^kE_r^{pq}(X)$ respectively and called the *spectral sequences* of involution c. (Here and in other similar notations space X will often be omitted.)

Note that if X is a trivial $\mathbb{Z}_2$-space, i.e. $c = \mathrm{id}_X$, then $\mathcal{S}^k X = \mathbb{R}\mathrm{p}^k \times X$, and $H_*(\mathcal{S}^k X) = H_*(\mathbb{R}\mathrm{p}^k) \otimes H_*(X)$. In particular, both the spectral sequences degenerate in term E^1.

2.2.2. Proposition.

(1) ${}^kE^1_{**} = H_*(\mathbb{R}\mathrm{p}^k) \otimes H_*(X)$, and ${}^kE_1^{**} = H^*(\mathbb{R}\mathrm{p}^k) \otimes H^*(X)$;

(2) $d^1(h_p \otimes x) = h_{p-1} \otimes (1+c_*)x$, and $d_1(h^p \otimes x) = h^{p+1} \otimes (1+c^*)x$;

(3) if $r \geqslant 2$, both the sequences coincide with the Serre spectral sequences of fibration $\mathcal{S}^k X \to \mathbb{R}\mathrm{p}^k = \mathcal{S}^k\mathrm{pt}$;

(4) ${}^kE_r^{**}$ is a bigraded $H^*(\mathbb{R}\mathrm{p}^k)$-algebra, and ${}^kE^r_{**}$ is a bigraded ${}^kE_r^{**}$-module; if $r \geqslant 2$, differentials d^r and d_r are derivatives;

(5) ${}^kE^r_{pq}$ and ${}^kE_r^{pq}$ converge to $H_{p+q}(\mathcal{S}^k X)$ and $H^{p+q}(\mathcal{S}^k X)$ respectively; the convergence observes both the multiplicative structures;

(6) if $l \leqslant k$, the inclusion homomorphism ${}^{l}E^{r}_{pq} \to {}^{k}E^{r}_{pq}$ (resp., ${}^{k}E^{pq}_{r} \to {}^{l}E^{pq}_{r}$) is surjective (resp., injective) for $p \leqslant l$ and one-to-one for $p \leqslant l+1-r$.

Proof. By definition,

$$\begin{aligned} {}^{k}E^{1}_{pq} &= H_{p+q}(S^{p}X, S^{p-1}X) \\ &= H_{p+q}(S^{p}\times X, D^{p}_{-}\times X) \\ &= H_{p+q}(D^{p}_{+}\times X, S^{p-1}\times X) \\ &= H_{q}(X). \end{aligned}$$

This proves (1). Statement (2) follows from 2.4.4(1) below. (3) is just a rewording of the construction of the Serre spectral sequence using the skeleton filtration of the base of the fibration. (4) and (5) are well known properties of the Serre spectral sequence. (6) follows from the fact that ${}^{l}E^{1}_{**}$ (resp., ${}^{l}E^{**}_{1}$) is a truncation of ${}^{k}E^{1}_{**}$ (resp., ${}^{k}E^{**}_{1}$). □

2.3. Stabilized spectral sequence. Denote by $\varphi^{r}_{p} : {}^{\infty}E^{r}_{p+1,*} \to {}^{\infty}E^{r}_{p,*}$ and $\varphi^{p}_{r} : {}^{\infty}E^{p,*}_{r} \to {}^{\infty}E^{p+1,*}_{r}$ the maps $x \mapsto h \cap x$ and $x \mapsto h \cup x$ respectively.

2.3.1. Definition (Kalinin [5]). Spectral sequences

$$E^{r}_{q}(X) = \varprojlim({}^{\infty}E^{r}_{pq}, \varphi^{r}_{p}) \quad\text{and}\quad E^{q}_{r}(X) = \varinjlim({}^{\infty}E^{pq}_{r}, \varphi^{p}_{r})$$

are called the ***stabilized spectral sequences*** of involution c.

Natural maps $E^{r}_{q} \to {}^{\infty}E^{r}_{pq}$ and ${}^{\infty}E^{pq}_{r} \to E^{q}_{r}$ will be denoted by ψ^{r}_{q} and ψ^{q}_{r} respectively.

2.3.2. Theorem (Kalinin [5]).

(1) $E^{1}_{q} = H_{q}(X)$, and $E^{q}_{1} = H^{q}(X)$;

(2) $d^{1} = 1 + c_{*}$, and $d_{1} = 1 + c^{*}$;

(3) E^{*}_{r} is a graded $\mathbb{Z}_2$-algebra, and E^{r}_{*} is a graded E^{*}_{r}-module; d_r and d^r are derivatives if $r \geqslant 2$;

(4) provided that X is a finite CW-complex, E^{r}_{*} and E^{*}_{r} converge to $H_{*}(\operatorname{Fix} c)$ and $H^{*}(\operatorname{Fix} c)$ respectively; the convergence observes the multiplicative structures, but it does not observe the grading of $H_{*}(\operatorname{Fix} c)$ and $H^{*}(\operatorname{Fix} c)$;

(5) homology maps ψ^{r}_{p} are injective; cohomology maps ψ^{p}_{r} are surjective.

2.3.3. Notation. The filtration on $H_{*}(\operatorname{Fix} c)$ which arises due to convergence $E^{r}_{*} \Rightarrow H_{*}(\operatorname{Fix} c)$ is denoted by $\mathcal{F}_{*}(X)$. The composed maps $\mathcal{F}_{q} \longrightarrow \mathcal{F}_{q}/\mathcal{F}_{q+1} \xrightarrow{\cong} E^{\infty}_{q}$ are denoted by e_q.

Note that filtration $\mathcal{F}_{q}$ can as well be defined if E^{r}_{*} does not converge to $H_{*}(\operatorname{Fix} c)$ (Kalinin [5]). By definition, the class of cycle $x = \sum x_i$, $x_i \in S_i(\operatorname{Fix} c)$, belongs to $\mathcal{F}_{q}$ if for some p the image of $\sum h_{p+q-i} \otimes x_i \in H_{p+q}(\mathcal{S}^{\infty} \operatorname{Fix} c)$ in $H_{p+q}(\mathcal{S}^{\infty} X)$ comes from $H_{p+q}(\mathcal{S}^{p} X)$.

2.4. Geometric description. Let C_* (resp., C^*) be a chain (cochain) complex of Λ-modules. Define filtered complexes $\mathcal{S}^{k}C_{*}$ and $\mathcal{S}^{k}C^{*}$ as follows:

as a graduated group $\mathcal{S}^{k}C_{*} = H_{*}(\mathbb{R}\mathrm{p}^{k}) \otimes C_{*}$ (resp., $\mathcal{S}^{k}C^{*} = H^{*}(\mathbb{R}\mathrm{p}^{k}) \otimes C^{*}$);

the boundary (coboundary) operator is the map $h_i \otimes x \mapsto h_i \otimes \partial x + h_{i-1} \otimes (1+c)x$ (resp., $h^{i} \otimes x \mapsto h^{i} \otimes \delta x + h^{i+1} \otimes (1+c)x$);

the filtration is, respectively, $F_p = \text{Im}[\text{inclusion}_* \otimes \text{id} : \mathcal{S}^p C_* \to \mathcal{S}^k C_*]$, and $F^p = \text{Ker}[\text{inclusion}^* \otimes \text{id} : \mathcal{S}^k C^* \to \mathcal{S}^p C^*]$.

The spectral sequences of filtered complexes $\mathcal{S}^k C_*$ and $\mathcal{S}^k C^*$ are denoted by ${}^k E^r_{pq}(C_*)$ and ${}^k E_r^{pq}(C^*)$ respectively.

2.4.1. Theorem. *There exists a natural chain map* $\theta : \mathcal{S}^k S_*(X) \to S_*(\mathcal{S}^k X)$ *which induces isomorphisms of*

(1) *spectral sequences* ${}^k E^r_{**}(S_*(X)) \cong {}^k E^r_{**}(X)$ *and* ${}^k E_r^{**}(S^*(X)) \cong {}^k E_r^{**}(X)$;
(2) *homology groups* $H_*(\mathcal{S}^k S_*(X)) \cong H_*(\mathcal{S}^k X)$ *and* $H^*(\mathcal{S}^k S^*(X)) \cong H^*(\mathcal{S}^k X)$;
(3) *Smith homology groups* $H_*(\mathbb{R}\text{p}^k) \otimes SH_*(X) \cong SH_*(\mathcal{S}^k X)$.

Proof. Let p be the projection $S^k \times X \to \mathcal{S}^k X$, and $\tilde{c}$ be the diagonal involution on $S^k \times X$. Fix some natural chain homotopy equivalence $\mu : S_*(X) \otimes S_*(Y) \longrightarrow S_*(X \times Y)$ and define θ as the map $h_i \otimes x \mapsto p_\# \mu(\Sigma^i_+ \otimes x)$. Since $p_\# \tilde{c}_\# = p_\#$, and μ is natural,

$$\begin{aligned}
&\theta\partial(h_i \otimes x) - \partial\theta(h_i \otimes x) = \\
&= p_\#\mu(\Sigma^{i-1}_+ \otimes c_\# x + \Sigma^i_+ \otimes \partial x) - p_\#\mu(\Sigma^{i-1}_+ \otimes x + \Sigma^{i-1}_- \otimes x + \Sigma^i_+ \otimes \partial x) = \\
&= p_\#\mu(\Sigma^{i-1}_+ \otimes (1 + c_\#)x - \Sigma^{i-1}_- \otimes x) = p_\#(1 - \tilde{c}_\#)\mu(\Sigma^{i-1}_+ \otimes c_\# x) = 0.
\end{aligned}$$

This proves that θ is a chain map. The subquotient map

$$\theta_p : \mathcal{S}^p S_*(X)/\mathcal{S}^{p-1} S_*(X) = h_p \otimes S_*(X) \longrightarrow S_*(\mathcal{S}^p X, \mathcal{S}^{p-1} X)$$

can be decomposed through the map

$$\begin{gathered}
\tilde{\theta}_p : h_p \otimes S_*(X) \longrightarrow S_*(D^p_+ \times X, S^{p-1} \times X), \\
h_p \otimes x \mapsto \mu(\Sigma^i_+ \otimes x),
\end{gathered}$$

which induces an isomorphism of terms E^1 of the corresponding spectral sequences. Hence θ induces isomorphisms of the spectral sequences and homology groups.

To prove the last statement, we need the following lemma:

2.4.2. Lemma. *Let* $f : C_* \to D_*$ *be a chain map which induces isomorphism* $f_* : H_*(C_*) \to H_*(D_*)$. *Then* $(\text{id} \otimes f)_* : H_*(\mathcal{S}^k C_*) \to H_*(\mathcal{S}^k D_*)$ *is also an isomorphism for any* k.

Proof. $\text{id} \otimes f$ induces an isomorphism of terms E^1 of spectral sequences ${}^k E^r_{**}(C_*)$ and ${}^k E^r_{**}(D_*)$. □

Lemma 2.4.2 and Proposition 2.1.3(2) imply that θ induces isomorphism

$$H_*(\mathcal{S}^k(S_*(X)/Sm_*(X))) \cong H_*(S_*(\mathcal{S}^k X)/Sm_*(\mathcal{S}^k X)),$$

and, due to Five Lemma, isomorphism

$$H_*(\mathcal{S}^k Sm_*(X)) \cong SH_*(\mathcal{S}^k X).$$

Since $c_\#$ acts trivially on $Sm_*(X)$, the first group is equal to $H_*(\mathbb{R}\text{p}^k) \otimes SH_*(X)$. □

2.4.3. Corollary (of 2.4.1 and 2.4.2). *Theorem 2.4.1 remains valid if $S_*(X)$ is replaced with the cell complex corresponding to some c-invariant cell partition of pair $(X, \operatorname{Fix} c)$. (Note that for (1) and (2) it even is not necessary that* $\operatorname{Fix} c$ *be a cell subcomplex of* X.*)*

The remaining statements provide a clear geometric description of spectral sequences ${}^kE^r_{**}$ and E^r_*. (The dual description of cohomology spectral sequences is also valid.) The term "chain" ("cycle") means either singular or cell chain (resp., cycle).

2.4.4. Corollary. *Let x be a q-dimensional cycle in X. Then*

(1) $d^1(h_p \otimes x) = h_{p-1} \otimes (1 + c_\#)x$;

(2) let $r \leqslant p$. Then $d^r(h_p \otimes x)$ vanishes in ${}^kE^r_{p-r,q+r-1}(X)$ if and only if there exist some i-dimensional chains y_i in X, $q+1 \leqslant i \leqslant q+r$, such that $(1+c_\#)x = \partial y_{q+1}$, and $(1+c_\#)y_i = \partial y_{i+1}$ for $i \leqslant q+r-1$. In this case $d^{r+1}(h_p \otimes x) = h_{p-r-1} \otimes (1+c_\#)y_{q+r}$.

2.4.5. Corollary. *Let x be a q-dimensional cycle in X. Then*

(1) $d^1 x = (1 + c_\#)x$;

(2) $d^r x$ vanishes in $E^r_{q+r-1}(X)$ if and only if there exist some i-dimensional chains y_i in X, $q+1 \leqslant i \leqslant q+r$, such that $(1+c_\#)x = \partial y_{q+1}$, and $(1+c_\#)y_i = \partial y_{i+1}$ for $i \leqslant q+r-1$. In this case $d^{r+1} x = (1+c_\#)y_{q+r}$.

Proof. 2.4.4 is, in fact, the definition of the spectral sequence of complex $\mathcal{S}^k S_*(X)$. 2.4.5 is obtained by stabilizing differentials of ${}^\infty E^r_{**}$. □

2.4.6. Proposition. *Let $\operatorname{in} : \operatorname{Fix} c \hookrightarrow X$ be the inclusion. Consider some i-dimensional cycles x_i in $\operatorname{Fix} c$, and denote $x = \sum x_i$. Then $x \in \mathcal{F}_q(X)$ if and only if there exist some i-dimensional chains y_i in X, $i \leqslant q$, such that $\operatorname{in}_\# x_i = \partial y_{i+1} + (1+c_\#)y_i$ for $i \leqslant q-1$. In this case $e_q x = \operatorname{in}_\# x_q + (1+c_\#)y_q$.*

Proof. Replace $S_*(\mathcal{S}^k X)$ with $\mathcal{S}^k S_*(X)$. By definition (cf. 2.3.3) $x \in \mathcal{F}_q$ if and only if $\operatorname{in}_\#(\sum h_{p+q-i} \otimes x_i) = \bar{x} + \partial y$ for some $p \geqslant 0$, $\bar{x} \in \mathcal{S}^p S_{p+q}(X)$, and $y \in \mathcal{S}^\infty S_{p+q+1}(X)$. In this case $\psi_p e_q x$ is the class of $\bar{x}$ in ${}^\infty E^\infty_{pq}$. Expanding y in the sum $\sum h_{p+q+1-i} \otimes y_i$ we obtain that $\operatorname{in}_\# x_i = \partial y_{i+1} + (1+c_\#)y_i$ for $i \leqslant q-1$, and

$$\begin{aligned}\bar{x} &= h_p \otimes (\operatorname{in}_\# x_q + (1+c_\#)y_q) \\ &\quad + \partial\Big(\sum_{i \geqslant q+1} h_{p+q+1-i} \otimes y_i\Big) \qquad \text{(belongs to } \partial\mathcal{S}^p S_*(X)) \\ &\quad + \sum_{i \geqslant q+1} h_{p+q-i} \otimes \operatorname{in}_\# x_i \qquad \text{(belongs to } \mathcal{S}^{p-1} S_*(X)).\end{aligned}$$

Hence $\psi_p e_q x = h_p \otimes (\operatorname{in}_\# x + (1+c_\#)y_q)$. □

2.5. Higher inclusion maps. Define filtration $\mathcal{SF}^r_q(X)$ on $SH_*(X)$ and homomorphisms $e^r_q : \mathcal{SF}^r_q(X) \to E^r_q(X)$ as follows:

The class of a cycle $x = \sum_{i \geqslant q-r+1} x_i$, $x_i \in Sm_i(X)$, belongs to $\mathcal{SF}^r_q$ if and only if there exist some chains $y_i \in S_i(X)$, $q-r+2 \leqslant i \leqslant q$, such that $x_{q-r+1} = \partial y_{q-r+2}$, and $x_i = \partial y_{i+1} + (1+c_\#)y_i$ for $q-r+2 \leqslant i \leqslant q-1$. In this case $e^r_q x$ is the class of $x_q + (1+c_\#)y_q$ in $E^r_q(X)$.

Denote by $\mathcal{F}^r_q(X)$ the filtration $\mathcal{SF}^r_q \cap H_*(\operatorname{Fix} c)$ on $H_*(\operatorname{Fix} c)$ (cf. 2.1.3(3)). Restriction $e^r_q\big|_{\mathcal{F}^r_q}$ will also be denoted by e^r_q.

2.5.1. Definition. Homomorphisms e_q^r are called the *higher inclusion homomorphisms.*

Remark. In this definition singular chains can as well be replaced with cell chains corresponding to some c-invariant cell partition of pair $(X, \operatorname{Fix} c)$.

2.5.3. Proposition. *Homomorphisms e_q^r are well defined.*

Proof. We have to prove the following three statements: (1) $x_q + (1 + c_\#)y_q$ is a cycle; (2) $d^s(x_q + (1 + c_\#)y_q) = 0$ for $s \leqslant r$, so this cycle projects into E_q^r; (3) the class of $x_q + (1 + c_\#)y_q$ in E_q^r does not depend on the choice of y_i's.

(1) $\partial(x_q + (1+c_\#)y_q) = (1+c_\#)x_{q-1} + (1+c_\#)^2 y_{q-1} = 0$ since x_{q-1} is $c_\#$-invariant, and $(1 + c_\#)^2 = 0$;

(2) $d^s(x_q + (1 + c_\#)y_q) = 0$ for any s due to the fact that this cycle is $c_\#$-invariant (see Corollary 2.4.5);

(3) let $y_i' = y_i + z_i$ be some other chains. Then $\partial z_{q-r+2} = 0$, and $(1+c_\#)z_i = \partial z_{i+1}$ for $q - r + 2 \leqslant i \leqslant q - 1$. Hence $(1 + c_\#)y_q' = (1 + c_\#)y_q + d^{r-1} z_{q-r+2}$ (Corollary 2.4.5), and the image of $x_q + (1 + c_\#)y_q'$ in E_q^r coincides with that of $x_q + (1 + c_\#)y_q$. □

2.5.4. Theorem.

(1) $\mathcal{SF}_q^1 = \bigoplus_{i \geqslant q} SH_i(X)$, *and* $\mathcal{SF}_q^r = \operatorname{Ker} e_{q-1}^{r-1}$;

(2) $\mathcal{SF}_q^r \supset \bigoplus_{i \geqslant q} SH_i(X)$; *restriction of e_q^r to $SH_i(X)$ is trivial for $i > q$ and coincides (in the obvious sense) with sm_X for $i = q$;*

(3) $\mathcal{SF}_q^r \subset \mathcal{SF}_q^{r+1}$; *restriction of e_q^{r+1} to $\mathcal{SF}_q^r$ coincides with e_q^r;*

(4) *statements* (1)–(3) *remain valid if $\mathcal{SF}_q^r$ are replaced with $\mathcal{F}_q^r$;*

(5) $\mathcal{F}_q^r = \mathcal{F}_q$ *and* $e_q^r = e_q$ *for any $r \geqslant q + 2$ (note that in this case $E_q^r = E_q^\infty$.*

Proof. If $x \in \mathcal{SF}_q^r$, then obviously $x \in \mathcal{SF}_{q-1}^{r-1}$, and $e_{q-1}^{r-1}x = x_{q-1} + (1 + c_\#)y_{q-1} = \partial y_q$ vanishes in E_{q-1}^{r-1}. Hence $\mathcal{SF}_q^r \subset \operatorname{Ker} e_{q-1}^{r-1}$.

Let $x \in \operatorname{Ker} e_{q-1}^{r-1}$. This means that $x_{q-1} + (1 + c_\#)y_{q-1} = d^{r-2} z_{q-r+2} + \partial y_q'$ for some cycle $z_{q-r+2} \in S_{q-r+2}(X)$ and chain $y_q' \in S_q(X)$. Due to Corollary 2.4.5 this is equivalent to existence of some chains z_i, $q - r + 3 \leqslant i \leqslant q - 1$, such that $\partial z_i = (1 + c_\#)z_{i-1}$, and $x_{q-1} + (1 + c_\#)y_{q-1} = (1 + c_\#)z_{q-1} + \partial y_q'$. Denoting $y_i' = y_i + z_i$ for $q - r + 2 \leqslant i \leqslant q - 1$, we see that $x \in \mathcal{SF}_q^r$. This completes proof of (1).

Statements (2)–(4) immediately follow from the definitions, and (5) follows from comparing the definition of $\mathcal{SF}_q^r$ and Proposition 2.4.6. □

2.5.5. Definition. Define map $\bar{e}_q^r : \mathcal{SF}_q^r \to H_q(S^{r-1}X)$ as the product

$$\mathcal{SF}_q^r \xrightarrow{e_q^r} E_q^r \xrightarrow{\psi_0^r} {}^\infty E_{0,q}^r \xleftarrow{\cong} {}^{r-1}E_{0,q}^r \to H_q(S^{r-1}X)$$

(here the third map is invertible due to 2.2.2(6); the last map is the edge homomorphism of spectral sequence ${}^{r-1}E_{pq}^r$).

2.5.6. Proposition. *Let* $x = \sum_{i \geqslant q} x_i$, $x_i \in SH_i(X)$, *and* $r \geqslant 0$. *Then* $\mathrm{sm}_{S^r X}(\sum h_{q+r-i} \otimes x_i) = 0$ *in* $H_{q+r}(S^r X)$ *if and only if* $x \in S\mathcal{F}^{r+2}_{q+r+1}$. *In this case* $\mathrm{sm}_{S^{r+1}X}(\sum h_{q+r+1-i} \otimes x_i) = \bar{e}^{r+2}_{q+r+1}(x)$.

Proof. According to the definition the condition $x \in S\mathcal{F}^{r+2}_{q+r+1}$ is equivalent to the equality $\sum h_{q+r-i} \otimes x_i = \partial(\sum h_{q+r+1-i} \otimes y_i)$ for some $y_i \in S_i(X)$ (cf. Corollary 2.4.4). In this case $\sum h_{q+r+1-i} \otimes x_i = \partial(\sum h_{q+r+2-i} \otimes y_i) + h_0 \otimes (x_{q+r+1} + (1 + c_\#) y_{q+r+1})$; hence this element projects to $\bar{e}^{r+2}_{q+r+1} x \in H_{q+r+1}(S^{r+1}X)$. $\square$

Remark. Since ψ_0^r and the edge homomorphisms ${}^{r-1}E^r_{pq} \to H_q(S^{r-1}X)$ are injective, Proposition 2.5.6 can as well be taken for definition of $S\mathcal{F}^r_q$ and e^r_q.

3. The basic construction

3.1 Groups $H^{n-1}(S^k S(\xi))$. Throughout this section ξ denotes some fixed O_n-bundle over space X.

3.1.1 Definition. Denote by η_k the tautological linear bundle over $\mathbb{R}\mathrm{p}^k$, and by $\eta_k \otimes \xi$ the exterior tensor product, which is a bundle over $\mathbb{R}\mathrm{p}^k \times X$.

3.1.2. Proposition. *(1) There exists a natural fibrewise homeomorphism* $S(\eta_k \otimes \xi) = S^k S(\xi)$, $S(\xi)$ *being considered as a* $\mathbb{Z}_2$*-space via the standard antipodal involution;*

(2) if space X *is connected and* $k \leqslant n-1$, *there exists a natural exact sequence*

$$0 \to \sum_{i=n-k-1}^{n-1} H^i(X) \to H^{n-1}(S^k S(\xi)) \xrightarrow{i^*} H^{n-1}(S^{n-1}),$$

where i^* *is induced by the inclusion of a fibre* S^{n-1} *of fibration* $S(\eta_k \otimes \xi) \to \mathbb{R}\mathrm{p}^k \times X$;

(3) homomorphism i^* *is non-zero if and only if* $w_n(\xi) = \cdots = w_{n-k}(\xi) = 0$;

(4) if $l \leqslant k$, *the following diagram commutes*

$$\begin{array}{ccccccc} 0 & \longrightarrow & \bigoplus_{i=n-k-1}^{n-1} H^i(X) & \longrightarrow & H^{n-1}(S^k S(\xi)) & \xrightarrow{i^*} & H^{n-1}(S^{n-1}) \\ & & \big\downarrow \text{projection} & & \big\downarrow \text{restriction} & & \big\downarrow \text{identity} \\ 0 & \longrightarrow & \bigoplus_{i=n-l-1}^{n-1} H^i(X) & \longrightarrow & H^{n-1}(S^l S(\xi)) & \xrightarrow{i^*} & H^{n-1}(S^{n-1}), \end{array}$$

Proof. Statement (1) immediately follows from definitions. To prove (2) and (3) consider the Gisin exact sequence of the sphere bundle $S(\eta_k \otimes \xi) \to \mathbb{R}\mathrm{p}^k \times X$:

$$\begin{array}{ccccccc} H^{n-1}(\mathbb{R}\mathrm{p}^k \times X) & \hookrightarrow & H^{n-1}(S(\eta_k \otimes \xi)) & \longrightarrow & H^0(\mathbb{R}\mathrm{p}^k \times X) & \xrightarrow{e\cup} & H^n(\mathbb{R}\mathrm{p}^k \times X) \\ \Big\| & & & \searrow^{i^*} & \big\downarrow \cong & & \\ \bigoplus_{i=n-k-1}^{n-1} H^i(X) & & & & H^{n-1}(S^{n-1}) & & \end{array}$$

where $e = \sum h^{n-i} \otimes w_i(\xi)$ is the characteristic class of the bundle. It yields that $\operatorname{Im} i^* \neq 0$ if and only if $e = 0$. Statement (4) follows from the naturallity of the Gisin sequence. $\square$

3.2. Maps $\chi^k : H^{n-1}(\mathcal{S}^{k-1}S(\xi)) \longrightarrow \bigoplus_{i=1}^{k} H^{n-i}(V_{n,i}(\xi))$.

3.2.1. Definition. Let $f : Y \to V_{n,k}(\xi)$ be some map, $y \mapsto (x(y), \mathbf{e}_1(y), \ldots, \mathbf{e}_k(y))$, where $x(y) \in X$ and $(\mathbf{e}_1(y), \ldots, \mathbf{e}_k(y))$ is a k-frame in the fibre at $x(y)$ of the associated vector bundle $\mathbb{R}^n(\xi)$. Define the *coordinatewise lift* of f as the map $\tilde{f} : Y \times S^{k-1} \to S(\xi)$, $y \times \sum t^i \mathbf{e}_i \longmapsto (x(y), \sum t^i \mathbf{e}_i(y))$. (Here S^{k-1} and $S(\xi)$ are considered as subspaces of $\mathbb{R}^k$ and $\mathbb{R}^n(\xi)$ respectively.)

3.2.2. Definition. Fix some natural chain homotopy equivalence $\mu : S_*(X) \otimes S_*(Y) \longrightarrow S_*(X \times Y)$ and define the following maps:

(1) $\theta_{ij} : S_i(V_{n,k}(\xi)) \longrightarrow S_{i+j}(S(\xi))$, $j \leqslant k-1$. Let $x = \sum \sigma_\alpha$ be a singular chain, i.e. a sum of singular simplexes $\sigma_\alpha : \Delta^i \to V_{n,k}(\xi)$. Then $\theta_{ij}(x) = 0$ for $j < 0$, and $\theta_{ij}(x) = \sum (\tilde{\sigma}_\alpha)_\# \mu(\mathrm{id}_{\Delta^i} \otimes \Sigma^j_+)$ otherwise;

(2) $\bar{\varphi}_{k,i} : S_i(V_{n,k}(\xi)) \longrightarrow \mathcal{S}^{k-1} S_{i+k-1}(S(\xi))$. By definition, $\bar{\varphi}_{k,i}(x) = \sum h_{k-j-1} \otimes \theta_{ij}(x)$.

3.2.3. Proposition. *(1) $\bar{\varphi}_{k,i}$ is a chain map, so it induces homomorphisms*

$$\varphi_k = (\bar{\varphi}_{k,n-k})_* : H_{n-k}(V_{n,k}(\xi)) \longrightarrow H_{n-1}(\mathcal{S}^{k-1}S(\xi)), \qquad \text{and}$$
$$\varphi^k : H^{n-1}(\mathcal{S}^{k-1}S(\xi)) \longrightarrow H^{n-k}(V_{n,k}(\xi));$$

(2) if $X = \mathrm{pt}$, homomorphism $\varphi^k : H^{n-1}(\mathcal{S}^{k-1}S^{n-1}) \longrightarrow H^{n-k}(V_{n,k})$ is one-to-one;
(3) the following diagram commutes (cf. 3.1.2(2))

$$\begin{array}{ccc} H^{n-k}(X) & \longrightarrow & H^{n-1}(\mathcal{S}^{k-1}S(\xi)) \\ & \searrow & \downarrow \varphi^k \\ & & H^{n-k}(V_{n,k}(\xi)). \end{array}$$

Proof. We prove the dual statements about homology maps φ_k. Denote by c the antipodal involution on $S(\xi)$.

Let $f : Y \to V_{n,k}(\xi)$ be some map, and $x \in S_i(Y)$ be a singular chain in Y. Then obviously $\theta_{ij}(f_\# x) = \tilde{f}_\# \mu(x \otimes \Sigma^j_+)$. In particular, for a singular simplex $\sigma : \Delta^i \to V_{n,k}(\xi)$ this implies that $\partial \theta_{ij}(\sigma) = \tilde{\sigma}_\# \mu(\partial\, \mathrm{id}_{\Delta^i} \otimes \Sigma^j_+ + \mathrm{id}_{\Delta^i} \otimes (\Sigma^{j-1}_+ + \Sigma^{j-1}_-)) = \theta_{i-1,j}(\partial\sigma) + (1 + c_\#)\theta_{ij}(\sigma)$, and $\partial \bar{\varphi}_{k,i}(\sigma) = \bar{\varphi}_{k,i}(\partial\sigma)$. This proves statement (1).

To prove (2), consider the inclusion $\iota : S^{n-k} \to V_{n,k}$:

$$\sum t^i \mathbf{e}_i \longmapsto (\mathbf{e}_1, \ldots, \mathbf{e}_{k-1}, \sum t^i \mathbf{e}_{i+k-1}),$$

which induces isomorphism $\iota_* : H_{n-k}(S^{n-k}) \to H_{n-k}(V_{n,k})$ (cf. Steenrod, Epstein [7]); we suppose that $k < n$). It is easy to see that the coordinatewise lift $\tilde{\iota} : S^{n-k} \times S^{k-1} \to S^{n-1}$ is a degree 1 map, whose restriction to $S^{n-k} \times S^{k-2}$ coincides with the product const $\times$ inclusion $: S^{n-k} \times S^{k-2} \longrightarrow \mathrm{pt} \times S^{n-1}$. Hence $\theta_{n-k,j}(\Sigma^{n-k}) = \mu(\mathrm{const}_\# \Sigma^{n-k} \otimes \mathrm{inclusion}_\# \Sigma^j_+) = 0$ for $j \leqslant k-2$, and $\theta_{n-k,k-1}(\Sigma^{n-k})$ is a fundamental cycle of S^{n-1}, which generates $H_{n-1}(S^{n-1})$ (cf.3.1.2). If $k=n$, the same construction applied separately to each cycle $\Sigma^0_\pm$ in S^0 gives both the generators of $H_{n-1}(S^{n-1}S^{n-1}) = \mathbb{Z}_2 \oplus \mathbb{Z}_2$.

Prove statement (3). Let $\sigma : \Delta^i \to V_{n,k}(\xi)$ be a singular simplex. Denote by p the projection $V_{n,k}(\xi) \to X$ and by σ_X the simplex $p \circ \sigma : \Delta^i \to X$, and consider the following commutative diagram

$$\begin{array}{ccc} \Delta^i \times S^{n-1} & \xrightarrow{\tilde{\sigma}} & S(\xi) \\ {\scriptstyle \mathrm{id}\times\mathrm{const}}\big\downarrow & & \big\downarrow{\scriptstyle p_S} \\ \Delta^i \times \mathrm{pt} & \xrightarrow{\sigma_X \times \mathrm{id}} & X \times \mathrm{pt} \end{array}$$

Since μ is natural, $(p_S)_\# \theta_{ij}(\sigma) = \mu(\sigma_X \otimes \mathrm{const}_\# \Sigma^j_+)$. If $j>0$, $\mathrm{const}_\# \Sigma^j_+ = 0$. Hence $\bar{\varphi}_{k,n-k}$ projects to the map $\sigma \mapsto h_{k-1} \otimes \mu(p_\# \sigma \otimes \mathrm{id}_{\mathrm{pt}})$, which induces isomorphism $H_{n-k}(X) \longrightarrow h_{k-1} \otimes H_{n-k}(X)$.

3.2.4. Definition. Define $\chi_k : H^{n-1}(S^{k-1}S(\xi)) \longrightarrow \bigoplus_{i=1}^{k} H^{n-i}(V_{n,i}(\xi))$ as the direct sum of composed maps

$$H^{n-1}(S^{k-1}S(\xi)) \xrightarrow{\text{restriction}} H^{n-1}(S^{i-1}S(\xi)) \xrightarrow{\varphi^i} H^{n-i}(V_{n,i}(\xi)).$$

3.2.5. Proposition. *For any $l \leqslant k \leqslant n$ the following diagram commutes*

$$\begin{array}{ccccc} \bigoplus_{i=1}^{k} H^{n-1}(X) & \longrightarrow & H^{n-1}(S^{k-1}S(\xi)) & \xrightarrow{\text{inclusion}^*} & H^{n-1}(S^{l-1}S(\xi)) \\ & \searrow & \big\downarrow{\scriptstyle \chi^k} & & \big\downarrow{\scriptstyle \chi^l} \\ & & \bigoplus_{i=1}^{k} H^{n-i}(V_{n,i}(\xi)) & \xrightarrow{\text{projection}} & \bigoplus_{i=1}^{l} H^{n-i}(V_{n,i}(\xi)) \end{array}$$

Proof. Immediately follows from 3.2.3(3). □

3.3. Flag structures.

3.3.1. Definition. A *k-flag structure* on bundle ξ is a class $\varkappa \in H^{n-1}(S^{k-1}S(\xi))$ whose restriction to every fibre S^{n-1} is non-zero. The set of all k-flag structures on ξ is denoted by $\mathrm{Flag}^k(\xi)$.

3.3.2. Theorem. *(1) $\mathrm{Flag}^k(\xi)$ is not empty if and only if $w_n(\xi) = \cdots = w_{n-k+1}(\xi) = 0$. In this case $\mathrm{Flag}^k(\xi)$ is an affine space over $\bigoplus_{i=1}^{k} H^{n-i}(X)$;*

(2) if $l<k$, the inclusion homomorphism $H^{n-1}(S^{k-1}S(\xi)) \longrightarrow H^{n-1}(S^{l-1}S(\xi))$ restricts to an affine map $\mathrm{Flag}^k(\xi) \longrightarrow \mathrm{Flag}^l(\xi)$;

(3) if $\mathrm{Flag}^k(\xi)$ *is not empty,* χ^k *restricts to an affine isomorphism* $\mathrm{Flag}^k(\xi) \longrightarrow \prod_{i=1}^{k} \mathrm{St}_{n-i}(\xi)$*. The following diagram commutes:* $(l \leqslant k)$

$$\begin{array}{ccc} \mathrm{Flag}^k(\xi) & \xrightarrow{\text{restriction}} & \mathrm{Flag}^l(\xi) \\ \downarrow \chi^k & & \downarrow \chi^l \\ \prod_{i=1}^{k} \mathrm{St}_{n-i}(\xi) & \xrightarrow{\text{projection}} & \prod_{i=1}^{l} \mathrm{St}_{n-i}(\xi) \end{array}$$

Proof. Statements (1) and (2) follow from 3.1.2; statement (3) follows from 3.2.5. □

3.3.3. Definition. Denote by p the projection $S^{k-1}S(\xi) \longrightarrow \mathbb{R}p^{k-1} \times X$ and fix a subgroup (not necessary graded) $G \subset \bigoplus_{i=1}^{k} H_{n-i}(X) = H_{n-1}(\mathbb{R}p^{k-1} \times X)$. A *partial k-flag structure* (defined over G) is a class $\varkappa \in H^{n-1}(S^{k-1}S(\xi))/\operatorname{Ann}(p_*^{-1}G)$ whose restriction to every fibre S^{n-1} is non-zero. The set of all partial k-flag structures on ξ is denoted by $\mathrm{Flag}^k(\xi, G)$. (Remark: a k-flag structure can be thought of as a functional $H_{n-1}(S^{k-1}S(\xi)) \longrightarrow \mathbb{Z}_2$. A partial structure is a partial functional defined on $p_*^{-1}G$. This explains the term.)

In the case of partial structures Theorem 3.3.2 should be reworded as follows:

3.3.4. Theorem. *(1)* $\mathrm{Flag}^k(\xi, G)$ *is not empty if and only if* $w_i(\xi) = 0$ *for* $i \geqslant n-k+1$. *(Note that this condition does not depend on G.) This set is an affine space over* $\left[\bigoplus_{i=1}^{k} H^{n-i}(X)\right] / \operatorname{Ann} G$*;*

(2) if $l \leqslant k$*, the inclusion homomorphism induces an affine map* $\mathrm{Flag}^k(\xi, G) \longrightarrow \mathrm{Flag}^l\left(\xi, G \cap \bigoplus_{i=1}^{l} H^{n-i}(X)\right)$.

4. Flag structures on the fixed point set of an involution

4.1. Construction. Let X be an $(n+m)$-dimensional closed smooth manifold and c be a smooth involution on X. Fix some c-invariant m-dimensional closed smooth submanifold $V \in X$ and denote by F_V the intersection $V \cap \operatorname{Fix} c$, by $i : F_V \hookrightarrow \operatorname{Fix} c$, $j : F_V \hookrightarrow V$ and $i_V : V \hookrightarrow X$ the inclusions, and by ν_V the normal bundle of V in X.

4.1.1. Theorem. *Let for some* $k \geqslant 1$ *class* $(i_V)_*[V]$ *belongs to* $\mathcal{SF}^{k+1}_{m+k}(X)$*, or, equivalently,* $e^1_m[V] = \cdots = e^k_{m+k-1}[V] = 0$*. Then there exists a natural k-flag structure* $\varkappa_k$ *on* $j^*\nu_V$ *defined over* $i_*^{-1}\mathcal{F}^{k+1}_n(X)$*. In particular,* $w_p(j^*\nu_V) = 0$ *for* $p > n-k$.

Proof. Consider an equivariant tubular neighborhood T_{k-1} of $S^{k-1}V$ in $S^{k-1}X$ and denote by ∂T_{k-1} its boundary, by $\mathrm{pr} : \partial T_{k-1} \to S^{k-1}V$ the projection, and by $\mathrm{in} : T_{k-1} \hookrightarrow S^{k-1}X$ the inclusion. The restriction $\mathrm{pr}\,|_{\,\mathrm{pr}^{-1}F_V}$ obviously coincides with the sphere bundle associated with $\eta^{k-1} \otimes j^*\nu_V$, so the orientation $\varkappa_k$ that is to be constructed is a partial functional $H_{n-1}(\partial T_{k-1}) \dashrightarrow \mathbb{Z}_2$, which is non-trivial on the fundamental class of every fibre S^{n-1} of pr. We let $\varkappa_k$ be the linking coefficient map. More precisely, consider the following commutative diagram with exact rows: (D^{-1} denotes the Poincaris duality isomorphism)

$$\begin{array}{ccccccc}
 & & \bar{\varkappa}_k & \longmapsto & [\mathcal{S}^{k-1}V] & & \\
 & & \cap & & \cap & & \\
H_{m+k}(\mathcal{S}^{k-1}X) & \xrightarrow{\text{rel}} & H_{m+k}(\mathcal{S}^{k-1}X, T_{k-1}) & \xrightarrow{\partial} & H_{m+k-1}(T_{k-1}) & \xrightarrow{\text{in}_*} & \\
\downarrow D^{-1} & & \downarrow D^{-1} & & \downarrow D^{-1} & & \\
H^{n-1}(\mathcal{S}^{k-1}X) & \longrightarrow & H^{n-1}(\mathcal{S}^{k-1}X \setminus T_{k-1}) & \xrightarrow{\delta} & H^{n}(T_{k-1}, \partial T_{k-1}) & \longrightarrow &
\end{array}$$

Due to 2.5.5 the hypothesis of the theorem implies that $\text{in}_*[\mathcal{S}^{k-1}V] = 0$. Hence there exists an element $\bar{\varkappa}_k \in H_{m+k}(\mathcal{S}^{k-1}X, T_{k-1})$ such that $\partial\bar{\varkappa}_k = [\mathcal{S}^{k-1}V]$; this element is unique up to Im rel. Consider $D^{-1}\bar{\varkappa}_k$ as an element of $\text{Hom}(H_{n-1}(\mathcal{S}^{k-1}X \setminus T_{k-1}), \mathbb{Z}_2)$ and define $\varkappa_k$ as the composed homomorphism

$$H_{n-1}(\text{pr}^{-1} F_V) \xrightarrow{\text{inclusion}} H_{n-1}(\mathcal{S}^{k-1}X \setminus T_{k-1}) \xrightarrow{D^{-1}\bar{\varkappa}_k} \mathbb{Z}_2$$

restricted to

$$\text{Ker}[\text{inclusion}_* : H_{n-1}(\text{pr}^{-1} F_V) \to H_{n-1}(\mathcal{S}^{k-1}X)].$$

Since $\delta D^{-1}\bar{\varkappa}_k = D^{-1}[\mathcal{S}^{k-1}V]$ is the Thom class of pr, the restriction of $\varkappa_k$ to every fibre is non-zero, and $\varkappa_k$ is a k-flag structure. The domain of $\varkappa_k$ is the pull back of $\text{Ker}[\text{in}_* : H_{n-1}(\mathcal{S}^{k-1}F_V) \to H_{n-1}(\mathcal{S}^{k-1}X)]$; due to 2.5.5 the latter group coincided with $\mathcal{F}_n^{k+1}$. $\square$

4.1.3. Let $(i_V)_*[V] \in \mathcal{SF}_{m+k}^{k+1}$. Then due to 2.5.3 $(i_V)_*[V] \in \mathcal{SF}_{m+l}^{l+1}$ for any $l \leqslant k$. This gives rise to series of natural partial l-flag structures $\varkappa_l$ on $j^*\nu_V$, $1 \leqslant l \leqslant k$, defined over $\mathcal{F}_n^{l+1}$. On the other hand, due to 2.3.4 $\varkappa_k$ can be restricted to an l-flag structure $\varkappa_l'$ defined over subgroup $\mathcal{F}_n^{k+1} \cap \bigoplus_{i=1}^{l} H_{n-1}(X)$, which contains $\mathcal{F}_n^{l+1}$ (see 2.5.3).

4.1.4. Proposition. *The restriction of $\varkappa_k$ to $\mathcal{F}_n^{l+1}$ coincides with $\varkappa_l$.*

Proof. Denote by in : $(\mathcal{S}^{l-1}X, T_{l-1}) \hookrightarrow (\mathcal{S}^{k-1}X, T_{k-1})$ the inclusion and consider the following diagram

$$\begin{array}{ccc}
H_{m+k}(\mathcal{S}^{k-1}X, T_{k-1}) & \xrightarrow{\partial} & H_{m+k-1}(T_{k-1}) \\
\downarrow \text{in}^! & & \downarrow \text{in}^! \\
H_{m+l}(\mathcal{S}^{l-1}X, T_{l-1}) & \xrightarrow{\partial} & H_{m+l-1}(T_{l-1})
\end{array}$$

Since $\text{in}^![\mathcal{S}^{k-1}V] = [\mathcal{S}^{l-1}V]$, we can choose $\bar{\varkappa}_l = \text{in}^!\,\bar{\varkappa}_k$. Then $D^{-1}\bar{\varkappa}_l = in^*D^{-1}\bar{\varkappa}_k$, that shows that $\varkappa_l$ coincides with the restriction of $\varkappa_k$. $\square$

4.2. Geometric description.

4.2.1. The definition of map χ^k (3.2.4), filtration $\mathcal{F}^r_q$ and homomorphisms e^r_q, and the geometric construction of the linking coefficient map yield the following clear geometric description of flag structures $\varkappa_k$: (For simplicity we replace $\mathcal{F}^{k+1}_n$ with its maximal graded subgroup. This allows to consider orientations of different dimensions separately.)

Suppose that the total space of the sphere bundle $S(\nu_V)$ is embedded as the boundary of an equivariant tubular neighborhood of V in X. Let $\sigma : P \to V_{n,k}(j^*\nu_V)$ be an $(n-k)$-dimensional cycle (i.e. a singular polyhedron, a singular manifold, a cell cycle, etc.) Consider the composed map $\tilde{\sigma}_i : P \times D^i_+ \hookrightarrow P \times S^{k-1} \xrightarrow{\tilde{\sigma}} S(j^*\nu_V) \hookrightarrow X$ as a chain in X (of the same nature as σ), and suppose that there exist some j-dimensional chains y_j in$X$, $n-k+1 \leqslant j \leqslant n$, such that $\partial y_{n-k+1} = \tilde{\sigma}_0$, and $\partial y_j = (1+c_\#)y_{j-1} + \tilde{\sigma}_{j+k-n-1}$ for $j > n-k+1$. Suppose that y_n is transversal to V. Then $\varkappa_k(\sigma) = \#(y_n \cap F) \mod 2$.

Remark. It is this construction that was originally proposed by O.Viro [8].

4.3. Projective complete intersections.

4.3.1. Let A be an n-dimensional non-singular real algebraic variety. Denote by $\mathbb{C}A$ the set of complex points of A, by $\mathbb{R}A$ the real part of A, and by conj the involution of complex conjugation on $\mathbb{C}A$. Multiplication by $\sqrt{-1}$ induces an isomorphism between the tangent bundle $\tau_{\mathbb{R}A}$ of $\mathbb{R}A$ and its normal bundle $\nu_{\mathbb{R}A}$ in $\mathbb{C}A$. Since $\mathbb{R}A = \operatorname{Fix}\operatorname{conj}$, Theorem 4.1.1 yields:

4.3.2. Theorem. *If $[\mathbb{R}A] \in \mathcal{F}^{k+1}_{n+k}(\mathbb{C}A)$, there exists a natural partial k-flag structure on $\tau_{\mathbb{R}A}$ defined over $\mathcal{F}^{k+1}_n(\mathbb{C}A)$. In particular, $w_p(\mathbb{R}A) = 0$ for $p > n-k$.*

4.3.3. Let A be a projective complete intersection. Denote by $\mathbb{C}h$ (resp, $\mathbb{R}h$) the complex (resp,real) hyperplane section class in $H^2(\mathbb{C}A, \mathbb{Z})$ (resp, $H^1(\mathbb{R}A, \mathbb{Z}_2)$), and by $\widetilde{\mathbb{R}A}$ the double cover of $\mathbb{R}A$ corresponding to $\mathbb{R}h$.

4.3.4. Theorem. *Let $[\mathbb{R}A]$ is either equal to zero or dual to $\mathbb{C}h^{n/2} \mod 2$ in $H_n(\mathbb{C}A)$. Then for every k, $0 \leqslant k \leqslant n-1$, $\widetilde{\mathbb{R}A}$ is naturally supplied with a reduced k-dimensional Stiefel orientation.*

Remark. In particular, 4.3.4 implies that $w_1(\widetilde{\mathbb{R}A}) = \cdots = w_n(\widetilde{\mathbb{R}A}) = 0$. Note, however, that this holds for every projective complete intersection, no matter whether $[\mathbb{R}A]$ vanishes or not, since $\widetilde{\mathbb{R}A}$ is a complete intersection in sphere $S^N = \widetilde{\mathbb{R}p^N}$.

Proof. Consider the following diagram

$$\begin{array}{ccccc}
\widetilde{\mathbb{R}A} & \xrightarrow{\text{in}} & \operatorname{pr}_{\mathbb{C}}^{-1}\mathbb{R}A & \xrightarrow{\text{inclusion}} & \widetilde{\mathbb{C}A} \\
& \searrow^{\operatorname{pr}_{\mathbb{R}}} & \downarrow & & \downarrow \operatorname{pr}_{\mathbb{C}} \\
& & \mathbb{R}A & \xrightarrow{\text{inclusion}} & \mathbb{C}A
\end{array}$$

where $\operatorname{pr}_{\mathbb{C}} : \widetilde{\mathbb{C}A} \to \mathbb{C}A$ is the S^1-bundle corresponding to $\mathbb{C}h$. $\widetilde{\mathbb{C}A}$ can be constructed as the pull back of $\mathbb{C}A$ under the standard projection $S^{2N+1} \to \mathbb{C}p^N \supset \mathbb{C}A$. The complex conjugation on $S^{2N+1} \subset \mathbb{C}^{N+1}$ restricts to involution $\widetilde{\operatorname{conj}}$ on $\mathbb{C}A$ such that $\widetilde{\operatorname{conj}}$ covers conj, $\operatorname{pr}_{\mathbb{C}}^{-1}\mathbb{R}A$ is $\widetilde{\operatorname{conj}}$-invariant, and $\mathbb{R}A = \operatorname{Fix}\widetilde{\operatorname{conj}}$.

Class $[\mathrm{pr}_{\mathbb{C}}^{-1}\,\mathbb{R}A] = \mathrm{pr}_{\mathbb{C}}^{!}[\mathbb{R}A]$ vanishes in $H_{n+1}(\mathbb{C}A)$. Since $H^*(\mathbb{C}A)$ is generated by powers of $\mathbb{C}h$ up to dimension $n-1$, $H_i(\widetilde{\mathbb{C}A}) = H^{2n+1-i}(\widetilde{\mathbb{C}A}) = 0$ for $n+2 \leqslant i \leqslant 2n$, and all maps e_q^r are trivial for $n+2 \leqslant q \leqslant 2n$. Hence $[\mathrm{pr}_{\mathbb{C}}^{-1}\,\mathbb{R}A] \in \mathcal{SF}_{2n+1}^{n+1}(\widetilde{\mathbb{C}A})$, and Theorem 4.1.1 applied to $V = \mathrm{pr}_{\mathbb{C}}^{-1}\,\mathbb{R}A$ yields a partial k-flag structure $\varkappa_n$ on the bundle $\mathrm{in}^*\,\nu_V = \mathrm{pr}_{\mathbb{R}}^*\,\nu_{\mathbb{R}A} \cong \mathrm{pr}_{\mathbb{R}}^*\,\tau_{\mathbb{R}A} = \tau_{\widetilde{\mathbb{R}A}}$. This structure is defined over subgroup $\mathcal{F}_{n+1}^{n+1}$, which coincides with $\widetilde{H}_*(\widetilde{\mathbb{R}A})$ since $\widetilde{H}_i(\widetilde{\mathbb{C}A}) = 0$ for $i \leqslant n$. Due to Theorem 3.3.2 $\chi^n\varkappa_n$ is a collection of reduced k-dimensional Stiefel orientations, $0 \leqslant k \leqslant n-1$. □

References

1. G. E. Bredon, *Introduction to compact transformation groups*, Academic Press, New York, London, 1972.
2. A. Degtyarev, *Cohomology approach to structures on G-bundles* (to appear).
3. ———, *Screwed Steenrod squares and some applications* (to appear).
4. W. I. Hsiang, *Cohomology theory of topological transformation groups*, Springer-Verlag, Berlin, Heidelberg, New York, 1975.
5. I. Kalinin, *Cohomology characteristics of real algebraic hypersurfaces*, Algebra i Analis 3 (1991). (Russian)
6. V. A. Rokhlin, *Complex orientations of real algebraic curves*, Funkz. Analis 8 (1974), 71–75. (Russian)
7. N. E. Steenrod, D. B. A. Epstein, *Cohomology operations*, Princeton University Press, Princeton, 1962.
8. O. Ya. Viro, *Progress in the topology of real algebraic varieties over the last six years*, Russian Math. Surveys 41 (1986), 55–82.

A. I. Degtyarev
Steklov Mathematical Institute
St Petersburg Branch qqq (LOMI)
27, Fontanka 191011, St Petersburg USSR
E-mail: degt@lomi.spb.su

SUBANALITICITY AND THE SECOND PART OF HILBERT'S 16th PROBLEM

Zofia Denkowska

Summary : The aim of this paper is to clarify the role of subanalytic sets in the proofs concerning the second part of Hilbert's 16th problem.

After a brief introduction (§ 1), we give (§ 2) a somehow modified version of the subanalytic proofs contained, among other results, in the work of J.-P. Françoise and C.C. Pugh ([FP], 1986).

Finite cyclicity of elliptic points is proved in § 2 both by the methods of [FP] and as a consequence of the subanalyticity of the Poincaré map (this is our main modification).

In § 3, which is based mostly on the remarks of R. Roussarie, we give the reasons for which singular points other than elliptic do not admit the same subanalytic treatment.

The author of this paper had worked with subanalytic sets for over 10 years, so naturally the paper of J.-P. Françoise and C.C. Pugh ([FP]), published in 1986, immediately drew her attention, as it contains, among other results, some very ingenious applications of the subanalyticity to Hilbert's 16th problem have their limitations and why.

The author gathered together both the subanalytic results (§ 2) and the calculus that explains why direct generalization was impossible and what obstacles appear for singular points other than elliptic one (§ 3).

The author is very grateful to P. Biler (Wroclaw, visiting Orsay) for numerous suggestions concerning the proofs in § 3 as well as for his invaluable help in finding the references.

§ 1 Introduction

This paper is written in such a way that is could be read both by people having no knowledge of subanalytic sets and by those who know little about limit cycles, dynamical systems and Hilbert's 16th problem in particular (although O.D.E. are more essential to the subject than the subanalyticity).

This is why we begin with this introduction, which gives essential facts both about Hilbert's 16th problem and about subanalytic sets.

A. Hilbert's 16th problem

(see [R_1], [FP] ou [D] for more details)

We use the following notation equivalently:

A system of analytic differential equations

$$(*)\quad \begin{cases} x' = f(x,y,\lambda) \\ y' = g(x,y,\lambda), \end{cases}$$

with $(x,y,\lambda)\in \Omega\times U$, Ω open in $\mathbb{R}^2$, U open in $\mathbb{R}^l$, f, g analytic in $\Omega\times U$ (for simplicity sake we assume that Ω, U are cubes and $0\in \Omega\times U$) can be also regarded as an analytic deformation X_λ of an analytic vector field X_o, i.e. an analytically parametrized (by λ) family of planar analytic vector fields: $X_\lambda = f(x,y,\lambda)\partial/\partial x + g(x,y,\lambda)\partial/\partial y$ with f and g as above.

Definition 1.1. A limit cycle of X_λ is a periodic trajectory isolated in the set of all periodic trajectories of X_λ.

Problem 1.2. (a reformulation of Hilbert's 16th problem in the terms of analytic functions instead of polynomials, see [R_1]).

Is it true that the number of limit cycles of X_λ is locally uniformly bounded (with a bound independent of λ) for every analytic deformation X_λ?

To put it more precisely, we need

Definition 1.3. ([FP]) A compact subset Γ of $\mathbb{R}^2$ is called a limit periodic set of the analytic deformation X_λ if there is a sequence γ_n of periodic orbits of X_{λ_n} such that γ_n converges to Γ in the Hausdorff metric.

Remark 1.4. ([FP], [G, p.128])

A limit periodic set of an analytic deformation X_λ is:

either a singular point,

or a periodic trajectory,

or else a polycycle (a graphic) i.e. a closed, piecewise analytic curve in $\mathbb{R}^2$ consisting of singular points (vertices) and trajectories for which these vertices are, alternately,stable and unstable limit points and possibly of lines of zeroes.

We will not deal with polycycles in this paper, as subanalyticity cannot appear for them (cf. [D] or [I_1]).

Conjecture 1.5. Finite Cyclicity Conjecture ([R_1], [FP]).

Suppose that Γ is a compact subset of $\mathbb{R}^2$, invariant by X_o, which is a limit periodic set for the analytic deformation X_λ of X_o. Then it has finite cyclicity in X_λ, i.e. there exist $\varepsilon>0$, $n\in\mathbb{N}$ and a neighbourhood V of $\lambda=0$ such that: for all $\lambda\in V$ the number of limit cycles of X_λ contained in the ε-neighbourhood of Γ is not greater than N.

Remark. The Finite Cyclicity Conjecture amounts to saying that limit cycles of X_λ cannot accumulate (i.e. cannot form a convergent, nonconstant sequence), because they cannot accumulate on sets other than limit periodic sets.

Definition 1.6. ([AP] for example). Let Γ be a periodic orbit of the planar field X_o and γ its local cross section i.e. an analytic curve transverse to Γ and intersecting Γ in the unique point Q. Suppose $\gamma:(-\varepsilon,\varepsilon) \longrightarrow \mathbb{R}^2$, $\gamma(0)=Q$, is an analytic parametrization of γ.
For $t\in(-\varepsilon,\varepsilon)$ and λ in a neighbourhood of 0 take the solution of the Cauchy problem

$$(*)\quad \begin{cases} x' = f(x,y,\lambda) \\ y' = g(x,y,\lambda), \quad (x(0),y(0)) = \gamma(t) \end{cases}$$

and let $P = P(t,\lambda)$ be the point of the first return of this trajectory to the curve γ, i.e. the point of intersection of γ and the trajectory whose parameter $\tau\in(-\varepsilon,\varepsilon)$ has the smallest length $|\tau|$. Shrinking possibly the interval $(-\varepsilon,\varepsilon)$ to a smaller one $(-\upsilon,\upsilon)$ and using the Implicit Function Theorem we obtain an analytic map

$$P:(-\upsilon,\upsilon)\times\text{neighbourhood of } \lambda=0 \ni (t,\lambda) \longmapsto \tau=P(t,\lambda) \in \mathbb{R},$$

which is called the Poincaré map of Γ in X_λ.

The notion of a Poincaré map can be easily generalized to the case of $\Gamma = Q$ equal to an elliptic point of X_o (a singular point with purely

imaginary eigenvalues), because for such a point a ray transverse to all trajectories near Q, for λ small enough, can be chosen and it is shown that all trajectories close to Q return to this ray after a finite time ([FP]). The only difference is that in this case P is defined only in $[0,\upsilon)$:
$P:[0,\upsilon)\times$neighbourhood of $\lambda=0 \to \mathbb{R}$. For simplicity sake we will assume here that $Q=(0,0)$ and the ray is a segment of the x-axis.

We will see later that the Poincaré map for an elliptic point is subanalytic (while the one for a periodic trajectory is analytic, by the Implicit Function Theorem).

Remark. Periodic trajectories of X_λ correspond to the zeroes of the map $\delta(t,\lambda) = t-P(t,\lambda)$.

The notion of the first return map can be generalized to polycycles (see [Du], [I_1], [EMMR]), but then it becomes truly difficult and subtle and not subanalytic. This is why we do not even attempt to speak about polycycles in this work.

B. Subanalytic sets

We will need only some basic facts concerning them. Detailed references are to be found in [DŁS_1], [DŁS_2] or [BM].

Definition 1.7. (cf. above and [Ga]) Subanalytic germs in $a\in\mathbb{R}^n$ form the smallest family of germs closed under the finite union, taking the complement and projection and containing all the germs $\{f>0\}$, $\{f=0\}$, for f a germ f an analytic function in a. A set $B\subset\mathbb{R}^n$ is said to be subanalytic in $\mathbb{R}^n$ if for each $a\in\mathbb{R}^n$ its germ in a is a subanalytic germ.

Lemma 1.8. ([DŁS_1], [Ga]) Let B be a relatively compact subanalytic subset of $\mathbb{R}^k\times\mathbb{R}^n$ and let B_y, $y\in\mathbb{R}^n$, denote the section $\{x\in\mathbb{R}^k;\ (x,y)\in B\}$. Then there exists a constant $C\in\mathbb{N}$ such that for all $y\in\mathbb{R}^n$ the number of connected components of $B_y = \#_{cc}(B_y)$ is not greater than C (i.e. is uniformly bounded).

Proposition 1.9. ([DŁS$_1$])

If a subanalytic set $B \subseteq \mathbb{R}^k \times \mathbb{R}^n$ is $\mathbb{R}^n$-relatively compact (i.e. for each $K \subseteq \mathbb{R}^k$ relatively compact, the intersection $(K \times \mathbb{R}^n) \cap B$ is relatively compact) and $\pi: \mathbb{R}^k \times \mathbb{R}^n \to \mathbb{R}^k$ is the usual projection, then $\pi(B)$ is subanalytic in $\mathbb{R}$.

§2 Subanalyticity in Hilbert's 16th problem

In this paragraph we deal with limit periodic sets which are either periodic orbits or algebraically isolated singular points of X_o. In the case where x_o is a singular point of X_o we suppose, with no loss of generality, that $x_o=(0,0)$ and that for λ small enough, if there is a singular point near (0,0), it is (0,0) itself.

Proposition 2.1. If for a given analytic deformation X_λ and its limit periodic set Γ the Poincaré map is well defined in $I \times U$, where $I=(-\varepsilon,\varepsilon)$ if Γ a periodic orbit and $I=[0,\varepsilon)$ if Γ a singular point, and U is a small neighbourhood of $\lambda=0$, and it is subanalytic (its graph is subanalytic in $\mathbb{R} \times \mathbb{R}^l \times \mathbb{R}$), then the limit cycles of X_λ cannot accumulate on Γ.

P r o o f. Observe that $\delta(t,\lambda)=t-P(t,\lambda)$ is also subanalytic and therefore the germ in $t=0$, $\lambda=0$ of the set $A=\{(t,\lambda);\ \delta(t,\lambda)=0\}$ is subanalytic. Taking its subanalytic, relatively compact representant and projecting it on $\mathbb{R}$ we get the subanalyticity of the set $B = \{t \in I';\ \exists\ \lambda \in U'\ \ \delta(t,\lambda) = 0\}$, where $I' \subseteq I$, $U' \subseteq U$ are, respectively, the interval $(-\varepsilon',\varepsilon')$ or $[0,\varepsilon')$ and a neighbourhood of $\lambda=0$.
Now it is enough to remark that each point of B corresponds to a periodic orbit of one of the X_λ, $\lambda \in U'$, and each isolated point of B (its 0-dimensional connected component) corresponds to a limit cycle of one of the X_λ, $\lambda \in U'$. Since $\#_{cc}(B)$ is locally finite (cf. [DŁS$_1$]) the set B cannot contain a convergent sequence of isolated points, hence limit cycles cannot accumulate on Γ.

Proposition 2.2. Limit cycles of an analytic deformation X_λ cannot accumulate on an elliptic point of X_o.

P r o o f. It is well known that for such a point the Poincaré map is defined in $[0,\varepsilon)\times W$ small enough (cf. [FP]). Shrinking possibly $[0,\varepsilon)$ and U, take the analytic flow $\phi:[0,T]\times W\times U \to Q$, where U is a neighbourhood of $\lambda=0$, W a neighbourhood of (0,0) and Q a cube in $\mathbb{R}^2$ (cf. [L] for example) such that ϕ is analytic in a neighbourhood of $[0,T]\times\bar{W}\times\bar{U}$. Let us denote $\phi = (\phi_1,\phi_2)$. We have: graph $P = \{(x,\lambda,x')\in [0,T]\times U\times\mathbb{R};\ x'=P(x,\lambda)\}$. Observe that $x'=P(x,\lambda) \Leftrightarrow \exists\ t\in[0,T]$ such that $\phi_2(0,x,0,\lambda) = \phi_2(t,x',0,\lambda) = 0$ and for all $\tau>0$, $\tau<t$, $\phi_2(\tau,x',0,\lambda)\neq 0$.

Take the set

$$C = \Big\{(t,\tau,x,x',\lambda) \in [0,T]\times[0,T]\times W\times W\times U;\ 0<\tau<t \text{ and } \phi_2(\tau,x',0,\lambda) = \phi_2(0,x,0,\lambda) = \phi_2(t,x',0,\lambda) = 0\Big\}$$

and denote by p the projection $p(t,\tau,x,x',\lambda) = (t,x,x',\lambda)$. Since C is subanalytic and τ is bounded, by Prop. 1.9 we get the subanalyticity of p(C) and therefore of the complement $\setminus p(C)$ of p(C).

Take now the projection $\pi(t,x,x',\lambda) = (x,x',\lambda)$ and observe that, since t is bounded, the projection $\pi(\setminus p(C))$ is subanalytic. It is enough to remark that graph $P = \pi(\setminus p(C))$ and use Prop. 2.1.

In this proof taking t and τ bounded is essential, otherwise the projections might not be subanalytic.

Proposition 2.3. Limit cycles of X_λ cannot accumulate on periodic orbits.

P r o o f. The same as that of Prop. 2.2, only now [-T,T] will replace [0,T]. Of course, the result for periodic orbits is classical and can be obtained directly from the analyticity of P (Implicit Function Theorem) without ever using the subanalyticity.

Without the hypothesis about the periods being bounded from above, the set A need not be relatively compact and Lemma 1.8 would not apply.

Proposition 2.4. Given an analytic deformation X_λ of X_o and its limit periodic set Γ suppose that:
(i) there exists a neighbourhood W of Γ in $\mathbb{R}^2$ and a T>0 such that all the periodic orbits of X_λ, $\lambda \in U$, contained in W have their prime periods < T,
(ii) the Poincaré map $P: I \times U \to \mathbb{R}$ can be defined for I and U small enough (with $I=[0,\varepsilon)$ for Γ a singular point and $I=(-\varepsilon,\varepsilon)$ for Γ a periodic orbit).
Then P is subanalytic in $\mathbb{R}\times\mathbb{R}^l$.

P r o o f. The same as that of Prop. 2.2 Observe that the condition (i) plays a very important role in it.

Remark. It may seem a priori that Prop. 2.1 together with Prop. 2.4 above might give more than Prop. 2.2 and 2.3, i.e. more than just the finite cyclicity of elliptic points and periodic orbits; e.g. that they might apply to isolated periodic points other than elliptic points. We will see in §3 that this is not true.

On the other hand, Prop. 2.2 and 2.3 are so simple that they have many other proofs. As said above, Prop. 2.3 is classical. As to the finite cyclicity of elliptic points, we repeat below another proof of it, also subanalytic, given in [FP]. There is a third proof also, in $[R_2]$. not subanalytic, but much more useful because admitting generalization to the hyperbolic loops, which is not the case of the subanalytic methods.

We now give, after [FP], another subanalytic proof of Prop. 2.3 and 2.4, based on the following:

Theorem 2.5. ([FP]) For any analytic deformation X_λ of an analytic vector field X_o and any T>0 the number of limit cycles of X_λ of period < T is locally uniformly bounded.

P r o o f. The proof is the same as that of [FP], more detailed here because of the explanatory character of this paper.

Take the flow $\phi:[0,\tau]\times W\times U \to Q$, with $\tau<T$, W a neighbourhood of (0,0), U a neighbourhood of $\lambda=0$, both small enough, and Q a cube in $\mathbb{R}^2$, such that ϕ is analytic in a neighbourhood of $[0,\tau]\times\bar{W}\times\bar{U}$ ([L], [G]). For each $\lambda\in U$, $\phi(.,.,.,\lambda)$ denotes the flow of X_λ in $[0,\tau]\times W$.
Take the set $A = \{(t,x,y,\lambda)\in[0,\tau]\times W\times U;\ \phi(t,x,y,\lambda) = (x,y)\}$. Because of the hypothesis above, this is a relatively compact, subanalytic subset of $\mathbb{R}\times\mathbb{R}^2\times\mathbb{R}^l$. Applying Lemma 1.8 to the sections $A_{(t,\lambda)}=\{(x,y)\in W;\ \phi(t,x,y,\lambda)=(x,y)\}$ we get: there exists a constant N such that $\#_{cc}(A_{(t,\lambda)})\le N$ for all $(t,\lambda)\in [0,\tau]\times U$.

Observe that $A_{(t,\lambda)}$ is composed of all periodic orbits of X_λ for which t is a period. We assign now to each limit cycle of X_λ, $\lambda\in U$, its prime period t_o. If $t_o\le\tau$, we can assign to t_o the unique connected component of $A_{(t_o,\lambda)}$ in which this limit cycle lies (in fact each limit cycle is in itself a connected component of $A_{(t_o,\lambda)}$, where t_o its period).

We therefore have the injection
limit cycle of $X_\lambda \to (t_o,\lambda) \to$ connected component of $A_{(t_o,\lambda)}$ and therefore $\#\{\text{limit cycles of } X_\lambda,\ \lambda\in U,\ \text{prime period}\le\tau<T\} \le$ $\le \#_{cc}(A_{(t,\lambda)}) \le N$. Since τ was chosen arbitrarily among numbers $< T$, we obtain the theorem.

Remark that, as for any reasoning that involves subanalyticity, we need the existence of the bound T on the periods, otherwise Lemma 1.8 and Prop. 1.9 will not apply.

Now all we need to prove Prop. 2.2 (Prop. 2.3 is an obvious consequence of Thm 2.5) is the following:

Proposition 2.6. If (0,0) is an elliptic point of X_o and a limit periodic set of X_λ then there exists a neighbourhood W of (0,0) in $\mathbb{R}^2$ and T>0 such that all the periodic orbits of X_λ contained in W have the prime period < T.

P r o o f. (cf. [A] *Chapitre* 1 and also [BN], problems 1.668 and 1.558)

Observe that an elliptic point of

$$(*)\quad \begin{cases} x'=f(x,y) \\ y'=g(x,y) \qquad f,\ g \text{ analytic,} \end{cases}$$

is either a focus or a centre (cf. [G] for instance), while it is of course always a centre for the linearization

$$(**)\quad \begin{cases} x'=d_{(0,0)}f(x,y) \\ y'=d_{(0,0)}g(x,y) \end{cases}$$

and the periods of periodic orbits of (**) are constant, the angular velocity being equal to $\omega=|\mathrm{Im}\ \mu|$, where μ is one of the complex eigenvalues of (0,0).

We will first show that the angular velocities on periodic trajectories of (*) tend to ω as the trajectories tend to the elliptic point (0,0). To this end, take the simplest blowing up i.e.

$$\phi: S^1\times\mathbb{R} \ni (\bar{x},\bar{y},r) \to (r\bar{x},r\bar{y}) \in \mathbb{R}^2,\quad \bar{x}^2+\bar{y}^2=1,$$

which in local coordinates becomes:

$$(^\circ)\quad \begin{cases} x=r\cos\alpha \\ y=r\sin\alpha. \end{cases}$$

The blow up of the system (*) takes the form:

$$(\#)\quad \begin{cases} r'=f(r\cos\alpha,r\sin\alpha)\cos\alpha + g(r\cos\alpha,r\sin\alpha)\sin\alpha \\ \alpha'=r^{-1}[g(r\cos\alpha,r\sin\alpha)\cos\alpha - f(r\cos\alpha,r\sin\alpha)\sin\alpha] \end{cases}$$

We calculate the angular velocity α' using the Taylor expansions of f and g in (0,0) and substituting ($^\circ$):

$$f(r\cos\alpha,r\sin\alpha) = r\partial f/\partial x(0,0)\cos\alpha + r\partial f/\partial y(0,0)\sin\alpha + r^2P_1(\alpha) + \dots$$

$$g(r\cos\alpha,r\sin\alpha) = r\partial g/\partial x(0,0)\cos\alpha + r\partial g/\partial y(0,0)\sin\alpha + r^2Q_1(\alpha) + \dots$$

where $P_i(\alpha)$, $Q_i(\alpha)$ are homogeneous polynomials of variables $\cos\alpha$, $\sin\alpha$, of degree i.

Up to an analytic conjugation the Jacobian matrix of (f,g) has the form:

$$\begin{pmatrix} 0 & -\omega \\ \omega & 0 \end{pmatrix}$$

so we obtain $\alpha' = r^{-1}[f\cos\alpha - g\sin\alpha] = \omega + rF(r,\alpha)$. Therefore α' tends to ω as r tends to 0. All the periodic orbits of (*) remain periodic after the polar blowing up. Their periods tend to the double period of the closed orbit $S^1\times(0)=\{r=0\}$ as the orbits approach $S^1\times(0)$. Since ϕ is proper analytic and a two-leaf covering outside $S^1\times(0)=\phi^{-1}(0,0)$, we obtain that the periods of periodic orbits of (*) tend to $4\pi/\omega$, which is the double period of $S^1\times(0)$ in (#).

It is not as much their limit that interests us, it is the fact that their periods are bounded by, say, $D=4\pi/\omega+1/2$ in a small neighbourhood of (0,0).

Take now the system (*) with parameter:

$$(*_\lambda)\begin{cases} x'=f(x,y,\lambda) \\ y'=g(x,y,\lambda) \end{cases}$$

i.e. take our analytic deformation of X_0. From the continuous dependence of the solutions on parameters we obtain that ω_λ defined as the angular velocity for linearized $(*_\lambda)$ when X_λ has a centre in (0,0) and as ω_0 otherwise, where ω_0 is the angular velocity for linearized $(*_0)$, tends to ω_0 as λ tends to 0. This implies that periodic orbits of $(*_\lambda)$ near (0,0) have the periods uniformly bounded (by $D=4\pi/\omega+1/2$, for example).

A different proof of Prop. 2.5 is to be found in [FP].

§3 The reasons for which the use of subanalyticity stops here

We repeat again that we do not discuss limit periodic sets other than periodic orbits and singular points because (cf. [Du], [EMMR] or $[I_1]$, $[I_2]$) no subanalytic approach can be successful for polycycles.

As was said above, it may seem a priori that hypothesis (i) and (ii) of Prop. 2.4 might be verified for singular points other than elliptic ones. To show that this is not true, we make the following:

Observation 3.1. Suppose (0,0) is a singular point of the analytic field X_0 and a limit periodic set of its analytic deformation X_λ.
If (0,0) is:

(a) nilpotent (two eigenvalues zero, linear part non-trivial),
(b) semi-hyperbolic (one eigenvalue zero, one non-zero),
(c) completely degenerate (linear part vanishing),
(d) hyperbolic,

then there is no uniform bound on the periods of periodic orbits of X_λ converging to (0,0).

P r o o f. <u>Case</u> <u>(a)</u> The point (0,0) being a limit periodic set, there is a sequence γ_n of periodic orbits of X_λ converging to (0,0) in the Hausdorff metric. Let T_n denote the prime period of γ_n. For each γ_n take the smallest rectangle containing it, of length R_n (variation of x) and height L_n (variation of y). We will consider R and L two variables tending both to zero and will estimate T as R and L tend to 0.

Take the simplest normal form for nilpotent points ([AP], [GH]):

$x' = y + o(x,y)$

$y' = o(x,y)$,

where the Landau symbol $o(x,y)$ is understood as (x,y,λ) tends to 0. Suppose T is bounded for periodic orbits near (0,0) i.e. $T = \mathcal{O}(1)$ as R and L tend to 0.

We have: By the first equation, R = variation of x $\leq \max|x'|T = LT$, hence $R = \mathcal{O}(L)$ as L tends to 0. By the second equation, $L \leq (o(R)+o(L))T$. Since $T = \mathcal{O}(1)$ and $R = \mathcal{O}(L)$, we obtain thus $L \leq o(L)T$, which is impossible.

<u>Case</u> <u>(b)</u> The normal form for semi-hyperbolic points:

$x' = cx + o(x,y)$

$y' = o(x,y)$,

where $c \neq 0$, $o(x,y)$ are understood as above.

Suppose that the minimal period T of converging orbits is bounded near (0,0). Observe that, for a periodic trajectory $((x(t,\lambda), y(t,\lambda))$, there are two possibilities:

either

(1) $\max_{t \in [0,T]} x(t,\lambda) > 0$

or

(2) $\min_{t \in [0,T]} x(t,\lambda) < 0$.

Suppose $c>0$. Take case (1). The first equation implies that, at the point (t_o, λ_o) which realizes the maximum, we have $x'(t_o,\lambda_o) > 0$, which is impossible. This reasoning is valid for small λ, of course.

Case (2) is altogether impossible when $c>0$ and we are in a small neighbourhood of (0,0), with λ small enough.

If $c<0$, the reasoning is analogous, the cases (1) and (2) being inversed: case(1) impossible near (0,0), for λ small; case (2) excluded by the first equation, which would imply $x'(t_o,\lambda_o)<0$ at the minimum point of $x(t,\lambda)$.

<u>Case</u> <u>(c)</u> Normal form is now:

$$x'=o(x,y)$$
$$y'=o(x,y)$$

with o understood as above.

On a periodic orbit of diameter r the velocity $v=o(r)$ as r tends to 0 (as a consequence of the equations above for λ small enough). Therefore their periods T satisfy: $T\geq\pi r/o(r)$ and cannot be bounded as r tends to 0.

<u>Case</u> <u>(d)</u> This case is of course immediately excluded by the Grobman-Hartman theorem, but we can also reason as above, in a completely elementary way, taking the normal form:

$$x'=cx+o(x,y)$$
$$y'=dy+o(x,y), \quad c\neq0,\ d\neq0 \quad \text{(real eigenvalues)}$$

or

$$x'=ax+by+o(x,y)$$
$$y'=-bx+ay+o(x,y),\ a\neq0,\ b\neq0 \quad \text{(complex eigenvalues, real part}\neq0).$$

The reasoning is analogous to that of case (b), except that now we consider only the case (1) $\max(x^2+y^2)>0$ on $[0,T_o]$.

It is not hard to construct examples illustrating Observation 3.1 (a), (c), where arbitrarily small periodic orbits do exist and, of course, their periods are unbounded.

References

[A] Arnold V.I., *Chapitres supplémentaires de la théorie des équations différentielles ordinaires*, Editions Mir, Moscou (1980)

[AP] Arrowsmith D.K., Place C.M., *An Introduction to Dynamical Systems*, Cambridge University Press (1990)

[BM] Bierstone E., Milman P., Semianalytic and subanalytic sets, *Publications IHES* n° 67, pp. 1-42 (1988)

[BN] Biler P., Nadzieja T., *Differential Equations: A Collection of Exercises, Problems, Theorems, Examples and Counterexamples*, Wrocław (1990)

[D] Denkowska Z., What has happened to the Hilbert 16th Problem, University of Lodz Publications, Workshop in Sielpia (1988)

[DŁS$_1$] Denkowska Z., Łojasiewicz S., Stasica J., Certaines propriétés élémentaires des ensembles sous-analytiques, *Bull. Acad. Pol. Sci. sér. math.* vol. XXVII, n° 7-8, (1979), pp. 529-536

[DŁS$_2$] Denkowska Z., Łojasiewicz S., Stasica J., Sur le théorème de complémentaire d'un sous-analytique, *ibidem*, pp. 537-539

[DŁS$_3$] Denkowska Z., Lojasiewicz S., Stasica J., Sur le nombre de composantes connexes de la section d'un sous-analytique, *Bull. Acad. Pol. Sci. sér. math.* vol. XXX, n° 7-8, (1982) pp. 333-336

[Du] Dulac H., Sur les cycles limites, *Bull. Soc. Math. France* 51 (1923) pp. 45-88

[EMMR] Ecalle J., Martinet J., Moussu R., Ramis J.P., Non-accumulation des cycles limites, Publications IRMA, Strasbourg (1987)

[FP] Françoise J.P., Pugh C.C., Keeping track of limit cycles, *J. Diff. Equations*, 65 (1986) pp. 139-157

[G] Godbillon C., Systèmes dynamiques sur les surfaces, Publications IRMA, Strasbourg (1979)

[Ga] Gabrielov A.M., Projections of semi-analytic sets, *Funct. Anal. Appl.* 2 (1968) pp. 282-291

[GH] Guckenheimer J., Holmes Ph., *Nonlinear Oscillations, Dynamical Systems and Bifurcations of Vector Fields*, Springer Verlag (1983)

[I_1] Il'yashenko Yu.S., Limit cycles of polynomial vector fields with nondegenerate singular points on the real plane, *Funct. Anal. Appl.* 18 (1984) pp. 199-209

[I_2] Il'yashenko Yu.S., Finiteness theorems for limit cycles (in Russian), *Uspekhi Mat. Nauk* t.45, vyp.2, (1990) pp. 144-200

[L] Lefschetz S., *Differential Equations: Geometric Theory*, Dover Publications, New York (1977)

[R_1] Roussarie R., Note on the Finite Cyclicity Property and Hilbert 16th Problem, preprint Dijon (1987)

[R_2] Roussarie, R., Cyclicité finie des lacets et des points cuspidaux, Nonlinearity 2 (1989) pp. 73-117

The decidability of real algebraic sets by the index formula

J.-P. Françoise and F. Ronga

Abstract. We propose an algorithm to decide if a real algebraic set defined by polynomials with integer coefficients has a non empty intersection with a given ball by the index formula of Kronecker. We approximate the integral by a Riemann sum and we give an estimate of the time of computation which is needed. The method is well adapted to the use of parallel time computations.

In a recent article, L. Blum, M. Shub and S. Smale [B-S-S] introduced the classes of P-problems and NP-problems over the real numbers. They proved that the decidability problem of real algebraic sets is NP-complete and they raised the question: is $P = NP$? analogous to the well-known problem over the integers. This motivates the effective understanding of the geometry of real algebraic sets, to which we wish to contribute with this work. This subject has been recently considered by several authors (G. Collins [Co], J. Canny [C], M.F. Coste-Roy, Heintz, Solernò [H-R-S], J. Renegar [Re], D. Trotman [T]). Our aim is more modest, since we do not pretend to find points on an algebraic set but merely want to decide wether it is empty or not.

We use the index formula to count the geometric number of points at minimal distance from the origin in the given ball of a suitable perturbation of a polynomial equation. The difficulty is to find an explicit perturbation. Our approach is straightforward and in principle our result can be used directly to make explicit calculations, however the numbers involved may become extremely large as the dimension and the degree increase. It is well adapted to parallel computation as it needs the same computation to be done several times.

G. Stengle has made several suggestions to improve our results, that we plan to develop in a joint paper.

1. The algorithm.

We work in a given ball $B(0,r) = \{x \in \mathbf{R}^{n-1} \mid \sum x_i^2 \leq r^2\}$. Let $f(x) \in \mathbf{Z}[x_1, \dots, x_{n-1}]$ be a polynomial of degree at least 2 with integer coefficients and let

$$V(f) = \{x \in B(0,r) \mid f(x) = 0\} .$$

We want to decide whether V is empty or not.

Let $h(x) = x_1^\ell + \cdots + x_{n-1}^\ell - 1$, where ℓ is even and $\ell \geq sup\,\{d, 4\}$; set

$$f_\varepsilon = (1-\varepsilon)f(x) + \varepsilon h(x) \quad , \quad \varepsilon \in \mathbf{R}$$

and set $V_\varepsilon = V(f_\varepsilon)$. Consider the map $\delta : V_\varepsilon \to \mathbf{R}$, $\delta(x_1, \dots, x_{n-1}) = x_1^2 + \cdots + x_{n-1}^2$. We build an explicit ε such that $V(f_\varepsilon)$ is smooth, δ is a Morse function on $V(f_\varepsilon)$ and $V(f_\varepsilon)$

Typeset by $\mathcal{A}\mathcal{M}\mathcal{S}$-TEX

is empty if and only if $V(f)$ is empty. Let $g^{\varepsilon} : \mathbf{R}^n \to \mathbf{R}^n$ be the algebraic map whose coordinates are:

$$f_\varepsilon(x_1, \dots, x_{n-1})$$
$$\frac{\partial f_\varepsilon}{\partial x_i} - x_n x_i\,, \quad i = 1 \dots n-1$$

and consider the set of equations:

$$(1) \qquad \begin{aligned} g^{\varepsilon} &= 0 \\ D(x) &= 0 \end{aligned}$$

where $D(x) = det\left(\frac{\partial g_i^{\varepsilon}}{\partial x_j}\right)_{i,j=1\dots n}$. It is readily verified that if ε is such that (1) has no solution, then V_ε is smooth and δ is a Morse function on V_ε.

Consider the set of equations:

$$(2) \qquad \sum_{i=1}^{n-1} x_i^2 - r^2 = 0 \quad , \quad g^{\varepsilon}(x_1, \dots, x_n) = 0$$

where r is fixed. They express the fact that the sphere of radius r centered at the origin is tangent to V_ε at the point x.

We can assume without loss of generality that $f(0) > 0$ and that $r > 1$. It can be verified that the assumption on ℓ and r make sure that the resultants of (1) and (2) obtained by eliminating $x_1, \dots, x_n$ do not vanish identically. We can therefore apply [**Ro**, prop.1.1] to the set of equations (1) (with substitutions $p \to n+1$, $q \to n$, $m_1 \to d'$, $m_2 \dots m_n \to d'-1$, $m_{n+1} \to (d'-2)(n-2)+d'$, $H \to H' = n|(d')^{n-2}(d'(d'-1)H)^n$) and (2) (with substitutions $p \to n+1$, $q \to 1$, $m_1 \to 2$, $m_2 \to d'$, $m_3 \dots m_{n+1} \to d'-1$), where $d' = sup\,\{d, 4\}$. In fact, careful reading of the proof of [Ro, prop. 1.1] shows that the estimates that they provide can be improved by replacing the integer N by 1; we will make use of this.

Proposition. *Let $f(x) = \sum_{\alpha \in S} f_\alpha x^\alpha$, with $1 \le |f_\alpha| \le H$ and $|\alpha| \le d$ for $\alpha \in S$. Let*

$$\varepsilon_1 = \frac{1}{1 + (n+1)^{s-1}(H')^s s!}$$

where $s = \binom{2d' + (n-2)(d'-2) + d'}{n}$ and let

$$\varepsilon_2 = \frac{1}{1 + 2^{t-1}(dH)^t t!}$$

where $t = \binom{2 + d' + (n-1)(d'-1)}{n}$. Then if $0 < \varepsilon < inf\{\varepsilon_1, \varepsilon_2\}$, the algebraic variety $V_{\pm\varepsilon}$ is smooth, δ is a Morse function on $V_{\pm\varepsilon}$ and $V(f)$ is empty if and only if $V_{-\varepsilon}$ is empty.

The last assertion of this proposition uses the assumption $f(0) > 0$.

It follows that if we take for instance $\varepsilon = -\frac{1}{2}inf\{\varepsilon_1, \varepsilon_2\}$, then $V(f)$ is cmpty if and only if V_ε is empty. Since we have to work with polynomials with integer coefficients, we replace f_ε by $\left(\frac{1}{\varepsilon} - 1\right) f + h$.

We want to compute the geometric number of solutions of the equation $g^\varepsilon(x) = 0$ with $(x_1, \ldots, x_{n-1}) \in B(0, r)$. A simple calculation shows that for such a solution we have:

$$|x_n| \leq \frac{\ell H^2 N}{f(0)}$$

where $N = \#\{\alpha \mid a_\alpha \neq 0\}$. Therefore we can replace the last coordinates x_n by $x_n \frac{f(0)}{\ell H^2 N} r$ (that we again denote by x_n) so that we can work in the ball

$$B = \left\{ x \in \mathbf{R}^n \mid \sum_{i=1\ldots n} x_i^2 \leq r^2 \right\} .$$

Let us introduce $\rho = \sqrt{x_1^2 + \cdots + x_n^2}$, the vector field:

$$X = \sum_{i=1\ldots n} (-1)^i \frac{x_i}{\rho^n} \frac{\partial}{\partial x_i}$$

and the volume form $\Omega = dx_1 \wedge \cdots \wedge dx_n$. The index formula reads :

$$\int_{\partial B} g^* (i_X \Omega) = vol(B)(N_+ - N_-)$$

with $N_+ = \#(D^{-1}(\mathbf{R}_+) \cap g^{-1}(0))$ and $N_- = \#(D^{-1}(\mathbf{R}_-) \cap g^{-1}(0))$. It is not possible to decide whether V_ε is empty or not with this formula. But, following [P] a slight change will do it. We introduce a $n+1^{th}$ variable x_{n+1} and define $g' : \mathbf{R}^{n+1} \to \mathbf{R}^{n+1}$ by:

$$g_i'(x_1, \ldots, x_n, x_{n+1}) = \begin{cases} g_i^\varepsilon(x_1, \ldots, x_n) \text{ , if } i < n+1, \\ x_{n+1} D(x_1, \ldots, x_n) \text{ , if } i = n+1 \end{cases}$$

the vector field $X' = \sum_{i=1\ldots n+1} (-1)^i \frac{x_i}{\rho'^{(n+1)}} \frac{\partial}{\partial x_i}$, with $\rho'^2 = x_1^2 + \ldots + x_{n+1}^2$ and the volume form $\Omega' = dx_1 \wedge \cdots \wedge dx_{n+1}$. Let B' be a ball in $\mathbf{R}^{n+1}$; it is easy to show now that $V(f_\varepsilon) \times \{0\} \cap B'$ is empty if and only if:

$$I = \int_{\partial B'} g'^* (i_{X'} \Omega') = vol(B') N_+' = 0.$$

We can approximate the integral I by a Riemann sum Σ such that : $|I - \Sigma| < vol(B')/3$. It is possible to decide if $V \cap B$ is empty or not on the Riemann sum.

2. Estimate of the number of operations

We estimate the number of operations necessary to compute the integral I.

THEOREM. *Given the algebraic mapping* $g' : \mathbf{R}^{n+1} \to \mathbf{R}^{n+1}$ *and a ball* $B'(0,r)$ *of fixed radius* r *it is possible to decide if* $g'^{-1}(0) \cap B(0,r)$ *is empty or not in a polynomial time by parallel computer.*

PROOF: Let us analyse more carefully the integral involved in the index formula. It splits into $n+1$ integrals:

$$I = \sum_{i=1}^{n+1} \int_{\partial B'} (-1)^i \frac{A_i}{(g_1'^2 + \cdots + g_{n+1}'^2)^{(n+1)/2}} dx_1 \wedge \cdots \wedge \hat{dx_i} \wedge \cdots \wedge dx_{n+1}$$

with

$$A_i = \begin{vmatrix} g_1' & \partial g_1'/\partial x_1 & \cdots & \widehat{\partial g_1'/\partial x_i} & \cdots & \partial g_1'/\partial x_{n+1} \\ \vdots & \vdots & \ddots & \vdots & \ddots & \vdots \\ g_{n+1}' & \partial g_{n+1}'/\partial x_1 & \cdots & \widehat{\partial g_{n+1}'/\partial x_i} & \cdots & \partial g_{n+1}'/\partial x_{n+1} \end{vmatrix}$$

We can assume that n is odd, because if not we can replace g' by $g'' : \mathbf{R}^{n+2} \to \mathbf{R}^{n+2}$, $g''(x_1, \ldots, x_{n+2}) = (g'(x_1, \ldots, x_{n+1}), x_{n+2})$.

Now we have to repeat N times the same computation, namely evaluate the function inside the integral:

$$\phi_i = \frac{A_i}{(g_1'^2 + \cdots + g_{n+1}'^2)^{(n+1)/2}}$$

at each point of a Riemann subdivision.

It is well-known that the computation of a determinant can be done in polynomial time (as function of the size of the matrix). And so we obtain that this algorithm needs a parallel polynomial time.

It is quite easy to get the number of terms which are necessary for the Riemann sum. We assume that the functions $\phi_i(x)$ are normalized so that

$$sup\left\{\left|\frac{\partial \phi_i}{\partial x_k}\right|, x \in B(0,r), k = 1, \ldots, n+1\right\} < C.$$

Let B_j be a Riemann subdivision of $\partial B'(0,r)$ with N elements centered at $P_j, j = 1, \ldots, N$.

$$\left| \int_{\partial B'} \phi_i dx_1 \wedge \cdots \wedge \hat{dx_i} \wedge \cdots \wedge dx_{n+1} - \left(\frac{vol(B')}{N}\right) \sum_j \phi_i(P_j)\right| <$$

$$\sum \left| \int_{B_j} \phi_i dx_1 \wedge \cdots \wedge \hat{dx_i} \wedge \cdots \wedge dx_{n+1} - \left(\frac{vol(B')}{N}\right) \phi(P_j)\right| <$$

$$vol(B')r^{d-1}\left(\frac{vol(B')}{N}\right)^{1/n}$$

If we want to have :

$$\left|\int_{\partial B'}\phi_i\,dx_1\wedge\cdots\wedge\widehat{dx_i}\wedge\cdots\wedge dx_{n+1}-\left(\frac{vol(B')}{N}\right)\sum\phi_i(P_j)\right|<vol(B')/3,$$

we need :

$$N>(3Cr)^n.$$

So we have obtained that N is simply exponential in the dimension n.

S. Smale [S] computed the average complexity of several approximation methods of computation of 1-dimensional integrals. It would be interesting to extend his results to higher dimensional cases for further studies of the index formula.

References

[B-S-S] L. Blum, M. Shub, S.Smale., *On a theory of computation and complexity over the real numbers: NP-completeness, recursive functions, and universal machines.*, Bull. Amer. Math. Soc. 21 (July 1989), 1–46.

[C] J. Canny, *Some algebraic and geometric computations in PSPACE*, 20th annual ACM Symposium on Theory of Computing. (1988), 460–467.

[Co] G.E. Collins, *Quantifier elimination for real closed fields by cylindrical algebraic decomposition*, Lecture Notes in Computer Science 33, 134–183.

[H-S-R] J. Heintz-M.-F. Roy-P.Solernó, *Sur la complexité du principe de Tarski-Seidenberg*, Bull. Soc. Math. France t. 118 (1990), 101–126.

[P] E. Picard, *Sur le nombre des racines communes à plusieurs équations simultanées*, Journal de mathématiques pures et appliquées (1892), 5–24.

[Re] J. Renegar, *A faster PSPACE algorithm for deciding the existencial theory of the reals*, 29th Annual Symposium on Foundations of Computer Science (1988), 291–295.

[Ro] F. Ronga, *Recherche des solutions d'inéquations polynômiales*, Astérisque 192 (1990), 11-16.

[S] S. Smale, *On the efficiency of algorithms of analysis*, Bull. Amer. Math. Soc. 13 (October 1985), 87-121.

[T] D. Trotman, *On Canny's roadmap algorithm : orienteering in semi-algebraic sets*, "Proceedings of the Warwick Singularities Symposium," Springer Lecture Notes in Mathematics (to appear).

Keywords. Real algebraic geometry, index formula, *NP*-problems.
1980 *Mathematics subject classifications*: Primary 32C05. Secondary 12-04.

J.-P. F. : Université de Paris VI, UFR 920, Mathématiques, 45-46, 5ème étage, 4 Pl. Jussieu, 75252 Paris, France.
F. R. : Section de Mathématiques, Université de Genève, C.P. 240, CH-1211 Genève 24, Switzerland.
E-mail:Ronga@ibm.unige.ch

PROPER POLYNOMIAL MAPS: THE REAL CASE

J.M. GAMBOA(*): Dpto. Algebra
Fac. Matemáticas. Univ. Complutense
28040 Madrid. Spain.

C. UENO: Dpto Geometría y Topología
Fac. Matemáticas. Univ. Complutense.
28040. Madrid. Spain

<u>Introduction</u>. Let R be a real closed field. If $x = (x_1,\dots,x_n) \in R^n$, $r \in R$, $r>0$, we denote

$\|x\| = \sqrt{x_1^2+\dots+x_n^2}$; $B_n(x,r) = \{y \in R^n\colon \|y-x\| < r\}$ (open ball);

$\overline{B}_n(x,r) = \{y \in R^n\colon \|y-x\| \le r\}$ (closed ball) ; $R^+ = \{r \in R\colon r > 0\}$;

$S^{n-1}(x,r) = \{y \in R^n\colon \|y-x\| = r\}$ (n-1-sphere); $p_N^n = (0,\dots,0,1)$

The <u>euclidean topology</u> on R^n is the topology in which the open balls are a basis of open sets. In what follows R^n will be always considered endowed with its euclidean topology. The <u>semialgebraic</u> subsets of R^n form the smallest collection of subsets of R^n containing those of type

$$U(f) = \{x \in R^n\colon f(x) > 0\}$$

where $f \in R[X_1,\dots,X_n]$ is a polynomial, and stable under finite union and intersection, and complementation. A continuous map $f\colon X \longrightarrow Y$ between semialgebraic subsets $X \subset R^n$ and $Y \subset R^m$ is <u>semialgebraic</u> if its graph $Gr(f)$ is a semialgebraic subset of R^{n+m}. For a semialgebraic function $f\colon U \longrightarrow R$, where U is an open semialgebraic subset of R^n, we copy from the case $R = \mathbb{R}$ the notion of derivability. The ring $S^\infty(U)$ is

(*) Partially supported by C.I.C.Y.T PB 89/0379/C02/01 and Science Plan ERB 4002 PL 910021 - (91100021)

the set of semialgebraic functions $f:U\longrightarrow R$ which have partial derivatives of each order and all of them are (continuous) semialgebraic functions. An element f in $S^{\infty}(U)$ is called a <u>Nash function</u>. A map $f=(f_1,\dots,f_m):U\longrightarrow R^m$ is a <u>Nash map</u> if each coordinate function f_j is a Nash function. If each f_j is a polynomial we say that f is a <u>polynomial map</u>. The notion of proper semialgebraic map was introduced in [DK_1,pg. 192]: a semialgebraic map $f:X\longrightarrow Y$ is called <u>semialgebraically proper</u> if for every semialgebraic map $g:Z\longrightarrow Y$, the canonical projection

$$p:X{\times}_Y Z=\{(x,z)\in X{\times}Z:\ f(x)=g(y)\}\longrightarrow Z$$

is <u>semialgebraically closed</u>, i.e. p maps every closed semialgebraic subset C of $X{\times}_Y Z$ onto a closed semialgebraic subset p(C) of Z. Taking Z = Y and g the identity map on Y, it follows immediately that semialgebraically proper maps are semialgebraically closed. Also, from 9.1 and 9.4 in [D-K_1] -see also [Br., 8.13.5]- it follows that the constant map on a closed semialgebraic subset X of R^n is proper if and only if X is bounded. This together with [D-K_1, Thm 12.5] provide us the following characterization of semialgebraically proper maps, which is the starting point of our considerations in this note:

<u>Theorem 1</u>.- Let $f:X\longrightarrow Y$ be a semialgebraic map between the semialgebraic subsets $X\subset R^n$, $Y\subset R^m$. The following statements are equivalent:

(1) f is semialgebraically proper.
(2) f is semialgebraically closed and its fibers are closed in R^n and bounded.

Our main results are refinements of this theorem for Nash or polynomial mappings:

Theorem 2.- Let $f:R^n \longrightarrow R^m$ be a non-constant semialgebraically closed Nash map. Then f is semialgebraically proper.

Theorem 3.- Let $f:R^n \longrightarrow R$ be a non-constant polynomial map. Then f is semialgebraically proper if and only if there exists some $M \in R^+$ such that the fibers $f^{-1}(t)$ are bounded for every $t \in R$ with $|t| > M$.

We obtain some consequences of these results. For example, from Thm 3 it follows:

Corollary 1.- Let $f:R^n \longrightarrow R^m$ be a non-constant polynomial map. Then f is semialgebraically proper if and only if there exists $M \in R^+$ such that for every $t \in R^+$ with $t>M$, the inverse image $f^{-1}(S^{m-1}(0,t))$ is bounded.

The results.- For the convenience of the reader we begin with the proof of an easy but useful result stated in [D-K$_1$, pg. 192], whose simple proof is omitted there:

Proposition 1.- Let $f:X \longrightarrow Y$, $h:Y \longrightarrow T$ be semialgebraic maps between semialgebraic subsets $X \subset R^n$, $Y \subset R^m$, $T \subset R^p$, such that $h \circ f$ is semialgebraically proper. Then, f is also semialgebraically proper.

Proof.- We must prove that if $Z \subset R^q$ is a semialgebraic subset, $g:Z \longrightarrow Y$ is a semialgebraic map, $p:Xx_YZ \longrightarrow Z$ is the canonical projection and C is a closed semialgebraic subset of Xx_YZ, then p(C) is closed in Z. But Xx_YZ is a closed subset of $Xx_TZ = \{(x,z) \in XxZ: h\circ f(x) = h\circ g(z)\}$ and so C is also closed in Xx_TZ. Our hypothesis says that the projection $p':Xx_TZ \longrightarrow Z$ is semialgebraically closed. Hence $p(C) = p'(C)$ is closed in Z.

Proof of theorem 2.- Using Thm. 1 we must only prove that the fibers of f are bounded. Let us assume that $f^{-1}(b)$ is unbounded for some $b \in R^m$. Then, for every $r \in R^+$, the semialgebraic set $M_r = f^{-1}(B_m(b,1/r))\setminus\bar{B}_n(0,r)$ is open and non-empty. On the other hand, since f is non-constant, $R^n\setminus f^{-1}(b)$ is a dense subset of R^n by [B-C-R, Prop. 8.1.13]. Thus, the intersection $M_r\cap(R^n\setminus f^{-1}(b))$ is non-empty for every $r \in R^+$. In other words, the semialgebraic subset

$$S = \{(r,x) \in RxR^n\colon r>0,\ x \in M_r,\ f(x)\neq b\}$$

projects onto R^+ under the semialgebraic map $\Pi\colon S\longrightarrow R^+\colon (r,x)\longrightarrow r$. We apply now Hardt's theorem or, better, its corollary [B-C-R, 9.3.2] to the triple (S,Π,R^+). Then, there exists a finite subset $V \subset R^+$ such that Π is semialgebraically trivial over each connected semialgebraic component of $R^+\setminus V$. In particular, let us take a point $r_0 \in R^+$ bigger than all points in V and let T be the connected semialgebraic component of $R^+\setminus V$ containing r_0. The triviality of Π over T means that there exists a semialgebraic set F and a semialgebraic homeomorphism $h\colon TxF\longrightarrow \Pi^{-1}(T)$ such that $\Pi\circ h$ is the projection $TxF\longrightarrow T$. The set F is non-empty since the map $\Pi\colon S\longrightarrow R^+$ is surjective, and we pick an arbitrary point $p \in F$. Let us define a map $i\colon T\longrightarrow TxF$ by $i(r) = (r,p)$ and consider the semialgebraic map

$$\sigma\colon [r_0,\rightarrow\ ) \subset T\xrightarrow{i} TxF\xrightarrow{h} \Pi^{-1}(T) \subset S \subset R^+xR^n$$

If $\Pi_2\colon R^+xR^n\longrightarrow R^n$ is the canonical projection and $\varphi = \pi_2\circ\sigma\colon [r_0,\rightarrow)\longrightarrow R^n$ we are going to show that $N = \varphi([r_0,\rightarrow))$ is a closed semialgebraic subset of R^n, but f(N) is not closed. This contradiction proves the theorem. The semialgebraicity of N follows from Tarski theorem, see e.g. [B-C-R, Prop. 2.2.7]. Let us assume that there is some point $a \in Adh(N)\setminus N$. For some $r>r_0$, $a \in B_n(0,r)$. Then, if

$s>r$, the point $(s,\varphi(s)) = \sigma(s) \in S$ and so $\varphi(s) \in M_s$. In particular $\varphi(s)\notin \overline{B}_n(0,s)$. Thus $\varphi(s)\notin B_n(0,r)$. Consequently

$$B_n(0,r) \cap N \subset \varphi([r_0,r])$$

But $\varphi([r_0,r])$ is a closed semialgebraic subset of R^n, by [B-C-R, Thm. 2.5.8] and it does not contain a. Thus $A = B_n(0,r)\setminus\varphi([r_0,r])$ is an open semialgebraic neighborhood of a which does not meet N, i.e., $a\notin \mathrm{Adh}(N)$. To prove that f(N) is not closed we shall check that $b \notin f(N)$, but $b \in \mathrm{Adh}\, f(N)$.. For every $r \in [r_0,\rightarrow)$ it is $f(\varphi(r))\neq b$, since $(r,\varphi(r)) \in S$. Hence $b\notin f(N)$. On the other hand, if $\varepsilon \in R^+$ there exists $r \in [r_0,\rightarrow)$ such that $1/r<\varepsilon$ and since $(r,\varphi(r)) \in S$, we get

$$f(\varphi(r)) \in B_m(b,1/r) \subset B_m(b,\varepsilon)$$

and so the intersection $f(N)\cap B_m(b,\varepsilon)$ is non-empty as desired.

Remark.- Thm 2 is also true for semialgebraically closed maps whose fibers have dense complement in R^n, but it is false for arbitrary semialgebraically closed (non-constant) semialgebraic maps. For instance the semialgebraic map

$$f:\mathbb{R}\longrightarrow \mathbb{R}: t\longrightarrow \begin{cases} 0 \text{ if } t\leq 0 \\ t \text{ if } t>0 \end{cases}$$

is a topologically closed map but $f^{-1}(0)$ is unbounded.

In what follows we shall be mainly concerned with polynomial mappings. However we state first an easy proposition of very general nature to be used later -compare with Example 7.7 in [D-K_2]- We denote by $e_n:R^n \hookrightarrow S^n$ the inverse of the stereographic projection from the north pole

$$e_n(x_1,\ldots,x_n) = \left(\frac{2x_1}{\|x\|^2+1},\ldots,\frac{2x_n}{\|x\|^2+1}, \frac{\|x\|^2-1}{\|x\|^2+1}\right).$$

Proposition 2.- Let $f:R^n\longrightarrow R^m$ be a semialgebraic map. Then f is semialgebraically proper if and only if the map $\hat{f}:S^n \longrightarrow S^m$ defined by $\hat{f}(P^n_N) = P^m_N$ and $\hat{f}|e_n(R^n) = e_m \circ f$ is semialgebraic.

<u>Proof</u>.- Consider the cartesian square

$$\begin{array}{ccc} S^n & \xrightarrow{\hat{f}} & S^m \\ \uparrow e_n & & \uparrow e_m \\ R^n & \xrightarrow{f} & R^m \end{array}$$

If $\hat{f}$ is a semialgebraic map it is also semialgebraically closed since S^n is a closed and bounded semialgebraic subset of R^{n+1}-use Thm 1 and [B-C-R, Thm 2.5.8]. Then, also f is semialgebraically closed. In fact, if C is a closed semialgebraic subset of R^n and K is the closure of $e_n(C)$ in S^n, then $f(C) = e_m^{-1}(\hat{f}(K))$ is closed in R^m. Also, let b be a point in R^m and let $\Pi^n_N:S^n\setminus\{P^n_N\} \longrightarrow R^n$ be the stereographic projection from p^n_N. Since $f^{-1}(b) = \Pi^n_N(e_n(f^{-1}(b)))$ and $e_n(f^{-1}(b)) = (\hat{f})^{-1}(e_m(b))$ is closed and bounded in S^n, we get that also $f^{-1}(b)$ is bounded and by Thm.1, f is semialgebraically proper. For the converse, note that $Gr(\hat{f})\setminus Gr(f)$ is a unique point, and so $Gr(\hat{f})$ is semialgebraic. Hence we must only check the continuity of $\hat{f}$. Let U be an open semialgebraic subset of S^m. If $p^m_N \notin U$, then $(\hat{f})^{-1}(U) = e_n(f^{-1}(e_m^{-1}(U)))$ is open. If $p^m_N \in U$, then $e_m^{-1}(S^m\setminus U)$ is a closed and bounded subset of R^m. Therefore, by [A, Prop. 1.3], $f^{-1}(e_m^{-1}(S^m\setminus U))$ is a closed and bounded subset of R^n and so $(\hat{f})^{-1}(U) = S^n\setminus e_n(f^{-1}(e_m^{-1}(S^m\setminus U)))$ is open in S^n.

Let us denote by $\mathbb{P}_n(R)$ the projective space over R of dimension n, whose points are denoted $x = (x_0:x_1:...:x_n)$. We fix the embedding

$$u_n:R^n \hookrightarrow \mathbb{P}_n(R):x = (x_1,...,x_n) \longrightarrow (1:x_1:...:x_n).$$

We can relate semialgebraic properness with "projective extendibility".

Proposition 3.- Let $f:R^n \longrightarrow R^m$ be a semialgebraic map. Let us assume that there exists a semialgebraic map $F:\mathbb{P}_n(R) \longrightarrow \mathbb{P}_m(R)$ such that:

(i) $F(H^n_\infty) \subset H^m_\infty$, where $H^n_\infty = \{x = (x_0:x_1:\ldots:x_n) \in \mathbb{P}_n(R): x_0 = 0\}$.

(ii) $u_m \circ f = F \circ u_n$.

Then f is semialgebraically proper.

Proof.- First we check that the fibers of f are bounded. Let $b \in R^m$. By [B-C-R, 3.4.6], $\mathbb{P}_n(R)$ is a closed and bounded semialgebraic (in fact algebraic) subset of some R^M. Thus, $e_n(f^{-1}(b)) = F^{-1}(u_m(b))$ is a closed and bounded subset of $\mathbb{P}_n(R)$, and so $f^{-1}(b)$ is bounded. Secondly, if C is a closed semialgebraic subset of R^n and K denotes the closure of $u_n(C)$ in $\mathbb{P}_n(R)$, it follows from conditions i) and ii) that $f(C) = u_m^{-1}(F(K))$, and this is a closed subset in R^m since F is semialgebraic and $\mathbb{P}_n(R)$ is closed and bounded. Now we apply Thm. 1.

Corollary 2. Let $f:R \longrightarrow R^m$ be a non constant polynomial map. Then f is semialgebraically proper.

Proof.- We can assume m = 1. In fact, for some $1 \le i \le m$, the composition $f_i = \Pi_i \circ f$ is non constant, where $\Pi_i: R^m \longrightarrow R$ is the canonical i^{th}-projection. If f_i is semialgebraically proper, the same is true for f by Prop. 1. Thus we can write $f(t) = a_0t^n + a_1t^{n-1} + \ldots + a_n$ for some natural number $n \ge 1$ and some $a_0, \ldots, a_n \in R$, $a_0 \ne 0$. Now the semialgebraic map

$$F:\mathbb{P}_1(R) \longrightarrow \mathbb{P}_1(R): x = (x_0:x_1) \longrightarrow (x_0^n : a_0x_1^n + a_1x_1^{n-1}x_0 + \ldots + a_nx_0^n)$$

verifies conditions i) and ii) in Prop. 3, and so f is semialgebraically proper.

Remarks.- (1) The converse of proposition 3 is not true. Consider for example the polynomial map $f:\mathbb{R}^2 \longrightarrow \mathbb{R}^2:(x,y)\longrightarrow (x^2+y^2, x(x^2+y^2))$. There is no extension $F:\mathbb{P}_2(\mathbb{R})\longrightarrow \mathbb{P}_2(\mathbb{R})$ verifying conditions i) and ii) in Prop. 3. Otherwise we take the point $a = (0:0:1) \in \mathbb{P}_2(\mathbb{R})$ and the sequences $a_n = (\frac{1}{n}:\frac{1}{n}:1) \in \mathbb{P}_2(\mathbb{R})$, $b_n = (\frac{1}{n}:0:1)$, both convergent to a. Then

$$\lim_{n\to\infty} u_2 f(1,n) = \lim_{n\to\infty} F\circ u_2(1,n) = \lim_{n\to\infty} F(a_n) = F(a), \text{ and also}$$

$$\lim_{n\to\infty} u_2 f(0,n) = \lim_{n\to\infty} F\circ u_2(0,n) = \lim_{n\to\infty} F(b_n) = F(a). \text{ Consequently}$$

$$\lim_{n\to\infty} (1:1+n^2:1+n^2) = \lim_{n\to\infty} (1:n^2:0), \text{ i.e., } (0:1:1) = (0:1:0) \text{ a}$$

contradiction.

On the other hand, f is semialgebraically proper. In fact, let us consider $g:\mathbb{R}^2\longrightarrow \mathbb{R}^2:(x,y)\longrightarrow (x^3,y^2)$ and $h = g\circ f$. By Prop. 1 it is enough to prove that h is semialgebraically proper. But it is clear that the semialgebraic map

$$H:\mathbb{P}_2(\mathbb{R})\longrightarrow \mathbb{P}_2(\mathbb{R}):(x_0:x_1:x_2)\longrightarrow (x_0^6:(x_1^2+x_2^2)^3:x_1^2(x_1^2+x_2^2)^2)$$

fulfills conditions i) and ii) with respect to h, and so we are done.

(2) Corollary 2 is not true for larger classes of semialgebraic mappings. For example, the regular function $f:\mathbb{R}\longrightarrow \mathbb{R}: t\longrightarrow \dfrac{1}{1+t^2}$ is not semialgebraically proper.

<u>Corollary 3</u>.- Let $f = (f_1,\dots,f_m):\mathbb{R}^n\longrightarrow \mathbb{R}^m$ be a polynomial map. For every $1\le i\le m$, let P_i be the form of highest degree of f_i. Let us assume that each f_i is non-constant and that $P_1,\dots,P_m$ have not non-trivial common zeros in $\mathbb{R}^n$. Then f is semialgebraically proper.

<u>Proof</u>.- Let $d_i = \deg f_i$, $e_i = \prod_{j\ne i} d_i$, for $1\le i\le m$. Let us define a semialgebraic map $\varphi:\mathbb{R}^m\longrightarrow \mathbb{R}^m$ by $\varphi(y_1,\dots,y_m) = (y_1^{e_1},\dots,y_m^{e_m})$. By Prop. 1 it suffices to check that the composition $g = \varphi\circ f$ is

semialgebraically proper. For that we apply Prop. 3. Let us write $g = (g_1,\dots,g_m)$ and $d = \prod_{i=1}^{m} d_i$. Then $\deg g_i = d > 0$, for every $1 \le i \le m$, and we write

$$g_i = g_{i,d}(x)+g_{i,d-1}(x)+\dots+g_{i,0}(x),\ x = (x_1,\dots,x_n)$$

where each g_{ij} is the homogeneous component of degree j of g_i. Moreover, $g_{i,d} = P_i^{e_i}$, and so the set of commmon zeros of $g_{1,d},\dots,g_{m,d}$ is the origin. Therefore, the semialgebraic map

$$G: \mathbb{P}_n(R) \longrightarrow \mathbb{P}_m(R): (x_0:x_1:\dots:x_n) \longrightarrow (x_0^d: \sum_{j=0}^{d} g_{1j}(x)x_0^{d-j}:\dots: \sum_{j=0}^{d} g_{mj}(x)x_0^{d-j})$$

satisfies conditions i) and ii) in Prop. 3 with respect to g, and so g is semialgebraically proper.

Remark. The converse of the last corollary is not true in general. For example, the map $f: \mathbb{R}^2 \longrightarrow \mathbb{R}^2: (x,y) \longrightarrow (x^4+x^2+y^2, x^4+x^2+y^2)$ does not verify the hypothesis, but since $\|(x,y)\|^2 \le \|f(x,y)\|$ for every point $(x,y) \in R^2$, it follows that f is in fact topologically proper. However we get a converse if we restrict ourselves to homogeneous polynomial mappings:

Corollary 4. Let $f_1,\dots,f_m$ be non-constant homogeneous polynomials in $R[x_1,\dots,x_n]$, let Z be the set of common zeros in R^n of $f_1,\dots,f_m$ and let $f = (f_1,\dots,f_m): R^n \longrightarrow R^m$. Then, f is semialgebraically proper if and only if $Z = \{0\}$.

Proof. The "if" part follows from Cor. 3, and for the "only if" part note that the cone $Z = f^{-1}(0)$ is unbounded if $Z \ne \{0\}$.

Proof of theorem 3. We must prove the "if" part and we may assume that f is a polynomial of even degree and non-negative on R^n. In fact, let us define $g = \varphi \cdot f$, where $\varphi: R \longrightarrow R: t \longrightarrow t^2$. Then g is a polynomial map of even degree, $g(x) \geq 0$ for every point $x \in R^n$ and if $s \in R$ with $|s| > M^2$, the fiber $g^{-1}(s)$ is bounded. Hence, if we prove the theorem for g, and using Prop. 1, we get that f is semialgebraically proper. If $n = 1$, it is enough to apply Cor. 2. Thus we assume $n \geq 2$ and so, for each $r \in R^+$, $R^n \setminus B_n(0,r)$ is a semialgebraically connected semialgebraic subset of R^n. Thus the set $M_r = g(R^n \setminus B(0,r))$ is an interval (perhaps unbounded) in R, with $M_r \subset [0,\rightarrow)$. Consequently we have a well defined map

$$d: R^+ \longrightarrow R: r \longrightarrow \text{ infimum of } M_r.$$

The map d has semialgebraic graph and so it exists $b = \sup d(R^+) \in R \cup \{\infty\}$. We are going to prove that in fact $b = \infty$. Otherwise, choose a point $c \in R$, $c > b$, $c > M$. We shall see that for every $r \in R^+$, the fiber $g^{-1}(c)$ has points with norm bigger than r which contradicts the hypothesis. Since $c > b \geq d(r) = \inf M_r$, there exists some point $x_1 \in R^n \setminus B_n(0,r)$ such that $g(x_1) < c$. On the other hand, g being a square, there exists a point $y_1 \in R^n \setminus B_n(0,r)$ verifying $c < g(y_1)$. Thus $g(x_1) < c < g(y_1)$ for some points x_1, y_1 in the semialgebraically connected set $R^n \setminus B_n(0,r)$ and by Bolzano's theorem, $c = g(z_1)$ for some point $z_1 \in R^n \setminus B_n(0,r)$, as desired. Now, to finish, it suffices to apply proposition 2, i.e., we must prove that the extension of g,

$$\hat{g}: S^n \longrightarrow S^1 : p_N^n \longrightarrow p_N^1, \ \hat{g}|R^n = g,$$

is continuous. For the sake of simplicity we identify the embeddings $e_n: R^n \hookrightarrow S^n$ and $e_1: R \hookrightarrow S^1$ with set theoretic inclusions. We must check that the inverse image under $\hat{g}$ of every neighbourhood of p_N^1 contains a neighbourhood of p_N^n, i.e., for every $a \in R$ there exists $r \in R^+$ such that $R^n \setminus B_n(0,r) \subset g^{-1}((a,\rightarrow))$, which follows immediately from what precedes.

Now we are ready for the proof of corollary 1:

Proof of corollary 1. Let $f = (f_1, \dots, f_m)$, $F = f_1^2 + \dots + f_m^2 : R^n \longrightarrow R$. The polynomial map $\varphi: R^m \longrightarrow R: (y_1, \dots, y_m) \longrightarrow y_1^2 + \dots + y_m^2$ is semialgebraically proper since its fibers are bounded. Hence, $F = \varphi \circ f$ is semialgebraically proper if and only if f is so, by Prop. 1. But, by Thm. 3, F is semialgebraically proper if and only if there exists $M \in R^+$ such that $f^{-1}(S^{m-1}(0,t)) = F^{-1}(t)$ is bounded for all $t \in R^+$, $t > M$.

Remark. With the notations in the proof above, let C_f be the set of critical values of F. It is a finite set, [B-C-R, Thm. 9.5.2], and so there exists $\mu(f) \in R^+$ such that $C_f \cap [\mu(f), \rightarrow) = \emptyset$. Since the fibers $F^{-1}(t)$ and $F^{-1}(t')$ are semialgebraically homeomorphic when $t, t' \in R \setminus C_f$ we can state Cor. 1 in the following way:

"f is semialgebraically proper if and only if $f^{-1}(S^{m-1}(0,t))$ is bounded for some (and so for all) $t \ge \mu(f)$".

Example. Let $P, Q \in R[y]$ be polynomials and $d \ge 1$ a positive integer. Let $f: R^2 \longrightarrow R^2: (x,y) \longrightarrow (x^d + P(x), x^d + Q(y))$. If the fiber $f^{-1}f(0)$ is bounded, then f is semialgebraically proper.

W.l.o.g. we may suppose that $f(0) = 0$ and it is enough to prove that for every $k \in R$, $M_k = \{(x,y) \in R^2: (x^d + P(y))^2 + (x^d + Q(y))^2 = k^2\}$ is bounded. If $P = Q$, then $f^{-1}(0) = \{(x,y) \in R^2: x^d + P(y) = 0\}$ is bounded and so d is even and $P(y) = \sum_{y=0}^{2e} a_j y^{2e-j}$ for some integer $e \ge 1$ and $a_j \in R$, $a_0 > 0$. Thus, the sets $A = \{(x,y) \in R^2: x^d + P(y) = \ell\}$ and $B = \{(x,y) \in R^2: x^d + P(y) = -\ell\}$ are bounded for $\ell = |k|/\sqrt{2}$, and $M_k = A \cup B$. If $P \ne Q$, the set $F_k = \{(x,y) \in R^2, (P-Q)^2(y) \le 2k^2, |x^d| \le |P(y) + Q(y)| +$

$+ \frac{|P^2(y)+Q^2(y)-k^2|}{2}\}$ is bounded and $M_k \subset F_k$.

Comment. In some cases, condition in Thm 3 is easily verified. Consider for instance a polynomial map $f:\mathbb{R}^2 \longrightarrow \mathbb{R}: (x,y) \longrightarrow \sum_{j=0}^{d} f_j(x,y)$, where each $f_j \in \mathbb{R}[x,y]$ is an homogeneous polynomial of degree j. To decide if the fiber $f^{-1}(t)$ is bounded we consider the projectivization

$$C \equiv \{(x:y:z) \in \mathbb{P}_2(\mathbb{R}): \sum_{j=0}^{d} f_j(x,y)z^{d-j} = tz^d\}$$

Then, $f^{-1}(t)$ is bounded if and only if the intersection $C\cap\{z=0\}$ consists on isolated points. Of course this is the case if f_d has not non-trivial zeros, according to our sufficient condition in Prop. 3. Otherwise, if for example the point (0:1:0) is in C, we deshomogeneize to get

$$C' = \{x,z) \in \mathbb{R}^2: \sum_{j=0}^{d} f_j(x,1)z^{d-j} = tz^d\}$$

and the question is to decide if the origin is an isolated point in C. But this is the case if and only if the Puiseux branches at the origin of the equation defining C' are non-real.

Corollary 5. Let $f \in \mathbb{R}[x_1,\dots,x_n]$ be a strictly positive definite polynomial such that the level curves $\{x \in \mathbb{R}^n: f(x) = t\}$ are bounded, for every $t \in \mathbb{R}^+$. Then there exists $r_0 \in \mathbb{R}^+$ such that $f(x) \geq r_0$ for all $x \in \mathbb{R}^n$.

Proof. By Thm. 3, the non-constant polynomial map $f:\mathbb{R}^n \longrightarrow \mathbb{R}$ is semialgebraically proper and so its image $f(\mathbb{R}^n) \subset \mathbb{R}^+$ is a closed interval. Thus, its lower bound r_0 does the work.

Remark. The boundness of the level curves is not a superfluous condition in the last corollary: the polynomial $f(x,y) = (xy-1)^2+y^2$ is strictly positive and it attains all positive real values since $f(n,1/n) = 1/n^2$.

Corollary 6. Let $f = (f_1,\dots,f_m): R^n \longrightarrow R^m$ be a non-constant polynomial map. Let A be the polynomial ring $A = R[x_1,\dots,x_n,T]$ and let us denote by $\sum$ the sums of squares of polynomials in A. Then, f is semialgebraically proper if and only if there exist $C \in R^+$, $p \in \mathbb{N}$, $g \in A$ and $S_1, S_2 \in \sum$ such that

$$g\left(\sum_{j=1}^{m} f_j^2(x) - T^2\right) = 1 + S_1 + S_2\left(\sum_{i=1}^{n} x_i^2 - C(1+T^2)^p\right)$$

Proof. Let $F = f_1^2 + \dots + f_m^2: R^n \longrightarrow R$. If such an identity exists, we get an inclusion $F^{-1}(t^2) \subset B_n(0, \sqrt{C(1+t^2)^p})$ for every $t \in R$, and we apply Cor. 1. Conversely, if f is semialgebraically proper, the same holds true for F and so $F(R^n) = [a,\rightarrow)$ for some point $a \in R$. Hence $F^{-1}(t)$ is a non-empty, bounded and closed algebraic subset of R^n for every $t \geq a$, and the well defined map

$$r: [a,\rightarrow): t \longrightarrow \max\ \{\|x\|: F(x) = t\}$$

has semialgebraic graph. By [B-C-R, Prop. 2.6.1] there exists $b \geq a$, $C_1 \in R^+$ and $p \in \mathbb{N}$ such that $r(t) \leq C_1 \cdot t^p$ for every $t \geq b$. By Thm. 1 the preimage $F^{-1}([0,b])$ is a bounded subset of R^n, i.e., $F^{-1}([0,t]) \subset B_n(0,C_2)$ for some $C_2 \in R^+$. Then, if $C = \max\ \{C_1, C_2\}$ we get $F^{-1}(t^2) \subset B_n(0, C(1+t^2)^p)$ for every $t \in R$, and so the set $\{(x,t) \in R^{n+1}: \sum_{i=1}^{n} x_i^2 - C(1+T^2)^p \geq 0,\ \ F(x) - T^2 = 0\}$ is empty. Applying [B-C-R, 4.4.2] we get the desired identity.

Let $i = \sqrt{-1}$ and $C = R(i)$ be the algebraic closure of R. In what follows we denote $x = (x_1,\dots,x_n)$, $y = (y_1,\dots,y_n)$, $z = (z_1,\dots,z_n) =$

$x+iy$, and we identify C^n with R^{2n} via $C^n \longrightarrow R^{2n}$, $z = x+iy \longrightarrow (x,y)$. We can get some insight on semialgebraically proper polynomial maps $f = (f_1,\dots,f_n): C^n \longrightarrow C^m$, $f_j \in C[z_1,\dots,z_m]$. There exists real polynomials $g_j, h_j \in R[x,y]$ such that $f_j(z) = g_j(x,y)+ih_j(x,y)$ and we identify f with the polynomial map $F: R^{2n} \longrightarrow R^{2m}$, given by $F=(g_1,\dots,g_m,h_1,\dots,h_m)$. Since

$$\sum_{j=1}^{m} (g_j^2(x,y)+h_j^2(x,y)) = \sum_{j=1}^{m} |f_j(z)|^2$$

we get from Thm. 3:

<u>Corollary 7</u>. The polynomial map $f = (f_1,\dots,f_m): C^n \longrightarrow C^m$ is semialgebraically proper if and only if there exists $M \in R^+$ such that for every $r \in R$, $r>M$, the subset

$$\{z \in C^n: \sum_{j=1}^{m} |f_j(z)|^2 = r\}$$

is bounded.

In the one variable case we have an analogous result to Cor. 2:

<u>Corollary 8</u>. Every non-constant polynomial map $f: C \longrightarrow C^m$ is semialgebraically proper.

<u>Proof.</u> As in the proof of Cor.2, we may assume $m = 1$. Then $f(z) = \sum_{j=0}^{d} a_j (z)^{d-j}$ with $a_j \in C$, $a_0 \neq 0$ and the induced map $F = (g,h): R^2 \longrightarrow R^2$ is given by real polynomials g and h of degree d whose homogeneous components g_d and h_d of degree d verify

$$g_d(x,y) + i\, h_d(x,y) = a_0(x+iy)^d.$$

Thus, g_d and h_d have not non-trivial common zeros and by Cor.3 the map F is semialgebraically proper.

On the other hand, for $n \geq 2$, there is a significant difference between the real and the complex cases:

<u>Proposition 4</u>. The fibers of a semialgebraically proper polynomial map $f:C^n \longrightarrow C^m$ are finite. If m = n, then f is semialgebraically open, i.e., f maps open semialgebraic subsets of C^n onto open (semialgebraic) subsets of C^n and in particular f is surjective.

<u>Proof</u>. For the first part it is enough to see that bounded algebraic subsets of C^n are finite. This is true for $C = \mathbb{C}$, the field of complex numbers since if $Z \subset \mathbb{C}^n$ is an irreducible algebraic subset of $\mathbb{C}^n$ which is not a point, its projective closure W verifies that $W \setminus Z \neq \emptyset$ and Z is a dense subset of W, [S, Ch.7, §2.1, lemma1] and so Z is not closed in W, hence it is not compact. To obtain the result in the general case we use Tarski Principle for real closed fields -[T] or [B-C-R, 5.2.4]. Each algebraic subset of C^n can be seen as an algebraic subset of R^{2n} by separating the real and imaginary parts of its equations. For every integer d>0, let B_d be the collection of bounded algebraic subsets of R^{2n} "coming from" algebraic subsets of C^n and expressible as the set of real zeros of a polynomial in $R[x_1,\dots,x_n,y_1,\dots,y_m]$ of degree $\leq 2d$. Clearly, the fact "$V \in B_d$" is expressible as a formula in the first order language $\mathcal{L}(R)$ with parameters in R. Moreover, there exists a positive integer M(n,d) such that, if $C = \mathbb{C}$, and $V \in B_d$, then $card(V) \leq M(n,d)$. In fact, by [B-C-R, Thm. 9.9.4], there exists algebraic subsets $V_1,\dots,V_s$ of R^{2n} such that every V in B_d is semialgebraically homeomorphic to some V_j and we can take as M(n,d) the maximum of the number of semialgebraically connected components of $V_1,\dots,V_s$. Consequently, for every positive integer d, the following formula in $\mathcal{L}(R)$: "$V \in B_d \Rightarrow card(V) \leq M(n,d)$" is true for $R = \mathbb{R}$ and so it is also true for arbitrary real closed field R. In particular bounded algebraic subsets of C^n are finite.

Assume now n = m. Once the openess of f is proved, then $f(C^n)$ is an open and closed semialgebraic subset of C^n, i.e. $f(C^n) = C^n$. Let us denote by $F = (F_1,\dots,F_{2n}):R^{2n} \longrightarrow R^{2n}$ the polynomial map induced by F and let $\deg F = \max \{\deg F_1,\dots,\deg F_{2m}\}$. For every integer d>0, let $\mathcal{F}_d$ be the collection of polynomial maps $R^{2n} \longrightarrow R^{2n}$ "coming from" a polynomial map $C^n \longrightarrow C^n$ with degree ≤d. Since f has finite fibers, then F is an open map in case $C = \mathbb{C}$ -see [N, Prop 4, pp 132]- and so the following formula in $\mathcal{L}(R)$ is true for $R = \mathbb{R}$ and every d>0:

" For every $F \in \mathcal{F}_d$ such that card $F^{-1}(b) \leq M(n,d)$ for every $b \in R^{2n}$, the set $F(B_{2n}(a,r))$ is an open subset of R^{2n} for every $a \in R^{2n}$ and every $r \in R$, r>0".

Therefore the formula is also true for every d>0 and every real closed field R and since the open balls form a basis of open sets in the euclidean topology of R^{2n}, F is a semialgebraically open map.

<u>Remark</u>. Finiteness of the fibers is not a sufficient condition for a polynomial map $f:C^n \longrightarrow C^m$ to be semialgebraically proper. For example, the map $f:\mathbb{C}^2 \longrightarrow \mathbb{C}^2: (x,y) \longrightarrow (xy, x-y(xy-1))$ has finite fibers but it is not proper. In fact the point $(1,0) \notin f(\mathbb{C}^2)$ but for every natural number n, the point $(1,1/n) = f(1/n,n)$ is in $f(\mathbb{C}^2)$.

Each polynomial map $f:R^n \longrightarrow R^m$ induces another one $f_C:C^n \longrightarrow C^m$ defined by the same polynomials. Clearly, f_C semialgebraically proper implies f is semialgebraically proper, since R^n is a closed semialgebraic subset of $C^n = R^{2n}$, but the converse is not true: consider the map $f:\mathbb{R}^2 \longrightarrow \mathbb{R}: (x,y) \longrightarrow x^2+y^2$, which has not finite fibers but it is, by Thm 3, semialgebraically proper. To finish we show that semialgebraic maps with finite fibers -e.g. semialgebraically proper polynomial maps $C^n \longrightarrow C^m$ - have some nice properties:

Proposition 5. Let $f: R^n \longrightarrow R^m$ be a semialgebraic map with finite fibers. Let $F \subset \mathbb{R}^m$ be a semialgebraic set with dim $F \leq n-2$. Then $R^n \setminus f^{-1}(F)$ is semialgebraically connected.

Proof. From the finiteness of the fibers and Hardt's theorem [B-C-R 9.3.1] it follows that dim $f^{-1}(F) \leq n-2$ and now it suffices to apply Theorem 13.2 of [D-K] with $M = R^n$ and $N = f^{-1}(F)$.

Acknowledgement: We wish to thank the referee for his contribution in shortening the proofs of Thm.3, Cor.4 and Prop.5 and some other useful comments.

References

[A] Alonso, M.E. Real proper morphisms. Archiv der Math. Vol 43, pp. 237-243 (1984).

[B-C-R] Bochnak, J.; Coste, M.; Roy, M.F. *Géométrie Algébrique Réelle*. Ergebnisse der Math. 3 Folge, Band 12. Springer. (1987).

[Br] Brumfiel, G.W. *Partially ordered rings and semialgebraic geometry*. Cambridge: Cambridge Univ. Press (1979).

[D-K_1] Delfs, H.; Knebusch, M. Semialgebraic Topology over a Real Closed Field II: Basic Theory of Semialgebraic Spaces. Math. Zeit. 178, pp. 175-213 (1981).

[D-K_2] Delfs, H.; Knebusch, M. *Locally semialgebraic spaces*. L.N.M. 1173. Springer (1985).

[N] Narasimhan, R. *Introduction to the theory of analytic spaces*. L.N.M. 25, Springer (1966).

[S] Shafarevich, I.R. *Basic algebraic geometry*. Springer (1974).

[T] Tarski, A. *A decision method for elemmentary algebra and geometry*. Prepared for publication by J.C.C. Mac Kinsey, Berkeley (1951).

SUR LES ORDRES DE NIVEAU 2^n ET SUR UNE EXTENSION DU 17ème PROBLEME DE HILBERT

DANIELLE GONDARD-COZETTE

UNIVERSITE PARIS VI, 4 place Jussieu, 75252 PARIS cedex 05, FRANCE.

ABSTRACT. Let K be a chainable field, which means a field admitting orderings of exact level 2^n for any $n \geq 1$, from [G1] and [G2] we know that in such a field there exists $\alpha \in K$ such that $\alpha^2 \notin \sum K^4$; we then define the notion of α-chain of orderings and use it to obtain explicit formulas for orderings of exact level 2^n in some special fields. We also obtain for some chain-closed fields a new extension of Hilbert's 17th problem.

INTRODUCTION. Les corps ordonnables sont de deux sortes : les corps non chaînables (i.e. il n'existe pas de chaîne d'ordres de niveau supérieur de niveaux exacts 2^n) où toute somme de carrés est une somme de puissances quatrièmes (par exemple $\mathbb{R}$, $\mathbb{Q}$ et ses extensions algébriques ordonnables), et les corps chaînables où il existe un élément α tel que α^2 ne soit pas somme de puissances quatrièmes ; dans ce cas nous dirons que K est un *corps α-chaînable* (par exemple $\mathbb{Q}(X)$ et $\mathbb{R}((X))$ sont des corps X-chaînables) .

La nécessité de faire intervenir une constante α a été mise en évidence par les axiomatisations des théories des corps chaînables et des corps chaîne-clos que nous avons données dans [G1] (preuves dans [G2]). Ce point de vue nous a déjà permis de démontrer des résultats du type 17ème problème de Hilbert dans [D-G] et de créer un analogue à l'algèbre réelle dans [B-G] , et nous le conservons ici pour obtenir une expression des ordres de niveau 2^n de certains corps ($\mathbb{R}(X)$ par exemple) et une nouvelle forme de généralisation du 17ème problème de Hilbert ; l'introduction de la partie III expliquera comment la généralisation précédemment étudiée dans [D-G] et [B-B-D-G] est différente de celle présentée dans cet article.

I-NOYAUX DES CHAINES D'UN CORPS CHAINABLE.

Définition I-1. Un corps K sera dit α-chaînable s'il existe dans K un élément α tel que α^2 ne soit pas somme de puissances quatrièmes d'éléments de K.

Définitions I-2. Un préordre T de niveau 2^n (i.e. $T + T \subseteq T$, $T \cdot T \subseteq T$, $K^{2^n} \subseteq T$) sera dit un α-préordre si $\alpha^{2^{n-1}} \notin T$ (il est alors de niveau exact 2^n car $\sum K^{2^{n-1}}$ n'est pas contenu dans T). De même un ordre P de niveau exact 2^n sera dit un α-ordre si $\alpha^{2^{n-1}} \notin P$.

Definition I-3. Nous appelerons α-chaîne une chaîne d'ordres de niveau supérieur $(P_i)_{i \in \mathbb{N}}$ telle que $\alpha^2 \notin P_2$.

Proposition I-4. Un corps K est α-chaînable si et seulement s'il existe au moins une α-chaîne.

Ceci résulte des définitions I-1 et I-3 et du résultat de Becker-Harman qui donne l'expression suivantes des sommes de puissances : $\sum K^4 = (\bigcap_i P_{i2}) \cap \sum K^2$ où P_{i2} désigne un ordre de niveau exact 4 quelconque de K. Donc si α^2 n'est pas une somme de puissances quatrièmes dans K, il existe au moins un j tel que $\alpha^2 \notin P_{j2}$. D'après le corollaire I-4 de [H] il existe au moins une chaîne d'ordres de niveau supérieur passant par ce P_{j2}.

Proposition I-5. Dans un corps K α-chaînable tout ordre P_n de niveau exact 2^n, $n \geq 2$, d'une α-chaîne $(P_i)_{i \in \mathbb{N}}$ est un α-ordre.

En effet si $\alpha^{2^{n-1}} \notin P_n$, alors P_n est bien un α-ordre. Si $\alpha^{2^{n-1}} \in P_n$ P_n étant de α-chaîne $\alpha^2 \notin P_2$, on en déduit qu'il existe $i \geq 2$ tel que $\alpha^{2^{i-1}} \notin P_i$ et $\alpha^{2^i} \in P_{i+1}$ (puisque d'après la relation de chaîne dès que $\alpha^{2^p} \in P_l$ on a $\alpha^{2^{p+q}} \in P_{l+q}$ pour tout $q \geq 1$) ; or il est connu (cf. [Bel]) que si $x \in K^*$ et si x^2 appartient à un ordre P de niveau exact 2^n alors $x \in P \cup -P$: en effet si $x \notin P$ alors $P + Px$ est un préordre propre contenant strictement P et donc $P + Px = K$, d'où $-1 = a + bx$ avec a et b dans P et $x = (-1-a)/b \in -P$. On déduit de ce résultat que $\alpha^{2^l} \in P_{l+1}$, entraîne que $\alpha^{2^{l-1}} \in P_{l+1} \cup -P_{l+1}$, puis par la relation de chaîne $P_{l+1} \cup -P_{l+1} = (P_0 \cap P_l) \cup -(P_0 \cap P_l)$ que $\alpha^{2^{l-1}} \in P_l$ ce qui est contraire à l'hypothèse.

Lemme I-6. Soit K un corps α-chaînable. Pour tout $n \in \mathbb{N}^*$, $T_n = \sum K^{2^n} - \alpha^{2^{n-1}} \sum K^{2^n}$ est un α-préordre propre (i.e. $-1 \notin T_n$) de niveau exact 2^n.

C'est tout à fait clair pour $n = 1$ et $n = 2$; pour $n \geq 3$ K étant α-chaînable $\alpha^2 \notin \sum K^4$ et par la proposition I-4 il existe une α-chaîne $(P_l)_{l\in\mathbb{N}}$; par I-5 tout P_n de α-chaîne est un α-ordre donc $\alpha^{2^{n-1}} \notin P_n$ et la relation de chaîne $P_n \cup -P_n = (P_{n-1} \cap P_0) \cup -(P_{n-1} \cap P_0)$ jointe au fait que P_{n-1} contient toutes les puissances 2^{n-1}-èmes montre que $\alpha^{2^{n-1}} \in P_n \cup -P_n$ et donc que $-\alpha^{2^{n-1}} \in P_n$; P_n contenant toutes les puissances 2^n-èmes d'éléments de K contient donc T_n. T_n est clairement un préordre de niveau 2^n, et le niveau est exactement 2^n car $\alpha^{2^{n-1}}$ n'appartenant pas à P_n ne peut donc pas appartenir à T_n ; de même -1 ne peut appartenir à T_n qui est donc propre.

Proposition I-7. Dans un corps K α-chaînable pour toute α-chaîne $(P_l)_{l\in\mathbb{N}}$ on a $T_k = \sum K^{2^k} - \alpha^{2^{k-1}} \sum K^{2^k} \subseteq P_k$, pour tout $k \geq 2$.

C'est clair d'après la preuve du lemme I-6.

Proposition I-8. Dans un corps K α-chaînable, tout P_n, ordre de niveau exact 2^n $n \geq 2$, qui contient $T_n = \sum K^{2^n} - \alpha^{2^{n-1}} \sum K^{2^n}$ est un P_n de α-chaîne.

En effet seulement l'un des deux éléments $\alpha^{2^{n-1}}$ ou $-\alpha^{2^{n-1}}$ appartient à P_n. Donc $\alpha^{2^{n-1}} \notin P_n$ et on en déduit que toute chaîne passant par P_n est une α-chaîne puisque si $\alpha^2 \in P_2$, en utilisant la condition de chaîne (qui s'exprime au niveau 3 par : $P_3 \cup -P_3 = (P_2 \cap P_0) \cup -(P_2 \cap P_0))$ on obtiendrait $\alpha^2 \in P_3 \cup - P_3$ d'où $\alpha^4 \in P_3$; on déduirait en itérant que $\alpha^{2^{n-1}} \in P_{n-1}$ ce qui est impossible. Toute chaîne $(P_i)_{i\in\mathbb{N}}$ passant par P_n est donc une α-chaîne .

Théorème I-9. *Dans un corps α-chaînable* $T_n = \sum K^{2^n} - \alpha^{2^{n-1}} \sum K^{2^n}$ *est égal à l'intersection de tous les α-ordres de niveau exact* 2^n *(qui appartiennent à au moins une α-chaîne).*

D'après le théorème 1 de Becker [Be1], un préordre propre de niveau 2^n est égal à l'intersection de tous les ordres de même niveau, exact ou non, qui le contiennent ; un ordre qui contient T_n est un α-ordre car $-\alpha^{2^{n-1}} \in T_n \subseteq P_n$ ce qui entraîne que $\alpha^{2^{n-1}} \notin P_n$; enfin P_n est bien de niveau exact 2^n puisque $\sum K^{2^{n-1}}$ n'est pas contenu dans P_n.

Corollaire I-10. *Dans un corps* K *α-chaînable ,* $K_n = \sum K^{2^n} - \alpha^{2^{n-1}} \sum K^{2^n}$ *est égal à l'intersection des* P_n *, ordres de niveau exact* 2^n *des α-chaînes.*

Cela résulte immédiatement de I-7, I-8 et I-9.

Remarque. Dans un corps K α-chaînable $T'_n = \sum K^{2^n} + \alpha^{2^{n-1}} \sum K^{2^n}$ est toujours un préordre propre de niveau 2^n mais il n'est pas forcément de niveau exact 2^n .

Définition I-12. Appelons , pour $n \geq 2$, T_n noyau au niveau n des α-chaînes.

II-ON ORDERINGS OF SOME CHAINABLE FIELDS.

Théorème II-1. *Si un corps* K *α-chaînable n'admet qu'une seule α-chaîne* $(P_i)_{i\in\mathbb{N}}$ *, à échange de* P_0 *et* P_1 *près, alors l'ordre de niveau exact* 2^n *de celle-ci est pour tout* $n \geq 2$ *,* $P_n = T_n = \sum K^{2^n} - \alpha^{2^{n-1}} \sum K^{2^n}$.

En effet par la proposition I-7 toute α-chaîne $(Q_i)_{i\in\mathbb{N}}$ est telle que $Q_n \supseteq T_n$ pour tout $n \geq 2$, d'après l'hypothèse faite que K n'admet qu'une seule α-chaîne le corollaire I-10 donne alors $P_n = T_n$.

Exemple d'application : Dans [Bel] page 61 est donné un exemple de corps ayant exactement trois ordres, dont l'un P est archimédien, et admettant des ordres de niveau exact 2^n pour un $n \geq 2$ donc pour tout n : il s'agit d'un corps K intersection d'une clotûre réelle de $\mathbb{Q}(X)$ pour un ordre archimédien avec une clôture réelle généralisée pour un ordre de niveau exact 2^n avec $n \geq 2$. De [G4] il découle que les deux ordres (non archimédiens) P_0 et P_1 sont co-chaînables avec $A(P_0) = A(P_1)$ et $\bar{P}_0 = \bar{P}_1$. On en déduit (ce corps étant de Pasch) que d'après le théorème 17 (ii) de [Bel] ce corps K n'admet qu'un seul ordre de niveau exact 2^n pour chaque $n \geq 2$; d'après le théorème II-1 cet ordre est explicitement donné par : $P_n = \sum K^{2^n} - \alpha^{2^{n-1}} \sum K^{2^n}$ où α est un élément de K séparant les ordres P_0 et P_1.

Théorème II-2. *Soit K un corps α-chaînable et P_0, P_1 deux ordres qui soient le début d'une α-chaîne $(P_i)_{i\in\mathbb{N}}$; pour $n \geq 2$, soit $C_n = \sum (P_0 \cap P_1)^{2^{n-1}} - \alpha^{2^{n-1}} \sum (P_0 \cap P_1)^{2^{n-1}}$, alors C_n est égal à l'intersection de tous les α-ordres P_n le contenant (de tels P_n sont alors de niveau exact 2^n et sont de α-chaîne).*

La démonstration est analogue à celle de I-10 dès que l'on a montré que C_n est un préordre propre de niveau exact 2^n. Si $(P_i)_{i\in\mathbb{N}}$ est une α-chaîne commençant par $P_0 \cap P_1$ alors $(P_0 \cap P_1)^{2^{n-1}} \subseteq P_n$: c'est vrai pour $n = 2$ d'après la relation $P_2 \cup -P_2 = (P_0 \cap P_1) \cup -(P_0 \cap P_1)$, et par récurrence si $(P_0 \cap P_1)^{2^{n-1}} \subseteq P_n$ alors par la relation de chaîne, au niveau $n+1$, $P_{n+1} \cup -P_{n+1} = (P_0 \cap P_n) \cup -(P_0 \cap P_n)$ on déduit que $(P_0 \cap P_1)^{2^n}$ est contenu dans P_{n+1}. P_n étant de α-chaîne $\alpha^{2^{n-1}} \notin P_n$ et $-\alpha^{2^{n-1}} \in P_n$ donc on a $C_n \subseteq P_n$. On en déduit que C_n, qui est clairement un préordre de niveau au plus 2^n ($\forall x \;\; x^{2^n} = (x^2)^{2^{n-1}} \in (P_0 \cap P_1)^{2^{n-1}}$, est un préordre propre de niveau exact 2^n. Il est donc égal à l'intersection de tous les ordres P qui le contiennent, ceux-ci sont de niveau au plus 2^n car si le niveau était supérieur on aurait $\sum K^{2^n}$ non contenu dans P ce qui est impossible puisque $(P_0 \cap P_1)^{2^{n-1}} \geq K^{2^n}$. Le niveau est évidemment au moins 2^n puisque $\sum K^{2^{n-1}}$ n'est pas contenu dans P qui contenant C_n ne peut contenir $\alpha^{2^{n-1}}$. Etant de niveau exact 2^n et tels que $-\alpha^{2^{n-1}} \in P_n$, ces ordres P contenant C_n sont des ordres de niveau exact 2^n de α-chaînes.

Exemple d'application : la proposition II-2 permet d'obtenir une expression des ordres de niveau supérieur de $\mathbb{R}(X)$:
En effet $\mathbb{R}(X)$ est un corps de Pasch et, comme déjà utilisé plus haut, il découle de [Bel] et de [G4] qu'un couple d'ordres co-chaînables ne l'est que par une seule chaîne. Enfin il est montré dans [Di] que les chaînes de ce corps sont construites sur des paires d'ordres "symétriques" P_{a+} et P_{a-}, qui placent X infiniment près de $a \in \mathbb{R}$ avec $X-a$ positif ou négatif respectivement, ou sur $P_{\infty+}$ et $P_{\infty-}$ qui rendent X infiniment grand positif ou négatif respectivement.
Il y a donc dans $\mathbb{R}(X)$ deux $(X-a)$-chaînes bâties sur les deux couples d'ordres P_{a+} et P_{a-} d'une part et $P_{\infty+}$ et $P_{\infty-}$ d'autre part. Les ordres de niveau exact 2^n de ces deux chaînes sont alors donnés par

$P_{na} = \sum (P_{a+} \cap P_{a-})^{2^{n-1}} - (X-a)^{2^{n-1}} \sum (P_{a+} \cap P_{a-})^{2^{n-1}}$ et par

$P_{n\infty} = \sum (P_{\infty+} \cap P_{\infty-})^{2^{n-1}} - (X-a)^{2^{n-1}} \sum (P_{\infty+} \cap P_{\infty-})^{2^{n-1}}$ respectivement ;

ceci résulte évidemment du fait que ces deux expressions sont des $(X-a)$-préordres propres de niveau exact 2^n distincts .

III-UNE AUTRE GENERALISATION DU 17ème PROBLEME DE HILBERT AU NIVEAU 2^n

A l'origine le 17ème problème de Hilbert se présentait sous la forme : *soit* $f \in \mathbb{R}[\bar{X}]$ *telle que* $\forall\ \bar{x} \in \mathbb{R}^p$ $f \geq 0$, *a-t-on* $f \in \sum \mathbb{R}(\bar{X})^2$? Artin le résolut en 1927 (cf. [R]).
$\mathbb{R}$ n'ayant qu'un seul ordre dont les éléments positifs sont $P = \mathbb{R}^2 = \sum \mathbb{R}^2$ une première extension possible est :
soit K *un corps ordonnable, soit* $f \in K[\bar{X}]$ *telle que* $\forall\ \bar{x} \in K^p$ $f \in \sum K^2$, *a-t-on* $f \in \sum K(\bar{X})^2$? Ceci a été étudié dans [G-R], [McK], [P2] et [Z-G].
Une autre extension est le passage aux sommes de puissances 2^n-èmes :

soit K *un corps ordonnable et soit* $f \in K[\bar{X}]$ *telle que* $\forall\ \bar{x} \in K^p$ $f \in \sum K^{2^n}$ *a-t-on* $f \in \sum K(\bar{X})^{2^n}$? Cette extension a été résolue pour une variable et certain corps chaîne-clos dans [D-G]. Puis dans [B-B-D-G] ce problème est résolu dans un cadre plus général, pour les puissances 2n-èmes et plusieurs variables.

Nous proposons ici une extension différente du 17-ème problème de Hilbert aux ordres de niveau supérieur sous la forme :

Soit K *un corps chaînable (i.e. ordonnable et tel que* $\sum K^2 \neq \sum K^4$ *), soit* P_n *un ordre de niveau exact* 2^n *(* $n \geq 2$ *) de* K*, soit* $f \in K[\bar{X}]$ *telle que* $\forall\ \bar{x} \in K^p$ $f(\bar{x}) \in P_n$*, peut-on caractériser* f *?*

Nous apportons une réponse dans le cas de K chaîne-clos (<2>-réel-clos dans la terminologie de [B-B-D-G]) n'admettant qu'une seule valuation henselienne à corps des restes réel-clos. Insistons sur le fait que dans [D-G] ou [B-B-D-G] l'obtention d'un résultat nécessite de considérer les $f(\bar{x})$ pour $\bar{x} \in L^p$, où L est une extension algébrique ordonnable de K, alors qu'ici on ne considère $f(\bar{x})$ que pour $\bar{x} \in K^p$.

Théorème III-1. *Soit* K *un corps chaîne-clos* α*-chaînable n'admettant qu'une seule valuation hensélienne à corps des restes réel-clos, et soit* $f \in K(\bar{X})$ *; alors pour* $n \geq 2$ *les propriétés suivantes sont équivalentes :*

(i) $f \in \sum K(\bar{X})^{2^n} - \alpha^{2^{n-1}} \sum K(\bar{X})^{2^n}$;

(ii) $\forall \bar{x} \in K^p$ *où* f *est définie* $f(\bar{x}) \in P_n$*, où* P_n *désigne l'unique ordre de niveau exact* 2^n *de* K *;*

(iii) $\forall \bar{x} \in K^p$ *où* f *est définie* $f(\bar{x}) \in \sum K^{2^n} - \alpha^{2^{n-1}} \sum K^{2^n}$;

(iv) $\forall \bar{x} \in K^p$ *où* f *est définie* $f(\bar{x}) \in K^{2^n} \cup - \alpha^{2^{n-1}} K^{2^n}$.

La preuve utilise le corollaire I-10 du présent article et le lemme suivant obtenu avec Delon dans [D-G] :

Lemme III-2 [D-G]. Soient K et L deux corps chaîne-clos tels que $K \subseteq L$ et K n'admet qu'une seule valuation hensélienne à corps des restes réel-clos alors sont équivalents :

(i) $K \cap L^2 = K^2$;

(ii) K est relativement algébriquement clos dans L.

(iii) $K \prec L$ (où "$\prec$" est une inclusion élémentaire).

De l'expression d'un corps chaîne clos α-chaînable sous la forme $K = K^2 \cup - K^2 \cup \alpha K^2 \cup - \alpha K^2$ et du lemme précédent il résulte que si K n'a qu'une seule valuation hensélienne à corps des restes réel-clos et est contenu dans un autre corps chaîne-clos α-chaînable L, alors on a $K \{ L$.

Preuve de III-1.

(ii) $\Leftrightarrow$ (iii) $\Leftrightarrow$ (iv) est immédiat : En effet dans un corps chaîne-clos α-chaînable K l'unique ordre de niveau exact 2^n est donné par les expressions suivantes : $P_n = \sum K^{2^n} - \alpha^{2^{n-1}} \sum K^{2^n} = K^{2^n} \cup - \alpha^{2^{n-1}} K^{2^n}$; la première forme résulte du fait qu'un corps chaîne-clos α-chaînable est uniquement α-chaînable et du théorème II-2, et la seconde vient de l'expression des ordres de niveau supérieur d'un corps n'admettant que deux ordres usuels donnée par Becker dans [B1].

Il suffit de montrer le théorème pour $f \in K[\overline{X}]$ car si $f = g / h$ alors $f = gh^{2^n-1} / h^{2^n}$ et l'on sait que d'une part $\sum K^{2^n} \subseteq P_n$ et que d'autre part $T_n(K(\overline{X})) = \sum K(\overline{X})^{2^n} - \alpha^{2^{n-1}} \sum K(\overline{X})^{2^n}$ est un préordre (voir § I).

(i) $\Rightarrow$ (iii) : il suffit de vérifier que si f, appartenant à $\sum K(\overline{X})^{2^n} - \alpha^{2^{n-1}} \sum K(\overline{X})^{2^n}$, est définie en x, alors $f(x)$ appartient à $\sum K^{2^n} - \alpha^{2^{n-1}} \sum K^{2^n}$. Cette preuve est due à Becker et nous le remercions de nous autoriser à la reproduire ici :

Notons $f \in \sum K(\overline{X})^{2^n} - \alpha^{2^{n-1}} \sum K(\overline{X})^{2^n}$ sous la forme $f = \sum r_i^{2^n} - \alpha^{2^{n-1}} \sum s_j^{2^n}$ où les r_i, $s_j \in K(\overline{X})$; soit $x = (x_1, \ldots, x_p) \in K^p$ et soit $\mathcal{O}_x = K[\overline{X}]_{(X_1 - x_1, \ldots, X_p - x_p)}$ l'anneau local en x ; soit $\lambda : K(\overline{X}) \longrightarrow K \cup \{\infty\}$ une place telle que $\lambda(X_i) \longmapsto x_i$ et soit V_λ l'anneau correspondant ; alors si f est définie en x, $f \in \mathcal{O}_x \subseteq V_\lambda$. Il suffit de montrer que r_i, $s_j \in V_\lambda$ d'où l'on déduit $\lambda(f) = f(x_1, \ldots, x_p) =$
$= \sum \lambda(r_i)^{2^n} - \alpha^{2^{n-1}} \sum \lambda(s_j)^{2^n} \in P_n$.

Soit par exemple r_1 tel que $v(r_1) = \min \{ v(r_i), v(s_j) \}$, si $r_1 \notin V_\lambda$, alors on a $f = r_1^{2^n} [1 + \sum (r_i / r_1)^{2^n} - \alpha^{2^{n-1}} \sum (s_j / r_1)^{2^n}]$; on sait que $f \in V_\lambda$ et le crochet, noté z dans la suite, dans l'expression ci-dessus est une unité ce qui donne une contradiction ; en effet si z n'était pas une unité alors dans le corps résiduel $V_\lambda / m_\lambda = K$ on aurait, puisque les r_i / r_1 et s_j / r_1 appartiennent à V_λ :

$\lambda(z) = 1 + \sum y_i^{2^n} - \alpha^{2^{n-1}} \sum y_j'^{2^n} = 0$ ce qui est impossible puisque $-1 \notin P_n$.

L'autre cas où l'on a par exemple s_1 défini par :

$v(s_1) = \min \{ v(r_1), v(s_j) \}$ se traite de manière analogue.

(iv) ⇒ (i) : On considère la théorie des corps chaîne-clos α-chaînables et on utilise le langage des anneaux augmenté d'un symbole de constante α. Dans le corps K chaîne-clos α-chaînable l'hypothèse (ii) se traduit, en notant f par P / Q avec $P, Q \in K[\overline{X}]$, par la formule suivante :

" $\forall \overline{x}\ \exists y\ \exists z\ (Q(\overline{x}) = 0 \ \vee\ f(\overline{x}) = y^{2^n} \ \vee\ f(\overline{x}) = -\alpha^{2^{n-1}} z^{2^n})$ " ; dans $K(\overline{X})$ qui est α-chaînable on fixe une α-chaîne $(P_i)_{i \in \mathbb{N}}$ et on considère L une clôture chaîne de $K(\overline{X})$ pour cette chaîne. D'après le lemme III-2 et l'hypothèse faite sur K on a $K \prec L$, donc dans L la même formule " $\forall \overline{x}\ \exists y\ \exists z\ (Q(\overline{x}) = 0 \ \vee\ f(\overline{x}) = y^{2^n} \ \vee\ f(\overline{x}) = -\alpha^{2^{n-1}} z^{2^n})$ " est satisfaite ; on choisit alors $\overline{x} = \overline{X}$ dans L et on obtient que f appartient à l'unique ordre de niveau exact 2^n de L ; cet ordre prolongeant l'ordre P_n de niveau 2^n de l' α-chaîne choisie sur $K(\overline{X})$, nous avons aussi que f appartient à P_n. Ce raisonnement est faisable pour toutes les α-chaîne de $K(\overline{X})$, par conséquent f appartient à l'intersection de tous les ordres de niveau exact 2^n des α-chaînes de $K(\overline{X})$ dont on sait par le corollaire I-10 qu'elle est égale à $\sum K(\overline{X})^{2^n} - \alpha^{2^{n-1}} \sum K(\overline{X})^{2^n}$.

REFERENCES

[Be1] E. Becker :*"Hereditarily pythagorean fields and orderings of higher" types"*, I.M.P.A., Lectures Notes # 29 (1978), Rio de Janeiro.

[Be2] E. Becker : *"The real holomorphy ring and sums of 2n-th powers"*, in Géométrie Algébrique réelle, Lecture Notes in Math. # 959, p.138-181, Springer-Verlag 1982.

[B-B-D-G] E. Becker, R. Berr, F. Delon et D. Gondard : *"Hilbert's 17th problem and sums of 2n powers"*, preprint.

[B-G] E. Becker et D. Gondard : *"On rings admitting orderings and 2-primary chains of orderings of higher level"*, Manuscripta Mathematica, 69, 267-274, 1990.

[D-G] F. Delon et D. Gondard :*"17ème problème de Hilbert au niveau n dans les corps chaîne-clos"*, The Journal of Symbolic Logic, vol. 56, # 3 Sept. 1991, p.853-861.

[Di] M. Dickmann :"*Couples d'ordres chaînables*", Exposé au Séminaire "Structures algébriques ordonnées" (D.D.G.), Univ. paris VII 1987-88.

[G1] D. Gondard :"*Théorie du premier ordre des corps chaînables et des corps chaîne-clos*", C. R. Acad. Sc. Paris, Tome 304, # 16, 1987.

[G2] D. Gondard : "*Chainable fields and real algabraic geometry*", in Proceedings "Real Algebraic and Analytic Geometry" (Trento Oct. 88) , Lectures Notes in Math. 1420, Springer-Verlag, 1990.

[G3] D. Gondard :"*Kernels of chains and chainable fields*", in Abstracts A.M.S., vol 10 #5 Oct. 1989.

[G4] D. Gondard :"*Sur l'espace des places réelles d'un corps chaînable*", preprint.

[G-R] D. Gondard et P. Ribenboim :"*Sur le 17ème problème de Hilbert*", C. R. Acad. Sc. Paris, tome 277, 20-08-1973, p. 303-304.

[H] J. Harman :"*Chains of higher level orderings*", Contemporary Mathematics, vol. 8, 1982, pp. 141-174, A.M.S..

[L1] T. Y. Lam :"*The theory of ordered fields*", Proceedings of Alg. Conference, pp 1-152, M. Dekker (1980).

[L2] T. Y. Lam :"*Orderings, Valuations and Quadratic Forms*", C.B.M.S. regional conference, # 52, 1983, A.M.S..

[McK] K. McKenna : "*New facts about Hilbert's 17th problem*", Lecture notes in Math. 498, Springer-Verlag, pp. 220-230.

[P1] A. Prestel :"*Lectures on formally real fields*", I.M.P.A., Monografias de Matematica, # 22, 1975, Rio de Janeiro.

[P2] A. Prestel : "*Sums of squares in fields*", Bol. Soc. Brasil., Rio de Janeiro, 1975.

[R] P. Ribenboim : "*Arithmétique des corps*", Herman, Paris 1972.

[Z-G] Zeng Guangxin : "*A characterization of preordered fields with the weak Hilbert property*", proc. of the A.M.S. 104 (1988), 335-342.

CURVES OF DEGREE 6 WITH ONE NON-DEGENERATE DOUBLE POINT AND GROUPS OF MONODROMY OF NON-SINGULAR CURVES

I.V. ITENBERG

Abstract.

The paper is devoted to the rigid isotopy classification of plane projective real algebraic curves of degree 6 with a non-degenerate double point and to the calculation of the groups of monodromy of non-singular curves of degree 6.

Key words and phrases :

curve of degree 6 with a non-degenerate double point, rigid isotopy classification, group of monodromy.

Address : Dep. of Math. and Mechanics
Leningrad State University
Bibliotechnaya pl. 2,
Leningrad, Petrodvoretz,
198904, USSR

e-mail address : degt@lomi.spb.su

Introduction

There are two main problems for every class of plane projective real algebraic curves : the classification up to real isotopy and the classification up to rigid isotopy.

Two curves A and B are really isotopic iff sets $\mathbb{R}A$ and $\mathbb{R}B$ of their real points are isotopic in $\mathbb{R}P^2$. Two curves of a fixed class are rigidly isotopic iff there exists a real isotopy connecting these curves and consisting of the sets of real points of the curves of the same class.

Real isotopy and rigid isotopy classifications of non-singular curves of degree ≤ 4 and of the curves with a non-degenerate double point of degree ≤ 4 were known in the last century. The classification up to real isotopy of curves of degree 5 was also known. In 1969 D.A.Budkov [1] obtained the real isotopy classification of non-singular curves of degree 6. The classification of non-singular curves of degree 5 and of degree 6 up to rigid isotopy was completed in 1978-1981 in works of V.A.Rokhlin [2], V.V.Nikulin [3] and V.M.Kharlamov [4]. The results of Kharlamov [4] also allow to obtain the rigid isotopy classification of curves of degree 5 with a non-degenerate double point.

In 1991 the author ([5], [6]) obtained the classifications up to real isotopy and up to rigid isotopy of the curves of degree 6 with one non-degenerate double point using Nikulin's scheme ([3]) and the results of E.B.Vinberg ([7], [8]) on the groups generated by reflections.

In the present paper we consider the scheme of obtaining these classifications. Besides, we discuss the classification up to rigid isotopy in the cases of the most interest.

Kharlamov noticed that the scheme developed by Nikulin ([3]) gives a method for calculation of the groups of monodromy of non-singular curves of degree 6 (if the beginning and the end of a rigid isotopy coincide with non-singular curve A, then this rigid isotopy defines some permutation of the components of the set of real points of A; we will call the group of all such permutations the group of monodromy of curve A). Kharlamov has calculated the groups of monodromy of non-singular M-curves of degree 6 (M-curve is a curve with maximal possible number of components of the real point set for given degree; this number is equal to 11 for degree 6)

In the present paper we consider the general scheme of

calculation of the group of monodromy of non-singular curves of degree 6. Only the cases of M-curves and (M-1)-curves are discussed in details.

1. Statements of results

We will consider the curves of degree 6 with a non-degenerate double point as the results of the simplest degenerations of non-singular curves.

Rigid isotopy classification of these curves is given by Theorem 2.1 and Propositions 3.4, 3.5 of the present paper.

Calculation of the groups of monodromy of non-singular curves of degree 6 can be done using Theorem 3.6.

In the present part of the paper we will discuss the cases of M-curves and (M-1)-curves only.

For the schemes of disposition of ovals of non-singular curves (an oval is a connected component of the set of real points of a curve which is homeomorphic to circle and embedded two-sidely into $\mathbb{R}P^2$) we will use the system of notations suggested by Viro [9]. The curve consisting of a single oval will be denoted by symbol $\langle 1\rangle$, the empty curve - by symbol $\langle 0\rangle$. If symbol $\langle A\rangle$ stands for some set of ovals then the set of ovals obtained by addition of an oval surrounding all old ovals will be denoted by symbol $\langle 1\langle A\rangle\rangle$. If the curve is the union of two non-intersecting sets of ovals denoted by $\langle A\rangle$ and $\langle B\rangle$ respectively with no oval of one set surrounding an oval of the other set, then this curve will be denoted by symbol $\langle A\amalg B\rangle$. Besides, if A is the notation for some set of ovals then part $A\amalg\ldots\amalg A$ of another notation where A repeats n times will be denoted by $n\times A$; parts $n\times 1$ of notation will be denoted by n.

Let us notice that the classifications up to real isotopy and up to rigid isotopy coincide in the cases of non-singular M-curves and (M-1)-curves of degree 6 ([3]). Thus in these cases it is enough to point out the scheme of disposition of ovals in order to fix the rigid isotopy type of a curve.

Now we will present the schemes of possible conjunctions and contractions of ovals of non-singular M-curves and (M-1)-curves of degree 6. Note that no oval of these curves can conjunct with itself.

An oval of a non-singular curve is called even (odd) if it lies inside of even (odd) number of ovals of this curve.

For convenience we first formulate a general statement :

each empty oval of a non-singular curve of degree 6 can be contracted and there exists only one rigid isotopy class of the results of such degeneration.

M-curves

There are three rigid isotopy types of non-singular M-curves of degree 6 : $\langle 1\langle 1\rangle \amalg 9\rangle$, $\langle 1\langle 5\rangle \amalg 5\rangle$, $\langle 1\langle 9\rangle \amalg 1\rangle$.

Scheme $\langle 1\langle 1\rangle \amalg 9\rangle$. The scheme of possible conjunctions and contractions of ovals of a curve of this isotopy type is shown in Fig. 1.

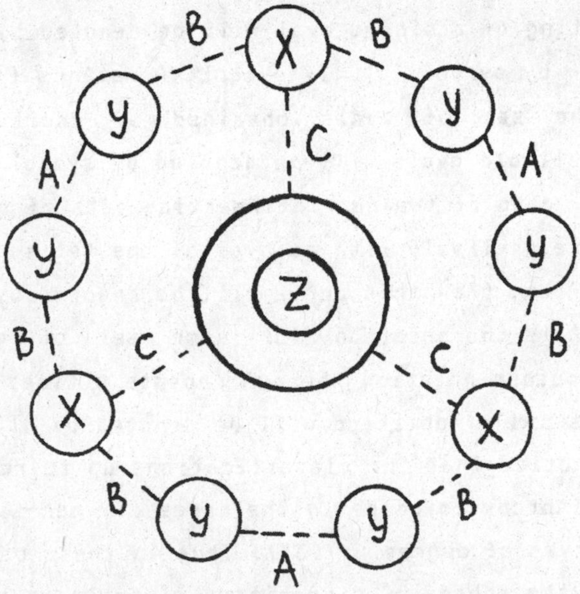

Fig.1

It is stated that the even empty ovals of the curve belonging to this rigid isotopy type can be numerated by numbers from 1 to 9 in such a way that the first oval can conjunct with the second one, the second oval can conjunct with the third one and so on with the last nineth oval conjuncting with the first one. Moreover, the first, the fourth and the seventh ovals can conjunct with a non-empty one while the other conjunctions of ovals are impossible.

The letters standing near two dotted lines coincide iff the results of the conjunctions corresponding to these lines are rigidly isotopic. The letters placed inside the empty ovals coincide iff the results of contractions of these ovals are rigidly isotopic.

We will name such schemes the marked schemes of conjunctions and contractions of ovals.

It is clear that the even empty ovals of a non-singular M-curve of degree 6 with the scheme $\langle 1\langle 1\rangle \amalg 9\rangle$ can conjunct only in a cyclic order. There is natural cyclic order for these ovals (this order is defined up to inversion). Let us take a point inside the odd empty oval and a line containing this point; we can rotate the chosen line around the chosen point and mark the order in which the line intersects the even empty ovals. It is the corollary from Bezout theorem that the cyclic order obtained cannot be changed by a rigid isotopy. Thus non-neighbouring (in sense of the given cyclic order) ovals cannot conjunct.

Rigid isotopy classification of the results of the simplest degenerations of curves with scheme $\langle 1\langle 1\rangle \amalg 9\rangle$ can be formulated as follows :

the numbers of rigid isotopy classes contained in isotopy classes represented in Fig. 2 are:

2a) - 1, 2b) - 1, 2c) - 2, 2d) - 2.

But in this statement some information is lost.

The group of monodromy of a curve of the rigid isotopy type under discussion is equal to S_3, where S_3 is the group of permutations of 3 elements.

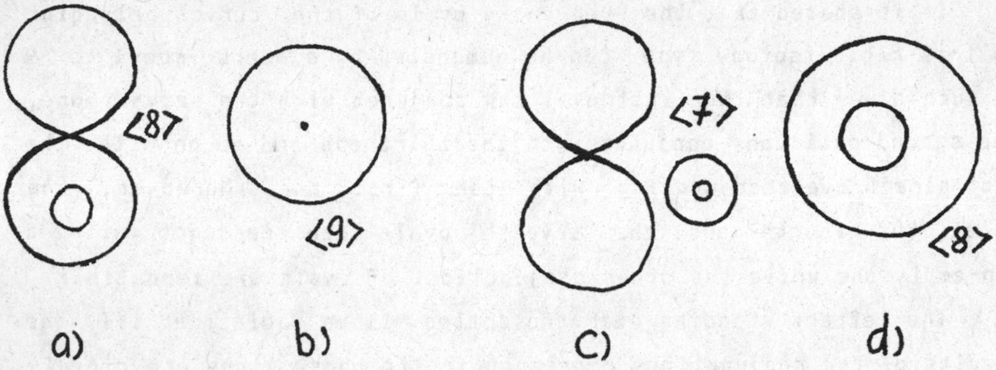

Fig.2

Scheme <1<5> ⨿ 5>. The scheme of conjunctions and contractions of ovals is given in Fig.3. This scheme is not marked because in this case it is convinient to say that the results of the contractions of non-coincident ovals are not rigidly isotopic and the results of conjunctions of non-coincident pairs of ovals are not rigidly isotopic as well. We will always present non-marked schemes in the cases like this.

It is necessary to note that there are linear orders both on the even empty and on the odd ovals of a curve of degree 6 having scheme <1<5> ⨿ 5>.

The group of monodromy of a curve of the rigid isotopy type under discussion is trivial.

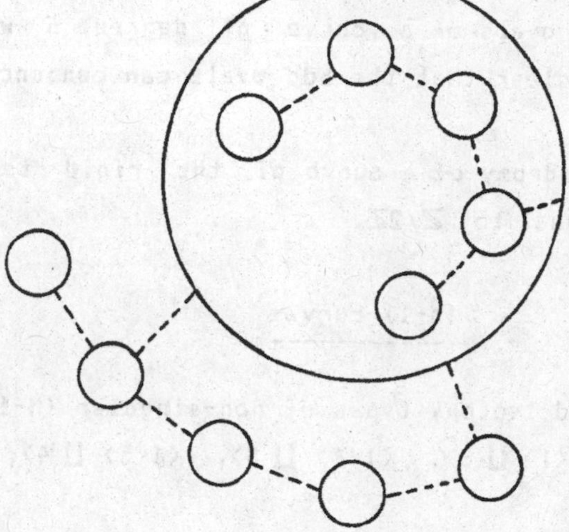

Fig.3

<u>Scheme $\langle 1\langle 9\rangle \amalg 1\rangle$.</u> The marked scheme of conjunctions and contractions of ovals is represented in Fig.4.

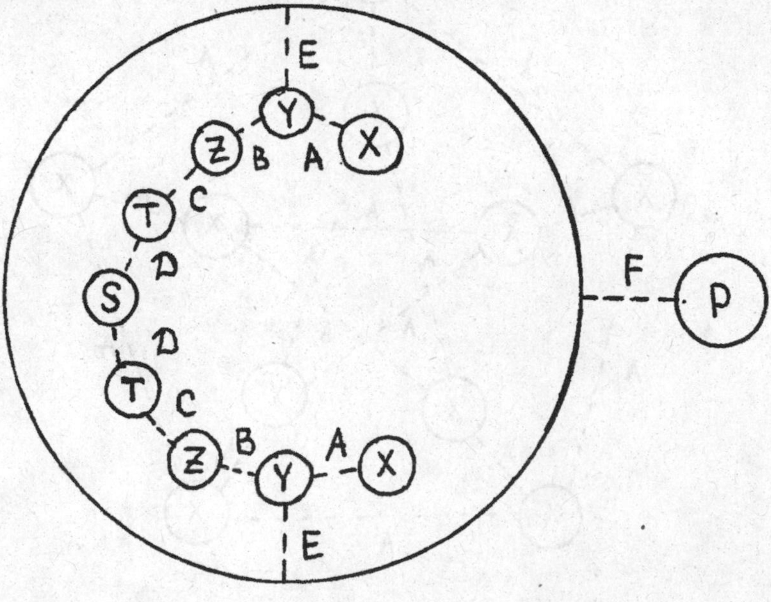

Fig.4

Note that there is a natural linear order (defined up to inversion) on the odd ovals of a curve of degree 6 with scheme ⟨1⟨9⟩ ∐ 1⟩. So it is clear that the odd ovals can conjunct only in chain.

The group of monodromy of a curve of the rigid isotopy type under discussion is equal to $\mathbb{Z}/2\mathbb{Z}$.

(M-1)-curves

There are 6 rigid isotopy types of non-singular (M-1)-curves of degree 6 : ⟨10⟩, ⟨1⟨1⟩ ∐ 8⟩, ⟨1⟨4⟩ ∐ 5⟩, ⟨1⟨5⟩ ∐ 4⟩, ⟨1⟨8⟩ ∐ 1⟩, ⟨1⟨9⟩⟩.

Only those degenerations are discussed whose results cannot be obtained as the results of a degeneration of an M-curve.

<u>Scheme ⟨10⟩</u>. The marked scheme of conjunctions and contractions of ovals is given in Fig.5.

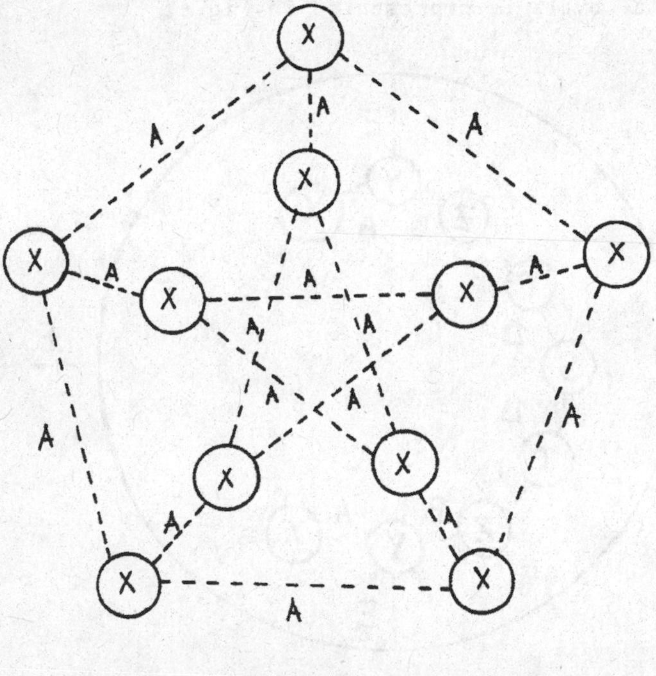

Fig.5

The group of monodromy of a curve of the rigid isotopy type under discussion is equal to A_5, where A_5 is the group of even permutations of 5 elements.

Scheme ⟨1⟨1⟩ ∐ 8⟩. The marked scheme of conjunctions and contractions of ovals is given in Fig.6.

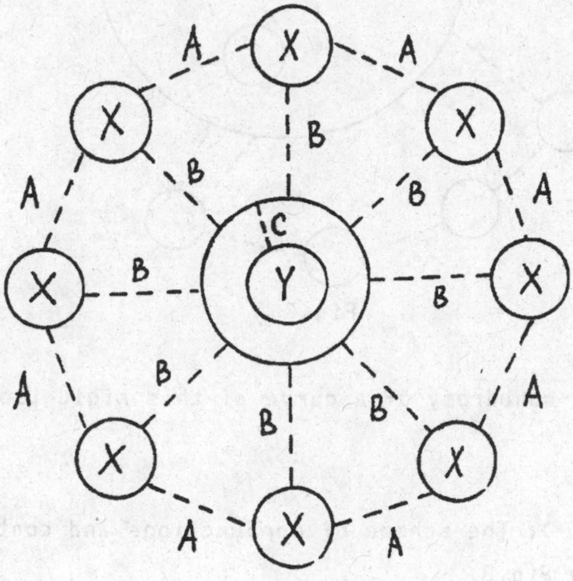

Fig.6

The group of monodromy of a curve of the rigid isotopy type discussed is equal to D_8, where D_8 is the group of symmetries of the regular 8-gon.

Scheme ⟨1⟨4⟩ ∐ 5⟩. The scheme of conjunctions and contractions of ovals is given in Fig.7.

The two dotted lines between the non-empty oval and one of the even empty ovals denote the existence of two rigid isotopy classes of the results of conjunctions of the corresponding ovals. Thus it is not enough to point out a pair of conjunct ovals for pointing out the rigid isotopy class of the results of conjunctions.

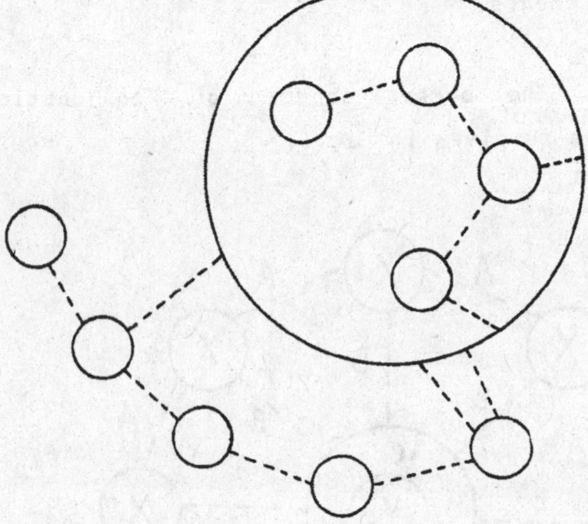

Fig.7

The group of monodromy of a curve of this rigid isotopy type is trivial.

Scheme $\langle 1\langle 5\rangle \amalg 4\rangle$. The scheme of conjunctions and contractions of ovals is given in Fig.8.

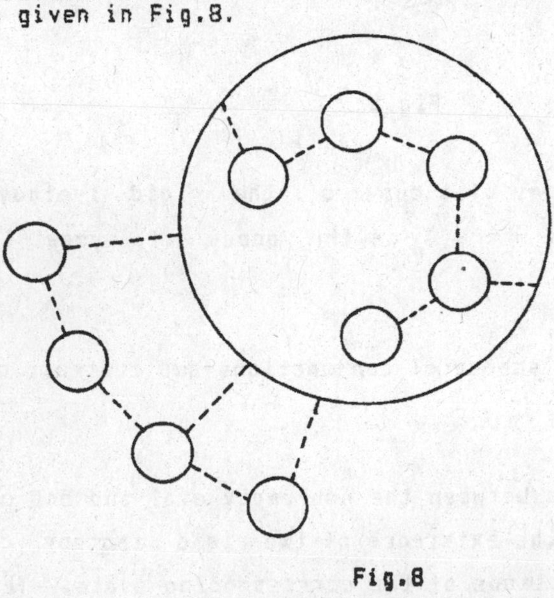

Fig.8

The group of monodromy of a curve of this rigid isotopy type is trivial.

Scheme $\langle 1\langle 8\rangle \amalg 1\rangle$. The marked scheme of conjunctions and contractions of ovals is given in Fig.9.

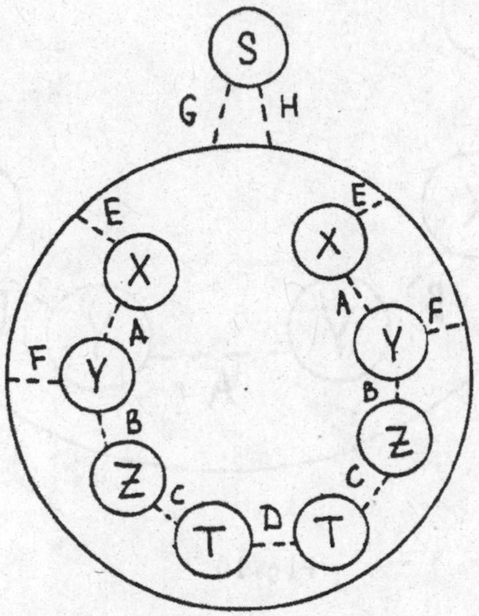

Fig.9

The group of monodromy of a curve of this rigid isotopy type is equal to $\mathbb{Z}/2\mathbb{Z}$.

Scheme $\langle 1\langle 9\rangle\rangle$. The marked scheme of conjunctions and contractions of ovals is given in Fig.10.

Thus one can introduce a natural cyclic order defined up to inversion on the odd ovals of a curve of degree 6 having scheme $\langle 1\langle 9\rangle\rangle$.

The group of monodromy of a curve of the rigid isotopy type discussed is equal to S_a.

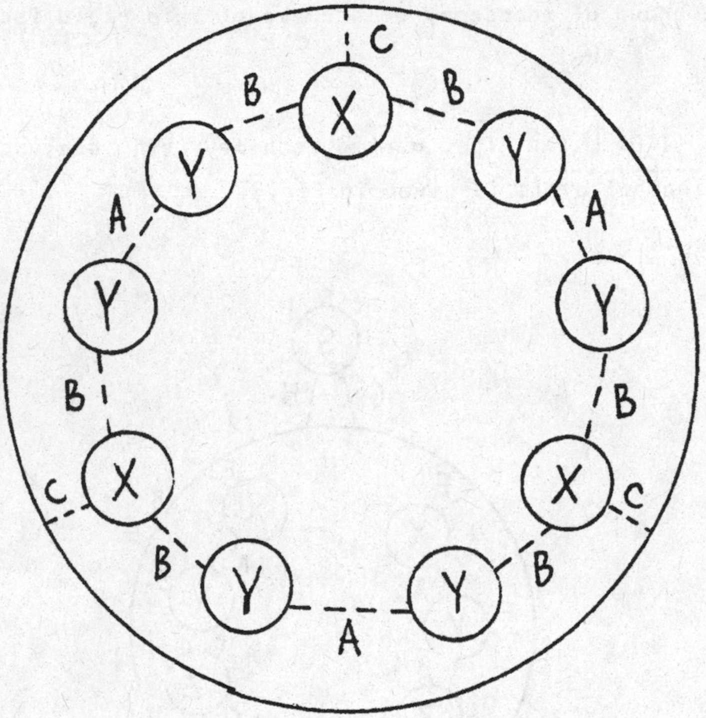

Fig.10

The second part of the present paper is devoted to obtaining the rigid isotopy classification of curves of degree 6 with a non-degenerate double point and to calculation of the groups of monodromy of non-singular curves of degree 6.

2. The reduction to a problem on K3-surfaces

Let A be a curve of degree 6 with a non-degenerate double point, $\mathbb{C}A$ be the set of complex points of this curve. Set $\mathbb{R}A$ of real points of curve A divides real projective plane $\mathbb{R}P^2$ in two sets with common boundary $\mathbb{R}A$. If one of them is non-orientable (both sets cannot be non-orientable) then let us denote it by $\mathbb{R}P^2_-$ and the other by $\mathbb{R}P^2_+$. If both sets are orientable then let us denote an arbitrary one by $\mathbb{R}P^2_-$, the other by $\mathbb{R}P^2_+$. Consider the two-sheeted branched covering space of complex projective plane $\mathbb{C}P^2$ with branch locus $\mathbb{C}A$. Variety X obtained as a result of resolution of the singular point of the branched covering space is a K3-surface.

The involution of complex conjugation acting on $\mathbb{C}P^2$ is lifted to X by two different ways. Let us denote the antiholomorphic involution on X with set of the fixed points lying above $\mathbb{R}P^2_-$ by $conj_-$. The other antiholomorphic involution on X (its set of the fixed points lies above $\mathbb{R}P^2_+$) will be denoted by $conj_+$. Choose one of these involutions and denote it by $conj$. We can identify group $H_2(X;\mathbb{Z})$ together with the intersection form and involuation $conj_*$ on it with lattice L having the rank equal to 22, the signature equal to -16 and involution φ. Denote by L_+ and L_- the eigenspaces of the involution φ corresponding to eigenvalues 1 and -1 respectively. Let B be the bilinear form on L. The diagonal forms over $\mathbb{Q}$ of the restrictions of B on L_+ and L_- have one and two positive squares respectively . In $H_2(X;\mathbb{Z})$ we can point out two elements: class h of hyperplane section and class δ of exceptional divisor. Denote the images in L of classes h and δ by h' and δ' respectively.

The following statements hold true:

$$B(h',h')=2, \quad B(\delta',\delta')=-2, \quad B(h',\delta')=0,$$

$$\varphi(h')=-h', \quad \varphi(\delta')=-\delta', \quad h'\not\equiv\delta' \pmod{2L}$$

The isomorphism of lattice $f\colon L_1\to L_2$ we will call the isomorphism of quadruples $(L_1,\varphi_1,h_1{}'\delta_1{}')$ and $(L_2,\varphi_2,h_2{}'\delta_2{}')$ if

$$f\circ\varphi_1=\varphi_2\circ f, \quad f(h_1{}')=h_2{}', \quad f(\delta_1{}')=\delta_2{}'.$$

Let us fix quadruple

$$(L,\varphi,h',\delta') \qquad (1)$$

satisfying all conditions above. We will call K3-surface X with fixed isomorphism of the quadruples

$$\psi : (H_2(X;\mathbb{Z}),conj_*,h,\delta) \dashrightarrow (L,\varphi,h',\delta')$$

the marked K3-surface.

Let subspace $H^{2,0}(X)$ of complex vector space $H^2(X,\mathbb{C})$ is the image of the de Rham isomorphism of space $H^{2,0}(X)$ consisting of cohomologous classes of holomorphic 2-forms. The dimension of space $H^{2,0}(X)$ is equal to 1. Denote by H the real one-dimensional subspace of $H^{2,0}(X)$ consisting of vector x satisfying the following condition: the image of vector x under the involution of $H^2(X,\mathbb{C})$ induced by the involution of complex conjugation of coefficients coincides with $conj^*(x)$. We will denote by H' the subspace of $H_2(X,\mathbb{C})$ generated by a class dual (in the Poincaré sense) to a class generating space H.

Define (as was done in [3],[10]) the space

$$\Omega = \{0 \neq \omega \in L \otimes \mathbb{C} \mid B_{\mathbb{C}}(\omega,\omega)=0,\ B_{\mathbb{C}}(\omega,\bar{\omega}) > 0,$$
$$B_{\mathbb{C}}(\omega,h')=0,\ B_{\mathbb{C}}(\omega,\delta')=0,\ \varphi(\omega)=\bar{\omega}\ \}/\mathbb{R}^*,$$

where $B_{\mathbb{C}}$ is the extension of bilinear form B on L to bilinear form on $L \otimes \mathbb{C}$, $\mathbb{R}^*$ acts by multiplication. The period mapping bringing one-dimentional subspace $\psi(H')$ of $L \otimes \mathbb{C}$ in correspondance with the marked K3- surface makes space Ω the space of periods of marked K3-surfaces with a non-degenerate double point with involution equivalent to φ and with linear system $|h|$.

We will call equivalent periods the periods which can be transformed one into another by automorphism of quadruple (1). Let us describe the image of the period mapping up to the equivalence. Let element $\omega \in L \otimes \mathbb{C}$ generate class $[\omega] \in \Omega$. Let $\omega = \omega_+ + i\omega_-$ where $\omega_+ \in L_+ \otimes \mathbb{R}$, $\omega_- \in L_- \otimes \mathbb{R}$. Denote the orthogonal complement in L_- of lattice generated by vectors h' and δ' by $L_{-h,\delta}$. Note that $\omega_- \in L_{-h,\delta} \otimes \mathbb{R}$, $\omega_+^2 = \omega_-^2 > 0$. The diagonal forms over $\mathbb{Q}$ of the restrictions of B on L_+ and $L_{-h,\delta}$ have one positive square. Thus the space Ω consists of two connected components. Each of these components can be transformed to one another by involution $-\varphi$. We are interested in periods only up to equivalence and thus we can consider the quotient space $\Omega/(1,-\varphi)$ as the space of periods. Let $\mathfrak{L}_+$ and $\mathfrak{L}_{-h,\delta}$ be the spaces of Lobachevsky obtained by the standard procedure from spaces $L_+ \otimes \mathbb{R}$ and $L_{-h,\delta} \otimes \mathbb{R}$ respectively. It is easy to verify that

$$\Omega/(1,-\varphi) = \mathfrak{L}_+ \times \mathfrak{L}_{-h,\delta}$$

Periods $([\omega_+],[\omega_-])$ belonging to $\mathfrak{L}_+ \times \mathfrak{L}_{-h,\delta}$ are the periods of the marked k3-surfaces with a non-degenerate double point if there is no vector v (*) such that $B_{\mathbb{C}}(\omega,v)=0$, $B(v,h')=0$, $B(v,v)=-2$ in lattice L. Consider a reflection with respect to the hyperplane in $\mathfrak{L}_+$ orthogonal to vector $v \in L_+$ with square -2 (this reflection transforms x to $x+B(x,v)\cdot v$). It can be transposed with φ leaving vectors h' and δ' in their places. Thus the reflection described gives us equivalent periods. It also holds true for a reflection with respect to hyperplane in $\mathfrak{L}_{-h,\delta}$ orthogonal to vector $v' \in L_{-h,\delta}$ having square -2. Let Ω_+ and $\Omega_{-h,\delta}$ be the fundamental domains of groups generated by reflections with respect to hyperplanes orthogonal to vectors with squares -2 in $\mathfrak{L}_+$ and $\mathfrak{L}_-$ respectively. We have shown that the image of the period mapping belongs to $\Omega_+ \times \Omega_{-h,\delta}$ up to equivalence.

Consider element $\omega \in L \otimes \mathbb{C}$ such that there is vector $v \in L$ with the properties:

$$B_{\mathbb{C}}(\omega,v)=0, \quad B(v,h')=0, \quad B(v,v)=-2.$$

Let $v=v_+ + v_-$ where $v_+ \in L_+ \otimes \mathbb{Q}$, $v_- \in L_- \otimes \mathbb{Q}$. We will denote the orthogonal complement in L_- to vector h' by L_{-h}. Then we have $2v_- \in L_{-h}$. Note that if $[\omega] \in \Omega_+ \times \Omega_{-h,\delta}$, then $v_- \neq 0$. Now we consider two cases:

a) $v_+ = 0$. If vector v_- does not belong to space $L_{-h,\delta} \otimes \mathbb{R}$, then the projection of vector v_- on $L_{-h,\delta} \otimes \mathbb{R}$ will be denoted by v'_-. It is easy to notice that

$$2v'_- \in L_{-h,\delta}, \; B(2v'_-,2v'_-) \geq 0 \text{ or } B(2v'_-,2v'_-) = -6, \quad 2v_- \equiv \delta' \pmod{2L}.$$

Consider hyperplanes in $L_{-h,\delta}$ orthogonal to vectors $v'_- \in L_{-h,\delta}$ where $B(2v'_-,2v'_-)=-6$ and $2v \equiv \delta' \pmod{2L}$. Let these hyperplanes divide open polytope $\Omega_{-h,\delta}$ into polytopes $\Omega^i_{-h,\delta}$ (i=1,2,...) (the number of these polytopes may be infinite).

b) $v_+ \neq 0$. It should be noted that

$$B(v_+ - v_-, v_+ - v_-) = B(v_+ + v_-, v_+ + v_-) = -2.$$

If vectors $v_+ - v_-$, $v_+ + v_-$, δ' are linear independent, then class $[\omega]$

(*) except δ'

belongs to the sertain subset of codimention 2 of space $\mathfrak{L}_+ \times \mathfrak{L}_{-h,\delta}$ This subset is the intersection of two hyperplanes orthogonal to vectors $v_+ + v_-$ and $v_+ - v_-$. Linear dependence of vectors $v_+ + v_-$, $v_+ - v_-$, δ' is possible only if $v = \delta'/2$. But then $B(2v_+, 2v_+) = -6$ and $2v_+ \equiv \delta'$ (mod 2L). Consider the hyperplanes in $\mathfrak{L}_+$ orthogonal to vectors $2v_+ \in \mathfrak{L}_+$ where $B(2v_+, 2v_+) = -6$ and $2v_+ \equiv \delta'$ (mod 2L) similar to the previous case. Let these hyperplanes divide open polytope Ω_+ into polytopes Ω_+^j (j=1,2,...) (the number of these polytopes may be infinite as well).

We have proved that the image of mapping of periods up to equivalence can be obtained by removing the closed subset with condimension at least 2 (this subset will be denoted by Z) from the union of components $\Omega_+^j \times \Omega_{-h,\delta}^i$ (i=1,...;j=1,...). It should be noted that some components $\Omega_+^j \times \Omega_{-h,\delta}^i$ can be transformed to other ones by the automorphisms of quadruple (1).

Similar to [3] it is possible to show that every class of $[\omega] \in (\bigcup_{i,j} \Omega_+^j \times \Omega_{-h,\delta}^i) \setminus Z$ defines only one class of the projective equivalence of real K3-surfaces being two-sheeted covering spaces of $\mathbb{C}P^2$ branched over a curve of degree 6 with a non-degenerate double point.

It is easy to prove that one of lattices L_+ and $L_{-h,\delta}$ has no vectors v such that $v \equiv \delta'$ (mod 2L). Thus one of collections Ω_+^j and $\Omega_{-h,\delta}^i$ consists of the single polytope. Let Ω_* stand for collection Ω_+^j if there exists vector v in L such that $v \equiv \delta'$ (mod 2L) and for collection $\Omega_{-h,\delta}^i$ in the other case.

We have proved the theorem.

Theorem 2.1. Consider the curves of degree 6 with a non-degenerate double point having quadruple $(H_2(X,\mathbb{Z}), conj_*, h, \delta)$ isomorphic to quadruple (1). The rigid isotopy types of these curves are in one-to-one correspondence with the elements of collection Ω_* considered up to the automorphisms of (1).

We can choose for each curve of degree 6 with a non-degenerate double point involution *conj* in such a way that lattice L_+ would not have such vectors v that $v \equiv \delta'$ (mod 2L), i.e. collection Ω_* would coincide with collection $\Omega^{i}_{-h,\delta}$. We suggest that involution *conj* is chosen in this way.

3. Corollaries of theorem 2.1.

Let $\mathfrak{L}_{-h}$ be the space of Lobachevsky obtained in a standard way from space L_{-h}. The group generated by the reflections with respect to the hyperplanes orthogonal to the vectors in L_{-h} having square -2 acts in $\mathfrak{L}_{-h}$. Let $\tilde{\Omega}_-$ stand for a polytope being a fundamental domain of this group and having a face orthogonal to δ'.

The Coxeter scheme (see for example [7]) of polytope $\tilde{\Omega}_-$ appears to contain all information required about collection Ω_* (let us remind that the involution *conj* was chosen in such a way that collection Ω_* coincides with collection $\Omega^{i}_{-h,\tilde{\delta}}$).

Let C be the Coxeter scheme of polytope $\tilde{\Omega}_-$, e be the vertex of scheme C corresponding to δ'. Remove all thick and dotted edges from scheme C (two vertices of a scheme are connected by thick (or dotted) edge if bilinear form B has the value equal (or more than) 2 on corresponding vectors). Let C′ stand for the scheme obtained. Consider the symmetries of scheme C′ generated by the automorphisms of (L,φ,h'). The group of these symmetries acts on C′. Let C′′ be the corresponding quotient-scheme and e' for the class containing vertex e. Let K stand for the connected component of vertex e' in C′′. It is easy to prove the following statement.

Proposition 3.1. The number of the polytopes of collection Ω_* which cannot be transformed one into another by the automorphisms of (1) is equal to the number of vertices of scheme K.

Corollary 3.2. The number of the rigid isotopy types of the curves of degree 6 with a non-degenerate double point having quadruple $(H_2(X,\mathbb{Z}), conj_*, h, \delta)$ isomorphic to (1) is equal to the number of vertices of scheme K.

The construction of schemes K for all possible classes of isomorphism of quadruples $(H_2(X,\mathbb{Z}), conj_*, h, \delta)$ corresponding to the curves of degree 6 with a non-degenerate double point allows to obtain rigid isotopy classification of the curves mentioned. It is necessary to note that scheme K (unlike collection Ω_*) is always finite.

We will reformulate now the above rigid isotopy classification in terms of conjunctions and contractions of ovals of non-singular curves of degree 6.

But let us first make some remarks.

Let A_0 be a curve of degree 6 with a non-degenerate double point. There are two types of smoothings of singular point of this curve. Let A_t ($t\geq 0$) be a smoothing of one of these types and A_t ($t\leq 0$) be a smoothing of the other type. Let the number of ovals of non-singular curve A_{t_0} ($t_0<0$) be no less than the number of ovals of non-singular curve A_{t_1} ($t_1>0$). Then the family A_t ($t_0\leq t\leq 0$) will be called the non-increasing simplest degeneration of curve A_{t_0} and the family A_t ($0\leq t\leq t_1$) will be called the non-decreasing simplest degeneration of curve A_{t_1}.

We will consider the curves of degree 6 with a non-degenerate double point as the results of the non-increasing simplest degenerations of non-singular curves of degree 6.

There are 8 cases of the non-increasing degeneration of the non-singular curve of degree 6 (if its scheme of disposition of ovals does not coincide with ⟨1⟨1⟨1⟩⟩⟩):

1) conjunction of an even empty oval with the non-empty one;

1') conjunction of an odd oval with the non-empty one;

2) conjunction of two even empty ovals;

2') conjunction of two odd ovals;

3) contraction of an even empty oval;

3') contraction of an odd oval;

4) conjunction of the non-empty oval with itself in such a way that the component obtained of the set of real points of the curve is embedded in $\mathbb{R}P^2$ like the union of two lines.

5) conjunction of an even empty oval with itself in such a way that the component obtained of the set of real points of the curve is embedded in $\mathbb{R}P^2$ like the union of two lines.

Cases 4) and 5) are impossible for M-curves and (M-1)-curves of degree 6.

<u>Proposition 3.3.</u> Two curves being the results of degeneration of types 1-3 of the same non-singular curve of degree 6 have isomorphic quadruples $(H_2(X,\mathbb{Z}), conj_*, h, \delta)$ and involution $conj$ coincides with $conj_-$. Similarly two curves being the results of degenerations of types 1'-3' of the same non-singular curve of degree 6 have isomorphic quadruples $(H_2(X,\mathbb{Z}), conj_*, h, \delta)$ and involution $conj$ coincides with $conj_+$.

Define two graphs P and P' for an arbitrary non-singular curve of degree 6. The vertices of graph P (P' respectively) correspond to the rigid isotopy classes of the results of degenerations of types 1-3 (1'-3' respectively) of the given curve. Two vertices are connected by edge if the class corresponding to one of these vertices can be realized as conjunction of ovals A and B and the class corresponding to the other vertex can be realized as contraction of oval A.

<u>Proposition 3.4.</u> The graph P (P' respectively) of a non-singular curve of degree 6 is isomorphic to scheme K constructed for a result of an arbitrary degeneration of types 1-3 (1'-3' respectively) of the given curve.

<u>Proposition 3.5.</u> Two curves being the results of degenerations of type 4 (type 5 respectively) of the same non-singular curve of

degree 6 are rigidly isotopic.

Proposition 3.5 is the corollary of Theorem 2.1 and the fact that in the cases of such degenerations both lattices $L_{-h,\delta}$ and L_+ do not have vectors which are congruent to δ' modulo 2L.

Construction of graphs P and P' for all rigid isotopy classes of the non-singular curves of degree 6 together with the statement of Proposition 3.5 gives the rigid isotopy classification of the curves of degree 6 with a non-degenerate double point in terms of conjunctions and contractions of ovals of the non-singular curves of degree 6.

The schemes K for all possible classes of isomorphism of quadruples $(H_2(X,\mathbb{Z}), conj_*, h, \delta)$ corresponding to the curves of degree 6 with a non-degenerate double point (or, as is just the same, graphs P and P' for all rigid isotopy classes of the non-singular curves of degree 6) have been calculated by the auther of the present paper. The results of Vinberg ([7], [8]) on the groups generated by reflections are used.

If $\tilde{\Omega}_-$ has infinite number of faces it is difficult to calculate its Coxeter scheme. In this case it is possible to calculate scheme K directly.

Now we will discuss the question on groups of monodromy of non-singular curves of degree 6.

Consider a non-singular curve of degree 6 and polytope $\tilde{\Omega}_-$ constructed for the result of an arbitrary degeneration of types 1-3 (types 1'-3' respectively). Consider vertices $a_1, a_2, \ldots, a_n$ of Coxeter scheme C of polytope $\tilde{\Omega}_-$ corresponding to the classes of contractions of even empty ovals (odd ovals respectively) of the given curve

Let S be the group of symmetries of C generated by automorphisms of (L, φ, h'), S' be the subgroup of S consisting of the symmetries leaving vertices $a_1, a_2, \ldots, a_n$ on their places.

Theorem 3.6. Quotient group S/S' is isomorphic to the group of monodromy of even ovals (odd ovals respectively) of non-singular curve.

It is easy to calculate the group of monodromy of an arbitrary non-singular curve using Theorem 3.6.

References

1. Gudkov, D.A., Utkin, G.A.: Topology of the curves of degree 6 and of the surfaces of degree 4 (in Russian). Uch. Zap. Gor'kovskogo Universiteta, v. 87, (1969)

2. Rokhlin, V.A.: Complex topological characteristics of the real algebraic curves (in Russian). Usp. Mat. Nauk, v. 33, N 5, 77-89 (1978)

3. Nikulin, V.V.: Integer quadratic forms and some geometrical applications (in Russian). Izv. Akad. Nauk SSSR Ser. Mat. v. 43, N 1, 111-177 (1979)

4. Kharlamov, V.M.: Rigid isotopy classification of the real plane curves of degree 5 (in Russian). Funk. Analis i ego Pril . v. 15, N 1, 88-89 (1981)

5. Itenberg, I.V.: Rigid isotopy classification of the simplest degenerations of M-curves of degree 6 (in Russian). Vestn. Len. Universiteta. Ser. Mat. v. 3, 32-37 (1991)

6. Itenberg, I.V.: Rigid isotopy classification of curves of degree 6 with a non-degenerate double point (in Russian). Zap. Nauch. Sem. LOMI AN SSSR. v. 193, 72-89 (1991)

7. Vinberg, E.B.: On the groups of units of some quadratic forms (in Russian). Mat. Zb. v. 87, N 1, 18-36 (1972)

8. Vinberg, E.B.: The two most algebraic K3 surfaces. Math. Ann. v. 265, 1-21 (1983)

9. Viro, O.Ya.: The curves of degree 7, the curves of degree 8 and hypothesis of Ragsdale (in Russian). Dokl. Akad. Nauk SSSR v. 254, N 6, 1305 -1310 (1980)

10. Nikulin, V.V.: Involutions of integer quadratic forms and applications to real algebraic geometry (in Russian). Izv. Akad. Nauk SSSR Ser. Mat. v. 47, N 1, 109-188 (1983)

The 17-th Hilbert problem for noncompact real analytic manifolds

Piotr Jaworski, University of Warsaw, Institute of Mathematics,
ul.Banacha 2, 00-913 Warszawae, Poland

Abstract

In this paper we deal with the ring of real analytic functions on a real analytic manifold. We investigate under what assumptions this ring and its field of quotients posses the Artin-Lang property and give the positive answer to Hilbert's 17 problem.

1 Introduction

Let Q be a real analytic connected paracompact manifold . By A we shall denote the ring of (global) real analytic functions on Q, and by M the field of meromorphic functions on Q, i.e. the field of quotients of A. The investigation of these objects gives rise to certain basic problems; under what assumptions the following properties are valid:

Property 1 (Hilbert's 17 problem) *Every positive-definite analytic function on Q is a sum of squares of meromorphic functions i.e.*

$$(\forall x \in Q \ \ f(x) \geq 0) \Longrightarrow \exists h_1, \ldots, h_k \in M \ \ f = \sum_{i=1}^{k} h_i^2.$$

The field of meromorphic functions is formally real hence it may be ordered (see [9] §XI.2).

Property 2 (Artin-Lang property) *Let $f_1, \ldots, f_k$ be analytic functions on Q. If in every point of Q at least one f_i is not negative then in every ordering σ on M at least one f_i is not negative;*

$$(\forall x \in Q \ \ \exists i \ \ f_i(x) \geq 0) \Longrightarrow \forall \sigma \ \exists i \ \ f_i \geq 0.$$

We remark that the first property is a consequence of the second one.

It is known that the answer is "yes" if:

i) Q is compact ([10],[8] both properties);

ii) Q is lowdimensional:

property 1 is valid for dim $Q \leq 2$ ([3], [2], [7]);

property 2 is valid for dim $Q = 1$ ([1]),

iii) there are additional conditions on functions:

property 1 is valid if $f^{-1}(0)$ is discrete ([2]).

The aim of this paper is to show that it is enough to assume that certain preimages are compact. Namely the property 2 (resp. 1) is valid if $\bigcap_{i=1}^{k} f_i^{-1}((-\infty, 0])$ (resp. $f^{-1}(0)$) is compact. Moreover we shall show that the property 1 is valid if $f^{-1}(0)$ is a union of a discrete set and a compact set.

We record here that independently the similar results were recently obtained by Ana Castilla from the University of Madrid.

2 Orderings and filter bases

Let σ be any ordering on the field M. Let m_σ (resp. W_σ) be the set of all infinitely small (resp. finite) elements of M:

$$m_\sigma = \{f : \forall n \in N \quad | f |< 1/n \},$$

$$W_\sigma = \{f : \exists n \in N \quad | f |< n \}.$$

We remark that W_σ is a valuation ring and m_σ is its maximal ideal (see [9] §XI.1). Moreover all bounded analytic functions belong to W_σ (between the others the constant ones) - see [1]. Hence the residue field is isomorphic to the field of the real numbers:

$$W_\sigma / m_\sigma = R.$$

Every equivalence class $[f] \; mod \, m_\sigma$ contains a constant (a constant function). Hence we put $[f] = r$ where $r \in R$ and $f - r \in m_\sigma$.

Let $\varphi : M \longrightarrow R \cup \infty$ be the place associated to σ;

$$\varphi(f) = \begin{cases} 0 & for \quad f \in m_\sigma \\ [f] & for \quad f \in W_\sigma \\ \infty & for \quad f \in M \setminus W_\sigma \end{cases}$$

Obviously φ is compatible with the ordering σ:

$$f \geq 0 \Longrightarrow \varphi(f) \geq 0 \;\; or \;\; \varphi(f) = \infty.$$

We identify

$$R \cup \infty = RP_1.$$

We shall consider both the place φ and the meromorphic function f as a mapping to RP_1.

We shall describe the orderings on M with the help of filter bases of closed semianalytic sets (compare [1]).

We associate to the ordering σ the family of sets:

$$G_\sigma = \{f^{-1}([-1, 1]) \; : \; f \in m_\sigma \cap A \}$$

Proposition 1 *The family G_σ has the following properties:*
a) The empty set does not belong to G_σ

$$\neg \emptyset \in G_\sigma;$$

b) The intersection of any two elements of G_σ contains the third one

$$A, B \in G_\sigma \Longrightarrow \exists C \in G_\sigma \quad C \subset A \cap B;$$

c) If the closed subsets of Q - V_1, V_2 intersect every element of G_σ then their intersection is not empty

$$(\forall V \in G_\sigma \quad V \cap V_1 \neq \emptyset \text{ and } V \cap V_2 \neq \emptyset) \Longrightarrow V_1 \cap V_2 \neq \emptyset;$$

d) If the compact subset of Q - V_0 intersects every element of G_σ then the intersection of all elements of G_σ is not empty and consists of one point which belongs to V_0

$$(\forall V \in G_\sigma \quad V \cap V_0 \neq \emptyset) \Longrightarrow \bigcap_{V \in G_\sigma} V = \{p_\sigma\} \subset V_0;$$

e) The intersection of the closures (in RP_1) of the images of elements of G_σ under the analytic function is not empty and consists of one point equal to the value of the associated place φ at this function

$$\forall f \in A \quad \bigcap_{V \in G_\sigma} Cl\, f(V) = \{\varphi(f)\};$$

f) If the analytic function maps a set from G_σ to an interval then the value of the associated place φ at this function is contained in the same interval

$$\forall f \in A\,(\,\exists V \in G_\sigma \;\; \forall x \in V \;\; a \leq f(x) \leq b\,) \Longrightarrow a \leq \varphi(f) \leq b;$$

g) If the analytic function is positive on some set from G_σ then it is positive in the ordering σ

$$\forall f \in A\,(\,\exists V \in G_\sigma \;\; \forall x \in V \;\; 0 < f(x)\,) \Longrightarrow 0 <_\sigma f.$$

Remark.
Conditions a and b imply that the family G_σ is a filter base. Condition c implies that there is an unique ultrafilter G_σ^c of closed subsets of Q which contains G_σ; namely

$$G_\sigma^c = \{W \overset{cl}{\subset} Q : \forall V \in G_\sigma \;\; V \cap W \neq \emptyset\}.$$

Hence the filter base G_σ determines a point in the Čech-Stone compactification of Q. Moreover $\varphi(f)$ equals to the limit of f at this point (cond. "e"). More details about filters and their connection with Čech-Stone compactification the reader may find in [5] §1.6 and 3.6.

Proof.
a. If the set $f^{-1}([-1,1])$ is empty then either f is positive on the whole domain either $-f$. In the first case

$$\forall x \quad f(x) > 1$$

hence the function $f - 1$ is a square in A. Thus f is greater then 1 in the ordering σ and does not belong to m_σ. Analogicaly in the second case. Therefore the family G_σ does not

contain the empty set.
b. If $A = f^{-1}([-1,1])$ and $B = g^{-1}([-1,1])$ then we put $C = (f^2+g^2)^{-1}([-1,1])$.
c. If the sets V_1 and V_2 were disjoint then there would exist a bounded analytic function f which would separate them.

$$f_{|V_1} > 2 \quad f_{|V_2} < -2$$

Let $r = \varphi(f)$ then the set

$$V = (f-r)^{-1}([-1,1])$$

would belong to G_σ and intersect at most one of V_i's; the contradiction.
d. The family of closet sets $V \cap V_0$ has the finite intersection property and is contained in the compact set V_0 hence the intersection is nonempty ([5] Th.3.1.1). Moreover it follows from point c that the intersection of all elements of the family G_σ may contain at most one point. Hence this point must be contained in V_0.
e. The manifold RP_1 is compact, the family of closures of the images of the sets V has the finite intersection property hence the intersection is nonempty ([5] Th.3.1.1). Moreover. Let $\varphi(f) = r \neq \infty$ (resp. $= \infty$) then the functions $n(f-r), n \in N$ (resp. $(1+n^2)/(1+f^2)$) belong to m_σ. Therefore the sets $V_n = f^{-1}([r-1/n, r+1/n])$ (resp. $V_n = f^{-1}((-\infty,-n] \cup [n,\infty))$), $n \in N$, belong to the family G_σ. Obviously $\bigcap Cl\, f(V_n) = \{r\}$.
f. Point f is a direct corollary of e. Indeed if one image is contained in the interval then the intersection of the closures of all of them is also.
g. Let

$$W = f^{-1}((-\infty, 0]).$$

The function f is positive on the set V hence the sets V and W are disjoint. Let

$$\chi : Q \longrightarrow [0,1]$$

be a C^∞ function which equals 1 on W and 0 on V. We put

$$\psi = \chi + (1-\chi)f.$$

Obviously ψ is positive on Q and equals to f on V. We approximate ψ by an analytic function g such that

$$\forall x \in Q \quad | g(x) - \psi(x) | \leq \frac{1}{2}\psi(x)$$

(see [6], also [2]). The analytic function g is positive on Q hence it is invertible and moreover it is a square of an analytic function, so it is positive in every ordering on M. On the other hand

$$\forall x \in V \quad | g(x) - f(x) | \leq \frac{1}{2} f(x)$$

hence

$$\forall x \in V \quad \frac{1}{2} f(x) \leq g(x) \leq \frac{3}{2} f(x)$$

i.e.

$$\forall x \in V \quad \frac{2}{3} \leq \frac{f(x)}{g(x)} \leq 2.$$

The fraction f/g is an analytic function, thus from the point "f" we obtain that

$$\frac{2}{3} \leq \varphi(\frac{f}{g}) \leq 2.$$

Therefore the analytic function f/g is positive in σ (compatibility of the place and the ordering), g is a square hence f is positive in σ.

3 The Artin-Lang property with a compact assumption

Theorem 1 *Let $f_1, \ldots, f_k$ be analytic functions on Q. If*

$$a) \quad \bigcap_{i=1}^{k} f_i^{-1}((-\infty, 0)) = \emptyset,$$

$$b) \quad \bigcap_{i=1}^{k} f_i^{-1}((-\infty, 0]) \text{ is compact}$$

then in every ordering on the field of meromorphic functions M at least one f_i is positive.

Proof.
Let us assume that the thesis of the theorem is not valid; i.e. there exists an ordering σ on M such that all f_i are negative.

Case 1. The filter base G_σ is free i.e.

$$\bigcap_{V \in G_\sigma} V = \emptyset.$$

The set

$$W = \bigcap_{i=1}^{k} f_i^{-1}((-\infty, 0])$$

is compact hence there exists a set V in G_σ which is disjoint with W (property "d").

$$\emptyset = W \cap V = \bigcap_{i=1}^{k} (V \cap f_i^{-1}((-\infty, 0])).$$

The sets

$$V \cap f_i^{-1}((-\infty, 0])$$

are closed therefore for some i there exists a set $V_1 \in G_\sigma$ disjoint with $V \cap f_i^{-1}((-\infty, 0])$. Indeed, it follows by induction from "c" that the intersection of sets $V \cap f_i^{-1}((-\infty, 0])$ would be not empty if each of them intersected all elements of G_σ.
f_i is positive on $V \cap V_1$ hence the properties "g" and "b" assure us that f_i is positive in σ - a contradiction. Thus in any "free" ordering at least one f_i is positive.

Case 2. The filter base G_σ is fixed i.e.

$$\bigcap_{V \in G_\sigma} V = \{p_\sigma\}.$$

In this case the value of the associated place φ at an analytic function f equals to the value of f at the point p_σ (property "e "). Hence the valuation ring W_σ contains the localization of the ring A at the point p_σ. Next one may either apply the method of desingularisation - see [8] Th.C, or the method of excellent rings - see [10] Pr.2.2. Both lead to a contradiction.
This finishes the proof of the theorem. We remark that the hypothesis "a" is used only in the second case, while the hypothesis "b" only in the first one.

Corollary 1 *Let f be a positive-definite analytic function on Q. If f has compact zero set then f is a sum of squares of meromorphic functions.*

Proof.
From theorem we have that f is positive in every ordering on M. Thus f is a sum of squares (see [9] §XI.2).

4 17th Hilbert problem with compact+discrete assumption

Theorem 2 *Let f be a positive-definite analytic function on Q such that $f^{-1}(0)$ is a union of a discrete set and a compact set then f is a sum of squares of meromorphic functions.*

Proof.
The theorem follows directly from the following lemma:

Lemma 1 *Let f be a positive-definite analytic function on Q such that $f^{-1}(0)$ is a union of a discrete set and a compact set then f is a product of two positive-definite analytic functions on Q*

$$f = gh$$

where $g^{-1}(0)$ is discrete and $h^{-1}(0)$ is compact.

Indeed, both g and h are sums of meromorphic functions (see [2] for g and corollary 1 for h), hence f is also a sum of squares.
Proof of the lemma.
Let D be a compact component of the zero set of f. We consider a subsheaf $\mathcal{J}$ of the structure sheaf $\mathcal{O}$ defined by:

$$\begin{array}{ll} \mathcal{J}_p = f_p \cdot \mathcal{O}_p & \text{if } p \in D \\ \mathcal{J}_p = \mathcal{O}_p & \text{if } p \notin D \end{array}$$

$\mathcal{J}$ is coherent hence there is a finite number of global sections $h_1, \ldots, h_t$ which generate each stalk $\mathcal{J}_p$, $p \in Q$ (see [4]). We put

$$h = h_1^2 + \cdots + h_t^2 + f.$$

The analytic funtion h is positive-definite and $h^{-1}(0)$ equals to D. Moreover at each point $p \in D$

$$h_p = f_p + q_p f_p^2$$

where q_p is a germ of an analytic function.
Thus the fraction

$$g = \frac{f}{h}$$

is an analytic function which does not vanish on D.

References

[1] C.Andradas,E.Becker, A note on the real spectrum of analytic functions on an analytic manifold of dimension one. in: Real Analytic and Algebraic Geometry. L.N. in Math. 1420. Springer V. 1990.

[2] J.Bochnak,W.Kucharz, M.Shiota, On equivalence of ideals of real global analytic functions and the 17th Hilbert problem. Inventiones Math.63 (1981) 403-421.

[3] J.Bochnak,J.J.Risler, Sur le theoreme des zeros pour les varietes analytiques reelles de dimension 2. Ann.Sci.Ec.Norm. Super.8 (1975) 353-364.

[4] S.Coen, Sul rango dei fasci coerenti. Boll.Un.Mat.Ital.22 (1967) 373-383.

[5] R.Engelking, General Topology. PWN Polish Sc. Publ. 1977.

[6] M.Hirsch, Differential Topology. Springer V. 1976.

[7] P.Jaworski, Positive Definite Analytic Functions and Vector Bundles. Bull. Acad.Pol.Sc. serie math. 30:11-12 (1982) 501-506. 329-339.

[8] P.Jaworski, Extensions of Orderings on Fields of Quotients of Rings of Real Analytic Functions. Math.Nachr.125 (1986) 329-339.

[9] S.Lang, Algebra, Reading Mass. 1965.

[10] J.M.Ruiz, On Hilbert's 17th problem and real Nullstellensatz for global Analytic Functions, Math.Z. 190 (1985) 447-459.

CONSTRUCTION OF NEW M-CURVES OF 9-th DEGREE.

A.B.Korchagin
Nizhny Novgorod University, USSR

This paper continues investigations in the 1-st part of Hilbert 16-th problem. I construct 6 new M-curves of 9-th degree (Theorem 5), give all the information (Th.6 and 7) about M-curves of the 9-th degree that I know and formulate five conjectures about ovals arrangement of odd degree nonsingular algebraic curves. I use Viro techniques of gluing of charts of polynomials [1,2].

Theorem 1. For any two numbers $k,k' \in \mathbb{R}$ ($0<k<k'$) there exists neighbourhood $U \subset \mathbb{R}$ of number k' ($k \notin U$) such that for any different $k_1,k_2,k_3 \in U$ any plane curve N_{28} germ defined by equation

$$x_1(x_0x_2-kx_1^2)(x_0x_2-k_1x_1^2)(x_0x_2-k_2x_1^2)(x_0x_2-k_3x_1^2)=0 \qquad (1)$$

can be smoothed by gluing the charts shown on fig.1.

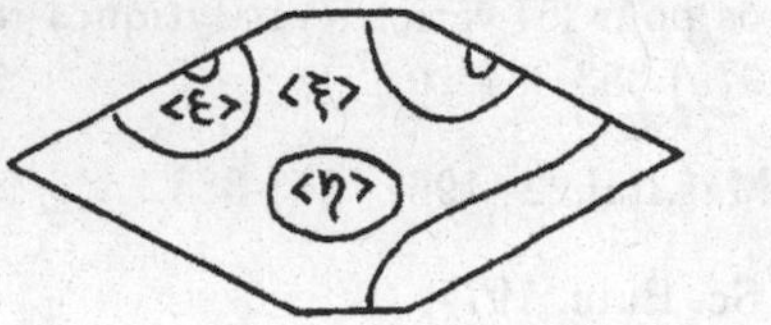

ε	4	10	4	6
ξ	0	0	4	4
η	7	1	3	1

Fig. 1

Proof. 1) There exists the 4-th degree curve F=0 placed in projective system of coordinates as it shown on fig.2.

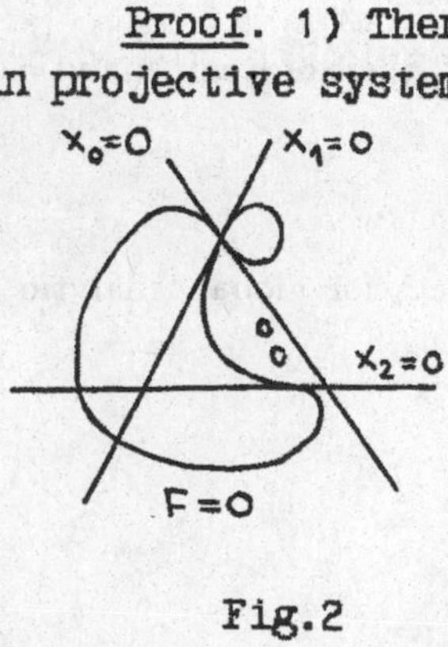

Fig.2

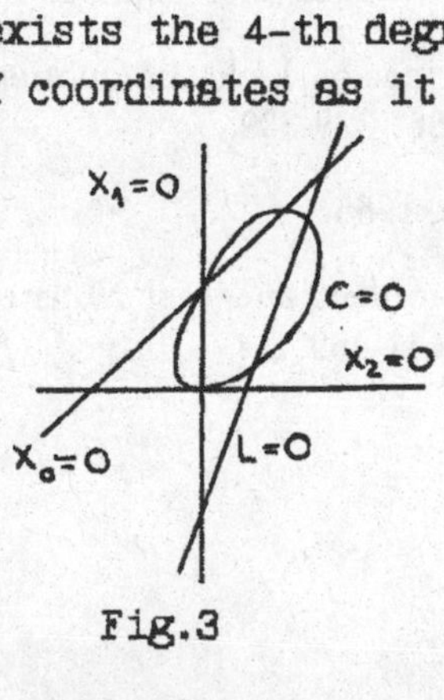

Fig.3

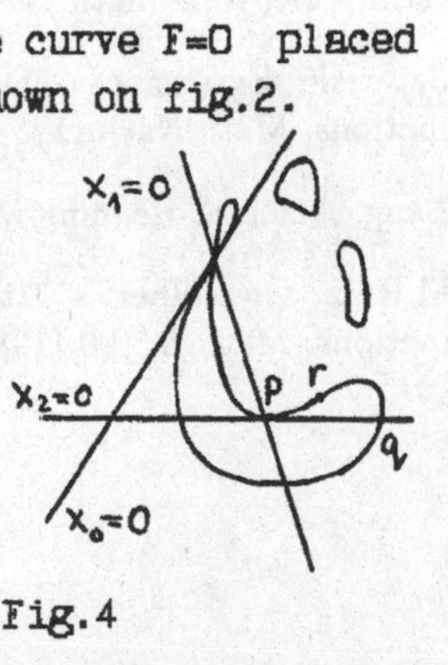

Fig.4

Let us consider conics C=0 and line L=0 placed in projective system of coordinates as you see on fig.3. The 4-th degree curve $CLx_0 + tx_1^4 = 0$ for sufficiently small $|t|$ and suitable sign of t is located such as shown on fig.4. Obvously the arc pq of this curve has a flex point r. Now we choose the new system of coorginates. Let the new axes $x_0=0$ and $x_1=0$ be the tangents at the double point of this curve (one of the branches crossing through double point has a flex point). And let the axis $x_2=0$ be a tangent to this curve at flex point r. So we obtain the 4-th degree curve F=0 shown on fig.2.

2) Let us consider the 6-th degree curve

$$G(x_0:x_1:x_2) = \frac{1}{x_0^4 x_1^2 x_2^4}(F\circ qu\circ l\circ qu)(x_0:x_1:x_2) = 0$$

where $qu(x_0:x_1:x_2) = (x_0x_2:x_1x_2:x_0^2)$ is quadratic projective involution and $l(x_0:x_1:x_2)=(x_0:x_2:x_1)$ is linear projective involution, $qu^{-1}=qu$, $l^{-1}=l$. The curve G=0 has double point in (0:0:1), cusp in (1:0:0) and degenerate 4-fold point in (0:1:0). It's shown on fig.5.

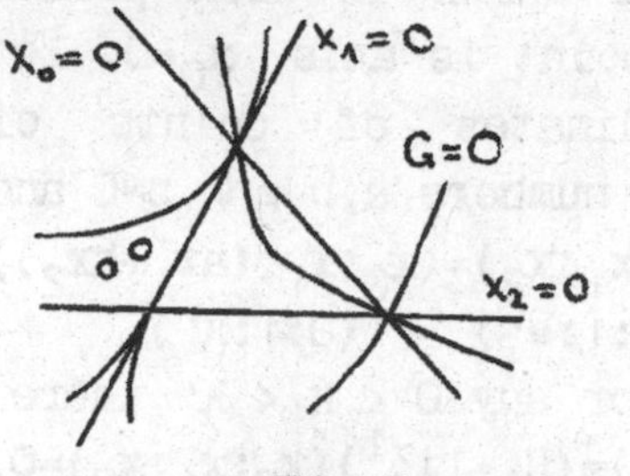

Fig.5

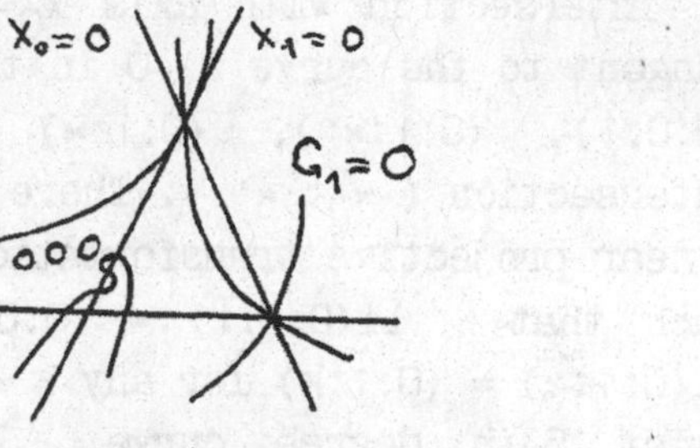

Fig.6

3) Keeping double point and degenerate 4-fold point we can obtain the 6-th degree curve $G_1=0$ shown on fig.6 as a result of sequence of small perturbations shown on fig.7.

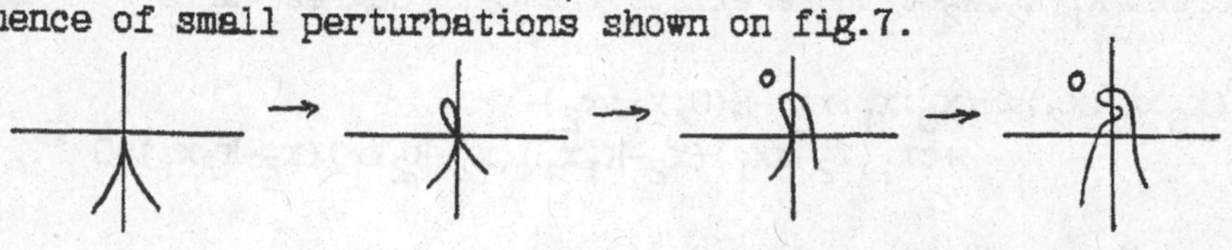

Fig.7

4) Let us consider the 5-th degree curve

$$H_1(x_0:x_1:x_2) \equiv \frac{1}{x_0{}^4 x_1{}^3}(G_1 \circ hy)(x_0:x_1:x_2) = 0$$

where $hy(x_0:x_1:x_2) = (x_0x_1:x_1{}^2:x_0x_2)$ is birational quadratic transformation and $hy^{-1}(x_0:x_1:x_2)=(x_0{}^2:x_0x_1:x_1x_2)$. It is shown on fig.8. This curve has 3-fold point in (1:0:0), and 3 points

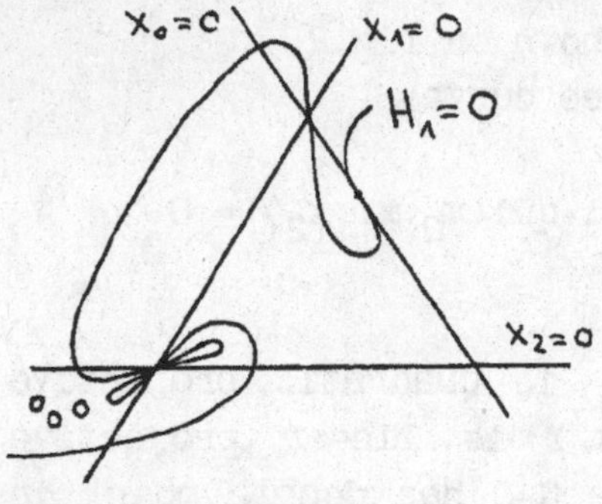

Fig.8

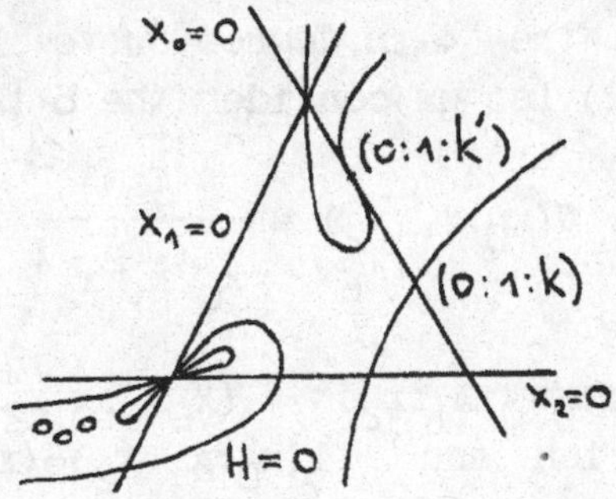

Fig.9

of intersection with axis $x_0=0$ one of which is flex point. Tangent to the curve $H_1=0$ in this flex point is axis $x_0=0$. Let (0:0:1), $(0:1:\varkappa')$, $(0:1:\varkappa)$ be coordinates of points of intersection ($\varkappa < \varkappa'$). There exist two numbers $a,b \in \mathbb{R}, b\neq 0$ and linear projective transformation $li(x_0:x_1:x_2)=(x_0:x_1:(ax_1+bx_2))$ such that $li(0:0:1) = (0:0:1)$, $li(0:1:\varkappa') = (0:1:k')$, $li(0:1:\varkappa) = (0:1:k)$ for any $k < k'$. So for any $0 < k < k'$ there exist 5-th degree curve $H(x_0:x_1:x_2)\equiv(H_1 \circ li^{-1})(x_0:x_1:x_2)=0$, located in $\mathbb{R}P^2$ as shown on fig.9 and $H(0:x_1:x_2)= cx_1(x_2-k'x_1)(x_2-kx_1)^3$, where $c\neq 0$. It is not difficult to see there exists neighbourhood $U \in \mathbb{R}$ of number k' such that for any different $k_1,k_2,k_3 \in U$ there exists the 5-th degree curve

$$\widetilde{H}(x_0:x_1:x_2)\equiv H(x_0:x_1:x_2)-H(0:x_1:x_2)+ +cx_1(x_2-kx_1)(x_2-k_1x_1)(x_2-k_2x_1)(x_2-k_3x_1)=0$$

shown with its chart on fig.10.

5) Under transformation hy^{-1} the curve $\widetilde{H}(x_0:x_1:x_2)=0$ maps onto 9-th degree curve $S(x_0:x_1:x_2)\equiv(\widetilde{H}\circ hy)(x_0:x_1:x_2)/cx_1=0$ that

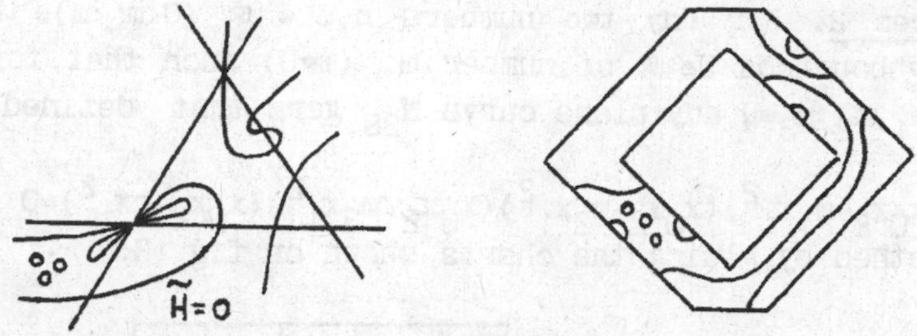

Fig.10

has a chart shown on fig.11. This curve has N_{28} germ defined by equation (1) in point (0:0:1) and has Z_{15} germ defined by equation

$$x_1(x_0x_2-m_1x_1^2)(x_0x_2-m_2x_1^2)(x_0x_2-m_3x_1^2)=0 \qquad (2)$$

in the point (1:0:0), where $0<m_1<m_2<m_3$.

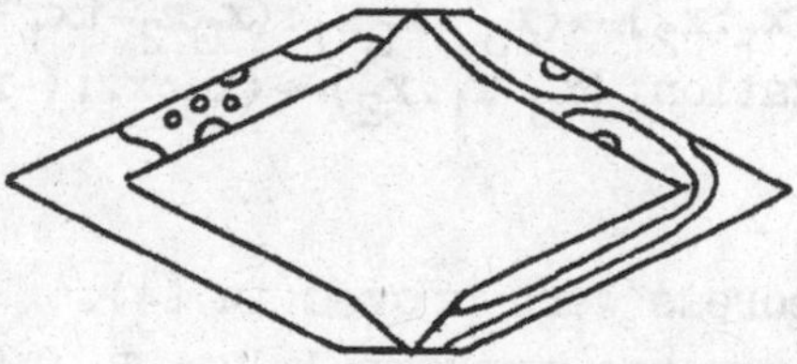

Fig.11

6) In [3] it was proved that for any three numbers $0<m_1<m_2<m_3$ any Z_{15} germ of plane curve defined by equation (2) can be smoothed by gluing the charts shown on fig.12.

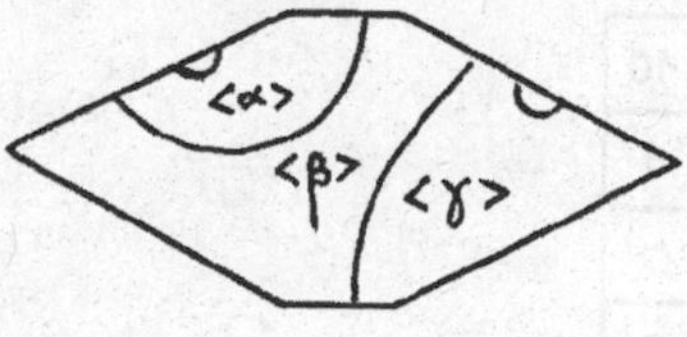

α	6	2	0	0
β	0	4	4	0
γ	0	0	2	6

Fig.12

7) Let us glue by Viro method the charts shown on fig.11 with the charts shown on fig.12 we obtain the charts shown on fig.1.

Theorem 2. For any two numbers $m, m' \in \mathbb{R}$ $(0<m'<m)$ there exists neighbourhood $U \in \mathbb{R}$ of number m' $(m \notin U)$ such that for any different $m_1, m_2, m_3 \in U$ any plane curve N_{28} germ that defined by equation

$$x_1(x_0x_2-m_1x_1^2)(x_0x_2-m_2x_1^2)(x_0x_2-m_3x_1^2)(x_0x_2-mx_1^2)=0$$

can be smoothed by gluing the charts shown on fig.13.

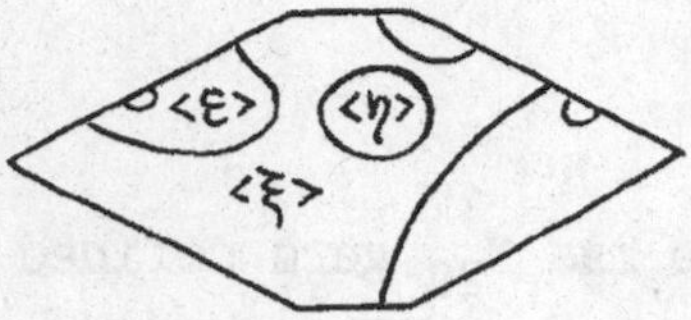

ε	4	10	4	6
ξ	0	0	4	4
η	7	1	3	1

Fig.13

Proof. Let us transform the curves provided with theorem 1 to transformation $(x_0:x_1:x_2)\mapsto(x_0^2:x_0x_1:(x_0x_2-tx_1^2))$, where $t>k'$ and then to symmetrization $(x_0:x_1:x_2)\mapsto(x_0:x_1:(-x_2))$ we obtain the charts we need.

The next two theorems were proved in [4].

Theorem 3. For any two numbers $k, k' \in \mathbb{R}$ $(0<k<k')$ there exists neighbourhood $U \in \mathbb{R}$ of number k' $(k \notin U)$ such that for any different $k_1, k_2, k_3 \in U$ any plane curve N_{28} germ defined by equation

$$x_1(x_0x_2-kx_1^2)(x_0x_1-k_1x_1^2)(x_0x_2-k_2x_1^2)(x_0x_2-k_3x_1^2)=0$$

can be smoothed by gluing the charts shown on fig.14.

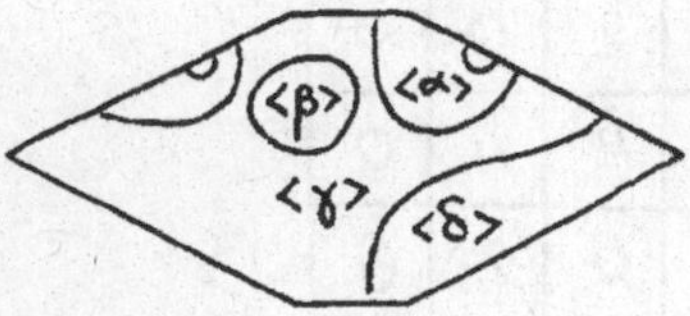

α	4	10
β	7	1
γ	0	0
δ	0	0

Fig.14

Theorem 4. For any two numbers $m, m' \in \mathbb{R}$ $(0<m'<m)$ there exists neighbourhood $U \in \mathbb{R}$ of number m' $(m \notin U)$ such that for any different $m_1, m_2, m_3 \in U$ any plane curve N_{28} germ defined by

equation

$$x_1(x_0x_2-m_1x_1^2)(x_0x_2-m_2x_1^2)(x_0x_2-m_3x_1^2)(x_0x_2-mx_1^2)=0$$

can be smoothed by gluing the charts shown on fig.15.

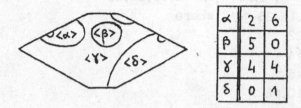

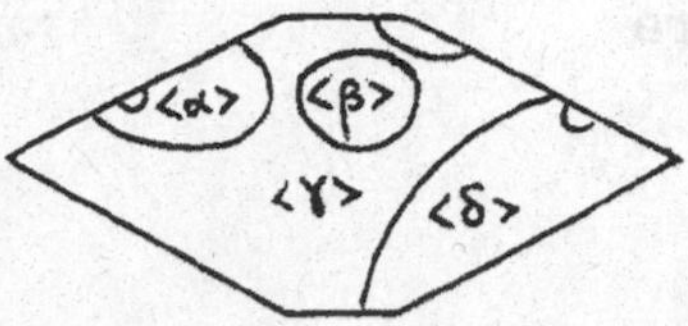

α	2	6
β	5	0
γ	4	4
δ	0	1

Fig.15

Theorem 5. The schemes contained in the last column of Table 1 can be realizable by M-curves of 9-th degree.

Table 1

N	Gluing				Scheme of M-curve
	fig.	α,β,γ,δ	fig.	ε,ξ,η	
1	14	4,7,0,0	1	4,0,7	<2<5>⊔2<7>⊔J>
2	14	10,1,0,0	1	4,0,7	<1<1>⊔1<5>⊔1<7>⊔1<11>⊔J>
3	14	10,1,0,0	1	10,0,1	<2<1>⊔2<11>⊔J>
4	15	6,0,4,1	13	10,0,1	<8⊔1<1>⊔1<17>⊔J>
5	15	2,5,4,0	13	10,0,1	<6⊔1<1>⊔1<5>⊔1<13>⊔J>
6	15	2,5,4,0	13	4,0,7	<6⊔1<5>⊔2<7>⊔J>

Proof. As it followed from theorems 1-4 the germs of curve

$$x_1(x_0x_2-kx_1^2)(x_0x_2-k_1x_1^2)(x_0x-k_2x_1^2)(x_0x_2-k_3x_1^2)=0$$

in points (1:0:0) and (0:0:1) can be smoothed by gluing the charts shown on fig.1,13,14,15. Let us do that according to Table 1 and we obtain M-curves that we need.

There are 1723 admissible (satisfying to Bezout theorem) schemes for M-curves of 9-th degree [12]. All the information that I know about 9-th degree M-curves is contained in the following two theorems.

<u>Theorem 6</u> ([4-9] + Th.5). There exist M-curves of 9-th degree which have the following 396 admissible schemes

1) $\langle J\sqcup 28\rangle$

2) $\langle J\sqcup\alpha\sqcup 1\langle\beta\rangle\rangle$, $\alpha=27-\beta$, $1\le\beta\le 16$, $\beta=18,19,22,23$

3) $\langle J\sqcup\alpha\sqcup 1\langle\beta\rangle\sqcup 1\langle\gamma\rangle\rangle$, $\alpha=26-\beta-\gamma$, where

$\beta=1$, $1\le\gamma\le 19$, $\gamma=22$
$\beta=2$, $2\le\gamma\le 10$, $\gamma=14$
$\beta=3$, $3\le\gamma\le 13$, $\gamma=15$
$\beta=4$, $4\le\gamma\le 8$, $\gamma=10,12$
$\beta=5$, $5\le\gamma\le 11$
$\beta=6$, $6\le\gamma\le 10$
$\beta=7$, $7\le\gamma\le 13$
$\beta=8$, $\gamma=8,10$

4) $\langle J\sqcup\alpha\sqcup 1\langle\beta\rangle\sqcup 1\langle\gamma\rangle\sqcup 1\langle\delta\rangle\rangle$, $\alpha=25-\beta-\gamma-\delta$, where

$(\beta,\gamma)=(1,1)$, $1\le\delta\le 7$, $\delta=9,10,11,13,14,18$
$(\beta,\gamma)=(1,2)$, $\delta=3,5,7,9,11,13,17$
$(\beta,\gamma)=(1,3)$, $3\le\delta\le 12$
$(\beta,\gamma)=(1,4)$, $\delta=5,7,9,11,15$
$(\beta,\gamma)=(1,5)$, $\delta=5,6,7,9,10,11,13,14$
$(\beta,\gamma)=(1,6)$, $\delta=7,9,11$
$(\beta,\gamma)=(1,7)$, $7\le\delta\le 12$
$(\beta,\gamma)=(1,8)$, $\delta=11$
$(\beta,\gamma)=(1,9)$, $\delta=9,10$
$(\beta,\gamma)=(1,10)$, $\delta=11$
$(\beta,\gamma)=(2,3)$, $\delta=3,5,7,11$
$(\beta,\gamma)=(2,5)$, $\delta=5,7,9$
$(\beta,\gamma)=(2,7)$, $\delta=7,11$
$(\beta,\gamma)=(3,3)$, $3\le\delta\le 7$, $\delta=9,10$
$(\beta,\gamma)=(3,4)$, $\delta=5,7,9$
$(\beta,\gamma)=(3,5)$, $5\le\delta\le 8$, $\delta=10$
$(\beta,\gamma)=(3,6)$, $\delta=7$
$(\beta,\gamma)=(3,7)$, $\delta=7,9,10$
$(\beta,\gamma)=(4,5)$, $\delta=7$
$(\beta,\gamma)=(4,7)$, $\delta=9$
$(\beta,\gamma)=(5,5)$, $\delta=5,6,7$
$(\beta,\gamma)=(5,6)$, $\delta=7$
$(\beta,\gamma)=(5,7)$, $\delta=7$
$(\beta,\gamma)=(5,8)$, $\delta=8,10$
$(\beta,\gamma)=(6,7)$, $\delta=7$

5) $\langle J\sqcup\alpha\sqcup 1\langle\beta\rangle\sqcup 1\langle\gamma\rangle\sqcup 1\langle\delta\rangle\sqcup 1\langle\varepsilon\rangle\rangle$, $\alpha=24-\beta-\gamma-\delta-\varepsilon$, where

$(\beta,\gamma,\delta)=(1,1,1)$, $\varepsilon=1,5,9,13,17,21$

$(\beta,\gamma,\delta)=(1,1,3)$, $\varepsilon=3,7,11,13$

$(\beta,\gamma,\delta)=(1,1,5)$, $\varepsilon=5,9,13$

$(\beta,\gamma,\delta)=(1,1,7)$, $\varepsilon=7,11,15$

$(\beta,\gamma,\delta)=(1,1,9)$, $\varepsilon=9$

$(\beta,\gamma,\delta)=(1,1,11)$, $\varepsilon=11$

$(\beta,\gamma,\delta)=(1,3,3)$, $\varepsilon=5,9$

$(\beta,\gamma,\delta)=(1,3,5)$, $\varepsilon=7,11$

$(\beta,\gamma,\delta)=(1,3,7)$, $\varepsilon=9$

$(\beta,\gamma,\delta)=(1,5,5)$, $\varepsilon=5$

$(\beta,\gamma,\delta)=(1,5,7)$, $\varepsilon=7,11$

$(\beta,\gamma,\delta)=(1,7,7)$, $\varepsilon=9$

$(\beta,\gamma,\delta)=(3,3,3)$, $\varepsilon=3,7$

$(\beta,\gamma,\delta)=(3,3,5)$, $\varepsilon=5$

$(\beta,\gamma,\delta)=(3,3,7)$, $\varepsilon=7$

$(\beta,\gamma,\delta)=(3,5,5)$, $\varepsilon=7$

$(\beta,\gamma,\delta)=(5,5,7)$, $\varepsilon=7$

6) $\langle J\sqcup\alpha\sqcup 1\langle\beta\sqcup 1\langle\gamma\rangle\rangle\rangle$, $\alpha=26-\beta-\gamma$, where

$\gamma=1$, $\beta=0$, $2\le\beta\le16$, $\beta=18,19$

$\gamma=2$, $\beta=2,3,6\ 7$, $9\le\beta\le19$

$\gamma=3$, $\beta=2,3,6,7$, $9\le\beta\le19$

$\gamma=4$, $\beta=2,3,6,7,8$, $10\le\beta\le19$, $\beta=21$

$\gamma=5$, $\beta=4,6,7$, $10\le\beta\le18$, $\beta=20$

$\gamma=6$, $\beta=2$, $6\le\beta\le13$, $15\le\beta\le18$

$\gamma=7$, $\beta=2,6,7,8,10,11,12$, $14\le\beta\le17$

$\gamma=8$, $\beta=4,6,8,9,10,12,13,14,16,17$

$\gamma=9$, $\beta=2$, $6\le\beta\le14$, $\beta=16$

$\gamma=10$, $2\le\beta\le9$, $11\le\beta\le14$

$\gamma=11$, $\beta=2,3,4,6,7,8,10,12,13$

$\gamma=12$, $\beta=2,6,8,10,13$

$\gamma=13$, $\beta=2,3,6,7,10,11,12$

$\gamma=14$, $\beta=3,4$, $7\le\beta\le11$

$\gamma=15$, $\beta=9$

$\gamma=16$, $\beta=6,9$

$\gamma=17$, $\beta=4,6,7,8$

$\gamma=18$, $\beta=6,7$

$\gamma=20$, $\beta=2,5$

$\gamma=21$, $\beta=2,3,4$

$\gamma=22$, $\beta=2,3$

These 396 M-curves contain 22 M-curves constructed by G.M. Polotovskii in [5].

<u>Theorem 7</u> ([10-15]). There not exist M-curves of 9-th degree which have the following 496 admissible schemes:

1) 2 schemes $\langle J\sqcup 1\langle 27\rangle\rangle$, $\langle J\sqcup 1\sqcup 1\langle 26\rangle\rangle$
2) 13 schemes $\langle J\sqcup 1\langle\alpha\rangle\sqcup 1\langle\beta\rangle\rangle$, where $\alpha+\beta=26$, $1\leq\alpha\leq\beta$
3) 52 schemes $\langle J\sqcup 1\langle\alpha\rangle\sqcup 1\langle\beta\rangle\sqcup 1\langle\gamma\rangle\rangle$, where $\alpha+\beta+\gamma=25$, $1\leq\alpha\leq\beta\leq\gamma$
4) 12 schemes $\langle J\sqcup 1\sqcup 1\langle\alpha\rangle\sqcup 1\langle\beta\rangle\sqcup 1\langle\gamma\rangle\rangle$, where $\alpha+\beta+\gamma=24$, $1\leq\alpha\leq\beta\leq\gamma$ and α,β,γ - even
5) 225 schemes $\langle J\sqcup\alpha\sqcup 1\langle\beta\rangle\sqcup 1\langle\gamma\rangle\sqcup 1\langle\delta\rangle\sqcup 1\langle\varepsilon\rangle\rangle$, where $\alpha+\beta+\gamma+\delta+\varepsilon=24$, $1\leq\alpha\leq\beta\leq\gamma\leq\delta\leq\varepsilon$ and α - odd
6) 85 schemes $\langle J\sqcup 1\langle\alpha\rangle\sqcup 1\langle\beta\rangle\sqcup 1\langle\gamma\rangle\sqcup 1\langle\delta\rangle\rangle$, where $\alpha+\beta+\gamma+\delta=2$, $1\leq\alpha\leq\beta\leq\gamma\leq\delta$ and $\alpha,\beta,\gamma,\delta$ - even.
7) 7 schemes $\langle J\sqcup 1\langle 1\langle 26\rangle\rangle\rangle$, $\langle J\sqcup 1\sqcup 1\langle 1\langle 25\rangle\rangle\rangle$, $\langle J\sqcup 1\langle 1\sqcup 1\langle 25\rangle\rangle\rangle$, $\langle J\sqcup 1\sqcup 1\langle 1\sqcup 1\langle 24\rangle\rangle\rangle$, $\langle J\sqcup 1\langle 2\sqcup 1\langle 24\rangle\rangle\rangle$, $\langle J\sqcup 1\langle 3\sqcup 1\langle 23\rangle\rangle\rangle$, $\langle J\sqcup 1\langle 25\sqcup 1\langle 1\rangle\rangle\rangle$.

These 496 admissible schemes contain 52 schemes which were restricted by T.Fiedler [10].

So the 9-th degree M-curves classification is not finished. We have inverstigated about 51% admissible schemes. Let us remind of the 7-th degree nonsingular curves classification.

<u>Theorem 8</u> (Viro [16,17]). There exist 7-th degree nonsingular plane curves which have the following schemes:

1) $\langle J\sqcup\alpha\rangle$, where $0\leq\alpha\leq 15$
2) $\langle J\sqcup\alpha\sqcup 1\langle\beta\rangle\rangle$, where $\alpha+\beta\leq 14$, $0\leq\alpha\leq 13$, $1\leq\beta\leq 13$
3) $\langle J\sqcup 1\langle 1\langle 1\rangle\rangle\rangle$.

There not exist M-curves which have the scheme $\langle J\sqcup 1\langle 14\rangle\rangle$.

These theorems 6,7,8 and my checking of nonsingular curves of 11 and 13-th degrees allowed to formulate the following conjectures.

Let $F=0$ be nonsingular real algebraic curve of odd degree and its oval S is surrounded by other r ovals of the curve. If r is odd (even) then the oval S is called odd (even) oval. Let $n(p)$ be the number of odd (even) ovals of the curve $F=0$.

Conjecture 1. If F=0 is nonsingular curve of odd degree d then

$$n-p \le \begin{cases} (3d^2-6d+3)/8 & \text{if } d\equiv 1 \bmod 4 \\ (3d^2-6d-1)/8 & \text{if } d\equiv 3 \bmod 4 \end{cases}$$

Conjecture 2. If F=0 is nonsingular curve of odd degree d then

$$n \le \begin{cases} (3d^2-6d+3)/8 & \text{if } d\equiv 1 \bmod 4 \\ (3d^2-6d-1)/8 & \text{if } d\equiv 3 \bmod 4. \end{cases}$$

Conjecture 1 is analoguos to Petrovsky theorem about even degree nonsingular curves [14]. Conjecture 2. is analogous to famous Ragsdale conjecture [15]. By analisis of theorems 6 and 7 I have also formulated the following conjectures.

Conjecture 3. If $\langle J\amalg\alpha\amalg 1\langle\beta\rangle\amalg 1\langle\gamma\rangle\amalg 1\langle\delta\rangle\rangle$, $\alpha+\beta+\gamma+\delta=25$, $\beta\ge 1$, $\gamma\ge 1$, $\delta\ge 1$ is the scheme of M-curve of 9-th degree then

1) β,γ,δ are odd

or 2) two of β,γ,δ are odd and one of them is even.

Conjecture 4. If $\langle J\amalg\alpha\amalg 1\langle\beta\rangle\amalg 1\langle\gamma\rangle\amalg 1\langle\delta\rangle\amalg 1\langle\varepsilon\rangle\rangle$, $\alpha+\beta+\gamma+\delta+\varepsilon=24$, $\beta\ge 1$, $\gamma\ge 1$, $\delta\ge 1$, $\varepsilon\ge 1$ is the scheme of M-curve of 9-th degree then $\alpha\equiv 0 \bmod 4$ and $\beta,\gamma,\delta,\varepsilon$ are odd.

Conjecture 5. If $\langle J\amalg\alpha\amalg 1\langle\beta\amalg 1\langle\gamma\rangle\rangle\rangle$, $\alpha+\beta+\gamma+=26$, $\gamma\ge 1$ is the scheme of M-curve of 9-th degree then $\alpha\ne 0$.

1. Viro O.Ya. Gluing of plane real algebraic curves and constructions of curves of degrees 6 and 7. - Lect. Notes in Math., 1984, N 1060, pp. 187-200.
2. Viro O.Ya. Gluing of algebraic hypersurfaces, smoothing of singularities and constructions of curves. - Proceedings of Leningrad Int. Topological Conf., 1983, pp. 149-197 (in Russian).

3. Korchagin A.B. Isotopy classification of plane seventh degree curves with the only singular point Z_{15}. - Lect.Notes in Math., 1988, v. 1346, pp. 407-426.
4. Korchagin A.B. M-curves of the 9-th degree: realizability of 167 types. - Paper placed in VINITI 10.11.87, N7884-B87 (in Russian).
5. Polotovskii G.M. To the problem of nonsingular algebraic cureve ovals arrangement topology classification in projective plane. - Methods of qualitative theory of diff. equations, Gorky State Univ., 1975, pp. 101-128 (in Rassian).
6. Korchagin A.B. New capacity of Bruzotti method for construction of M-curves of degree ≥ 8. - Methods of qualitative theory of diff. equations, Gorky State Univ., 1978, pp. 149-159,(in Russian).
7. Korchagin A.B. M-curves of the 9-th degree: constructions of 141 curves. - Paper placed in VINITI 29.10.86, N7459-B86 (in Russian).
8. Korchagin A.B. M-curves of the 9-th degree: realizability of 32 types. - Paper placed in VINITI 14.04.87, N2566-B87 (in Russian).
9. Korchagin A.B. M-curves of the 9-th degree: realizability of 24 types. - Paper placed in VINITI 29.11.87, N3049-B87 (in Russian).
10. Fiedler T. Sheats of lines and real algebraic curves topology. - Izvestia Acad. of Sc. USSR, math., v.46, N 4, 1982, pp.853-863 (in Rassian).
11. Korchagin A.B. M-curves of the 9-th degree: nonrealizability of 12 types. - Methods of qualitative theory of diff. equations, Gorky State Univ., 1985, pp.72-76,(in Russian).
12. Korchagin A.B. M-curves of the 9-th degree: the new prohibitions. - Math. Zametki, v.39, N2, 1986, pp.277-283, (in Russian).
13. Korchagin A.B. M-curves of the 9-th degree: nonrealizability of 12 types with two nests. - Paper placed in VINITI 04.03.88, N1832-B88 (in Russian).
14. Petrovsky I.G. On the topology of real plane algebraic curves.- Ann. of Math., v.39, N 1, 1938, pp. 187-209.

15. Ragsdale V. On the arrangment of the real branches of plane algebraic curves. - Amer. Jour. Math.,v.28, 1906, pp.377-404.
16. Viro O.Ya. Curves of degree 7, curves of degree 8 and Ragsdale conjecture. Soviet Dokl., v.254, N6, 1980, pp. 1306-1310 (in Russian).
17. Viro O.Ya. Curves of degrees 7 and 8: new restrictions. Izvestia Acad.of Sc. USSR, math.,V.47, N5, 1983, pp.1135-1150 (in Russian).

ON LINEAR DIFFERENTIAL OPERATORS RELATED TO THE n-DIMENSIONAL JACOBIAN CONJECTURE

Tadeusz KRASIŃSKI
Stanisław SPODZIEJA

Institute of Mathematics, University of Łódź, ul. S. Banacha 22, 90-238 Łódź, Poland

Introduction

Let E_n be the ring of entire functions on $\mathbb{C}^n$ with the standard topology of uniform convergence on compact subsets of $\mathbb{C}^n$, and let $F = (f_1,\dots,f_n) \in E_n^n$ be a fixed mapping from $\mathbb{C}^n$ into $\mathbb{C}^n$. In the paper we consider the linear differential operators

$$\Delta_i^F : E_n \ni f \longmapsto \mathrm{Jac}(f_1,\dots,f_{i-1},f,f_{i+1},\dots,f_n) \in E_n,$$

$i = 1,\dots,n$, where $\mathrm{Jac}(\cdot)$ means the usual jacobian. The operators Δ_i^F were studied many times in connection with the Jacobian Conjecture ([1],[3],[9],[10],[11],[12],[13]).

In the paper we prove the following theorem

Theorem. Let $F = (P_1,\dots,P_n) : \mathbb{C}^n \to \mathbb{C}^n$ be a polynomial mapping such that $\mathrm{Jac}F = 1$. Then the following conditions are equivalent:

(i) F is a polynomial automorphism of $\mathbb{C}^n$,

(ii) $\Delta_i^F(E_n)$ is dense in E_n for $i = 1,\dots,n-1$,

(iii) $\Delta_1^F(E_n)$ is dense in E_n and $\ker\Delta_1^F = \{ f(P_2,\dots,P_n) : f \in E_{n-1}\}$.

For $n = 2$ the equivalence (i) $\Leftrightarrow$ (ii) was proved by Y. Stein in [10]. He called the implication $\mathrm{Jac}F = 1 \Rightarrow$ (ii) the Analytic Jacobian Conjecture.

As corollaries we obtain similar theorems if we replace E_n by the rings of polynomials $\mathbb{C}[X] := \mathbb{C}[X_1,\dots,X_n]$ or $\mathbb{R}[X] := \mathbb{R}[X_1,\dots,X_n]$ (in the latter case for a real polynomial mapping F).

1. Operators Δ_i^F

Let $F = (f_1,\dots,f_n) : \mathbb{C}^n \longrightarrow \mathbb{C}^n$ be an entire mapping, i.e. $f_i \in E_n$, $i = 1,\dots,n$. The basic properties of Δ_i^F (defined in

Introduction) are given in the following propositions

Proposition 1.1. The functions Δ_i^F, $i = 1,\dots,n$, are linear differential operators on E_n, continuous in the topology of E_n. They are also derivations of the ring E_n. □

Proposition 1.2 (cf.[3] for n=2, [9]). If JacF is a nonzero constant, then, for any $x \in \mathbb{C}^n$ and a neighbourhood U of x such that $F|U$ is a biholomorphism, we have in U

$$\Delta_{i_1}^F \circ \dots \circ \Delta_{i_k}^F(f) = (\mathrm{Jac}F)^k \frac{\partial^k (f\circ(F|U)^{-1})}{\partial Y_{i_1}\dots\partial Y_{i_k}} \circ F$$

for any $i_1,\dots,i_k \in \{1,\dots,n\}$ and $f \in E_n$. □

Proposition 1.3 (cf.[1], p.297, [3],[9]). The operators Δ_i^F, $i = 1,\dots,n$, commute i.e. $\Delta_i^F \circ \Delta_j^F = \Delta_j^F \circ \Delta_i^F$ for any $i,j = 1,\dots,n$, if and only if JacF is constant. □

2. Operators Δ_i^F and level sets of polynomial mappings

Let $F = (P_1,\dots,P_n) : \mathbb{C}^n \longrightarrow \mathbb{C}^n$ be a polynomial mapping with $\mathrm{Jac}F = 1$. Fix $i \in \{1,\dots,n\}$ and numbers $y_1,\dots,y_{i-1},y_{i+1},\dots,y_n \in \mathbb{C}$. Consider the algebraic set

$$S := \{ x \in \mathbb{C}^n : P_j(x) = y_j,\ j = 1,\dots,i-1,i+1,\dots,n\}$$

and its decomposition into irreducible components

$$S = S_1 \cup \dots \cup S_k .$$

Since $\mathrm{Jac}F = 1$, therefore $S_1,\dots,S_k$ are nonsingular, one-dimensional and disjoint algebraic sets. They are also open Riemann surfaces.

Proposition 2.1. If $\Delta_i(E_n)$ (if F is fixed we shall omit the superscript F in Δ_i^F) is dense in E_n, then the algebraic curves S_j are biregular to $\mathbb{C}$ for $j = 1,\dots,k$. Moreover, $P_i|S_j$, $j = 1,\dots,k$, are biregular.

Proof. (cf.[10] for n = 2). Since the proof is analogous for each of the irreducible component S_j of S, we may assume that S is irreducible. Let us denote by $\mathcal{O}(S)$ the ring of holomorphic functions on S with the topology of uniform convergence on compact subsets of S, and by $I(S)$ the ideal in E_n of all entire functions vanishing on S. By Cartan's theorem $\mathcal{O}(S) \cong$

$E_n/I(S)$. Since $\Delta_i(I(S)) \subset I(S)$, Δ_i induces, in the canonical way, a unique continuous derivation D of the ring $\mathcal{O}(S)$.

The proof will be divided in several steps:

1. $D(\mathcal{O}(S))$ is dense in $\mathcal{O}(S)$. It follows from the assumption that $\Delta_i(E_n)$ is dense in E_n.

2. For any $f \in \mathcal{O}(S)$ we have $df = D(f)d(P_i|S)$ on S. Locally, $(P_i|S)$ is a holomorphic chart on S. So, we have locally (i.e. where $(P_i|S)$ is invertible) $df = (f\circ(P_i|S)^{-1})'\circ(P_i|S)d(P_i|S)$. On the other hand, we have also locally $D(f) = D(f\circ(P_i|S)^{-1}\circ(P_i|S))$ $= (f\circ(P_i|S)^{-1})'\circ(P_i|S)D(P_i|S) = (f\circ(P_i|S)^{-1})'\circ(P_i|S)$ because $D(P_i|S) = 1$. This gives the desired equality locally and hence globally.

3. Each holomorphic form ω on S is exact. Since $(P_i|S)$ is locally a holomorphic chart on S, therefore, for a given ω there exists $g \in \mathcal{O}(S)$ such that $\omega = gd(P_i|S)$. By step 1 there exist $g_n, h_n \in \mathcal{O}(S)$ such that $g_n \longrightarrow g$ in the topology of $\mathcal{O}(S)$ and $g_n = D(h_n)$. Hence, by step 2, $dh = D(h_n)d(P_i|S) = g_n d(P_i|S)$. Fix $z_o \in S$. We may assume that $h_n(z_o) = 0$ for $n \in \mathbb{N}$. Then, for any continuous piecewise-differentiable path γ joining z_o with $z \in S$ we have

$$h_n(z) = \int_\gamma dh_n = \int_\gamma g_n d(P_i|S).$$

Hence, there exists the limit $h(z) := \lim_{n\to\infty} h_n(z) = \int_\gamma gd(P_i|S)$ and it does not depend on the choice of γ. Of course, h is holomorphic on S and $dh = \omega$.

4. S is biregular to $\mathbb{C}$. From the above step and the Riemann theorem S is biholomorphic to the unit disc or to $\mathbb{C}$. Since S is an algebraic set, then by the Liouville property of algebraic sets, S is biholomorphic to $\mathbb{C}$. Let $\Phi : \mathbb{C} \longrightarrow S$ be such a biholomorphism. Again, since S is an algebraic set, Φ is a biregular mapping (see [7], Th.4).

5. $(P_i|S)$ is a biregular mapping. Obviously, $(P_i|S)$ is regular, locally invertible mapping. Hence, for any biregular mapping $\Phi : \mathbb{C} \longrightarrow S$ we have that $(P_i|S)\circ\Phi : \mathbb{C} \longrightarrow \mathbb{C}$ is regular and locally invertible. So, this composition has the form $L(z) = az + b$, $a,b \in \mathbb{C}$, $a \neq 0$. Hence $(P_i|S) = L\circ\Phi^{-1}$ is biregular. □

3. The main theorem

Let $F = (P_1,\dots,P_n) : \mathbb{C}^n \longrightarrow \mathbb{C}^n$ be a polynomial mapping such that $\mathrm{Jac}F = 1$. Then F is a dominating mapping, i.e. $\overline{F(\mathbb{C}^n)} = \mathbb{C}^n$. So, the degree $\deg F$ of the mapping F is well-defined ([6], §3A). Put $d := \deg F$ and define the set

$$B_F := \{y \in \mathbb{C}^n : \#F^{-1}(y) \neq d\}$$

which will be called the set of bifurcation points of F. It is known that B_F is a constructible set ([5], Lemma in I.8.4). On the other hand, from the fact that F is a local biholomorphism we have from Proposition 3.17 in [6] that $\mathbb{C}^n \setminus B_F$ is open and non-empty. Hence B_F is an algebraic set different from $\mathbb{C}^n$.

For any $i \in \{1,\dots,n\}$, we shall denote by $\pi_i : \mathbb{C}^n \longrightarrow \mathbb{C}^{n-1}$ the canonical projection $\pi_i(y_1,\dots,y_n) := (y_1,\dots,y_{i-1},y_{i+1},\dots,y_n)$.

Lemma 3.1. Fix $i \in \{1,\dots,n\}$. If $\Delta_i(E_n)$ is dense in E_n, then there exists an algebraic set $B_i \subset \mathbb{C}^{n-1}$ such that

(3.2) $\qquad B_F = \pi_i^{-1}(B_i)$.

Proof. We shall first show that the complement $\mathbb{C}^n_F \setminus B$ has the following property:

(3.3) $\qquad$ if $y \in \mathbb{C}^n \setminus B_F$, then $\pi_i^{-1}(\pi_i(y)) \subset \mathbb{C}^n \setminus B_F$.

Indeed, let us take $y = (y_1,\dots,y_n) \notin B_F$. By Proposition 2.1, the set $S = \{x \in \mathbb{C}^n : P_1(x) = y_1,\dots,P_{i-1}(x) = y_{i-1}, P_{i+1}(x) = y_{i+1},\dots,P_n(x) = y_n\}$ decomposes into disjoint irreducible components $S_1,\dots,S_k$ and, for any such S_j, the function $P_i|S_j$ is a biregular mapping of S_j onto $\mathbb{C}$. Since $y \notin B_F$, therefore $k = d$. Hence, from the bijectivity of $P_i|S_j$ and the fact that $S_1,\dots,S_d$ are disjoint, we obtain that, for any $z \in \mathbb{C}$, $\#F^{-1}(y_1,\dots,y_{i-1},z,y_{i+1},\dots,y_n) = d$. This gives property (3.3).

From (3.3) it easily follows that B_F has the property:

if $y \in B_F$, then $\pi_i^{-1}(\pi_i(y)) \subset B_F$.

Hence we obtain without difficulty that the set $B_i := \pi_i(B_F)$ is an algebraic set in $\mathbb{C}^{n-1}$. For the B_i so defined, equality (3.2) is obvious.□

Theorem 3.4. Under the above assumptions on F, the following conditions are equivalent:

(i) F is a polynomial automorphism of $\mathbb{C}^n$,

(ii) $\Delta_i(E_n) = E_n$, $i = 1,\dots,n-1$,

(iii) $\Delta_i(E_n)$ is dense in E_n, $i = 1,\dots,n-1$,

(iv) $\Delta_1(E_n)$ is dense in E_n and $\ker\Delta_1 = \{f(P_2,\dots,P_n) \in E_n : f \in E_{n-1}\}$.

Proof. 1. (i) ⇒ (ii). Take $f \in E_n$. Let g be an entire function such that $\partial g/\partial Y_i = f \circ F^{-1}$. Put $h := g \circ F$. Then, by Propositon 1.2,

$$\Delta_i(h) = \frac{\partial(h\circ F^{-1})}{\partial Y_i}\circ F = \frac{\partial g}{\partial Y_i}\circ F = f.$$

2. (ii) ⇒ (iii). Obvious.

3. (iii) ⇒ (i). Let d be the degree of F and let B_F be the bifurcation set of F. By Lemma 3.1, there exists an algebraic set $B^* \subset \mathbb{C}$ such that $B_F = \mathbb{C}^{n-1}\times B^*$. Since $B_F \neq \mathbb{C}^n$, therefore $B^* \neq \mathbb{C}$. Take $z \notin B^*$ and consider the nonsingular algebraic set

$$V := P_n^{-1}(z) = F^{-1}(\mathbb{C}^{n-1}\times\{z\}).$$

Since $(\mathbb{C}^{n-1}\times\{z\}) \cap B_F = \emptyset$, the mapping $F|V : V \longrightarrow \mathbb{C}^{n-1}\times\{z\}$ has the same number, equal to d, of elements in each fibre. Moreover, $F|V$ is a local biholomorphism. Hence we easily get that $F|V$ is a covering. Since $\mathbb{C}^{n-1}\times\{z\}$ is simply connected, then F is a biholomorphism on each topological component of V. So, the number of them is equal to d. Let $V_1,\ldots,V_d$ be these components. Since $\mathrm{grad}P_n := (\partial P_n/\partial X_1,\ldots,\partial P_n/\partial X_n)$ vanishes nowhere, therefore, from the connectivity of the irreducible components of algebraic sets ([6], Cor.4.16) we obtain that $V_1,\ldots,V_d$ are also the irreducible components of V. The mappings $F|V_i$, $i = 1,\ldots,d$, are regular and biholomorphic. Hence and from Zariski's Main Theorem ([6], Th.3.20) we get that $F|V_i$ is a biregular mapping of V_i onto $\mathbb{C}^{n-1}\times\{z\}$, $i = 1,\ldots,d$.

Let $Q_1,\ldots,Q_d$ be polynomials such that $V_i = \{x \in \mathbb{C}^n: Q_i(x) = 0\}$. Since $\mathrm{grad}P_n$ vanishes nowhere, therefore, by the above, $P_n - z = \alpha Q_1\ldots Q_d$, $\alpha \in \mathbb{C}$, and $Q_i|V_j = \mathrm{const.}$ for any i,j. Hence $Q_i = T_{ij}Q_j + \alpha_{ij}$ where $T_{ij} \in \mathbb{C}[X]$, $\alpha_{ij} \in \mathbb{C}$. Then we easily get $\deg T_{ij} = 0$. In consequence, there exists a polynomial $H \in \mathbb{C}[Z]$ such that $\deg H = d$ and $P_n - z = H\circ Q_1$. Hence $d = 1$ because, otherwise, there would exist points in $\mathbb{C}^n$ at which $\mathrm{grad}P_n$ vanishes.

Since $d = 1$, therefore F is injective. Hence F is a polynomial automorphism (see [8], Th.1.4, [1], Th.2.1).

4. (i) ⇒ (iv). By the implication (i) ⇒ (ii) it sufficies to show only the equality $\ker\Delta_1 = \{ f(P_2,\ldots,P_n) \in E_n : f \in E_{n-1}\}$. Denote the set on the right hand side of this equality by $\tilde{E}_n$.

Obviously, we have $\tilde{E}_n \subset \ker\Delta_1$. Let $g \in \ker\Delta_1$. By Proposition 1.2,

$$\frac{\partial(g\circ F^{-1})}{\partial Y_1}\circ F = 0$$

Then $\partial(g\circ F^{-1})/\partial Y_1 = 0$, and hence the entire function $g\circ F^{-1}$ depends only on the variables $Y_2,\ldots,Y_n$. Denoting $g\circ F^{-1}$ by f, we have $g = f(P_2,\dots,P_n)$. So, $g \in \tilde{E}_n$.

5. (iv) $\to$ (i). Put $\tilde{F} := (P_2,\dots,P_n) : \mathbb{C}^n \to \mathbb{C}^{n-1}$. It is a dominating mapping. So, it induces a monomorphism $\tilde{F}^*$ of the ring $\mathbb{C}[Y_1,\dots,Y_{n-1}]$ of regular functions on $\mathbb{C}^{n-1}$ into the ring $\mathbb{C}[X]$ of regular functions on $\mathbb{C}^n$. We claim that the assumption $\ker\Delta_1 = \tilde{E}_n$ implies that the ring $\tilde{F}^*(\mathbb{C}[Y_1,\dots,Y_{n-1}]) = \mathbb{C}[P_2,\dots,P_n]$ is integrally closed in $\mathbb{C}[X]$. Indeed, if an element $P \in \mathbb{C}[X]$ is integral over $\mathbb{C}[P_2,\dots,P_n] \subset \mathbb{C}[X]$ then $P \in \ker\Delta_1$ (because P is constant on each irreducible component of any fibre of $\tilde{F}$). Hence $P = f(P_2,\dots,P_n)$ for some $f \in E_{n-1}$. Since $P,P_2,\dots,P_n$ are polynomials, therefore, f is also a polynomial. This means that $P \in \mathbb{C}[P_2,\dots,P_n]$.

From the fact that $\mathbb{C}[P_2,\dots,P_n]$ is integrally closed in $\mathbb{C}[X]$ it easily follows that the field $\mathbb{C}(P_2,\dots,P_n)$ is algebraically closed in $\mathbb{C}(X_1,\dots,X_n)$. Hence $\tilde{F}$ is primitive, i.e. there exists an algebraic set $V \subset \mathbb{C}^{n-1}$, $V \neq \mathbb{C}^{n-1}$, such that, for any $y \in \mathbb{C}^{n-1}\setminus V$, the algebraic set $\tilde{F}^{-1}(y)$ is irreducible ([4]). So, for any $y \in \mathbb{C}^{n-1}\setminus V$, the set $\tilde{F}^{-1}(y)$ is connected ([6],Cor.4.16).

On the other hand, the assumption that $\Delta_1(E_n)$ is dense in E_n implies (by Proposition 2.1) that for each $y \in \mathbb{C}^{n-1}$, the mapping $P_1|\tilde{F}^{-1}(y)$ is biregular on each topological component of $\tilde{F}^{-1}(y)$. Hence, for each $y \in \mathbb{C}^{n-1}\setminus V$, the mapping $P_1|\tilde{F}^{-1}(y)$ is injective. This implies that in the open set $F^{-1}(\mathbb{C}\times(\mathbb{C}^{n-1}\setminus V))$ the mapping F is injective. Also, $\deg F = 1$. This gives that F is a polynomial automorphism ([8],Th.1.4,[1],Th.2.1).□

Remark.3.5. In the above theorem, for $n = 2$ the condition $\ker\Delta_1 = \tilde{E}_n$ in (iv) can be omitted because it follows from the assumption that $\mathrm{Jac}F = 1$. □

4. Operators Δ_i^F in the polynomial rings

Let $k = \mathbb{R}$ or $\mathbb{C}$. Let $F = (P_1,\dots,P_n) : k^n \to k^n$ be a polynomial mapping such that $\mathrm{Jac}F = 1$. The restriction of the operators Δ_i^F to the ring $k[X_1,\dots,X_n]$ we shall also denote by Δ_i^F.

Theorem 4.1. Under the above assumptions, the following conditions are equivalent:

(i) F is a polynomial automorphism of k^n,

(ii) $\Delta_i^F(k[X_1,\dots,X_n]) = k[X_1,\dots,X_n]$, $i = 1,\dots,n-1$,

(iii) $\Delta_1^F(k[X_1,\dots,X_n]) = k[X_1,\dots,X_n]$ and $\ker\Delta_1^F =$

$k[P_2,\dots,P_n]$.

Proof. 1. (i) ⇒ (ii). Analogously as the implication (i) ⇒ (ii) in theorem 3.4.

2. (i) ⇒ (iii). Analogously as the implication (i) ⇒ (iv) in theorem 3.4.

3. (ii) ⇒ (i). a) $k = \mathbb{C}$. From (ii) it follows that $\Delta_i^F(\mathbb{C}[X])$ is dense in E_n for $i = 1,\dots,n-1$. Hence, by theorem 3.4., F is a polynomial automorphism.

b) $k = \mathbb{R}$. F has a canonical extension to the mapping $\tilde{F}:\mathbb{C}^n \longrightarrow \mathbb{C}^n$. We shall show that $\Delta_i^{\tilde{F}}(\mathbb{C}[X]) = \mathbb{C}[X]$ for $i = 1,\dots,n-1$. Take $i \in \{1,\dots,n-1\}$ and $P \in \mathbb{C}[X]$. Let $P = P' + iP''$, $P',P'' \in \mathbb{R}[X]$. By the assumption there exist $Q',Q'' \in \mathbb{R}[X]$ such that $\Delta_i^F(Q') = P'$ and $\Delta_i^F(Q'') = P''$. Hence $\Delta_i^{\tilde{F}}(Q' + iQ'') = P$. From the case a) we obtain that $\tilde{F}$ is a polynomial automorphism of $\mathbb{C}^n$ Hence $F = \tilde{F}|\mathbb{R}^n$ is an injection. By the theorem in [2] F is also a surjection. So, F is a bijection. Since $F^{-1} = \tilde{F}^{-1}|\mathbb{R}^n$, therefore F^{-1} is a real polynomial mapping.

4. (iii) ⇒ (i). a) $k = \mathbb{C}$. Analogously as the implication (iv) ⇒ (i) in theorem 3.4.

b) $k = \mathbb{R}$. As above, let $\tilde{F} : \mathbb{C}^n \longrightarrow \mathbb{C}^n$ be a canonical extension of F. From the assumptions it follows easily that $\Delta_1^{\tilde{F}}(\mathbb{C}[X]) = \mathbb{C}[X]$ and $\ker\Delta_1^{\tilde{F}} = \mathbb{C}[P_2,\dots,P_n]$. From the case a) we have that $\tilde{F}$ is a polynomial automorphism. Proceeding analogously as in 3b) we obtain that F is a real polynomial automorphism. □

Acknowledgement. We thank Mr K. Rusek for his having called our attention to the paper by Y. Stein [10].

References

[1] H.Bass, E.H.Connel and D.Wright, The Jacobian Conjecture : Reduction degree and formal expansion of the inverse, Bull.Amer. Math.Soc.7(2)(1982) 287-330.

[2] A.Białynicki-Birula and M. Rosenlicht, Injective morphisms of real algebraic varietties, Proc.Amer.Math.Soc.13(1962), 200-203.

[3] Z.Charzyński, J.Chądzyński and P.Skibiński, A contribution to Keller's Jacobian Conjecture III, Bull.Soc.Sci.Lettres Łódź 39, NO.4(1989),1-8.

[4] T.Krasiński and S.Spodzieja, On the irreducibility of fibres of complex polynomial mappings (to appear).

[5] S.Lojasiewicz, Introduction to Complex Analytic Geometry (PWN, Warszawa, 1988)(in Polish).

[6] D.Mumford, Algebraic Geometry I, Complex Projective Varieties (Springer, Berlin-Heidelberg-New York, 1976).

[7] K.Rusek a nd T.Winiarski, Criteria for regularity of holomorphic mappings, Bull.Ac.Pol.:Math.28(1980), 471-475.

[8] K.Rusek and T.Winiarski, Polynomial Automorphisms of $\mathbb{C}^n$, Univ. Iagell. Acta Math.24(1984),143-149.

[9] S.Spodzieja, On commutativity of the composition of Whitney operators, Bull.Soc.Sci.Lettres Łódź 39, No.13(1989),1-6.

[10] Y.Stein, On linear differential operators related to the

Jacobian Conjecture, J.Pure Appl.Algebra 57(1989),175-186.
[11] Y.Stein, On the density of image of differential operators generated by polynomials, J.Analyse Math. 52(1989),291-300.
[12] Y.Stein, Linear differential operators related to the Jacobian Conjecture have a closed image, J.Analyse Math. 54 (1990),237-245.
[13] D.Wright, On the Jacobian Conjecture, Illinois J.Math.25 (1981),423-440.

On a subanalytic stratification satisfying a Whitney property with exponent 1

by Krzysztof Kurdyka

Instytut Matematyki, Uniwersytet Jagielloński, Reymonta 4, Kraków PL-30059 (Poland)

If A is a subanalytic and closed set in $\mathbf{R}^n$, then A satisfies following Whitney property (see [St]): every point of A has a neighbourhood U such that, if x and y are two points in $A \cap U$ then there exists an arc λ joining x and y in $A \cap U$ such that length$\lambda \leq C \mid x - y \mid^\alpha$,where C and α are constants depending only on U. Actually we have the same property for strata of a stratification obtained by projections (e.g. triangulations or cellular decompositions, see [Ha], [Ło3]). In this paper we construct a stratification compatible with a given family of subanalytic sets such that on each stratum the above property holds with exponent α=1 (corollary B). Actually the strata we obtain (theorem A) are L-regular in the sense of Parusiński [Pa1] i.e. are of the form $\{(x', x_n) \in \mathbf{R}^{n-1} \times \mathbf{R};\ f(x') < x_n < g(x'), x' \in A'\}$,where A' is L-regular in $\mathbf{R}^{n-1}$, f and g are subanalytic and continous in $\overline{A'}$, analytic in A', $f(x') < g(x')$ for $x' \in A'$. Moreover $\mid d_{x'}f \mid \leq M$,$\mid d_{x'}g \mid \leq M$ for each $x' \in A'$,where M is a constant depending only on n. Other strata are the graphs of analytic functions (with bounded differentials) defined on sets of previous form. Our result is related to the papers of Parusiński ([Pa1],[Pa2]). We use only elementary facts on subanalytic sets i.e. subanalycity of tangent mapping and stratifications of mappings. Hence our result holds in semialgebraic and semianalytic case.

This paper was inspired by a question of Prof. S.Łojasiewicz (corollary C), the author is very grateful to him for interest and encouragement. The author thanks G.Jasiński for helpful remarks. The paper was written when the author was associate member of Dept. of Math. of Université de Savoie at Chambéry and wants to acknowledge people working there (particulary P.Orro) for friendly ambiance and hospitality. The author thanks also the referee for valuables remarks.

0.Preliminary remarks. Let M be an analytic manifold. By a *stratification* of M we mean a locally finite family $\mathcal{T}$ of analytic, connected submanifolds of M such that: $M = \bigcup\{T : T \in \mathcal{T}\}$ (disjoint union), moreover if $S \cap (\overline{T} \setminus T) \neq \emptyset$, where $S, T \in \mathcal{T}$, then $S \subset (\overline{T} \setminus T)$ and $dimS < dimT$. We say that stratifcation $\mathcal{T}$ is subanalytic if all strata (i.e. members of $\mathcal{T}$) are subanalytic in M. Let $\mathcal{A}$ be a family of subsets of M, we say that stratification $\mathcal{T}$ is *compatible* with family $\mathcal{A}$ if $T \cap A \neq \emptyset$ implies $T \subset A$, for each $T \in \mathcal{T}, A \in \mathcal{A}$. Let $\mathcal{C}$ be a family of analytic, connected submanifolds of M, subanalytic in M. We say that family $\mathcal{C}$ is *stratifying* if for every locally finite family $\mathcal{A}$ of subanalytic subsets of M there exists a subanalytic stratification $\mathcal{T}$ of M, compatible with $\mathcal{A}$ such that $\mathcal{T} \subset \mathcal{C}$. It follows from the general method of constructing of subanalytic stratification (see e.g. [DS],[Lo3]) that $\mathcal{C}$ is stratifying if and only if :

(*)*For every A analytic submanifold of M, subanalytic in M, there exists F subanalytic in M, closed,nowhere dense subset of A such that every connected component of $A \setminus F$ belongs to the family $\mathcal{C}$.*

We recall a definiton of angle between linear subspaces.

1.Definition. Let X be a linear subspace in $\mathbf{R}^n$, let P be a line in $\mathbf{R}^n$.We define angle between P and X as

$$\delta(P,X) = \inf\{\sin(P,S);\, S \textit{ is a line in } X\}$$

where $\sin(P,S)$ is a sine of the angle between P and S. Let Y be a linear subspace in $\mathbf{R}^n$,we put

$$\delta(Y,X) = \sup\{\delta(P,X);\, P \textit{ is a line in } Y\}$$

If $Y = 0$ we put $\delta(0,X) = 0$. Function δ takes values in $[0,1]$. In general δ is not symmetric, however we have:

(1.0) if $dimX = dimY$ then $\delta(X,Y) = \delta(Y,X)$.

(1.1) if $Y \subset X$ then $\delta(Y,X) = 0$.

(1.2) if $dimZ \leq dimY \leq dimX$ then $\delta(Z,X) \leq \delta(Z,Y) + \delta(Y,X)$.

Let us denote by $\mathbf{G}_{i,n}$ the grassmanian space of all i-dimensional linear subspaces of $\mathbf{R}^n$ equiped with the natural structure of real algebraic variety. Then we have

(1.3) the mapping $\mathbf{G}_{i,n} \times \mathbf{G}_{j,n} \ni (Y,X) \longmapsto \delta(Y,X) \in \mathbf{R}$ is continous and semialgebraic (see [Ło4]).

2.Remark. Let X denotes a hyperplane and P a line in $\mathbf{R}^n$. Then for each $m > 0$ there exists $M > 0$ such that if $\delta(P,X) > m$ then X is a graph of linear mapping $\varphi : P^\perp \longrightarrow P$ satisfying $\| \varphi \| \leq M$ ($P^\perp$ denotes the orthogonal complement of P).

3.Lemma. *Given $r,n \in \mathbf{N}$, there exists constants $\varepsilon > 0$ and $m > 0$ such that for given $X_1,...X_r$ hyperplanes in $\mathbf{R}^n$,there exists a line P such that, if $Y_1,...Y_r$ are hyperplanes verifying $\delta(Y_i,X_i) < \varepsilon$, $i = 1,...r$, then*

$$\delta(P,Y_i) > m \quad \textit{for each } i = 1,...,r$$

Proof. Let us take a metric d on the sphere S^{n-1} defined as follows : $d(p,q) = \delta(\mathbf{R}p,\mathbf{R}q)$ for $p,q \in S^{n-1}$. Let us denote $X_i^\varepsilon = \{p \in S^{n-1} :\ dist(p, X_i \cap S^{n-1}) < \varepsilon\}$, where $dist(p,Z) = \inf\{d(p,q) : q \in Z\}$. It is enough and sufficient to prove the following fact :

(3.1) Given $r \in \mathbf{N}$, there exists $\varepsilon > 0$ and $m > 0$ such that the complement of $\bigcup_{i=1}^r X_i^\varepsilon$ in S^{n-1} contains a ball of radius m (in metric d).

We use induction on r to prove (3.1). The case $r = 1$ is obvious. Let us denote ε_r and m_r corresponding constants in (3.1) for r hyperplanes. Let $B(p,m_r)$ be a ball in

S^{n-1} disjoint with each $X_i^{\varepsilon_r}$, $i = 1, ...r$. Put $\varepsilon_{r+1} = m_{r+1} = \min\{\varepsilon_r, m_r\}/3$, then the set $B(p, m_r) \setminus X_{r+1}^{\varepsilon_{r+1}}$ contains a ball of radius m_{r+1}.

Let us recall some known facts on subanalytic sets. Let Γ be an analytic submanifold and subanalytic subset of $\mathbf{R}^n$, $dim\Gamma = k$, then the Gauss mapping

$$\tau : \Gamma \ni x \longmapsto T_x\Gamma \in \mathbf{G}_{k,n}$$

is subanalytic i.e. its graph is subanalytic in $\mathbf{R}^n \times \mathbf{G}_{k,n}$. As usual $T_x\Gamma$ denotes the tangent space to Γ at x. Moreover if Σ is a subanalytic subset of $\mathbf{G}_{k,n}$ then $\tau^{-1}(\Sigma)$ is subanalytic in $\mathbf{R}^n$ (see e.g. [DW],[Lo2]).

4.Definition. Let Γ be a C^1 submanifold of $\mathbf{R}^n$ and let $\varepsilon > 0$. We say that Γ is $\varepsilon - flat$, if for each $x, y \in \Gamma$ we have $\delta(T_x\Gamma, T_y\Gamma) < \varepsilon$. If $dim\Gamma = 0$ we assume that Γ is ε-flat for every $\varepsilon > 0$.

5.Proposition. *Let $\mathcal{A}$ be locally finite family of subanalytic sets in $\mathbf{R}^n$. Then for given $\varepsilon > 0$ there exists a subanalytic stratification $\mathcal{T}$ compatible with the family $\mathcal{A}$, such that each stratum of $\mathcal{T}$ is ε-flat.*

Proof. We use condition (*) of remark 0 to prove the existence of such stratifications. For every $k = 1, ..., n-1$ let us take a subanalytic finite partition of $\mathbf{G}_{k,n}$ into disjoint sets $\Sigma_i^k, i = 1, ..., r_k$ such that $\delta(X, Y) < \varepsilon$ for each $X, Y \in \Sigma_i^k$. Actually we can take a stratification of $\mathbf{G}_{k,n}$ compatible with a finite covering by δ-balls of radius ε.Let A be an analytic submanifold, subanalytic subset of $\mathbf{R}^n$, $dimA = k$. Let $\tau : A \longrightarrow \mathbf{G}_{k,n}$ be corresponding Gauss mapping, then the set $F = A \setminus \bigcup_{i=1}^{r_k} Int_A(\tau^{-1}(\Sigma_i^k))$ is closed and nowhere dense in A. Clearly every connected component of $A \setminus F$ is ε-flat. Hence by (*) of remark 0 the proposition follows.

Remark 5.1. Actually we çan require more for the stratifcation $\mathcal{T}$. By a theorem of Stasica [St] (see also [KR]), for given $M > 0$, we can refine $\mathcal{T}$ in a such way that every stratum T of $\mathcal{T}$, $dimT = k < n$, is a graph (in suitable coordinate system in $\mathbf{R}^n$) of an analytic mapping $\psi : U \longrightarrow \mathbf{R}^{n-k}$, where U is open in $\mathbf{R}^k$ and $| d_z\psi | \leq M$ for each $z \in U$. Suppose that $\psi = (\psi_1, ..., \psi_{n-k})$ and put $\varphi = \psi_{n-k}$, than T is also a graph of $\varphi : B \longrightarrow \mathbf{R}$, where B is a graph of $(\psi_1, ..., \psi_{n-k-1}) : U \longrightarrow \mathbf{R}^{n-k-1}$. Clearly B is an analytic submanifold of $\mathbf{R}^{n-1}$, we have also $| d_x\varphi | \leq M$ for each $x \in B$.

6. Definition.We say that subanalytic set A in $\mathbf{R}^n$ is a *s-cell*, if A is a point or in some coordinates in $\mathbf{R}^n$ A is of one of the following forms

α)
$$A = \{(x', x_n) \in \mathbf{R}^{n-1} \times \mathbf{R} : \; h(x') = x_n, \; x' \in B\}$$

β)
$$A = \{(x', x_n) \in \mathbf{R}^{n-1} \times \mathbf{R} : \; f(x') < x_n < g(x'), \; x' \in B\}$$

where B is a s-cell in $\mathbf{R}^{n-1}$; f, g and h are subanalytic continous functions on the closure of B, analytic on B. Moreover we assume that $f(x') < g(x')$ for each $x' \in B$.

Let A be a s-cell of dimension k, then A is an analytic submanifold of $\mathbf{R}^n$ homeomorphic to an open ball in $\mathbf{R}^k$. Moreover $\overline{A} \setminus A$ is homeomorphic to a sphere in $\mathbf{R}^k$.

7.Definition. We keep the notation of definition 6. We say that s-cell A in $\mathbf{R}^n$ is *L-regular with constant* $M > 0$, if $| d_{x'}h | \leq M$ (resp. $| d_{x'}f | \leq M$ and $| d_{x'}g | \leq M$) for each $x' \in B$, where B is a L-regular s-cell with constant M in $\mathbf{R}^{n-1}$. If A is a point we assume that A is L-regular with every constant $M > 0$.

Remark. The closure of our L-regular s-cell is a subanalytic version of the L-regular set of Parusiński (comp. def.3.2 of [Pa1]).

Proceeding by induction n we get:

8.Propostion. *Let $A \subset \mathbf{R}^n$ be a L-regular s-cell with constant $M > 0$. Then for every $x, y \in A$ there exists a smooth curve λ joining x and y such that*

$$length\lambda \leq C | x - y |,$$

where C is a constant depending only on n and M.

Now we state our main result.

Theorem A. *Let $\mathcal{A}$ be a locally finite family of subanalytic sets in $\mathbf{R}^n$. Then there exists a subanalytic stratification $\mathcal{T}$ of $\mathbf{R}^n$, compatible with the family $\mathcal{A}$, such that every stratum of $\mathcal{T}$ is a L-regular s-cell with a constant M, where M depends only on n.*

From theorem A and proposition 8 we get

Corollary B. *Let $\mathcal{A}$ be a locally finite family of subanalytic sets in $\mathbf{R}^n$. Then there exists a subanalytic stratifcation $\mathcal{T}$ of $\mathbf{R}^n$,compatible with the family $\mathcal{A}$, such that , if $x, y \in T$, where $T \in \mathcal{T}$, then there exists a smooth curve λ in T joining x and y, such that*

$$length\lambda \leq C | x - y |,$$

where C is a constant depending only on n.

We state also an immediate consequence of corollary B, which seems to be of its own interest.

Corollary C. *Let U be a subanalytic set, analytic submanifold of $\mathbf{R}^n$. Then there exist $M > 0$ and a subanalytic stratification $\mathcal{T}$ of $\mathbf{R}^n$, compatible with U such that, if $\varphi : U \longrightarrow \mathbf{R}$ is a differentiable function such that for some constant $C > 0$, $| d_x\varphi | < C$ for every $x \in U$, then*

$$| \varphi(x) - \varphi(y) | \leq CM | x - y |,$$

for every x and y belonging to the same stratum $T \in \mathcal{T}$, $T \subset U$.

Proof of theorem A. By (*) of remark 0 it is enough to prove the following:

(**) Let A be a subanalytic set, analytic submanifold of $\mathbf{R}^n$, then there exists F a subanalytic, closed, nowhere dense subset of A such that every connected component of $A \setminus F$ is a L-regular s-cell with constant M.

Constant M will be determined later on. We use induction on n to prove (**). We may add to $\mathcal{A}$ a family of cubes $[k_1, 1 + k_1] \times ... \times [k_n, 1 + k_n]$, $k_i \in \mathbf{Z}, k_i = 1, ..., n$. We call this new family $\mathcal{A}'$. Clearly every stratification compatible with $\mathcal{A}'$ will have only bounded strata. Hence it is enough to prove (**) for A bounded. Suppose now that

A is a bounded, subanalytic, analytic submanifold of $\mathbf{R}^n$, $\dim A = k < n$. Then by remark 5.1 for given $M > 0$ we have

$$A = F \cup \Gamma_1 \cup \ldots \cup \Gamma_s$$

where F is subanalytic in $\mathbf{R}^n$, closed in A, $\dim F < k$. Each Γ_i is a connected component of $A \setminus F$, moreover in suitable coordinate system every Γ_i is a graph of an analytic, subanalytic function $\varphi_i : B_i \longrightarrow \mathbf{R}$, where B_i is an analytic submanifold of $\mathbf{R}^{n-1}$ and $| d_x\varphi_i | \leq M$ for $x \in B_i$. Now applying induction hypothesis in $\mathbf{R}^{n-1}$ we may suppose that B_i is a s-cell L-regular with constant M. This ends the proof of (**) for A of dimension less than n.

Suppose now that A is subanalytic, open, bounded subset of $\mathbf{R}^n$. The following lemma is crucial for the proof of (**) in this case.

9.Lemma. *Let A be an open, bounded, subanalytic subset of $\mathbf{R}^n$, let $\varepsilon > 0$. Then there exists finitely many open and disjoint s-cells $A_i, i = 1, \ldots, p, A_i \subset A$ such that*
i) the set $A \setminus \bigcup_{i=1}^{p} A_i$ is closed and nowhere dense in A,
ii) for each $i = 1, \ldots, p$ there exists subanalytic subsets $B_1^i, \ldots, B_{k_i}^i$ such that
a) $k_i \leq 2n$
b) each B_j^i is an ε-flat analytic submanifold of $\mathbf{R}^n$, $\dim B_j^i = n-1$,
c) $\bigcup_{j=1}^{k_i} B_j^i$ is an open, dense subset of $\overline{A_i} \setminus A_i$.

Proof of the lemma 9. We prove the lemma by induction on $n = \dim A$. The case $n = 1$ is obvious. Let us assume the lemma holds true for open, bounded, subanalytic subsets of $\mathbf{R}^{n-1}$.We proceed by following steps:

Step 1. By Prop.5 there exists a subanalytic stratification $\mathcal{T}_1$ of $\mathbf{R}^n$, compatible with $\overline{A} \setminus A$, such that all strata of $\mathcal{T}_1$ are ε-flat.

Step 2. By Koopman-Brown theorem (see [Ło1],[Ło3]) we can find such coordinates in $\mathbf{R}^n$ that, if we denote by π a projection $\pi(x_1, .., x_{n-1}, x_n) = (x_1, .., x_{n-1}) = x'$, then π is finite on each stratum $T \in \mathcal{T}_1$, $T \subset \overline{A} \setminus A$.

By a theorem of Hardt (see [Ha1]) we can stratify the projection π restricted to $\overline{A}$, i.e. there exists a stratification $\mathcal{T}_2$ of $\mathbf{R}^n$ compatible with family $\{T \in \mathcal{T}_1 : T \in \overline{A}\}$, there exists a stratification $\mathcal{S}$ of $\mathbf{R}^{n-1}$, such that

i) each bounded stratum T of $\mathcal{T}_2$ is of one of the following form

α) $$T = \{(x', x_n) \in \mathbf{R}^n : \; x_n = h(x'), \; x' \in S\}$$

β) $$T = \{(x', x_n) \in \mathbf{R}^n : \; f(x') < x_n < g(x'), \; x' \in S\}$$

where S is a stratum of $\mathcal{S}$,$f(x') < g(x')$ for every $x' \in S$. Functions f, g and h are subanalytic and analytic in S.

ii) if T is stratum of $\mathcal{T}_2$ of the type α) and $T \subset \overline{A}$, $\dim T = n-1$, then $T \subset T_1$ for some $T_1 \in \mathcal{T}_1, \dim T_1 = n-1$ (this follows from the proof the theorem of Hardt). Hence T is also ε-flat.

iii) if stratum S satisfies condition (s) (i.e. every point of $\overline{S}$ has a basis of neighbourhoods $\{U_\nu\}$, such that $U_\nu \cap S$ is connected), then the function h (resp. f and g)

can be extended continously on $\overline{S}$ (see [Lo1],[Lo3]). Notice that every s-cell satisfies condition (s).

Step 3. For each open stratum $S \in \mathcal{S}, S \subset \pi(A)$ we apply induction hypothesis. Hence S is a union of open, disjoint s-cells $A'_1, ..., A'_q$ and a nowhere dense, closed subset of S. Every A'_i verifies conditions a), b), c) of *ii)* of lemma 9. Let us take a s-cell of the form

$$Q = \{(x', x_n) \in \mathbf{R}^n : f(x') < x_n < g(x'),\ x' \in A'_i\}$$

where f and g are those of Step 2. The boundary of Q is a union of the graphs of f and g (which are already ε-flat by Step 2) and the set

$$b(Q) = \{(x', x_n) \in \mathbf{R}^n :\ f(x') \leq x_n \leq g(x'),\ x' \in \overline{A'_i} \setminus A'_i\}$$

Let us denote by $B'^{\,i}_j$ the sets, contained in the boundary of A'_i, satisfying b) and c) of *ii)* in lemma 9. Then the sets (some of them perhaps empty)

$$B^i_j = \{(x', x_n) :\ f(x') < x_n < g(x'),\ x' \in B'^{\,i}_j\}$$

are ε-flat. Actually they are open in $B'^{\,i}_j \times \mathbf{R}$ which are ε-flat, since each $B'^{\,i}_j$ is ε-flat by induction hypothesis. Clearly the union of all B^i_j is dense in $b(Q) \setminus (\overline{f} \cup \overline{g})$ (we identify function with its graph). Hence making the induction step we get at most two sets more in *ii)* i.e. graphs of $f$and g. Thus our s-cell Q satisfies a), b), and c) of *ii)*. All s-cells constructed in the same way as Q i.e.

$$\{\pi^{-1}(A'_i) \cap T :\ i = 1, ..., q;\ T \in \mathcal{T}_2,\ T \subset A,\ dim T = n\}$$

gives our family of desired s-cells $A_1, ..., A_p$. This ends the proof of lemma 9.

Now we come back to the proof of (**) for A open and bounded in $\mathbf{R}^n$. Applying lemma 3 in $\mathbf{R}^n$ for $r = 2n$ we get corresponding positive constants ε and m. By lemma 9 we can suppose that A is a s-cell satisfying a), b) and c) of *ii)* with ε as above. Let us denote by $B_j, j = 1, ..., k,\ k \leq 2n$ open sets in the boundary of A, satisfying b) and c) of *ii)*. Let us take arbitrary points $x_j \in B_j, j = 1, ..., k$ and put $X_j = T_{x_j} B_j$. Since B_j is ε-flat, we have $\delta(X_j, T_{y_j} B_j) < \varepsilon$ for every $y_j \in B_j$. Hence by lemma 3 there exists a line P in $\mathbf{R}^n$ such that $\delta(P, T_{y_j} B_j) > m$. We take this line as $x_n - axis$, the orthogonal complement of P will be a $(x_1, ; ..., x_{n-1})$ -hyperplane denoted by $\mathbf{R}^{n-1}$. Locally each set B_j is a graph of an analytic mapping $\varphi : U \to \mathbf{R}$, where U is an open subset of $\mathbf{R}^{n-1}$. Moreover by Remark 2 there exists a constant $M > 0$ (depending only on m) such that $\mid d_{x'}\varphi \mid \leq M$ for each $x' \in U$. Since in the claim of the theorem of Koopman-Brown the set of admissible projections is dense , so we can assume that canonical projection $\pi : \mathbf{R}^n \to \mathbf{R}^{n-1}$ is finite on the boundary of A. Now we repeat the construction of Step 2 of the proof of lemma 9. Hence our A is a disjoint union of a nowhere dense, closed set F and sets of the form $\{(x', x_n) \in \mathbf{R}^n : f(x') < x_n < g(x'), x' \in A'\}$ where A' is an open and subanalytic subset of $\mathbf{R}^{n-1}$,f and g are analytic functions in A' such that $\mid d_{x'} f \mid \leq M,\ \mid d_{x'} g \mid \leq M$. By induction hypothesis we can assume that A' is a L-regular s-cell with constant M. This ends the proof of (**) for A open and bounded. Hence theorem A is proved.

Remark. Actually in the above step the th. of Koopman-Brown is superfluous, since the fact that π is finite on the boundary of A follows from Cor. 1.8 in [KR].

10.Remark To obtain a semialgebraic version of theorem A we suppose that $\mathcal{A}$ is a finite family of semialgebraic subsets of $\mathbf{R}^n$. The stratification we obtain will be also finite and semialgebraic. However we should change the definition of s-cell. We simply assume in definitions 6 and 7, case β), that f or g may be identically equal to ∞ or to $-\infty$. Clearly we can assume $d_x f = 0$ (resp. $d_x g = 0$) in that case. Notice that proposition 5 and the theorem of Hardt have semialgebraic versions, see respectively the proof in [Ło2] and chap.9 in [BCR]. Hence our proof holds in semialgebraic case.

References

[BCR] J. Bochnak, M. Coste, M-F. Roy; Géometrie algébrique réelle, (Ergeb. der Math., Folge 3, Bd.12), Springer, 1987.

[DS] Z. Denkowska, J. Stasica; Sur la stratifiction sous-analytique, Bull. Acad. Pol. Sci. Sér. Math., 30 (1982), 337-340.

[DW] Z. Denkowska, K. Wachta; Sur la sous-analycité de l'application tengente, Bull. Acad. Pol. Sci. Sér. Math., 30 (1982), 329-331.

[Ha1] R. Hardt; Stratification of real analytic mappings and images, Invent. Math. 28 (1975), 193-208.

[Ha2] R. Hardt; Triangulation of subanalytic sets and proper light subanalytic maps, Invent. Math. 38 (1977), 207-2170.

[KR] K. Kurdyka, G. Raby; Densité des ensembles sous-analytiques, Ann. Inst. Fourier, 39 (1989), 753-771.

[Ło1] S.Łojasiewicz; Triangulation of semi-analytic sets, Annali Scuola Norm. Sup., Sér.3, 8 (1964), 449-474.

[Ło2] S.Łojasiewicz; Sur la semi-analycité de l'application tangente, Bull. Acad. Pol. Sci. Sér. Math. 27 (1979) 525-527.

[Ło3] S.Łojasiewicz; Stratifications et triangulations sous-analytiques, Seminari di Geometria (Bologna), (1986) 83-97.

[Ło4] S.Łojasiewicz; Semianalytic and subanalytic geometry, book in prepartion.

[Pa1] A. Parusiński; Lipschitz properties of semianalytic sets, Ann. Inst. Fourier, 38 (1988) 189-213.

[Pa2] A. Parusiński; Regular projections for subanalytic sets, C. R. Acad. Sci. Paris, 307 Série I (1988) 343-347.

[St] J. Stasica; The Whitney condition for subanalytic sets, Zeszyty Naukowe Uniw. Jag. 23 (1982), 211-221.

UNE BORNE SUR LES DEGRES POUR LE THEOREME DES ZEROS REEL EFFECTIF

Henri LOMBARDI
Laboratoire de Mathématiques. UFR des Sciences et Techniques
Université de Franche-Comté. 25030 Besançon cédex France

Résumé Nous donnons les idées et résultats essentiels d'un calcul d'une majoration des degrés pour le théorème des zéros réels effectif.

Abstract We give the main ideas and results concerning a computation of a degree majoration for the real nullstellensatz.

1) Introduction

Nous rendons compte dans cet article du calcul d'une borne sur les degrés accompagnant la preuve constructive du théorème des zéros réel et de ses variantes (cf. [Lom d]). Les preuves sans les majorations de degré peuvent être trouvées dans [Lom b] .

Les résultats obtenus

Une formulation générale du théorème des zéros réel et de ses variantes peut être la suivante (cf [BCR] théorème 4.4.2) : on considère un système d'égalités et inégalités portant sur des polynomes de $K[X] = K[X_1,X_2,...,X_n]$, où K est un corps ordonné de clôture réelle R ; ce système définit une partie S semi-algébrique de R^n ; le théorème affirme que S est vide (fait géométrique) si et seulement si il y a une certaine identité algébrique construite à partir des polynomes donnés, identité qui donne une preuve de ce fait géométrique. Calculer une borne sur les degrés pour le théorème des zéros réels consiste à calculer une majoration sur les degrés des polynomes intervenant dans le résultat (l'identité algébrique construite) à partir de la taille de l'entrée (le système de conditions de signes portant sur la liste de polynomes donnée au départ). Les paramètres qui controlent la majoration des degrés dans le résultat sont en fait : le nombre k de polynomes dans l'entrée, le degré d des polynomes dans l'entrée, et le nombre n de variables.
Le calcul de majoration est obtenu en suivant pas à pas la preuve constructive d'existence de l'identité algébrique et en explicitant les majorations à chaque étape de la preuve.
C'est une majoration primitive récursive, donnée par une tour d'exponentielles : le nombre d'étages dans la tour est n+4 et en haut de la tour on trouve :

d.log(d) + loglog(k) + cte .

Ce résultat n'est pas trop mauvais, dans la mesure où la principale responsabilité de l'explosion est supportée par l'algorithme de Hörmander, à la base de la preuve effective. On peut espérer baser une autre preuve effective sur des algorithmes plus performants et néanmoins de conception très simple, et obtenir en conséquence une majoration où le paramètre n interviendrait de manière moins catastrophique, sans tour d'exponentielles. Il semble néanmoins improbable d'obtenir d'aussi bonnes bornes que dans les meilleures versions effectives du théorème des zéros de Hilbert (cf. [He] ,[FG] et [Ko]).

La preuve constructive du théorème des zéros réels

De manière générale un «théorème des zéros» affirme que certains faits «géométriques» ont une preuve purement «algébrique».
Un exemple simple est fourni par la formule de Taylor. Par exemple pour un polynome de degré ≤ 4 , on a l'identité algébrique : (avec $\Delta = U - V$)

$$P(U) = P(V) + \Delta.P'(V) + (1/2).\Delta^2.P''(V) + (1/6).\Delta^3.P^{(3)}(V) + (1/24).\Delta^4.P^{(4)}$$

Cette identité algébrique rend manifeste le fait géométrique suivant : si en un point v le polynome P a toutes ses dérivées positives, alors pour tout $u > v$ on a $P(u) > P(v)$. Ce fait géométrique, qui peut être rendu manifeste par un tableau de variation, est également clair par la formule de Taylor.
C'est un cas particulier du lemme de Thom, qui affirme (entre autres) que l'ensemble des points où un polynome et ses dérivées successives ont chacun un signe fixé, est convexe. La construction du nullstellensatz réel utilise une version "identité algébrique" de ce fait, donnée par ce que nous appelons les formules de Taylor mixtes et les formules de Taylor généralisées.

L'idée générale de notre preuve constructive est la suivante. Pour un corps ordonné **K** il y a un algorithme de conception très simple pour tester si un système de csg (conditions de signes généralisées) portant sur ces polynomes en plusieurs variables est possible ou impossible dans la clôture réelle de **K** . C'est l'algorithme de Hörmander (cf. la preuve du principe de Tarski-Seidenberg dans [**BCR**] chap. 1), appliqué de manière itérative pour diminuer par étapes le nombre de variables sur lesquelles portent les csg. Si on regarde les arguments sur lesquels est basée la preuve d'impossibilité (en cas d'impossibilité), on voit qu'il y a essentiellement des identités algébriques (traduisant la division euclidienne), le théorème des accroissements finis et l'existence d'une racine pour un polynome sur un intervalle où il change de signe.
Les ...-stellensatz réels effectifs doivent donc pouvoir être obtenus si on arrive à "algébriser" les arguments de base de la preuve d'incompatibilité et les méthodes de déduction impliquées.
Un pas important a déjà été réalisé avec la version algébrique du théorème des accroissements finis pour les polynomes (cf. [**LR**]), qui a été à l'origine des formules de Taylor mixtes et généralisées.
Un autre pas a consisté à traduire sous forme de *constructions d'identités algébriques* certains raisonnements élémentaires (du genre si $A \Rightarrow B$ et $B \Rightarrow C$ alors $A \Rightarrow C$).
Il fallait en outre trouver une version "identité algébrique" des axiomes d'existence dans la théorie des corps réels clos. C'est ce qui est fait à travers la notion d'*existence potentielle*.

Remarques sur l'article présent

Nous intoduisons dans cet article une problématique où le role central est tenu par les constructions d'identités algébriques (appelées «incompatibilités fortes») à partir d'autres identités algébriques, alors que dans les versions précédentes c'étaient plutôt les identités algébriques elles-mêmes qui jouaient le role central. Ce changement de point de vue a été motivé par le calcul de majoration lui-même. Ce qui, dans **[Lom d]**, apparaissait sous l'appellation peu plaisante d'«implication forte vue comme existence potentielle», s'appelle désormais «implication dynamique». Quant aux anciennes implications fortes, elles ne jouent pratiquement plus aucun role. Nous donnons dans ce nouveau cadre un traitement unifié pour les preuves cas par cas («disjonction dynamique»), l'implication («implication dynamique») et l'existence («existence potentielle»).

Significations de la preuve constructive pour différentes écoles philosophiques

Bien que nous nous placions a priori dans un cadre constructif "à la Bishop", tel que développé dans [MRR] pour ce qui concerne la théorie des corps discrets, comme nous ne précisons pas le sens du mot effectif ni celui du mot décidable, toutes les preuves peuvent être lues avec des lunettes adaptées à la philosophie ou au cadre de travail de chaque lecteur particulier.
Si on adopte un point de vue "classique" par exemple, les procédures effectives réclamées dans la structure du corps des coefficients par le mathématicien constructif peuvent être considérées comme données par des oracles. En conséquence, les preuves fournissent une preuve dans le cadre classique, *et sans recours à l'axiome du choix*, du théorème des zéros réels dans un corps ordonné arbitraire. En fait les preuves données fournissent des algorithmes uniformément primitifs récursifs, "uniformément" s'entendant par rapport à un oracle qui donne la structure du corps des coefficients du système de csg considéré :

si $(c_i)_{i=1,\ldots,m}$ est la famille des coefficients et si $P \in \mathbb{Z}[(C_i)_{i=1,\ldots,m}]$ l'oracle répond à la question « Quel est le signe de $P((c_i)_{i=1,\ldots,m})$? ».

2) Incompatibilités fortes

Incompatibilités fortes : définitions et notations

Nous considérons un corps ordonné K, et une liste de variables $X_1, X_2, \ldots, X_n$ désignée par $\mathbf{X}$.
Nous notons donc $K[\mathbf{X}]$ l'anneau des polynomes $K[X_1,X_2,\ldots,X_n]$.
Etant donnée une partie finie F de $K[\mathbf{X}]$:

nous notons F^{*2} l'ensemble des carrés d'éléments de F.

le *monoïde multiplicatif engendré* par F est l'ensemble des produits d'éléments de $F \cup \{1\}$, nous le noterons $\mathcal{M}(F)$.

le *cône positif engendré* par F est l'ensemble des sommes d'éléments du type $p.P.Q^2$ où p est positif dans K, P est dans $\mathcal{M}(F)$, Q est dans $K[\mathbf{X}]$. Nous le noterons $\mathcal{C}p(F)$.

enfin nous noterons $\mathcal{I}(F)$ l'idéal engendré par F.

Définition et notation 1 : Etant donnés 4 parties finies de $\mathbf{K}[X]$: $F_>$, $F_\geq$, $F_=$, $F_\neq$, contenant des polynomes auxquels on souhaite imposer respectivement les conditions de signes >0 , $\geqslant 0$, $=0$, $\neq 0$, on dira que $\mathbf{F} = [F_> ; F_\geq ; F_= ; F_\neq]$ est *fortement incompatible* dans $\mathbf{K}$ si on a une égalité dans $\mathbf{K}[X]$ du type suivant :

$S + P + Z = 0$ avec $S \in \mathcal{M}(F_> \cup F_\neq^{*2})$, $P \in \mathcal{Cp}(F_\geq \cup F_>)$, $Z \in \mathcal{I}(F_=)$

Nous utiliserons la notation suivante pour une incompatibilité forte:

$\downarrow [S_1 > 0, \ldots, S_i > 0, P_1 \geqslant 0, \ldots, P_j \geqslant 0, Z_1 = 0, \ldots, Z_k = 0, N_1 \neq 0, \ldots, N_h \neq 0] \downarrow$

Il est clair qu'une incompatibilité forte est une forme très forte d'incompatibilité. En particulier, elle implique l'impossibilité d'attribuer les signes indiqués aux polynomes souhaités, dans *n'importe quelle* extension ordonnée de $\mathbf{K}$.
Si on considère la clôture réelle $\mathbf{R}$ de $\mathbf{K}$, l'impossibilité ci-dessus est testable par l'algorithme de Hörmander, par exemple.

Le théorème des zéros réels et ses variantes

Les différentes variantes du théorème des zéros dans le cas réel sont conséquence du théorème général suivant :

Théorème : Soit $\mathbf{K}$ un corps ordonné et $\mathbf{R}$ une extension réelle close de $\mathbf{K}$. Les trois faits suivants, concernant un système de csg portant sur des polynomes de $\mathbf{K}[X]$, sont équivalents :

l'incompatibilité forte dans $\mathbf{K}$

l'impossibilité dans $\mathbf{R}$

l'impossibilité dans toutes les extensions ordonnées de $\mathbf{K}$

Ce théorème des zéros réels remonte à 1974 ([**Ste**]). Des variantes plus faibles ont été établies par Krivine ([**Kri**]), Dubois ([**Du**]), Risler ([**Ris**]), Efroymson ([**Efr**]). Toutes les preuves jusqu'à ([**Lom a**]) utilisaient l'axiome du choix.

Degré d'une incompatibilité forte

Si nous voulons préciser les majorations de degré fournis par notre preuve du théorème des zéros réel, nous devons préciser la terminologie.
Nous manipulons des incompatibilités fortes écrites *sous forme paire,* c.-à-d.:

$$S + P + Z = 0 \quad \text{avec} \quad S \in \mathcal{M}(F_>^{*2} \cup F_\neq^{*2}) \text{ , } P \in \mathcal{Cp}(F_\geq \cup F_>) \text{ , } Z \in \mathcal{I}(F_=)$$

(la considération des formes paires d'implications fortes a pour unique utilité de faciliter un peu le calcul de majoration des degrés).
Quand nous parlons de degré, sauf précision contraire, il s'agit du degré total maximum.
Le degré d'une incompatibilité forte est par convention au moins égal à 1, c'est le degré maximum des polynomes qui «composent» l'incompatibilité forte.
Par exemple, si nous avons une incompatibilité forte :

$$\downarrow [A > 0, B > 0, C \geqslant 0, D \geqslant 0, E = 0, F = 0] \downarrow$$

explicitée sous forme d'une identité algébrique :

$$A^2.B^6 + C.\sum_{i=1}^{h} p_i.P_i^2 \;+\; A.B.D.\sum_{j=1}^{k} q_j.Q_j^2 \;+\; E.U \;+\; F.V = 0$$

le degré de l'incompatibilité forte est :

$\sup \{\, d(A^2.B^6)\,,\, d(C.P_i^2)\ (i = 1,\ldots,h)\,,\, d(A.B.D.Q_j^2)\ (j = 1,\ldots,k)\,,\, d(E.U)\,,\, d(F.V)\, \}$.

Le calcul de majoration

Nous allons expliquer dans cet article comment peut être mené un calcul de majorations primitives récursives pour le théorème des zéros réels. Les détails des calculs sont dans [Lom d].

Les données sont trois entiers d, n, k qui majorent, dans un système de csg incompatible $\mathbb{H}$, respectivement les degrés des polynomes, le nombre des variables et le nombre de csg.

Le calcul doit aboutir à 3 fonctions primitives récursives explicites $\delta(d,n,k)$, $\sigma(d,n,k)$ et $\psi(d,n,k)$ qui donnent des majorants pour, dans une incompatibilité forte $\downarrow \mathbb{H} \downarrow$, respectivement le degré maximum, le nombre de termes dans la somme, et le nombre d'opérations arithmétiques dans $\mathbf{K}$ nécessaires pour calculer les coefficients dans l'incompatibilité forte à partir des coefficients donnés au départ.

En fait, chacun des théorèmes ou propositions qui conduit à la preuve constructive du théorème des zéros réel peut être accompagné d'une majoration primitive récursive du même type. Ces majorations s'enchainent les unes les autres, sans difficulté majeure.

Comme le calcul est très fastidieux, nous nous en sommes tenus aux majorations de degrés, laissant au lecteur courageux les deux autres majorations.

On notera que l'usage de l'algorithme de Hörmander 'sans raccourci', à la base de notre méthode, rend a priori les majorations obtenues sans intérêt pratique.

Constructions d'incompatibilités fortes

Définition 2 : Nous parlerons de construction d'une incompatibilité forte à partir d'autres incompatibilités fortes, lorsque nous avons un algorithme qui permet de construire la première à partir des autres.

Il s'agit donc d'une implication logique, au sens constructif, liant des incompatibilités fortes.

Notation 3 : Nous noterons cette implication logique (au sens constructif) par un signe de déduction "constructif". La notation

$$(\downarrow \mathbb{H}_1 \downarrow \text{ et } \downarrow \mathbb{H}_2 \downarrow) \;\vdash_{\text{cons}}\; \downarrow \mathbb{H}_3 \downarrow$$

signifie donc qu'on a un algorithme de construction d'une incompatibilité forte de type $\mathbb{H}_3$ à partir d'incompatibilités fortes de types $\mathbb{H}_1$ et $\mathbb{H}_2$

Cela n'a d'intérêt que lorsque les incompatibilités fortes désignées en hypothèse et en conclusion comportent des éléments variables.

Un exemple fondamental aidera à mieux comprendre.

Le raisonnement par séparation des cas (selon le signe d'un polynome)

Nous donnons ici un énoncé détaillé des «raisonnements cas par cas», incluant la propagation des majorations de degrés.

Proposition 4 : Soit H un système de csg portant sur des polynomes de K[X] , Q un élément de K[X] , alors:

$$[\downarrow(H, Q<0)\downarrow \text{ et } \downarrow(H, Q>0)\downarrow] \vdash_{cons} \downarrow(H, Q\neq 0)\downarrow \qquad (a)$$

$$[\downarrow(H, Q\leq 0)\downarrow \text{ et } \downarrow(H, Q\geq 0)\downarrow] \vdash_{cons} \downarrow H\downarrow \qquad (a')$$

De même :

$$[\downarrow(H, Q>0)\downarrow \text{ et } \downarrow(H, Q=0)\downarrow] \vdash_{cons} \downarrow(H, Q\geq 0)\downarrow \qquad (b)$$

$$[\downarrow(H, Q\neq 0)\downarrow \text{ et } \downarrow(H, Q=0)\downarrow] \vdash_{cons} \downarrow H\downarrow \qquad (c)$$

$$[\downarrow(H, Q>0)\downarrow \text{ et } \downarrow(H, Q\leq 0)\downarrow] \vdash_{cons} \downarrow H\downarrow \qquad (d)$$

Dans chacun de ces cas, notons d_1 et d_2 les degrés des deux incompatibilités fortes données dans l'hypothèse, le degré de l'incompatibilité forte construite est respectivement majoré par :

$$\varphi_a(d_1,d_2) = \varphi_{a'}(d_1,d_2) = d_1 + d_2$$

$$\varphi_b(d_1,d_2) = d_1.d_2$$

$$\varphi_c(d_1,d_2) = d_1.d_2$$

$$\varphi_d(d_1,d_2) = d_1.d_2 + d_2$$

Ces 4 fonctions sont majorées par $\varphi(d_1,d_2) = d_1.d_2 + d_1 + d_2$

Enfin, pour démontrer que H est fortement incompatible, on peut raisonner en séparant selon les 3 cas $Q>0$, $Q<0$, $Q=0$, et en construisant une incompatibilité forte dans chaque cas. Ce qui renvient à affirmer :

$$[\downarrow(H, Q>0)\downarrow \text{ et } \downarrow(H, Q<0) \text{ et } \downarrow(H, Q=0)\downarrow] \vdash_{cons} \downarrow H\downarrow \qquad (e)$$

Notons d_1 , d_2 et d_3 les degrés des trois incompatibilités fortes données dans l'hypothèse, le degré de l'incompatibilité forte construite est majoré par :

$$\varphi_e(d_1,d_2,d_3) = d_1.d_2 + d_1.d_3 + d_2.d_3$$

3) Versions algébriques dynamiques de l'implication et de la disjonction

La version algébrique dynamique de l'implication

Définition et notation 5 :

Soient H_1 et H_2 deux systèmes de csg portant sur des polynomes de K[X] . Nous dirons que *le système* H_1 *implique dynamiquement* H_2 lorsque, pour tout système de csg H portant sur des polynomes de K[X,Y] , on a la construction d'incompatibilité forte :

$$\downarrow [H_2(X), H(X,Y)]\downarrow \vdash_{cons} \downarrow [H_1(X), H(X,Y)]\downarrow$$

Nous noterons cette implication dynamique par :

$$^\bullet(\ \mathbb{H}_1(X) \Rightarrow \mathbb{H}_2(X)\ )^\bullet.$$

Lorsque le système $\mathbb{H}_1$ est vide, nous utilisons la notation $^\bullet(\ \mathbb{H}_2(X)\ )^\bullet$.

Remarques :

1) On a trivialement l'équivalence des affirmations :

$$\downarrow \mathbb{H}_1 \downarrow \quad \text{et} \quad ^\bullet(\ \mathbb{H}_1 \Rightarrow (1=0)\ )^\bullet$$

2) La vision dynamique de l'implication correspond, dans les références **[Lom x]** à l'«implication forte vue comme existence potentielle». En tant qu'implication forte «statique», c'était une liste d'identités algébriques. En tant qu'implication dynamique, cela devient un algorithme de manipulations d'identités algébriques. Dans la mise en oeuvre concrète d'algorithmes de construction du théorème des zéros réel, la vision dynamique est en fait beaucoup plus fructueuse que la vision statique. Certaines subtilités s'introduisent, comme le fait que deux implications qui ont la même signification peuvent avoir des dynamiques distinctes (c.-à-d. qu'elles se traduisent pas des algorithmes de manipulations d'identités algébriques distincts), et ont alors des coûts (en termes de temps de calcul) différents (pour plus de détails voir le paragraphe "Variations sur le thème des implications dynamiques").

Il apparaît en fin de compte que les "bonnes notions" sont celles d'incompatibilité forte et d'implication dynamique, tandis que la notion d'implication forte serait plutôt un incident de parcours. Nous verrons un peu plus loin que les raisonnements cas par cas peuvent être interprétés par une autre "bonne notion", la version dynamique du «ou».

La notation que nous utilisons ici est légèrement distincte de celle utilisée dans les précédentes références [Lom] pour noter les existences potentielles. Ceci nous permet de mieux insister sur la différence de signification entre une implication forte «statique» et une implication dynamique.

Fonction-degré d'une implication dynamique

Une implication dynamique $^\bullet(\ \mathbb{H}_1 \Rightarrow \mathbb{H}_2\ )^\bullet$ signifie par définition un algorithme fournissant la construction :

$$\downarrow [\ \mathbb{H}_2\ ,\ \mathbb{H}\ ] \downarrow \ \vdash_{\text{cons}} \ \downarrow [\ \mathbb{H}_1\ ,\ \mathbb{H}\ ] \downarrow$$

Chaque fois que nous établissons une implication dynamique particulière, nous devons établir des 'majorations primitives récursives de degré' pour cette construction d'incompatibilités fortes : le degré de l'incompatibilité forte construite est majoré par une fonction $\Delta(d,..;k,...)$ où d est le degré de l'incompatibilité forte initiale, k le nombre de csg dans $\mathbb{H}_2$ etc.... (le point-virgule isole les 'variables', qui dépendent de l'incompatibilité forte initiale, des 'paramètres', qui ne dépendent que de $\mathbb{H}_1$ et $\mathbb{H}_2$).

Nous disons qu'il s'agit d'une fonction-degré acceptable pour l'implication dynamique considérée, ou encore, (par abus) nous parlons de *la* fonction-degré attachée à l'implication dynamique.

Renforcement simultané de l'hypothèse et de la conclusion dans les implications dynamiques

Soient $\mathbb{H}_1$, $\mathbb{H}_2$, $\mathbb{H}_3$ des systèmes de csg portant sur des polynomes de $\mathbf{K[X]}$.

Si on a l'implication dynamique

$$^{\bullet}(\mathbb{H}_1 \Rightarrow \mathbb{H}_2)^{\bullet}$$

on a également l'implication dynamique

$$^{\bullet}([\mathbb{H}_1, \mathbb{H}_3] \Rightarrow [\mathbb{H}_2, \mathbb{H}_3])^{\bullet}$$

et cette dernière accepte la même fonction degré que la première (simple constatation).

La transitivité des implications dynamiques

La proposition suivante est immédiate : il suffit d'enchainer les deux algorithmes de constructions d'incompatibilités fortes.

Proposition 6 : Soient $\mathbb{H}_1$, $\mathbb{H}_2$, $\mathbb{H}_3$ trois systèmes de csg portant sur des polynomes de $\mathbf{K}[X]$. Alors:

$[\ ^{\bullet}(\mathbb{H}_1 \Rightarrow \mathbb{H}_2)^{\bullet}$ et $^{\bullet}([\mathbb{H}_1, \mathbb{H}_2] \Rightarrow \mathbb{H}_3)^{\bullet}\ ]$ impliquent $^{\bullet}(\mathbb{H}_1 \Rightarrow \mathbb{H}_3)^{\bullet}$

Supposons que la première implication dynamique admette comme fonction-degré acceptable $\Delta^1(d;p)$ où d est le degré de $\downarrow [\mathbb{H}_2, \mathbb{H}] \downarrow$ et $\mathbf{p}$ représente certains paramètres dépendant de $\mathbb{H}_1$ et $\mathbb{H}_2$, supposons de même une fonction-degré acceptable $\Delta^2(d;q)$ pour la deuxième implication dynamique, alors une fonction-degré pour l'implication dynamique construite est obtenue en composant les deux fonctions-degré précédentes :

$$\Delta(d;p,q) = \Delta^1(\Delta^2(d;q);p)$$

La proposition qui suit est un corollaire immédiat de la précédente.

Proposition 7 : Soient $\mathbb{H}_1$, $\mathbb{K}_1$, $\mathbb{K}_2$, ..., $\mathbb{K}_n$ des systèmes de csg portant sur des polynomes de $\mathbf{K}[X]$. Alors :

$$[\ ^{\bullet}(\mathbb{H}_1 \Rightarrow \mathbb{K}_1)^{\bullet}, ^{\bullet}(\mathbb{H}_1 \Rightarrow \mathbb{K}_2)^{\bullet}, ..., ^{\bullet}(\mathbb{H}_1 \Rightarrow \mathbb{K}_n)^{\bullet}\ ] \quad \vdash_{\text{cons}} \quad ^{\bullet}(\mathbb{H}_1 \Rightarrow [\mathbb{K}_1, \mathbb{K}_2, ..., \mathbb{K}_n])^{\bullet}$$

En outre, une fonction-degré pour l'implication dynamique construite est obtenue en composant (dans un ordre arbitraire) les fonctions-degré des implications dynamiques de l'hypothèse

Cas des implications dynamiques avec une seule condition de signe dans la conclusion

Combinée avec la proposition 7, la proposition qui suit permet de montrer l'équivalence d'une implication dynamique avec la donnée d'une liste d'incompatibilités fortes. Cette donnée était appelée une implication forte dans les articles **[Lom x]**.

Proposition 8 :

Soient $\mathbb{H}_1$ un système de csg portant sur des polynomes de $\mathbf{K}[X]$, Q un élément de $\mathbf{K}[X]$, σ un élément de $\{>, <, =, \geq, \leq, \neq\}$ et σ' l'élément opposé, alors :

$$\downarrow (\mathbb{H}_1, Q\,\sigma\,0) \downarrow \quad \text{si et seulement si} \quad ^{\bullet}(\mathbb{H}_1 \Rightarrow Q\,\sigma'\,0)^{\bullet}$$

Si d_1 est le degré d'une incompatibilité forte $\downarrow(\mathbb{H}_1, Q\,\sigma\,0)\downarrow$, alors une fonction-degré acceptable pour l'implication dynamique est doonée par

$$(d;d_1) \longmapsto \varphi(d,d_1) = d.d_1 + d + d_1 . \text{ (cf. prop. 4)}$$

Inversement si d_Q est le degré du polynome Q et si Δ^1 est une fonction-degré acceptable pour l'implication dynamique $^\bullet(\ \mathbb{H}_1 \Rightarrow Q\,\sigma'\,0\ )^\bullet$, le degré de l'incompatibilité forte peut être majoré par $\Delta^1(2.d_Q)$.

preuve> Dans le sens direct : soit $\mathbb{H}$ un système de csg et une incompatibilité forte $\downarrow(\mathbb{H}, Q\,\sigma'\,0)\downarrow$ de degré d, on peut construire l'incompatibilité forte $\downarrow(\mathbb{H}_1, \mathbb{H})\downarrow$ en raisonnant cas par cas. Dans le cas $Q\,\sigma\,0$ on utilise l'incompatibilité forte $\downarrow(\mathbb{H}_1, Q\,\sigma\,0)\downarrow$ de degré d_1 et dans le cas $Q\,\sigma'\,0$ on utilise l'incompatibilité forte $\downarrow \mathbb{H}, Q\,\sigma'\,0\downarrow$, on conclut en utilisant la proposition 4.
Réciproque : on a une incompatibilité forte sous forme paire de degré $2.d_Q$:

$$\downarrow(Q\,\sigma\,0, Q\,\sigma'\,0)\downarrow,$$

obtenue en lisant convenablement l'identité

$$Q^2 + Q.(-Q) = 0,$$

on applique alors la définition de l'implication dynamique en prenant pour $\mathbb{H}$ la seule condition $Q\,\sigma\,0$. □

La version algébrique dynamique de la disjonction

Définition et notation 9 :

Soient $\mathbb{H}_1, \mathbb{H}_2, \ldots, \mathbb{H}_k$ et $\mathbb{K}_1, \mathbb{K}_2, \ldots, \mathbb{K}_m$ des systèmes de csg portant sur des polynomes de $\mathbf{K}[X]$.

Nous disons que *le système* $\mathbb{H}_1$ *implique dynamiquement la disjonction* $\mathbb{K}_1 \vee \mathbb{K}_2 \vee \ldots \vee \mathbb{K}_m$ lorsque, pour tout système de csg $\mathbb{H}$ portant sur des polynomes de $\mathbf{K}[X,Y]$, on a la construction d'incompatibilité forte :

$$\{\downarrow[\ \mathbb{K}_1(X), \mathbb{H}(X,Y)\ ]\downarrow \text{ et} \ldots \text{et } \downarrow[\ \mathbb{K}_m(X), \mathbb{H}(X,Y)\ ]\downarrow\} \ \vdash_{\text{cons}} \ \downarrow[\ \mathbb{H}_1(X), \mathbb{H}(X,Y)\ ]\downarrow$$

Nous noterons cette implication-disjonction dynamique par :

$$^\bullet(\ \mathbb{H}_1(X) \Rightarrow [\ \mathbb{K}_1(X) \vee \mathbb{K}_2(X) \vee \ldots \vee \mathbb{K}_m(X)\ ]\ )^\bullet.$$

Lorsque le système $\mathbb{H}_1$ est vide, nous utilisons la notation

$$^\bullet(\ \mathbb{K}_1(X) \vee \mathbb{K}_2(X) \vee \ldots \vee \mathbb{K}_m(X)\ )^\bullet.$$

Enfin, la notation :

$$^\bullet(\ [\ \mathbb{H}_1 \vee \mathbb{H}_2 \vee \ldots \vee \mathbb{H}_k\ ] \Rightarrow [\ \mathbb{K}_1 \vee \mathbb{K}_2 \vee \ldots \vee \mathbb{K}_m\ ]\ )^\bullet$$

signifie que chacune des implications-disjonctions dynamiques

$$^\bullet(\ \mathbb{H}_i(X) \Rightarrow [\ \mathbb{K}_1(X) \vee \mathbb{K}_2(X) \vee \ldots \vee \mathbb{K}_m(X)\ ]\ )^\bullet \quad (i = 1, \ldots, k)$$

est vérifiée

Remarque : Toute formule sans quantificateur de la théorie du premier ordre des anneaux totalement ordonnés dicrets à paramètres dans $\mathbf{K}$ est équivalente à une formule en forme normale disjonctive et donc à une formule du type

$$\mathbb{K}_1(X) \vee \mathbb{K}_2(X) \vee \ldots \vee \mathbb{K}_m(X)$$

où les $\mathbb{K}_i(X)$ sont des systèmes de csg portant sur des polynomes de $\mathbf{K}[X]$.
Les implications-disjonctions dynamiques consituent une forme de raisonnement purement «identité algébrique» concernant les formules sans quantifi-

cateur, où la logique a été évacuée au profit d'algorithmes de constructions d'identités algébriques.

Fonction-degré d'une implication-disjonction dynamique

Une implication-disjonction dynamique

$$^{\bullet}(\ \mathbb{H}_1(X) \Rightarrow [\,\mathbb{K}_1(X) \vee \mathbb{K}_2(X) \vee \ldots \vee \mathbb{K}_m(X)\,]\ )^{\bullet}$$

signifie par définition un algorithme fournissant la construction :

$$\{\downarrow[\,\mathbb{K}_1(X)\,,\ \mathbb{H}(X,Y)\,]\downarrow \text{ et} \ldots \text{et } \downarrow[\,\mathbb{K}_m(X)\,,\ \mathbb{H}(X,Y)\,]\downarrow\,\} \ \vdash_{\text{cons}}\ \downarrow[\,\mathbb{H}_1(X)\,,\ \mathbb{H}(X,Y)\,]\downarrow$$

Chaque fois que nous établissons une implication-disjonction dynamique particulière, nous devons établir des 'majorations primitives récursives de degré' pour cette construction d'incompatibilités fortes : le degré de l'incompatibilité forte construite est majoré par une fonction $\Delta(d_1,..,d_m)$ où d_i est le degré de l'incompatibilité forte initiale n°i.

Nous disons qu'il s'agit d'une fonction-degré acceptable pour l'implication-disjonction dynamique considérée.

Exemples : La proposition 4 peut être relue comme affirmant des disjonctions ou implications-disjonctions dynamiques :

Proposition 4 bis : On a les implications-disjonctions dynamiques suivantes :

$^{\bullet}(\ Q \neq 0 \Rightarrow [\ Q > 0 \ \vee\ Q < 0\ ]\ )^{\bullet}$ (a)

$^{\bullet}(\ Q \leqslant 0 \ \vee\ Q \geqslant 0\ )^{\bullet}$ (a')

$^{\bullet}(\ Q \geqslant 0 \Rightarrow [\ Q > 0 \ \vee\ Q = 0\ ]\ )^{\bullet}$ (b)

$^{\bullet}(\ Q \neq 0 \ \vee\ Q = 0\ )^{\bullet}$ (c)

$^{\bullet}(\ Q > 0 \ \vee\ Q \leqslant 0)^{\bullet}$ (d)

$^{\bullet}(\ Q = 0 \ \vee\ Q > 0 \ \vee\ Q < 0\ )^{\bullet}$ (e)

Les fonctions-degré sont données par les fonctions φ_x décrites à la proposition 4

La transitivité des implications-disjonctions dynamiques

L'énoncé le plus général est le suivant :

les implications-disjonctions dynamiques

$$^{\bullet}(\ [\ \mathbb{H}_1 \vee \mathbb{H}_2 \vee \ldots \vee \mathbb{H}_k\,] \Rightarrow [\,\mathbb{K}_1 \vee \mathbb{K}_2 \vee \ldots \vee \mathbb{K}_m\,]\ )^{\bullet}$$

et

$$^{\bullet}(\ [\ \mathbb{K}_1 \vee \mathbb{K}_2 \vee \ldots \vee \mathbb{K}_m\,] \Rightarrow [\,\mathbb{L}_1 \vee \mathbb{L}_2 \vee \ldots \vee \mathbb{L}_n\,]\ )^{\bullet}$$

impliquent :

$$^{\bullet}(\ [\ \mathbb{H}_1 \vee \mathbb{H}_2 \vee \ldots \vee \mathbb{H}_k\,] \Rightarrow [\,\mathbb{L}_1 \vee \mathbb{L}_2 \vee \ldots \vee \mathbb{L}_n\,]\ )^{\bullet}$$

Cette transitivité s'obtient en enchainant les algorithmes de constructions d'incompatibilités fortes. Les fonctions-degré résultantes s'obtiennent donc par composition convenable des fonctions-degré initiales.

On démontrerait pour les implications-disjonctions dynamiques le principe de substitution analogue à celui démontré pour les implications dynamiques (cf. infra proposition 15), par la même méthode.

Cas avec une seule condition de signe

Proposition 10 : Supposons que dans une implication-disjonction dynamique

$$^\bullet(\ \mathbb{H}_1 \Rightarrow [\mathbb{K}_1 \vee \mathbb{K}_2 \vee \ldots \vee \mathbb{K}_m \,] \ )^\bullet$$

chaque système $\mathbb{K}_i$ du second membre soit une seule condition de signe $Q_i \,\sigma_i\, 0$, et notons $Q_i \,\tau_i\, 0$ la condition de signe opposée.

Alors on a l'implication-disjonction dynamique :

$$^\bullet(\ \mathbb{H}_1(X) \Rightarrow [\, Q_1 \sigma_1 0 \vee Q_2 \sigma_2 0 \vee \ldots \vee Q_m \sigma_m 0 \,] \ )^\bullet \qquad \text{(a)}$$

si et seulement si on a une incompatibilité forte :

$$\downarrow [\ \mathbb{H}_1(X) \,,\, Q_1 \tau_1 0 \,,\, Q_2 \tau_2 0 \,, \ldots ,\, Q_m \tau_m 0 \ ] \downarrow \qquad \text{(b)}$$

On obtient sans difficulté les précisions suivantes concernant les degrés :

Si (a) est vérifié avec une fonction-degré acceptable $\Delta^1(d_1 , \ldots , d_m)$ et si chaque Q_i a pour degré δ_i alors on a une incompatibilité forte (b) de degré $\Delta^1(2.\delta_1 , \ldots , 2.\delta_m)$.

Si on a une incompatibilité forte (b) de degré δ, alors l'implication dynamique (a) admet pour fonction-degré acceptable :

$$(d_1 , \ldots , d_m) \longmapsto \varphi(d_1 , \varphi(d_2 , \ldots , \varphi(d_m , \delta)\ldots) \qquad \text{(fonction } \varphi \text{ de la prop. 4)}$$

Variations sur le thème des implications dynamiques

Pour de nombreuses implications de base, on a des algorithmes plus rapides, et moins coûteux en degré, que celui donné en appliquant les propositions 7 et 8, lorsqu'on veut les traiter en implications dynamiques.

Implications triviales et implications simples

Définition 11 : (implications triviales)

Une implication $\mathbb{H}_1(X) \Rightarrow \mathbb{H}_2(X)$ est dite triviale lorsque toute incompatibilité forte $\downarrow [\ \mathbb{H}_2(X) \,,\, \mathbb{H}(X,Y)\ ] \downarrow$ fournit par simple relecture l'incompatibilité forte $\downarrow [\ \mathbb{H}_1(X) \,,\, \mathbb{H}(X,Y)\ ] \downarrow$.

L'implication dynamique $^\bullet(\mathbb{H}_1(X) \Rightarrow \mathbb{H}_2(X))^\bullet$ accepte alors pour fonction-degré : $\Delta_0(d) = d$.

Exemples : L'implication $[\, A > 0 \,,\, B > 0 \,] \Rightarrow AB > 0$ est triviale : dans l'incompatibilité forte

$$\downarrow [\, AB > 0 \,,\, \mathbb{H} \,] \downarrow$$

on relit chaque constituant AB (dans la partie «monoïde» ou dans la partie «cone») sous forme du produit de A par B pour obtenir l'incompatibilité forte

$$\downarrow [\, A > 0 \,,\, B > 0 ,\, \mathbb{H} \,] \downarrow$$

Notez que l'implication 'contrapposée' $[\, A > 0 \,,\, A.B \leq 0 \,] \Rightarrow B \leq 0$ n'est pas une implication simple. L'implication dynamique

$$^\bullet(\, [\, A > 0 \,,\, AB \leq 0 \,] \Rightarrow B \leq 0 \ )^\bullet$$

peut être obtenue par l'algorithme suivant : multipliez chaque terme de l'incompatibilité forte $\downarrow [\, B \leq 0 \,,\, \mathbb{H} \,] \downarrow$ par A et relire les termes où apparaît le produit A.B en considérant que AB est un seul bloc, provenant de l'hypothèse $AB \leq 0$; on obtient alors l'incompatibilité forte

$$\downarrow [\, A > 0 \,,\, A.B \leq 0 ,\, \mathbb{H} \,] \downarrow$$

De même, l'implication $B = 0 \Rightarrow A.B = 0$ est triviale, tandis que la contraposée ne l'est pas.
On a aussi l'implication triviale $[\, A \geq 0 \,,\, A \neq 0 \,] \Rightarrow A > 0$, tandis que l'implication $[\, A \geq 0 \,,\, A \leq 0 \,] \Rightarrow A = 0$ ne l'est pas.

Définitions 12 : (implications simples)

a) Une implication : $\mathbb{H}_1(X) \Rightarrow T(X) = 0$ est dite simple lorsqu'elle est donnée par une égalité $T = \Sigma\, N_i.V_i$ où les N_i sont les polynomes supposés nuls dans $\mathbb{H}_1$.
On appelle degré absolu d'une telle implication simple l'entier :
$\sup(d(N_i.V_i)) - d(T)$, et degré relatif le rationnel $\sup(\, d(N_i.V_i)\,) \,/\, d(T)$

b) Une implication : $\mathbb{H}_1(X) \Rightarrow T(X) \geqslant 0$ est dite simple lorsqu'elle est donnée par une égalité $T = \Sigma\, P_h.(\Sigma\, u_{h,j}\, U_{h,j}^{\,2}) + \Sigma\, N_i.V_i$ avec les mêmes hypothèses qu'en a), et où en outre les P_h sont des produits de polynomes supposés > 0 , ou ≥ 0 , dans $\mathbb{H}_1$. (les $u_{h,j}$ sont des positifs de K).
On appelle degré absolu d'une telle implication simple la différence :
$\sup(\, d(N_i.V_i)\,,\, d(P_h.U_{h,j}^{\,2})\,) - d(T)$, et degré relatif leur rapport.

c) Une implication : $\mathbb{H}_1(X) \Rightarrow T(X) > 0$ est dite simple lorsqu'elle est donnée par une égalité $T = S.R^2 + \Sigma\, P_h.(\Sigma\, u_{h,j}\, U_{h,j}^{\,2}) + \Sigma\, N_i.V_i$ avec les mêmes hypothèses qu'en b), et où en outre S (resp. R) est un produit de polynomes supposés > 0 (resp $\neq 0$) dans $\mathbb{H}_1$. On appelle degré relatif d'une telle implication simple le rationnel :
$\sup(\, d(S.R^2), d(N_i.V_i)\,,\, d(P_h.U_{h,j}^{\,2})\,) \,/\, d(T)$

d) Une implication : $\mathbb{H}_1(X) \Rightarrow T(X) \neq 0$ est dite simple lorsqu'elle est donnée par une égalité $T = S.R + \Sigma\, N_i.V_i$ avec les mêmes hypothèses qu'en c). On appelle degré relatif d'une telle implication simple le rationnel : $\sup(\, d(S.R), d(N_i.V_i)\,) \,/\, d(T)$

e) Une implication $\mathbb{H}_1(X) \Rightarrow \mathbb{H}_2(X)$ est dite simple lorsque chacune des csg du second membre, $Q_i\, \sigma_i\, 0$, résulte de $\mathbb{H}_1(X)$ par une implication simple. On appelle degré relatif de cette implication simple le sup des degrés relatifs des implications simples $\mathbb{H}_1(X) \Rightarrow Q_i\, \sigma_i\, 0$.
Lorsque le système $\mathbb{H}_2(X)$ ne comporte que des conditions de signes fermées ($= 0$, ≤ 0 , ≥ 0) on appelle degré absolu de l'implication simple, le sup des degrés absolus des implications simples $\mathbb{H}_1(X) \Rightarrow Q_i\, \sigma_i\, 0$.

Il y a un algorithme particulièrement simple pour expliciter l'implication dynamique correspondant à une implication simple donnée du type :

$$\mathbb{H}_1(X) \Rightarrow T(X) = 0$$

Dans l'incompatibilité forte :

$$\downarrow [\, \mathbb{H}(X,Y) \,,\, T(X) = 0 \,] \downarrow$$

on remplace T par $\Sigma\, N_i.V_i$.
Par exemple si T apparaissait sous forme T.W , on aura maintenant une

somme $\Sigma N_i.(W.V_i)$ où chaque terme a un rôle autonome dans la nouvelle implication forte :

$$\downarrow [\ \mathbb{H}(X,Y)\ ,\ \mathbb{H}_1(X)\ ] \downarrow\ .$$

On voit que le degré de cette dernière a augmenté au plus du degré absolu de l'implication simple, et on en déduit qu'il a été multiplié au plus par le degré relatif de l'implication simple.
Des considérations du même genre s'appliquent aux autres cas d'implications simples et on obtient :

Proposition 13 : (implications simples en tant qu'implications dynamiques)

a) Une implication simple : $\mathbb{H}_1(X) \Rightarrow \mathbb{H}_2(X)$, où $\mathbb{H}_2(X)$ ne comporte que des conditions de signes fermées, accepte pour fonction-degré :
$(d;\delta) \longmapsto d + \delta$, où δ est le degré absolu de l'implication simple.

b) Une implication simple : $\mathbb{H}_1(X) \Rightarrow \mathbb{H}_2(X)$ accepte pour fonction-degré :
$(d;\delta') \longmapsto d.\delta'$, où δ' est le degré relatif de l'implication simple.

Remarque : Souvent, une implication simple a un degré absolu nul et un degré relatif égal à 1, ce qui signifie que l'implication forte considérée ne coûte rien pour ce qui concerne les degrés. Nous dirons indifféremment 'implication simple de degré relatif égal à 1' ou 'implication simple qui ne coûte rien'. Dans une éventuelle mise en oeuvre de l'algorithme, il est *toujours* plus économique de traiter une implication simple en tant que telle.

Trois exemples :

Substitution d'égaux :
L'implication $U = V \Rightarrow P(X,U) = P(X,V)$ est une implication simple qui ne coûte rien. (U et V sont ici supposées être des variables et non des polynomes)

Somme de deux positifs :
L'implication $[A > 0, B \geq 0\ ] \Rightarrow A + B > 0$ est simple de degré relatif $\delta' = \sup(d(A),d(B)) / d(A+B)$ et accepte la fonction-degré : $d \longmapsto d.\delta'$.
L'implication $[A \geq 0, B \geq 0\ ] \Rightarrow A + B \geq 0$ est simple de degré absolu $\delta = \sup(d(A),d(B)) - d(A+B)$ et accepte la fonction-degré $\Delta : d \longmapsto d + \delta$.

Point où un polynome unitaire a le signe de son coefficient dominant :
Soit Q un polynome, unitaire en la variable U distincte des X_i :

$$Q(X,U) = U^s + C_{s-1}(X).U^{s-1} + ... + C_1(X).U + C_0(X)$$

Soit $V(X) = s + C_{s-1}(X)^2 + ... + C_1(X)^2 + C_0(X)^2$.
Alors on a des implications simples simultanées qui ne coutent rien :

$$[\ ] \Rightarrow Q(X,V(X)) > 0$$

$$[\ ] \Rightarrow Q^{(i)}(X,V(X)) > 0 \qquad \text{(dérivées par rapport à U)}$$

Signalons enfin quelques implications, qui sans être des implications simples, sont d'un traitement "rapide" en tant qu'implications dynamiques :

Proposition 14 : (fonctions-degré de quelques implications particulières)

a) L'implication $[\, A > 0\, , A.B \geq 0\,] \Rightarrow B \geq 0$ accepte pour fonction-degré : $(d;\delta) \longmapsto d + 2.\delta$ où $\delta = d(A)$. Même chose avec $=$ à la place de $\geq$.

b) L'implication $[\, A > 0\, , A.B > 0\,] \Rightarrow B > 0$ accepte pour fonction-degré : $(d;\delta,\delta') \longmapsto \sup(\, d.\delta', d + 2.\delta\,)$ où $\delta = d(A)$, $\delta' = d(A.B) \,/\, d(B)$.

c) L'implication $[\, A \geq 0\, , A.B > 0\,] \Rightarrow B \geq 0$ accepte pour fonction-degré : $(d;\delta) \longmapsto d + 2.\delta$ où $\delta = d(A.B)$.

d) L'implication $[\, A \geq 0\, , A.B > 0\,] \Rightarrow B > 0$ accepte pour fonction-degré : $(d;\delta,\delta') \longmapsto \sup(\, d.\delta', d + 2.\delta\,)$ où $\delta = d(A.B)$, $\delta' = d(A.B) \,/\, d(B)$.

e) L'implication $[\, A.B > 0\, ,\ A{+}B > 0\,] \Rightarrow [\, A > 0\, , B > 0\,]$ accepte pour fonction-degré : $(d;\delta,\delta') \longmapsto d.\delta' + 2\delta$ où $\delta' = d(AB)/\inf(d(A),d(B))$, $\delta = \sup(d(A),d(B))$.

f) L'implication $A^{2k} \leq 0 \Rightarrow A = 0$ accepte pour fonction-degré : $(d;k) \longmapsto 2k.d$

De même l'implication $[\, A \geq 0\, , A \leq 0\,] \Rightarrow A = 0$ accepte pour fonction-degré : $d \longmapsto 2d$.

g) L'implication $P(X,U) \neq P(X,V) \Rightarrow U \neq V$ accepte pour fonction-degré : $(d;\delta') \longmapsto d.\delta'$ où $\delta' = d(P(X,U) - P(X,V)) \,/\, d(U - V)$

Par exemple pour le a) : on multiplie, terme à terme, l'implication forte par A^2, en prenant soin de remplacer les $B.A^2$ par $(BA).A$.

Le principe de substitution

Proposition 15 :

On considère des variables $X_1,X_2,\ldots,X_n, U_1,U_2,\ldots,U_h, Z_1,Z_2,\ldots,Z_k$, et des polynomes $P_1,P_2,\ldots,P_n$ de $\mathbf{K}[Z]$. Notons $\mathbf{P}(Z)$ pour $P_1(Z), \ldots, P_n(Z)$.

Si on a $\quad {}^\bullet(\, \mathbb{H}_1(X,U) \Rightarrow \mathbb{H}_2(X,U)\,)^\bullet \quad$ (a)

alors on a aussi $\quad {}^\bullet(\, \mathbb{H}_1(\mathbf{P}(Z),U) \Rightarrow \mathbb{H}_2(\mathbf{P}(Z),U)\,)^\bullet \quad$ (b)

Si Δ^1 est une fonction-degré acceptable pour (a), une fonction-degré acceptable pour (b) est donnée par : $d \longmapsto \lambda + \Delta^1(\mu + \delta.d)$ où $\delta = \sup(\deg(P_i))$ et λ et μ sont explicités dans la preuve.

preuve>

Notons $\mathbf{X} = \mathbf{P}(Z)$ pour : $X_1 = P_1(Z), \ldots, X_n = P_n(Z)$. Soient $Y_1,Y_2,\ldots,Y_n$ des nouvelles variables.

Pour tout R figurant dans $\mathbb{H}_2(X,U)$ considérons l'égalité qui peut être obtenue par divisions successisves (les degrés des R_i sont tous inférieurs ou égaux au degré de R) :

$$R(X,U) = R(Y,U) + (X_1 - Y_1)\, R_1(X,Y,U) + \ldots + (X_n - Y_n)\, R_n(X,Y,U) \quad (1)$$

En substituant $P_i(Z)$ à Y_i on obtient une égalité :

$$R(\mathbf{P}(Z),U) = R(X,U) + (X_1 - P_1(Z))\, R_1(X,\mathbf{P}(Z),U) + \ldots + (X_n - P_n(Z))\, R_n(X,\mathbf{P}(Z),U) \quad (2)$$

Ces égalités fournissent une implication simple :

$$^\bullet(\ [\ \mathbb{H}_2(X,U)\ ,\ X = P(Z)\ ]\ \Rightarrow\ \mathbb{H}_2(P(Z),U)\ )^\bullet \quad (3)$$

dont le degré absolu est majoré par

$$\mu = \sup\ (\ 0\ ,\ \sup\{\ \deg((X_i - P_i(Z))\ R_i(X,P(Z),U)) - \deg(R(P(Z),U))\ ;\ R \text{ figure dans } \mathbb{H}_2\ ,\ R(P(Z),U) \neq \text{cte}\ \}$$

De la même manière, pour les R figurant dans $\mathbb{H}_1$ on a des égalités :

$$R(X,U) = R(P(Z),U) - (X_1 - P_1(Z))\ R_1(X,P(Z),U) - \ldots - (X_n - P_n(Z))\ R_n(X,P(Z),U) \quad (4)$$

qui fournissent une implication simple

$$^\bullet(\ [\ \mathbb{H}_1(P(Z),U)\ ,\ X = P(Z)\ ]\ \Rightarrow\ [\ \mathbb{H}_1(X,U)\ ,\ X = P(Z)\ ]\ )^\bullet \quad (5)$$

dont le degré absolu est majoré par

$$\lambda = \sup\ (\ 0\ ,\ \sup\{\ \deg((X_i - P_i(Z))\ R_i(X,P(Z),U)) - \deg(R(X,U))\ ;\ R \text{ figure dans } \mathbb{H}_1\ \}$$

Par ailleurs l'implication dynamique :

$$^\bullet(\ [\ \mathbb{H}_1(X,U)\ ,\ X = P(Z)\ ]\ \Rightarrow\ [\ \mathbb{H}_2(X,U)\ ,\ X = P(Z)\ ]\ )^\bullet \quad (6)$$

accepte la même fonction-degré que l'implication dynamique (a).
En composant (5), (6) et (3) on obtient :

$$^\bullet(\ [\ \mathbb{H}_1(P(Z),U)\ ,\ X = P(Z)\ ]\ \Rightarrow\ \mathbb{H}_2(P(Z),U)\ )^\bullet \quad (7)$$

Enfin, comme les variables X ne figurent pas dans $\mathbb{H}_1(P(Z),U)$, on a l'implication dynamique :

$$^\bullet(\ \mathbb{H}_1(P(Z),U) \Rightarrow\ [\ \mathbb{H}_1(P(Z),U)\ ,\ X = P(Z)\ ])^\bullet \quad (8)$$

obtenue en remplaçant les X_i par les P_i dans l'incompatibilité forte initiale. Elle accepte pour fonction-degré : $d \longmapsto d.\delta$ avec $\delta = \sup(\deg(P_i))$.
En résumé, si Δ^1 est une fonction-degré acceptable pour (a), une fonction-degré acceptable pour (b) est donc : $d \longmapsto \lambda + \Delta^1(\mu + \delta.d)$ ❑

Formules de Taylor mixtes

On considère deux variables U et V et on pose $\Delta := U - V$. On considère un polynome P à coefficients dans un corps ordonné K ou *plus généralement dans un anneau commutatif* **A** *qui est une* $\mathbb{Q}$ *-algèbre.*

Si $\deg(P) \leqslant 4$, on a les 8 formules de Taylor mixtes suivantes:

$$P(U) - P(V) = \Delta.P'(V) + (1/2).\Delta^2.P''(V) + (1/6).\Delta^3.P^{(3)}(V) + (1/24).\Delta^4.P^{(4)}$$
$$P(U) - P(V) = \Delta.P'(V) + (1/2).\Delta^2.P''(V) + (1/6).\Delta^3.P^{(3)}(U) - (1/8).\Delta^4.P^{(4)}$$
$$P(U) - P(V) = \Delta.P'(V) + (1/2).\Delta^2.P''(U) - (1/3).\Delta^3.P^{(3)}(V) - (5/24).\Delta^4.P^{(4)}$$
$$P(U) - P(V) = \Delta.P'(V) + (1/2).\Delta^2.P''(U) - (1/3).\Delta^3.P^{(3)}(U) + (1/8).\Delta^4.P^{(4)}$$
$$P(U) - P(V) = \Delta.P'(U) - (1/2).\Delta^2.P''(V) - (1/3).\Delta^3.P^{(3)}(V) - (1/8).\Delta^4.P^{(4)}$$
$$P(U) - P(V) = \Delta.P'(U) - (1/2).\Delta^2.P''(V) - (1/3).\Delta^3.P^{(3)}(U) + (5/24).\Delta^4.P^{(4)}$$
$$P(U) - P(V) = \Delta.P'(U) - (1/2).\Delta^2.P''(U) + (1/6).\Delta^3.P^{(3)}(V) + (1/8).\Delta^4.P^{(4)}$$
$$P(U) - P(V) = \Delta.P'(U) - (1/2).\Delta^2.P''(U) + (1/6).\Delta^3.P^{(3)}(U) - (1/24).\Delta^4.P^{(4)}$$

Comme toutes les combinaisons de signes possibles se présentent, on obtient :
– supposons que u et v attribuent la même suite de signes (au sens large) pour les dérivées successives d'un polynome P non constant de degré ≤ 4, notons $\varepsilon_1 = 1$ ou -1 selon que $P'(u)$ et $P'(v)$ sont tous deux ≥ 0 ou tous deux ≤ 0, alors le fait que $P(u) - P(v)$ a même signe que $\varepsilon_1.(u - v)$ est rendu évidnet

par l'une des formules ci-dessus, ce qui donne l'implication sous forme d'une implication simple (u et v peuvent être des éléments de **K** mais aussi des variables, ou des polynomes)

– si u et v n'attribuent pas la même suite de signes pour un polynome P de degré ≤ 4 et ses dérivées successives, alors on a une identité algébrique qui donne le signe de $u - v$ à partir des signes des $P^{(i)}(u)$ et des $P^{(i)}(v)$: la formule de Taylor mixte à utiliser est avec $P^{(i)}$ ($i = 0, 1, 2,$ ou 3) où i est le plus grand indice pour lequel les deux signes ne sont pas identiques

Plus généralement on a :

Proposition 16 : (formules de Taylor mixte)

Pour chaque degré s , il y a 2^{s-1} formules de Taylor mixtes et toutes les combinaisons de signes possibles apparaissent.

Formules de Taylor généralisées (le lemme de Thom sous forme d'identités algébriques)

Le lemme de Thom affirme (entre autres) que l'ensemble des points où un polynome et ses dérivées successives ont chacun un signe fixé, est un intervalle. Une preuve facile, par récurrence sur le degré du polynome, est basée sur le théorème des accroissements finis. Nous pouvons, grâce aux formules de Taylor mixtes, traduire ce fait géométrique sous forme d'identités algébriques, que nous appellerons des **formules de Taylor généralisées.** Plutôt que de risquer un énoncé, nous donnons un exemple.

Un exemple : Considérons le polynome générique de degré 4

$$P(X) = c_0 X^4 + c_1 X^3 + c_2 X^3 + c_3 X^2 + c_4 X^4 + c_5$$

Considérons le système de conditions de signe portant sur le polynome P et ses dérivées successives par rapport à la variable X :

$$\mathbb{H}(U) : \; P(U) > 0, \; P'(U) < 0, \; P^{(2)}(U) < 0, P^{(3)}(U) < 0, P^{(4)}(U) > 0.$$

Considérons également le système de conditions de signe généralisées obtenues en relachant toutes les inégalités, sauf la dernière :

$$\mathbb{H}'(U) : \; P(U) \geq 0, \; P'(U) \leq 0, \; P^{(2)}(U) \leq 0, P^{(3)}(U) \leq 0, P^{(4)}(U) > 0.$$

Le lemme de Thom affirme (entre autres) :

$$[\, \mathbb{H}'(U) \, , \mathbb{H}'(V) \, , \; U < Z < V \,] \;\; \Rightarrow \;\; \mathbb{H}(Z) \qquad (1)$$

Nous allons voir que ce fait géométrique est rendu évident par des identités algébriques.

On écrit les formules de Taylor mixtes suivantes :

α) $P^{(3)}(Z) = P^{(3)}(V) + P^{(4)}.(Z - V)$

β) $P^{(2)}(Z) = P^{(2)}(U) + P^{(3)}(Z).(Z - U) - 1/2\, P^{(4)}.(Z - U)^2$

γ) $P'(Z) = P'(U) + P^{(2)}(U).(Z - U) + 1/2\, P^{(3)}(Z).(Z - U)^2 - 1/3\, P^{(4)}.(Z - U)^3$

δ) $P(Z) = P(V) + P'(Z).(Z - V) - 1/2 P^{(2)}(Z).(Z - V)^2 + 1/6\, P^{(3)}(V).(Z - V)^3 + \ldots$
$1/8\, P^{(4)}.(Z - V)^4$

Posons $\boxed{\Delta_1 = Z - U \, , \; \Delta_2 = V - Z}$

Dans β) on remplace $P^{(3)}(Z)$ par son expression donnée dans α) et on obtient :

β') $\boxed{P^{(2)}(Z) = P^{(2)}(U) + P^{(3)}(V).\Delta_1 \; - \; P^{(4)}\, [\Delta_1.\Delta_2 + 1/2\, \Delta_1{}^2]}$

On obtient de la même manière, par substitutions :

$$\gamma')\quad P'(Z) = P'(U) + P^{(2)}(U).\Delta_1 + 1/2\, P^{(3)}(V).\Delta_1{}^2 - P^{(4)}.[\Delta_1{}^2.\Delta_2/2 + \Delta_1{}^3/3]$$

et enfin

$$\begin{aligned}\delta')\ P(Z) = {} & P(V) - P'(U).\Delta_2 - P^{(2)}(U).[\Delta_1.\Delta_2 + 1/2\Delta_2{}^2] \\ & - P^{(3)}(V).[\Delta_1{}^2.\Delta_2/2 + \Delta_1.\Delta_2{}^2/2 + \Delta_2{}^3/6] \\ & + P^{(4)}.[\Delta_1{}^3.\Delta_2/3 + \Delta_1{}^2.\Delta_2{}^2/2 + \Delta_1.\Delta_2{}^3/2 + \Delta_2{}^4/8]\end{aligned}$$

Les égalités α), β′), γ′), δ′) donnent l'implication (1) sous forme d'une implication simple. La première égalité est une formule de Taylor ordinaire portant sur le polynome $P^{(3)}$. Les trois dernières peuvent être vues comme des formules de Taylor généralisées portant sur les polynomes $P^{(2)}$, P' et P.

Plus généralement, on obtient:

Théorème 17 : (évidence forte du lemme de Thom)

Soit T une variable distincte des C_i. Soient $P \in K[C][T]$, de degré s en T, $\sigma_1, \sigma_2, \ldots, \sigma_s$ une liste formée de $<$ ou $>$. On note $H(C,T)$ ou $H(T)$ le système de csg : $P'(C,T)\ \sigma_1\ 0, \ldots, P^{(i)}(C,T)\ \sigma_i\ 0, \ldots, P^{(s)}(C,T)\ \sigma_s\ 0$ (les dérivées sont par rapport à T).

Soit $H'(T)$ le système de csg obtenu à partir de $H(T)$ en relachant toutes les conditions de signe sauf celle relative à $P^{(s)}$.

Soit $H_1(T)$ le système de csg : $P^{(s)}(C,T) > 0, P^{(i)}(C,T) \geqslant 0, i = 1, \ldots, s-1$.

Soient enfin trois variables U, V, Z distincte des C_i.

On a alors les implications dynamiques suivantes :

$${}^{\bullet}(\,[\,H'(U), H'(V), U\ \sigma_1\ V\,] \ \Rightarrow\ P(U) > P(V)\,)^{\bullet} \qquad \text{(a)}$$

$${}^{\bullet}(\,[\,H_1(U), V > U\,] \ \Rightarrow\ P(V) > P(U)\,)^{\bullet} \qquad \text{(b)}$$

$${}^{\bullet}(\,[\,H'(U), H'(V), U < Z < V\,] \ \Rightarrow\ H(Z)\,)^{\bullet} \qquad \text{(c)}$$

Ce sont des implications simples qui ne coûtent rien.

preuve> L'implication dynamique (a) résulte de formules de Taylor mixtes. L'implication dynamique (b) résulte de la formule de Taylor ordinaire au point U. Les formules de Taylor généralisées établies pour l'implication dynamique (c) résultent des formules de Taylor mixtes. On constate qu'il s'agit d'implications simples qui ne coûtent rien (ceci parce que U, V, Z sont des variables et non des polynomes). ❑

4) Existences potentielles

Notations et définitions

Elles sont tout à fait analogues à celles données pour les implications dynamiques.

Définition et notation 18 :

Soient $\mathbb{H}_1$ un système de csg portant sur des polynomes de $\mathbf{K}[X]$, $\mathbb{H}_2$ un système de csg portant sur des polynomes de $\mathbf{K}[X,T_1,T_2,\ldots,T_m] = \mathbf{K}[X,T]$.

Nous dirons que *les hypothèses* $\mathbb{H}_1$ *autorisent l'existence des* T_i *vérifiant* $\mathbb{H}_2$ lorsque, pour tout système de csg $\mathbb{H}$ portant sur des polynomes de $\mathbf{K}[X,Y]$, les variables Y_i et T_j étant deux à deux distinctes, on a la construction d'implication forte :

$$\downarrow [\ \mathbb{H}_2(X,T)\ ,\ \mathbb{H}(X,Y)\ ]\downarrow \ \vdash_{cons} \downarrow [\ \mathbb{H}_1(X)\ ,\ \mathbb{H}(X,Y)\ ]\downarrow .$$

Nous parlerons également *d'existence potentielle des* T_i *vérifiant* $\mathbb{H}_2$ *sous les hypothèses* $\mathbb{H}_1$

Nous noterons cette existence potentielle par :

$$^{\bullet}(\ \mathbb{H}_1(X) \Rightarrow \exists\, T\ \mathbb{H}_2(X,T)\ )^{\bullet}.$$

Lorsque le système $\mathbb{H}_1$ est vide, nous utilisons la notation $^{\bullet}(\ \exists\, T\ \mathbb{H}_2(X,T)\ )^{\bullet}$.

La notion de fonction-degré acceptable pour une existence potentielle peut être elle aussi directement recopiée du cas des implications dynamiques.

Remarques :

1) La notion d'existence potentielle est une notion d'existence faible. L'existence potentielle signifie qu'il n'est pas grave de faire comme si les T_i existaient vraiment, parce que cela n'introduit pas de contradiction: on peut paraphraser la définition en disant :

pour construire l'incompatibilité forte $\quad \downarrow [\ \mathbb{H}_1(X)\ ,\ \mathbb{H}(X,Y)\ ]\downarrow$

il suffit d'avoir construit $\quad \downarrow [\ \mathbb{H}_2(X,T)\ ,\ \mathbb{H}(X,Y)\ ]\downarrow$

2) On pourrait étendre la définition de l'existence potentielle en remplaçant le système de csg $\mathbb{H}_2(X,T)$ par une disjonction de systèmes de csg, comme on a fait avec la notion d'implication-disjonction dynamique.

Quelques règles de manipulation des énoncés d'existence potentielle

Transitivité

La transitivité des existences potentielles est immédiate. Voici l'énoncé précisé en termes de fonctions-degré acceptables.

Proposition 19 : (transitivité dans les existences potentielles)

On considère des variables $X_1,X_2,\ldots,X_n, T_1,T_2,\ldots,T_m, U_1,U_2,\ldots,U_k$ et des systèmes de csg $\mathbb{H}_1(X)$, $\mathbb{H}_2(X,T)$ et $\mathbb{H}_3(X,T,U)$.

Les existences potentielles

$$^{\bullet}(\ H_1(X) \Rightarrow \exists\, T\ \ H_2(X,T)\ )^{\bullet} \ \text{ et } \ ^{\bullet}(\ H_2(X,T) \Rightarrow \exists\, U\ \ H_3(X,T,U)\ )^{\bullet}$$

impliquent l'existence potentielle :

$$^{\bullet}(\ H_1(X) \Rightarrow \exists\, T, U\ \ H_3(X,T,U)\ )^{\bullet}$$

Supposons que la première existence potentielle admette comme fonction-degré acceptable $\Delta^1(d;p)$ où d est le degré de l'incompatibilité forte $\downarrow [\ H_2(X,T)\ ,\ H(X,Y)\ ] \downarrow$ et p représente certains paramètres dépendant de $H_1(X)$ et $H_2(X,T)$, supposons de même une fonction-degré acceptable $\Delta^2(d;q)$ pour la deuxième existence potentielle, alors une fonction-degré pour l'existence potentielle construite est donnée par :

$$\Delta(d;p,q) = \Delta^1(\Delta^2(d;q);p)$$

Preuves cas par cas

Voici maintenant un énoncé correspondant aux preuves cas par cas d'une existence potentielle, conséquence immédiate de la proposition 4 .

Proposition 20 : (raisonnement cas par cas)

Soit Q un polynome de $K[X]$.

a) Pour démontrer une existence potentielle

$$^{\bullet}(\ [\ H_1(X)\ ,\ Q \neq 0\ ) \Rightarrow \exists\, T\ \ H_2(X,T)\ )^{\bullet}$$

il suffit de démontrer chacune des existences potentielles

$$^{\bullet}(\ [\ H_1(X),\ Q > 0] \Rightarrow \exists\, T\ \ H_2(X,T)\ )^{\bullet} \ \text{ et } \ ^{\bullet}(\ [\ H_1(X),\ Q < 0] \Rightarrow \exists\, T\ \ H_2(X,T)\ )^{\bullet}$$

Si Δ^i $(i = 1,2)$ sont les deux fonctions-degré des existences potentielles supposées, une fonction-degré pour l'existence potentielle déduite est donnée par : $\Delta^1 + \Delta^2$

a'), b), c), d), e) : énoncés analogues décalqués de la proposition 4

Le principe de substitution

Le principe de substitution pour les existences potentielles se démontre comme pour les implications dynamiques.

L'existence implique l'existence potentielle

Un autre principe utile est le fait que l'existence implique l'existence potentielle. Il s'obtient facilement : on remplace les variables T_i «existentielles» par les polynomes concrets P_i qui réalisent l'existence. On reconnait là une analogie formelle avec la règle d'introduction du quantificateur existentiel en calcul naturel par exemple (cf. [Pra]).

Proposition 21 : (l'existence implique l'existence potentielle)

Soient $P_1,P_2,\ldots,P_m \in K[X]$ et notons $P(X)$ pour $P_1(X), \ldots, P_m(X)$. On a l'existence potentielle : $^{\bullet}(\ H_2(X,P(X)) \Rightarrow \exists\, T\ H_2(X,T)\ )^{\bullet}$.

Si δ majore les degrés des P_i , l'existence potentielle accepte pour fonction-degré : $(d;\delta) \longmapsto d.\sup(1,\delta)$

Corollaire : (mêmes hypothèses)

Si $^\bullet(\ H_1(X) \ \Rightarrow \ H_2(X,P(X))\ )^\bullet$ **alors** $^\bullet(H_1(X) \Rightarrow \exists\, T\ H_2(X,T)\ )^\bullet$

Si Δ^1 est une fonction-degré acceptable pour l'implication forte de l'hypothèse, une fonction-degré acceptable pour la conclusion est donnée par : $(d;\delta) \longmapsto \Delta^1(d.\sup(1,\delta))$ où δ majore les degrés des P_i .

Existences potentielles fondamentales

On sait démontrer les existences potentielles correspondant aux deux axiomes existentiels de la théorie des corps réels clos.

Théorème 22 : (autorisation de rajouter l'inverse d'un non nul)

On a l'existence potentielle de l'inverse d'un non nul. Ce qui s'écrit:

$$^\bullet(\ U \neq 0 \Rightarrow \exists\, T \quad 1 = U.T\ )^\bullet$$

Soit δ le degré de U, une fonction-degré acceptable pour l'existence potentielle est $(d;\delta) \longmapsto d + d.\delta + \delta$

Remarque: La preuve de cette existence potentielle recopie ce qu'on fait, dans la preuve du théorème des zéros de Hilbert, pour passer du théorème des zéros faible au théorème des zéros général (c'est le «Rabinovitch trick», par exemple dans l'exposé classique de van der Waerden). La notion d'existence potentielle de l'inverse d'un non nul est donc en filigrane dans les classiques.

Théorème 23 : (autorisation de rajouter une racine sur un intervalle où un polynome change de signe)

Soit $P(C,X)$ un polynome de degré s en X et de degré global δ .

On a l'existence potentielle d'une racine sur un intervalle où ce polynome change de signe. Ce qui s'écrit, en notant $P(X)$ pour $P(C,X)$:

$$^\bullet(\ [\ P(X).P(Y) < 0\ ,\ X < Y\ ] \ \Rightarrow \ \exists\, Z\ [\ P(Z) = 0\ ,\ X < Z < Y\ ]\ \ )^\bullet$$

et, si X , Y , Z désignent des variables, une fonction-degré acceptable est donnée par :

$$(d;\delta,s) \longmapsto \ (\ (2d+7)\,(\delta+1)\ )^{\gamma(s)} \quad \text{où} \quad \gamma(s) = 2^{(s+2)^2/2}$$

Remarque : La preuve du théorème précédent "recopie" la preuve classique, par récurrence sur le degré du polynome P , du théorème «si un corps est ordonné et si $P(u).P(v) < 0$ avec P irréductible, alors le corps $K[W]\,/\,P(W)$ est réel». Ceci donne l'existence potentielle d'une racine. Pour avoir la racine sur l'intervalle, il y a de nouveau une récurrence à faire. Tout ceci conduit à une relativement mauvaise fonction-degré. Le problème semble difficile à contourner. Dans le cas complexe (théorème des zéros de Hilbert), l'existence potentielle d'une racine d'un polynome non constant est au contraire extrêmement simple et conduit à une fonction-degré tout à fait raisonnable : par exemple si $P(X,Y)$ est un polynome unitaire en Y de degré s en Y et de degré δ en X , il suffit de tout réduire modulo P et une fonction-degré acceptable pour l'existence potentielle $^\bullet(\ \exists\, Y\ P(X,Y) = 0\ \ )^\bullet$ est donnée par : $(d;\delta) \longmapsto d.(\delta+1)$

5) Majorations finales

Tableaux de Hörmander

Nous donnons ici quelques majorations directement liées à l'algorithme de Hörmander lui-même (cf. [Hör] annexe, ou [BCR] chap. 1).

Proposition 24 : (Tableau de Hörmander pour des polynomes en une variable)
Soit **K** un corps ordonné, sous-corps d'un corps réel clos **R** .
Soit $L = [P_1, P_2, ..., P_k]$ une liste de polynomes de **K**[Y] .
Soit $\mathcal{P}$ la famille de polynomes engendrée par les éléments de L et par les opérations $P \longmapsto P'$, et $(P,Q) \longmapsto \mathrm{Rst}(P,Q)$. Alors :

1) $\mathcal{P}$ est finie.
2) On peut établir le tableau complet des signes pour $\mathcal{P}$ en utilisant les seules informations suivantes : le degré de chaque polynome de la famille; les diagrammes des opérations $P \longmapsto P'$, et $(P,Q) \longmapsto \mathrm{Rst}(P,Q)$ (où $\deg(P) \geqslant \deg(Q)$) dans $\mathcal{P}$; et les signes des constantes de $\mathcal{P}$.

Si s majore les degrés des P_i , le nombre de coefficients d'éléments de $\mathcal{P}$, et donc aussi le nombre de points du tableau de Hörmander est majoré par : $(k+1)^{2^s}$

L'algorithme de Hörmander traite des polynomes en n variables, en éliminant chaque variable l'une après l'autre. A chaque élimination d'une variable, le nombre de polynomes à considérer et leurs degrés croissent de manière impressionnante. Ceci est précisé dans la proposition suivante :

Proposition 25 : (Tableau de Hörmander paramétré)
Soit **K** un corps ordonné, sous-corps d'un corps réel clos **R** .
Soit $L = [Q_1, Q_2, ..., Q_k]$ une liste de polynomes de $K[X_1, X_2, ..., X_n][Y]$.
On peut construire une famille finie $\mathcal{F}$ de polynomes de $K[X_1, X_2, ..., X_n]$ telle que, pour tous $x_1, x_2, ..., x_n$ dans **R** , en posant $P_i(Y) = Q_i(x_1, x_2, ..., x_n;Y)$, le tableau complet des signes pour $L = [P_1, P_2, ..., P_k]$ est calculable à partir des signes des $S(x_1, x_2, ..., x_n)$ pour $S \in \mathcal{F}$.
Supposons que la liste L possède k éléments de degré en **X** majoré par δ et de degré en Y majoré par s . Considérons la famille $\mathcal{G}$, formée de tous les coefficients de tous les polynomes de tous les tableaux de Hörmander possibles, construits sur L , en remplaçant l'opération "reste" par l'opération "pseudo-reste". Une famille $\mathcal{F}$ convenable peut être extraite de $\mathcal{G}$. Alors :
le degré de chaque polynome de $\mathcal{G}$ et de chaque pseudo-division est majoré par : $\delta.(s+1)!$, (sauf si $n = 0$, donc $\delta = 0$, et les degrés sont majorés par s).
le nombre d'éléments de la famille $\mathcal{G}$ est majoré par : $(k+1)^{2^s}$

Mené jusqu'au bout, cet algorithme produit donc une explosion de degrés obtenue en itérant $n-1$ fois (n étant le nombre de variables) la fonction $s \longmapsto s!$. Ceci conduit à la majoration finale.

Nullstellensatz, positivestellensatz et nichtnegativestellensatz réels effectifs

Théorème 26 : Soit **K** un corps ordonné, sous-corps d'un corps réel clos **R**.
Soit $\mathbb{H}(X_1,X_2,...,X_n)$ un système de csg portant sur une famille finie de polynomes de $\mathbf{K}[X_1,X_2,...,X_n]$. Ce système est impossible dans **R** si et seulement si il est fortement incompatible dans **K**. En termes plus formalisés :

Si $\downarrow \mathbb{H}(X_1,X_2,...,X_n) \downarrow$ (dans **K**),

alors les csg $\mathbb{H}$ sont impossibles à réaliser dans n'importe quelle extension ordonnée de **K**.

Si $\forall\, x_1,x_2,...,x_n \in \mathbf{R}$ $\mathbb{H}(x_1,x_2,...,x_n)$ est absurde,

alors : $\downarrow \mathbb{H}(X_1,X_2,...,X_n) \downarrow$ (dans **K**).

Précisément, si k est le nombre de csg dans $\mathbb{H}(X_1,X_2,...,X_n)$ et d le degré maximum, on peut calculer une implication forte

$\downarrow \mathbb{H}(X_1,X_2,...,X_n) \downarrow$ (dans **K**) de degré majoré par le nombre $\mu_{26}(d,k,n)$ donné par la tour d'exponentielle à n+4 étages

$$2^{2^{\cdot^{\cdot^{\cdot^{d.\lg(d)+\lg\lg(k)+cte}}}}}$$

Remarque : La principale cause d'explosion des degrés dans la majoration finale actuelle réside dans l'utilisation de l'algorithme de Hörmander.
On peut donc espérer améliorer sensiblement ces majorations en se basant sur d'autres preuves, élémentaires mais moins longues, d'incompatibilité.

Remerciements : Je remercie Marie-Françoise Roy pour ses nombreux commentaires et suggestions.

Bibliographie :

[BCR] Bochnak, Coste M., Roy M.-F. : Géométrie Algébrique réelle. Springer-Verlag. A series of Modern Surveys in Mathematics n°11. 1987.

[Du] Dubois, D. W. : A nullstellensatz for ordered fields, Arkiv for Mat., Stockholm, t. 8, 1969, p. 111-114

[Efr] Efroymson, G. : Local reality on algebraic varieties, J. of Algebra, t. 29, 1974, p. 113-142.

[FG] Fitchas N., Galligo A. : Nullstellensatz effectif et Conjecture de Serre (Théorème de Quillen-Suslin) pour le Calcul Formel. Math. Nachr. 149 p 231-253 (1990)

[He] Hermann Greta : Die Frage der endlich vielen Schritte in der Theorie der Polinomideale, Math. ANN. 95 (1926), 736-788

[Hör] Hörmander, L. : The analysis of linear partial differential operators, vol 2, Berlin, Heidelberg, New-York, Springer (1983). 364-367.

[Ko] Kollár J. : Sharp effective Nulstellensatz. I. AMS 1 p 963-975 (1988)

[Kri] Krivine, J. L. : Anneaux préordonnés. Journal d'analyse mathématique, t.12, 1964, p. 307-326

[Lom a] Lombardi H. : Théorème effectif des zéros réel et variantes. Publications Mathématiques de l'Université (Besançon). 88-89. Fascicule 1.

[Lom b] Lombardi H. : Effective real nullstellensatz and variants, in «MEGA 90», mai 1991, chez Birkhaüser. (Version anglaise plus courte)

[Lom c] Lombardi H. : Nullstellensatz réel effectif et variantes. C.R.A.S. Paris, t. 310, Série I, p 635-640, 1990.

[Lom d] Lombardi H.: Théorème effectif des zéros réel et variantes, avec une majoration explicite des degrés. 1990. Mémoire d'habilitation.

[LR] Lombardi H., Roy M.-F. : Théorie constructive élémentaire des corps ordonnés. 1989. Publications Mathématiques de Besançon. Théorie des Nombres 1990-91.
Version anglaise moins détaillée «Constructive elementary theory of ordered fields» in «MEGA 90», mai 1991, chez Birkhaüser.

[MRR] R. Mines, F. Richman, W. Ruitenburg : A Course in Constructive Algebra. Springer-Verlag. Universitext. 1988.

[Pra] Prawitz D. : Ideas and results of proof theory. Proceedings of the second scandinavian logic symposium (juin 70). Studies in Logic and Foundations of Mathmatics n°63, 235-307. North Holland.

[Ris] Risler, J.-J. : Une caractérisation des idéaux des variétés algébriques réelles, C.R.A.S. Paris, t. 271, 1970, série A, p. 1171-1173.

[Ste] Stengle, G. : A Nullstellensatz and a Positivestellensatz in semialgebraic geometry. Math. Ann. 207, 87-97 (1974)

MINIMAL GENERATION OF BASIC SEMI–ALGEBRAIC SETS OVER AN ARBITRARY ORDERED FIELD

M. MARSHALL AND L. WALTER

If A is a ring (commutative with 1), $F(\mathfrak{p})$ denotes the residue field of A at a prime $\mathfrak{p} \subseteq A$ and $Sper\,A$ denotes the real spectrum of A [2], [3], [8], [9]. If A is an F-algebra and $T \subseteq F$ is a preordering, $Sper_T A := \{Q \in Sper\,A \mid Q \supseteq T\}$. This is a saturated set in $Sper\,A$ in the terminology of [13].

Fix a real closed field R and a subfield $F \subseteq R$. Let $P \in Sper\,F$ be the ordering induced by the inclusion. Fix an algebraic set $V = \{x \in R^n \mid f_i(x) = 0 \text{ for } i = 1, \dots, k\}$ where $f_1, \dots, f_k \in F[X_1, \dots, X_n]$ and let $A = \frac{F[X_1,\dots,X_n]}{(f_1,\dots,f_k)}$.

A semi-algebraic set $S \subseteq V$ is *basic over F* (or simply *F-basic*) if

$$S = \{x \in V \mid g_1(x) > 0 \wedge \cdots \wedge g_s(x) > 0 \wedge h_1(x) \geq 0 \wedge \cdots \wedge h_t(x) \geq 0\}$$

where $g_1, \dots, g_s, h_1, \dots, h_t \in F[X_1, \dots, X_n]$. If $t = 0$ (resp. $s = 0$), S is said to be *basic open over F* (resp. *basic closed over F*). A semi-algebraic set is *defined over F* if it is a finite union of F-basic sets.

Define $s_F(V)$ (resp. $\overline{s}_F(V)$) to be the least integer $s \geq 1$ such that each $S \subseteq V$ which is basic open over F (resp. basic closed over F) is expressible as

$$S = \{x \in V \mid g_1(x) > 0 \wedge \cdots \wedge g_s(x) > 0\}$$
$$(\text{resp. } S = \{x \in V \mid g_1(x) \geq 0 \wedge \cdots \wedge g_s(x) \geq 0\})$$

where $g_1, \dots, g_s \in F[X_1, \dots, X_n]$.

The object of this paper is to give bounds for $s_F(V)$ and $\overline{s}_F(V)$, thus generalizing results in [5], [16]. The result we obtain is stated in Theorem 1.2 below. In §2, we show, at least in the case $V = R^n$, that the bounds in Theorem 1.2 are best possible.

1. Bounds for $s_F(V)$ and $\overline{s}_F(V)$.

We need the following notation: for any $Q \in Sper\,F$, let $\lambda_Q : F \to \mathbb{R} \cup \infty$ be the associated real place and v_Q the valuation associated with λ_Q (see [10].) Define $\overline{m}(Q) := m(Q) + e(Q)$ where $m(Q) := dim_2 \frac{v_Q(F^*)}{2v_Q(F^*)}$ and $e(Q) \in \{0,1\}$ is defined as follows: let $\varphi_Q : F \twoheadrightarrow F(Q) \cup \infty$ be the coarsest place such that λ_Q factors through φ_Q and such that the induced place $F(Q) \to \mathbb{R} \cup \infty$ has 2-divisible value group. Then

$$e(Q) := \begin{cases} 0 & \text{if } F(Q) \text{ is hereditarily Euclidean} \\ 1 & \text{otherwise.} \end{cases}$$

Here a field K is called *hereditarily Euclidean* if K and all its formally real (finite) algebraic extensions are Euclidean [11].

Note 1.1. An ordered field (K, P) is hereditarily Euclidean iff P extends uniquely to each formally real finite extension of K. (The one implication is clear. For the other, let L be an extension of K and let $Q \in Sper\, L$ extend P. If $Q \neq L^2$, there exists $a \in Q$, $a \notin L^2$. Then Q (and therefore, P) extends in two ways to $L[\sqrt{a}]$.)

Theorem 1.2. *Let* $\overline{m} := \overline{m}(P)$ *and* $d = dim\, V$. *Then*

$$s_F(V) \leq d + \overline{m}$$
$$\overline{s}_F(V) \leq d(d+1)/2 + (d+1)\overline{m}.$$

Remarks 1.3. We are assuming $\overline{m} > 0$ or $d > 0$. If $d = \overline{m} = 0$ then a much simpler argument shows $s_F(V) = \overline{s}_F(V) = 1$. A similar remark applies to Theorem 1.7 below.

Remarks 1.4. (1) If F is real closed then $\overline{m} = 0$. In particular, if $F = R$ we get the usual bounds for $s(V) = s_F(V)$ and $\overline{s}(V) = \overline{s}_F(V)$ [5], [12], [16].

(2) If $R = \mathbb{R}$ then λ_P is trivial so $m(P) = 0$.

(3) If $m(P) = 0$ then $F(P) = F$, $\overline{m} = e(P) \in \{0, 1\}$ and $\overline{m} = 0$ iff F is hereditarily Euclidean.

For any Tychonoff closed $X \subseteq Sper\, A$, let $d_0(X)$ denote the supremum of all integers $l \geq 0$ such that there exists $Q_i \in X, i = 0, \ldots, l$, with $Q_0 \subsetneqq \cdots \subsetneqq Q_l$ [3] and let $d_1(X)$ denote the Krull dimension of the ring $A/\cap\{supp\, Q \mid Q \in X\}$ [1], [4].

Lemma 1.5. $dim\, V = d_0(Sper_P A) = d_1(Sper_P A)$.

Proof. Let $\overline{F}$ be the real closure of (F, P) in R. The natural map $A \to \frac{\overline{F}[X_1,\ldots,X_n]}{(f_1,\ldots,f_k)}$ induces a homeomorphism $Sper\, \frac{\overline{F}[X_1,\ldots,X_n]}{(f_1,\ldots,f_k)} \cong Sper_P A$ [**2**, 2.17] and therefore, the kernel of the composite map $A \to \frac{\overline{F}[X_1,\ldots,X_n]}{(f_1,\ldots,f_k)} \to \frac{\overline{F}[X_1,\ldots,X_n]}{I(V\cap\overline{F}^n)}$ is precisely the ideal $\cap\{supp\, Q \mid Q \in Sper_P A\}$. Since $F \subseteq \overline{F}$ is algebraic, $A \to \frac{\overline{F}[X_1,\ldots,X_n]}{I(V\cap\overline{F}^n)}$ is integral so $d_1(Sper_P A) = dim\, \frac{\overline{F}[X_1,\ldots,X_n]}{I(V\cap\overline{F}^n)} = dim\, V \cap \overline{F}^n$ and by [**3**, 7.5.6], $dim\, V \cap \overline{F}^n = d_0(Sper_P A)$. It remains only to show that $dim\, V = dim\, V \cap \overline{F}^n$. By [**3**, 2.3.6], $V \cap \overline{F}^n$ is a disjoint union of semi-algebraic sets $S_1, \ldots, S_m$ where each S_i is semi-algebraically homeomorphic to $(0,1)^{d_i} \subseteq \overline{F}^{d_i}$ and by [**3**, 2.8.9], $dim\, V \cap \overline{F}^n = max\{d_1, \ldots, d_m\}$. By the Transfer Principle [**3**, 5.2.3], V is the disjoint union of the semi-algebraic sets $(S_1)_R, \ldots, (S_m)_R$ and each $(S_i)_R$ is semi-algebraically homeomorphic to $(0,1)^{d_i} \subseteq R^{d_i}$. Therefore, by [**3**, 2.8.9], $dim\, V = max\{d_1, \ldots, d_m\} = dim\, V \cap \overline{F}^n$. This completes the proof.

Lemma 1.6. *The map* $\Phi : V \to Sper_P A$ *given by* $x \mapsto P_x = \{f \in F[X_1, \ldots, X_n] \mid f(x) \geq 0\}$ *induces an isomorphism* $C \mapsto \Phi^{-1}(C)$ *of the Boolean algebra of constructible sets in* $Sper_P A$ *onto the Boolean algebra of semi-algebraic sets in* V *which are defined over* F.

Proof. It is clear that the map $C \mapsto \Phi^{-1}(C)$ is a surjective homomorphism. The injectivity follows from the Transfer Principle [**3**, 5.2.3].

It follows from (1.6) that $s_F(V) = s(Sper_P A)$ and $\overline{s}_F(V) = \overline{s}(Sper_P A)$ where s, $\overline{s}$ are defined as in [**13**]. Thus, to prove (1.2), we need only show that $s(Sper_P A) \leq d + \overline{m}$

and $\overline{s}(Sper_P A) \leq d(d+1)/2 + (d+1)\overline{m}$, where $d = d_0(Sper_P A) = d_1(Sper_P A)$. In fact, we obtain a slightly more general result.

Theorem 1.7. *Let A be a finitely-generated F-algebra, $T \subseteq F$ a proper preordering and $X = Sper_T A$. Then*

(1) $\quad d_0(X) = d_1(X)$.

In this case, we call $d(X) := d_0(X) = d_1(X)$ the dimension of X. Also,

(2) $\quad s(X) \leq d + \overline{m}$

(3) $\quad \overline{s}(X) \leq d(d+1)/2 + (d+1)\overline{m}$

where $d := d(X)$ and $\overline{m} := sup\{\overline{m}(P) \mid P \in Sper_T F\}$.

Proof. (1) is true if T is an ordering (by (1.5) applied in the case R is the real closure of (F,T).) Also, it is clear that any chain of orderings in $Sper_T A$ contracts to an ordering $P \in Sper_T F$. Thus,

$$\begin{aligned} d_0(Sper_T A) &= sup\{d_0(Sper_P A) \mid P \in Sper_T F\} \\ &= sup\{d_1(Sper_P A) \mid P \in Sper_T F\} \\ &\leq d_1(Sper_T A). \end{aligned}$$

The remaining inequality $d_1(Sper_T A) \leq sup\{d_1(Sper_P A) \mid P \in Sper_T F\}$ is a consequence of the following:

Lemma 1.8. *Suppose A is a commutative ring, $X \subseteq Sper\, A$ is Tychonoff closed and $\mathfrak{p}$ is a minimal prime over $\cap\{supp\, P \mid P \in X\}$. Then $\mathfrak{p} = supp\, P$ for some $P \in X$.*

Proof. For each $f \in A\backslash\mathfrak{p}$, there exists $P \in X$ such that $f \notin supp\, P$. By the compactness of X in the Tychonoff topology,

$$\bigcap_{f \in A\backslash\mathfrak{p}} \{P \in X \mid f \notin suppP\} \neq \emptyset.$$

Thus, there exists $P \in X$ with $supp\, P \subseteq \mathfrak{p}$ and by the minimality of $\mathfrak{p}$, we have $supp\, P = \mathfrak{p}$.

Before we can complete the proof of (1.7), we must generalize a result in [**7**] on the behavior of the stability index under field extensions.

Theorem 1.9. *Let $F \subseteq K$ be fields, $T \subseteq F$ a proper preordering, $X = Sper_T K$. Then*

$$s(X) \leq trdeg\, K|F + \overline{m}$$

where $\overline{m} := sup\{\overline{m}(P) \mid P \in Sper_T F\}$.

Proof. We use the fan characterization of s(X) [**10**]:

$$s(X) = sup\{log_2|V| \mid V \subseteq X \text{ is a fan}\}.$$

Let $V \subseteq X$ be a fan. By Bröcker's theorem on trivialization of fans [**10**, 5.13], there exists a valuation v on K fully compatible with V such that the induced fan $\overline{V}$ on the residue field $\overline{K}$ is trivial. Let $T_1 = \sum K^2 T$. By the Krull-Baer Theorem [10, 3.12], we have

$$log_2|V| \leq dim_2 \frac{v(K^*)}{v(T_1{}^*)} + log_2|\overline{V}|$$

and $log_2|\overline{V}| \in \{0,1\}$. (Replacing V by a bigger fan if necessary, we may even assume we have equality.)

Fix an ordering $Q \in V$ and let $P = Q \cap F$. It suffices to show

$$(1) \qquad dim_2 \frac{v(K^*)}{v(T_1{}^*)} + log_2|\overline{V}| \leq trdeg\, K|F + \overline{m}(P).$$

Since $T_1 = \sum K^2 T \subseteq Q$ and Q is v-compatible, we have $v(T_1{}^*) = 2v(K^*) + v(T^*)$. Let

$$W = \{\alpha \in v(K^*) \mid 2^r \alpha \in v(F^*) \text{ for some } r \geq 0\}.$$

Then

$$(2) \qquad dim_2 \frac{v(K^*)}{v(T_1{}^*)} = dim_2 \frac{v(K^*)}{2v(K^*) + W} + dim_2 \frac{2v(K^*) + W}{2v(K^*) + v(T^*)}.$$

If $a_i \in K^*$, $k_i \in \mathbb{Z}$ not all zero are such that $\sum k_i v(a_i) \in v(F^*)$ then, dividing by the highest power of 2 common to the k_i, we have $\sum l_i v(a_i) \in W$, where the $l_i \in \mathbb{Z}$ and at least one of the l_i is odd. This shows $dim_{\mathbb{Z}} \frac{v(K^*)}{v(F^*)} \geq dim_2 \frac{v(K^*)}{2v(K^*)+W}$ and hence,

$$(3) \qquad \begin{aligned} trdeg\, K|F \geq & dim_{\mathbb{Z}} \frac{v(K^*)}{v(F^*)} + trdeg\, \overline{K}|\overline{F} \\ \geq & dim_2 \frac{v(K^*)}{2v(K^*) + W} \end{aligned}$$

where $\overline{F}$ is the residue field of $v|_F$. Also, $\frac{2v(K^*)+W}{2v(K^*)+v(T^*)} \cong \frac{W}{2W+v(T^*)}$ so we have

$$(4) \qquad dim_2 \frac{2v(K^*) + W}{2v(K^*) + v(T^*)} = dim_2 \frac{W}{2W + v(T^*)} \leq dim_2 \frac{W}{2W} \leq dim_2 \frac{v(F^*)}{2v(F^*)}$$

where the last inequality follows from Bröcker's Index Formula [7, 3.2].

Since Q is v-compatible, P is compatible with $v|_F$ so the place $\lambda_P : F \to \mathbb{R} \cup \infty$ factors through the place $\rho : F \to \overline{F} \cup \infty$ associated with $v|_F$. Thus, we have

$$(5) \qquad dim_2 \frac{v(F^*)}{2v(F^*)} \leq dim_2 \frac{v_P(F^*)}{2v_P(F^*)} = m(P).$$

Combining (2)–(5), we have

$$(6) \qquad dim_2 \frac{v(K^*)}{v(T_1{}^*)} \leq trdeg\, K|F + m(P).$$

This leaves the case where $|\overline{V}| = 2$ and we have equality at (6). To prove (1) in this case, we must show $e(P) = 1$.

Since we have equality at (6), we must have equality at (3) so $\overline{F} \subseteq \overline{K}$ is algebraic. Fix a finite extension $\overline{F}_1$ of $\overline{F}$ such that $\overline{F}_1 \subseteq \overline{K}$ and the two orderings in $\overline{V} \subseteq Sper\,\overline{K}$ remain distinct when restricted to $\overline{F}_1$.

We also have equality at (5) so $m(P) = dim_2 \frac{v(F^*)}{2v(F^*)}$. Thus the place $\overline{F} \to \mathbb{R} \cup \infty$ induced by factoring λ_P through ρ has 2-divisible value group. By the definition of the place $\varphi_P : F \to F(P) \cup \infty$, ρ factors through φ_P inducing a place $F(P) \to \overline{F} \cup \infty$. Let F_1 be a finite unramified extension of $F(P)$ having $\overline{F}_1$ as residue field. Then F_1 has at least two distinct orderings so $F(P)$ is not hereditarily Euclidean and therefore, $e(P) = 1$. This completes the proof.

We return to the proof of (1.7). By [**13**, 5.1],

$$s(X) = sup\{s(X(\mathfrak{p})) \mid \mathfrak{p} \subseteq A \text{ is a real prime with } X(\mathfrak{p}) \neq \emptyset\}.$$

Let $\mathfrak{p} \subseteq A$ be a real prime with $X(\mathfrak{p}) \neq \emptyset$. Since $X = Sper_T A$, $X(\mathfrak{p}) = Sper_T F(\mathfrak{p})$ so we have

$$(*) \qquad s(X(\mathfrak{p})) \leq trdeg\, F(\mathfrak{p})|F + \overline{m} = dim\, A/\mathfrak{p} + \overline{m}$$

by (1.9). Since $dim\, A/\mathfrak{p} \leq d(X) = d$, $s(X) \leq d + \overline{m}$. This proves (2).

Also, by [**13**, 5.3],

$$\overline{s}(X) \leq \sum_{i=0}^{d} s_i$$

where $s_i = sup\{s(X(\mathfrak{p})) \mid \mathfrak{p} \subseteq A \text{ is a real prime and } dim\, A/\mathfrak{p} \leq i\}$. By $(*)$, $s_i \leq i + \overline{m}$ for $0 \leq i \leq d$. Thus,

$$\overline{s}(X) \leq \sum_{i=0}^{d} (i + \overline{m}) = d(d+1)/2 + (d+1)\overline{m}$$

which proves (3). This completes the proof of (1.7).

2. $s_F(R^n)$ and $\overline{s}_F(R^n)$.

One would like to show the bound for $s_F(V)$ given in §1 is the best possible. Although we are not able to do this in general, we can in the special case where $V = R^n$. We use the following:

Theorem 2.1. *For each $P \in Sper\ F$ and for each finite $k \leq \overline{m}(P)$, there exists a finite extension K of F such that $s(Sper_P K) \geq k$.*

Proof. Let $\overline{P}$ denote the pushdown of P along the place $\varphi_P : F \twoheadrightarrow F(P) \cup \infty$. Fix a finite extension $\overline{K}$ of $F(P)$ so that $\overline{P}$ extends at least $2^{e(P)}$ ways to $\overline{K}$ ($\overline{K}$ exists by (1.1).) Let W be a group with $v(F^*) \subseteq W \subseteq \frac{1}{2}v(F^*)$ and $|\frac{W}{v(F^*)}| = 2^k$, where v is the valuation associated with φ_P. Choose $a_1, \ldots, a_k \in P^*$ such that $\frac{1}{2}v(a_1), \ldots, \frac{1}{2}v(a_k)$ is a $\mathbb{Z}_2$-basis of W modulo $v(F^*)$. Then v extends (in fact, uniquely) to $F[\sqrt{a_1}, \ldots, \sqrt{a_k}]$ and $F[\sqrt{a_1}, \ldots, \sqrt{a_k}]/F$ is totally ramified of degree 2^k. Let K be a finite unramified extension of $F[\sqrt{a_1}, \ldots, \sqrt{a_k}]$ having $\overline{K}$ as residue field. Since $W \subseteq v(K^*)$ and $|\frac{v(K^*)}{v(F^*)}| = 2^k$, $v(K^*) = W$.

Let $T = \sum K^2 P$ and $\overline{T} = \sum \overline{K}^2 \overline{P}$. Suppose $x_1, \ldots, x_s \in K^*$, $p_1, \ldots, p_s \in P^*$ are such that $t = \sum_{i=1}^{s} x_i^2 p_i \in T^*$. Since $v(K^*) = W$, $v(x_i) = \sum_{j=1}^{k} \frac{m_{ij}}{2} v(a_j) + v(y_i)$, where $m_{ij} \in \{0,1\}$ and $y_i \in F^*$, and therefore, there exists $u_i \in K^*$ with $v(u_i) = 0$ such that $x_i^2 p_i = u_i^2 a_1^{m_{i1}} \cdots a_k^{m_{ik}} y_i^2 p_i$. Let $q_i = a_1^{m_{i1}} \cdots a_k^{m_{ik}} y_i^2 p_i \in P^*$ and assume $v(q_1) \leq v(q_i)$ for all i. Then $t = q_1 u$, where $u = \sum_{i=1}^{s} u_i^2 q_i / q_1$. Clearly $v(u) \geq 0$, $\overline{u} \in \overline{T} = \sum \overline{K}^2 \overline{P}$ and since $-1 \notin \overline{T}$, we must have $v(u) = 0$. Thus, $v(t) = v(q_1) = v(x_1^2 p_1)$ and if $v(t) = 0$ then $v(q_1) = 0$ so $\overline{t} = \overline{q}_1 \overline{u} \in \overline{P} \cdot \overline{T} = \overline{T}$. This shows $\overline{T}$ is the pushdown of T and $v(T^*) = 2v(K^*) + v(P^*) = 2W + v(F^*) = v(F^*)$ so $|\frac{v(K^*)}{v(T^*)}| = |\frac{v(K^*)}{v(F^*)}| = 2^k$. It follows that $s(Sper_P K) \geq k + s(Sper_{\overline{P}} \overline{K}) \geq k + e(P)$, which completes the proof.

Let the notation be as in Theorem 1.2. Suppose, for each finite $k \leq \overline{m}$, there exist primes $\mathfrak{p}_0 \supseteq \cdots \supseteq \mathfrak{p}_d$ in A such that the natural maps $\lambda_i : A/\mathfrak{p}_i \to A/\mathfrak{p}_{i-1}$ extend to discrete rank 1 places $\lambda_i : F(\mathfrak{p}_i) \to F(\mathfrak{p}_{i-1}) \cup \infty$ and such that $s(Sper_P F(\mathfrak{p}_0)) \geq k$. Then, we have a fan Z_0 in $Sper_P F(\mathfrak{p}_0)$ of order 2^k and inductively, a fan Z_i in $Sper_P F(\mathfrak{p}_i)$ of order 2^{i+k} obtained by pulling back $Z_{i-1} \subseteq Sper_P F(\mathfrak{p}_{i-1})$ along λ_i, $i = 1, \ldots, d$. It follows that $s_F(V) \geq s(Sper_P F(\mathfrak{p}_d)) \geq s(Z_d) = d + k$.

For example, such a sequence of primes exists if $V = R^n$ (so $d = n$ and $A = F[X_1, \ldots, X_n]$.) Namely, one can take $\mathfrak{p}_d = (0)$, $\mathfrak{p}_{d-1} = (X_d), \ldots, \mathfrak{p}_1 = (X_2, \ldots, X_d)$, $\mathfrak{p}_0 = (f, X_2, \ldots, X_d)$ where $f \in F[X_1]$ such that $F[X_1]/(f) \cong K$, K as in (2.1). Thus we have the following:

Theorem 2.2. $s_F(R^n) = s(Sper_P F(X_1, \ldots, X_n)) = n + \overline{m}$.

That is, the bound for $s_F(V)$ given in (1.2) is best possible, at least if $V = R^n$.

The analogous problem for $\overline{s}_F(V)$ is more complicated. Motivated by Scheiderer's construction in [16] we look for a sequence of prime ideals $\mathfrak{p}_0 \supseteq \cdots \supseteq \mathfrak{p}_d$ in A such that $\lambda_i : A/\mathfrak{p}_i \to A/\mathfrak{p}_{i-1}$ extends to a discrete rank 1 place $\lambda_i : F(\mathfrak{p}_i) \to F(\mathfrak{p}_{i-1}) \cup \infty$, $i \geq 1$, and spaces of orderings

$$Z_i \oplus Z_i' \subseteq Sper_P F(\mathfrak{p}_i), \qquad i \geq 0,$$

such that Z_i, Z_i' are fans of order $2^{i+\overline{m}}$ and Z_i is the pull-back of Z_{i-1}' along λ_i if $i \geq 1$. (We assume $\overline{m} < \infty$.) Then we take

$$Y_i = \bigcup_{j=0}^{i} (Z_j \oplus Z_j'), \qquad S_i = \bigcup_{j=0}^{i} (\{P_j\} \oplus Z_j')$$

where P_j is a fixed element of Z_j.

To obtain Z_i', $i \geq 1$, we want to choose additional primes $\mathfrak{q}_{i-1} \supseteq \mathfrak{p}_i$, $i = 1, \ldots, d$, with $\mathfrak{q}_{i-1} \not\subseteq \mathfrak{p}_0$ such that the natural map $\rho_i : A/\mathfrak{p}_i \to A/\mathfrak{q}_{i-1}$ lifts to a discrete rank 1 place $\rho_i : F(\mathfrak{p}_i) \to F(\mathfrak{q}_{i-1}) \cup \infty$ and $s(Sper_P F(\mathfrak{q}_{i-1})) = i - 1 + \overline{m}$. Then Z_i' can be obtained by pulling back a fan of order $2^{i-1+\overline{m}}$ in $Sper_P F(\mathfrak{q}_{i-1})$ along ρ_i. By our construction, the places ρ_i, λ_i are independent so Z_i, Z_i' lie in distinct Bröcker classes and therefore $Z_i \cup Z_i' = Z_i \oplus Z_i'$ holds automatically if $i \geq 1$ (see [6], [14, 3.2].)

We use the notation and theory in [13] to show that Y_i is saturated and S_i is basic closed in Y_i. Let $\Sigma_i = \{a \in A \mid a > 0 \text{ on } Y_i\}$. Then $a^2 \in \Sigma_i$ for $a \in A \setminus \mathfrak{p}_0$. Thus, if $Q \in cl(Y_i)$, $Q \notin Y_i$, then $supp(Q) \supseteq \mathfrak{q}_{j-1}$ for some $j \leq i$ so $supp(Q) \not\subseteq \mathfrak{p}_0$, and consequently $Q \notin U(\Sigma_i^2)$. This proves that Y_i is closed in $U(\Sigma_i^2) = Sper\, \Sigma_i^{-1}A$. Since $Y_i(\mathfrak{p}_j) = Z_j \oplus Z'_j$ is saturated in $Sper_P F(\mathfrak{p}_j)$ for all $j \leq i$, this implies Y_i is saturated (see [**13**, 2.3].) Also, it is clear that S_i is closed in Y_i and $S(\mathfrak{p}_j) = \{P_j\} \oplus Z'_j$ is basic in $Y_i(\mathfrak{p}_j) = Z_j \oplus Z'_j$ for all $j \leq i$ so, by [**13**, 3.2], S_i is basic closed in Y_i.

The argument in [**16**] shows that at least $\sum_{j=0}^{i}(j + \overline{m}) = i(i+1)/2 + (i+1)\overline{m}$ inequalities ($\geq$) are required to describe S_i in Y_i. Namely, $i+\overline{m}$ of these inequalities, say $f_1 \geq 0, \ldots, f_{i+\overline{m}} \geq 0$, are required just to describe $\{P_i\} = S_i \cap Z_i$ in $Z_i = Y_i \cap Z_i$ (since Z_i is a fan of order $2^{i+\overline{m}}$.) Moreover, each f_k is strictly negative at some $Q \in Z_i$ and, of course, $f_k \geq 0$ at Q' where $Q' \supseteq Q$ is the specialization of Q in Z'_{i-1} (since $Z'_{i-1} \subseteq S_i$.) Thus $f_k = 0$ at Q', that is, $f_k \in \mathfrak{p}_{i-1}$. This is true for $k = 1, \ldots, i+\overline{m}$. Thus, none of the inequalities $f_1 \geq 0, \ldots, f_{i+\overline{m}} \geq 0$ can contribute to the description of $S_{i-1} = S_i \cap Y_{i-1}$ in $Y_{i-1} = Y_i \cap Y_{i-1}$. Therefore, by induction on i, $\sum_{j=0}^{i-1}(j + \overline{m})$ additional inequalities are required for the description of S_{i-1} in Y_{i-1} so, altogether, $\sum_{j=0}^{i}(j+\overline{m})$ inequalities are required to describe S_i in Y_i. Thus $\overline{s}(Y_i) \geq i(i+1)/2 + (i+1)\overline{m}$. In particular,

$$\overline{s}(Sper_P A) \geq \overline{s}(Y_d) \geq d(d+1)/2 + (d+1)\overline{m}.$$

To be able to apply this in the case $V = R^n$, we assume

(*) There exists a finite extension L of F with $Z_0 \oplus Z'_0 \subseteq Sper_P L$ where Z_0, Z'_0 are fans of order $2^{\overline{m}}$.

Assuming this, we can realize the above set-up in $A = F[X_1, \ldots, X_d]$ as follows: take $\mathfrak{p}_d = (0)$, $\mathfrak{p}_{d-1} = (X_d), \ldots, \mathfrak{p}_1 = (X_2, \ldots, X_d)$, $\mathfrak{p}_0 = (f, X_2, \ldots, X_d)$ where $f \in F[X_1]$ is such that $F[X_1]/(f) \cong L$. The choice of $\mathfrak{q}_{i-1}$, $i = 1, \ldots, d$, is fairly arbitrary, for example, we could take $\mathfrak{q}_{d-1} = (X_d + 1), \ldots, \mathfrak{q}_1 = (X_2 + 1, X_3, \ldots, X_d)$ and $\mathfrak{q}_0 = (f(X_1 + 1), X_2, \ldots, X_d)$. The point is, with this choice of q_i, $q_{i-1} \not\subseteq \mathfrak{p}_0$, $F(\mathfrak{q}_{i-1}) \cong F(X_1, \ldots, X_{i-1})$ for $i \geq 2$, and $F(\mathfrak{q}_0) \cong L$ so $s(Sper_P F(\mathfrak{q}_{i-1})) = i - 1 + \overline{m}$, $i = 1, \ldots, d$. This proves:

Theorem 2.3. *If (F, P) satisfies (*) then $\overline{s}_F(R^n) = n(n+1)/2 + (n+1)\overline{m}$.*

If (*) does not hold, we are still able to get a somewhat weaker result by replacing L by the field K constructed in (2.1) and taking $Z_0 = \emptyset$ (so we only require that $Sper_P F(\mathfrak{p}_0)$ contains a fan Z'_0 of order $2^{\overline{m}}$) and taking

$$Y_i = \bigcup_{j=1}^{i}(Z_j \oplus Z'_j), \qquad S_i = \bigcup_{j=1}^{i}(\{P_j\} \oplus Z'_j).$$

Then, starting our induction at $i = 1$ instead of $i = 0$ yields $\overline{s}(Y_i) \geq \sum_{j=1}^{i}(j + \overline{m}) = i(i+1)/2 + i\overline{m}$. Thus we have:

Theorem 2.4. *For any F, $n(n+1)/2 + n\overline{m} \leq \overline{s}_F(R^n) \leq n(n+1)/2 + (n+1)\overline{m}$.*

Remarks 2.5.

1) If $\overline{m} = 0$, (F, P) is hereditarily Euclidean so cannot satisfy $(*)$ but, nonetheless, by (2.4) the conclusion of (2.3) is valid.

2) We have been assuming $\overline{m} < \infty$ but this is unnecessary. If $\overline{m} = \infty$, then $\overline{s}_F(R^n) \geq s_F(R^n) = n + \overline{m} = \infty$, so the conclusion of (2.3) is true in this case as well.

3) For $0 < \overline{m} < \infty$, (F, P) may or may not satisfy $(*)$. Fixing an integer k, $0 < k < \infty$, we have the following examples:

   (*i*) Take F to be the iterated formal power series field $\mathbb{R}((t_1))\dots((t_k))$ and denote by $\lambda : F \to \mathbb{R} \cup \infty$ the naturally associated discrete rank k place. Any ordering $P \in \mathit{Sper}\, F$ is compatible with λ and $\overline{m}(P) = k$. But (F, P) does not satisfy $(*)$ since, for any finite extension L of F, L has at most 2^k orderings.

   (*ii*) Take F to be the rational function field $\mathbb{R}(t_1, \dots, t_k)$ and take $P \in \mathit{Sper}\, F$ to be the restriction to F of any ordering on $\mathbb{R}((t_1))\dots((t_k))$. Again, $\overline{m}(P) = k$, but now, according to [**15**, §4], (F, P) does satisfy $(*)$.

Of course, having $(*)$ fail does not necessarily imply that the conclusion of (2.3) fails. We know of no example where the conclusion of (2.3) fails.

References

1. C. Andradas, L. Bröcker, J.M. Ruiz, *Minimal generation of basic open semi-analytic sets*, Invent. Math. **92** (1988), 409–430.
2. E. Becker, *On the real spectrum of a ring and its application to semi-algebraic geometry*, Bull. Amer. Math. Soc. **15** (1986), 19–61.
3. J. Bochnak, M. Coste, M.-F. Roy, *Géométrie Algébrique Réelle*, Ergeb. der Math. **3** 12, Springer-Verlag, Berlin Heidelberg New York, 1987.
4. L. Bröcker, *On the stability index of Noetherian rings*, preprint.
5. ———, *On basic semi-algebraic sets*, Expositiones Math. (to appear).
6. ———, *Über die Anzahl der Anordnungen eines kommutativen Körpers*, Arch. Math. **29** (1977), 458–464.
7. ———, *Zur Theorie der quadratischen Formen über formal reelen Körpern*, Math. Ann. **210** (1974), 233–256.
8. M. Knebusch, C. Scheiderer, *Einführung in die reele Algebra*, Fiedr. Vieweg & Sohn, Braunschweig/Wiesbaden, 1989.
9. T.Y. Lam, *An introduction to real algebra*, Rocky Mountain J. Math. **14** (1984), 767–814.
10. ———, *Orderings, valuations, and quadratic forms*, CBMS Regional Conf. Ser. in Math. no. 52, Amer. Math. Soc., Providence, R.I., 1983.
11. ———, *The theory of ordered fields*, Ring Theory and Algebra III (ed. B. McDonald), Lecture Notes in Pure and Applied Math., vol. 55, Dekker, New York, 1980, pp. 1–152.
12. L. Mahé, *Une démonstration élementaire du théorème de Bröcker-Scheiderer*, preprint.
13. M. Marshall, *Minimal generation of basic sets in the real spectrum of a commutative ring*, preprint.
14. J. Merzel, *Quadratic forms over fields with finitely many orderings*, Ordered Fields and Real Algebraic Geometry (eds. D. Dubois and T. Récio), Contemporary Math., vol. 8, Amer. Math. Soc., Providence, R.I., 1982, pp. 185–229.
15. C. Scheiderer, *Spaces of orderings of fields under finite extensions*, Manuscripta Math. **72** (1991), 27–47.
16. ———, *Stability index of real varieties*, Invent. Math. **97** (1989), 467–483.

Department of Mathematics and Statistics, University of Saskatchewan, Saskatoon, Canada S7N 0W0

CONFIGURATIONS OF AT MOST 6 LINES OF $\mathbb{R}P^3$

V. F. MAZUROVSKII

Ivanovo Civil Engineering Institute

An unordered $(n;k)$-configuration of degree m is defined to be an unordered collection of m linear k-dimensional subspaces of $\mathbb{R}P^n$. We associate with each configuration its upper and lower ranks, i.e. the dimensions of the projective hull and intersection respectively of all the subspaces of the configuration. The combinatorial characteristic of a configuration is, by definition, the list of upper and lower ranks of all its subconfigurations. Two configurations are said to be rigidly isotopic if they can be joined by isotopy which consists of configurations with the same combinatorial characteristics. It is obvious that the property of being rigidly isotopic is equivalence relation. The equivalence class of a configuration by this relation is called its rigid isotopy type.

The space $SPC^m_{n,k}$ of unordered $(n;k)$- configurations of degree m is naturally isomorphic to m-th symmetric power of Grassmanian manifold $G_{n+1,k+1}$. A configuration is said to be non-singular if all its subspaces are in general position. The set $GSPC^m_{n,k}$ of non-singular configurations is an open subset of manifold $SPC^m_{n,k}$ in Zariski topology. The set of all non-singular configurations of the same rigid isotopy type forms a connected component of $GSPC^m_{n,k}$ (in strong topology). These connected components are called cameras of manifold $SPC^m_{n,k}$.

In the paper [6] O. Ya. Viro enumerated the cameras of the spaces $SPC^m_{3,1}$ for $m \leq 5$ and showed that non-singular $(3;1)$-configurations of degree ≤ 5 are determined up to rigid isotopy by the linking coefficients of the lines of a configuration. In [4] the author enumerated the cameras of $SPC^6_{3,1}$ and proved that non-singular configurations of 6 lines of $\mathbb{R}P^3$ are not determined up to rigid isotopy by the linking coefficients. The main purpose of the present paper is to describe the mutual position of the cameras in $SPC^m_{3,1}$ for $m \leq 6$. In particular we give a detailed proof of classification of non-singular $(3;1)$-configurations of degree 6 up to rigid isotopy (in [4] this proof was only outlined).

The author is grateful to O. Ya. Viro for posing the problem and fruitful discussions.

§ 1. BASIC CONSTRUCTIONS

1.1. Linking coefficients. In the next two sections we describe the constructions of O. Ya. Viro (see [6], [7]).

Two disjoint oriented lines L_1^*, L_2^* in the oriented space $\mathbb{R}P^3$ have linking coefficient $lk(L_1^*, L_2^*)$ equal to $+1$ or -1 (here the doubled linking coefficient of cycles L_1^* and L_2^* in the oriented manifold $\mathbb{R}P^3$ is considered). Let $L = \{L_1, L_2, L_3\}$ be an unordered non-singular configuration of three lines of the oriented space $\mathbb{R}P^3$, and let L_1^*, L_2^*, L_3^* be the same lines equiped with some orientations. The product $lk(L_1^*, L_2^*) \times$

$lk(L_1^*, L_3^*)lk(L_2^*, L_3^*)$ denoted by $lk(L_1, L_2, L_3)$ does not depend on the choise of the orientations of $L_i, i = 1, 2, 3$, is preserved under isotopies of L, and changes under reversal of the orientation of $\mathbb{R}P^3$. Unordered non-singular $(3;1)$-configurations $L = \{L_1, L_2, \dots, L_m\}$ and $L' = \{L'_1, L'_2 \dots L'_m\}$ are said to be homology equivalent if there exists a bijection $\varphi : L \to L'$ such that for a fixed orientation of $\mathbb{R}P^3$ $lk(L_i, L_j, L_k) = lk(\varphi(L_i), \varphi(L_j), \varphi(L_k))$ with any $i, j, k = 1, 2, \dots, m$, $i \neq j$, $i \neq k$, $j \neq k$.

1.2. Construction of the join of two configurations. Let $A = \{A_1, \dots, A_m\}$ be an unordered configuration of k-dimensional subspaces of $\mathbb{R}P^n$, and let $B = \{B_1, \dots, B_m\}$ be an unordered configuration of l-dimensional subspaces of $\mathbb{R}P^s$. We suppose that $\mathbb{R}P^n$ and $\mathbb{R}P^s$ are imbedded into $\mathbb{R}P^{n+s+1}$ as disjoint linear subspaces. If n and s are odd we suppose, in addition, that $\mathbb{R}P^n$, $\mathbb{R}P^s$, and $\mathbb{R}P^{n+s+1}$ are oriented, and linking coefficient of the images of $\mathbb{R}P^n$ and $\mathbb{R}P^s$ in $\mathbb{R}P^{n+s+1}$ equals $+1$. Let $f : \{1, \dots, m\} \to \{1, \dots, m\}$ be some bijection, and C_i be the projective hull of the images of A_i and $B_{f(i)}$ in $\mathbb{R}P^{n+s+1}$, $i = 1, \dots, m$. It is clear that $C = \{C_1, \dots, C_m\}$ is a $(n+s+1; k+l+1)$-configuration of degree m. The configuration C is called the join of A and B. A configuration is called an isotopy join if it is rigidly isotopic to the join of some two configurations.

1.2.1. Lemma. *The mirror image of an isototy join is also an isotopy join.*

1.3. Degeneration and perturbation. Let $s : [0,1] \to SPC^m_{n,k}$ be a path with the beginning at a point A such that the restriction $s\big|_{[0,1)}$ is a rigid isotopy of A. If configurations A and $A' = s(1)$ have distinct combinatorial characteristics, then s is called a degeneration of configuration A, and path s^{-1} is called a perturbation of A'. Two subspaces A_i and A_j of a configuration $A = \{A_1, \dots, A_m\}$ are said to be contiguous if either they coincide, or there exists a degeneration such that (1) its restrictions on the configurations $\{A_1, \dots, A_{i-1}, A_{i+1}, \dots, A_m\}$ and $\{A_1, \dots, A_{j-1}, A_{j+1}, \dots, A_m\}$ which are obtained by removing the elements A_i, A_j from A respectively are rigid isotopies, and (2) the subspaces corresponding to the subspaces A_i and A_j coincide in the result of this degeneration.

1.4. Adjacency graph. A configuration is said to be 1-singular if all configurations rigidly isotopic to it form a codimension 1 subset in the configuration space. The set of all 1-singular configurations of the same rigid isotopy type is called a wall. Two 1-singular configurations are said to be p-equivalent if they belong to walls which separate the same cameras. The set of all 1-singular configurations of $SPC^m_{n,k}$ will be denoted by $G^1SPC^m_{n,k}$.

The mutual position of the cameras in the configuration space can be described by means of the adjacency graph (see [2]), whose vertices and edges are in one-to-one correspondence with the cameras and walls respectively, and two vertices representing some cameras are connected by an edge if and only if these cameras are adjacent to the wall corresponding to this edge. It may happen that the beginning of an edge coincides with its end, as in the following cases:

a) if the configuration space has a boundary and the wall is contained in it;

b) if the wall is a one-sided subset of the configuration space;

c) if the wall is a two-sided subset, but has the same camera adjacent at each side.

Each of these cases corresponds to a loop in the adjacency graph. In cases b) and c) the wall is called inner one.

1.5. Affine (3; 1)–configurations. By an unordered affine (3; 1)–configuration of degree m we mean an unordered collection of m lines of $\mathbb{R}^3$. The canonical imbedding $\mathbb{R}^3 \hookrightarrow \mathbb{R}P^3$ induces a map of the set of affine (3; 1)–configurations of degree m into $SPC^m_{3,1}$. This map is called the projective completion, and the image of an affine (3; 1)–configuration K under this map is called the projective completion of K. A rigid isotopy of an affine (3; 1)–configuration is an isotopy of this configuration which induces a rigid isotopy of its projective completion.

Let $Oxyz$ be the canonical Cartesian coordinate system in $\mathbb{R}^3$. The planes of $\mathbb{R}^3$ defined by the equations $z = \text{const}$ will be called horizontal planes. The common line of the projective completions of horizontal planes is called the horizontal line of infinity.

Consider the affine (3; 1)-configuration S which consists of the following lines $\begin{cases} y=0 \\ z=1 \end{cases}$, $\begin{cases} x=0 \\ z=1 \end{cases}$, $\begin{cases} y=x \\ z=-1 \end{cases}$, $\begin{cases} y=-x \\ z=-1 \end{cases}$. The pairs of lines $\begin{cases} y=0 \\ z=1 \end{cases}$, $\begin{cases} x=0 \\ z=1 \end{cases}$ and $\begin{cases} y=x \\ z=-1 \end{cases}$, $\begin{cases} y=-x \\ z=-1 \end{cases}$ will be called the frames of configuration S. Let $\bar{S}$ be the projective completion of S. Consider a configuration obtained by adjoining to $\bar{S}$ several pairwise disjoint lines which have no common points with $\bar{S}$. Such configuration is said to be framed. The subconfiguration $\bar{S}$ of a framed configuration will be called the skeleton, the other lines will be called free lines.

We call the sliding translation the affine transformation given by the formulas $x' = x + az$, $y' = y + bz$, $z' = z$, where a and b are some numbers.

In what follows we suppose that $\mathbb{R}^3$ is canonically imbedded into $\mathbb{R}P^3$.

§ 2. Join configurations of lines of $\mathbb{R}P^3$

2.1. Classification of non-singular isotopy join configurations of 6 lines of $\mathbb{R}P^3$ up to rigid isotopy. Let L^1 and L^2 be two oriented disjoint lines in $\mathbb{R}P^3$ with positive linking coefficient. Let A^j_i be some pairwise different points of L^j, $j = 1, 2$, $i = 1, 2, \ldots, m$, such that the increase of index i agrees to the orientation of L^j. Consider a permutation σ of degree m. The lines passing through points A^1_i and $A^2_{\sigma(i)}$ form a non-singular join (3; 1)-configuration of degree m, which will be denoted by $jc(\sigma)$. It is easy to see that up to rigid isotopy this construction provides all isotopy join configurations of m lines of $\mathbb{R}P^3$. It is also clear that the map $S_m \longrightarrow \pi_0(GSPC^m_{3,1})$ which assigns to a permutation $\sigma \in S_m$ the rigid isotopy type of $jc(\sigma)$ is well defined. We denote this map by $\varkappa$.

In what follows $(\sigma_1, \sigma_2, \ldots, \sigma_m)$ denotes permutation $\begin{pmatrix} 1 & 2 & \ldots & m \\ \sigma_1 & \sigma_2 & \ldots & \sigma_m \end{pmatrix}$.

2.1.1. Lemma. *1) Let μ and ν be permutations of degree m preserving the natural cyclic order. Then $\varkappa(\mu \cdot \sigma \cdot \nu) = \varkappa(\sigma)$.*

2) $\varkappa(\sigma^{-1}) = \varkappa(\sigma)$.

3) $\varkappa((m, m-1, \ldots, 2, 1) \cdot \sigma \cdot (m, m-1, \ldots, 2, 1)) = \varkappa(\sigma)$.

4) Suppose that permutation $\sigma = (\sigma_1, \dots, \sigma_{i-1}, \sigma_i, \dots, \sigma_j, \sigma_{j+1}, \dots, \sigma_m)$ satisfies the following condition: for any integer p, with $\min_{i\le k\le j} \sigma_k \le p \le \max_{i\le k\le j} \sigma_k$, there exists an index $l (i \le l \le j)$ such that $\sigma_l = p$.

Let $\alpha = \min_{i\le k\le j} \sigma_k$, $(\bar\tau_i, \dots, \bar\tau_j) = (j-i, \dots, 2, 1)\cdot(\sigma_i - \alpha+1, \dots, \sigma_j - \alpha + 1)(j-i, \dots, 2, 1)$, and $\tau = (\sigma_1, \dots, \sigma_{i-1}, \tau_i, \dots, \tau_j, \sigma_{j+1}, \dots, \sigma_m)$, where $\tau_k = \bar\tau_k + \alpha - 1, k = i, i+1, \dots, j-1, j$. Then $\varkappa(\tau) = \varkappa(\sigma)$.

Proof. Statements 1) and 2) are obvious. Non-singular join configuration $jc(\sigma)$ is mapped to non-singular join configuration $jc((m, \dots, 1)\cdot\sigma\cdot(m, \dots, 1))$ by a projective transformation which reverses orientations of the lines L^1 and L^2 and preserves orientation of $\mathbb{R}P^3$. In case 4) the lines of non-singular join configuration $jc(\sigma)$, corresponding to elements $\sigma_i, \sigma_{i+1}, \dots, \sigma_{j-1}, \sigma_j$ of permutation σ, are in a regular neighbourhood which does not intersect the other lines of the configuration. The rigid isotopy, connecting the configurations $jc(\sigma)$ and $jc(\tau)$, is given by the projective transformation from 3) restricted to this regular neighbourhood.

A configuration is called a mirror configuration if it is rigidly isotopic to its mirror image.

2.1.2. Theorem. *Two non-singular isotopy join configurations of 6 lines of $\mathbb{R}P^3$ are rigidly isotopic if and only if they are homology equivalent. Any non-singular isotopy join (3; 1)-configuration of degree 6 is rigidly isotopic to one of the following 15 parwise non-isotopic non-singular join configurations:*

a)
$jc(1,2,3,4,5,6)$, $jc(1,2,3,4,6,5)$, $jc(1,2,3,5,6,4)$, $jc(1,2,4,3,6,5)$, $jc(1,2,4,6,3,5)$, $jc(1,2,5,6,3,4)$;

b) mirror configurations $jc(1,2,3,6,5,4)$, $jc(1,3,5,2,6,4)$, $jc(1,2,4,6,5,3)$;

c) the mirror images of the configurations of a): $jc(6,5,4,3,2,1)$, $jc(5,6,4,3,2,1)$, $jc(4,6,5,3,2,1)$, $jc(5,6,3,4,2,1)$, $jc(5,3,6,4,2,1)$, $jc(4,3,6,5,2,1)$.

Proof of Theorem 2.1.2 is reduced to sorting all permutations of S_6. First we split S_6 into union ofdisjoint classes of permutations with homology equivalentcorresponding join configurations. Then, using Lemma 2.1.1, weprove that elements of each of these classes define rigidly isotopic non-singular join configurations.

2.1.3. Remark. A similar consideration shows that any non-singular isotopy join (3; 1)-configuration of degree ≤ 5 is rigidly isotopic to one of the following pairwise non-isotopic non-singular join configurations:$jc(1)$, $jc(1,2)$, $jc(1,2,3)$, $jc(3,2,1)$, $jc(1,2,3,4)$, $jc(1,2,4,3)$, $jc(4,3,2,1)$, $jc(1,2,3,4,5)$, $jc(1,3,4,5,2)$, $jc(1,4,5,2,3)$, $jc(1,3,5,2,4)$, $jc(3,2,5,4,1)$, $jc(2,5,4,3,1)$, $jc(5,4,3,2,1)$ (compare with Viro's classification in [6]).

2.2. Classification of 1-singular isotopy join (3; 1)-configurations of degree ≤ 6 up to rigid isotopy. Consider again two oriented disjoint lines L^1 and L^2 in $\mathbb{R}P^3$ with positive linking coefficient. Let $A^1_1, \dots, A^1_{m-1}, A^1_m$ be different points of L^1, and $A^2_1, \dots A^2_{m-1}$ be different points of L^2 such that the increase of the lower indices of the points agrees to the orientations of L^1 and L^2. Consider a map f from $\{1, \dots, m-1, m\}$ onto $\{1, \dots, m-1\}$ and connect points A^1_i and $A^2_{f(i)}$ by lines, $i = 1, \dots, m-1, m$. The obtained 1-singular join (3; 1)-configuration of degree m is denoted by $sjc(f)$. The set of all surjections of $\{1,\dots,m-1,m\}$ onto $\{1,\dots,m-1\}$ will be denoted by $\hat{S}_m$. It is clear that the map $\hat{\varkappa}: \hat{S}_m \to \pi_0(G^1SPC^m_{3,1})$ which assigns to a map $f \in \hat{S}_m$ the rigid isotopy type of $sjc(f)$ is well defined. Since a surjection f can be represented by the table $\begin{pmatrix} 1 & \dots & m-1 & m \\ f(1) & \dots & f(m-1) & f(m) \end{pmatrix}$, we shall also use the symbol $(f(1),\dots,f(m-1),f(m))$ to denote f.

2.2.1. Lemma. *1) If μ and ν are permutations of degree m and $m-1$ respectively preserving the natural cyclic orders, then $\hat{\varkappa}(\mu \cdot f \cdot \nu) = \hat{\varkappa}(f)$;*

2) $\hat{\varkappa}((m,m-1,\dots,2,1)\cdot f \cdot (m-1,\dots,2,1)) = \hat{\varkappa}(f)$;

3) Suppose that surjection $f = (f(1),\dots,f(i-1), f(i),\dots,f(j), f(j+1),\dots,f(m))$ satisfies the following conditions: (1) the elements $f(i),\dots f(j)$ are parwise different, and (2) for any integer p, with $\min_{i\le k\le j} f(k) \le p \le \max_{i\le k\le j} f(k)$, there exists an index l $(i \le l \le j)$ such that $f(l) = p$.

Let $\alpha = \min_{i\le k\le j} f(k)$, $(\bar{\sigma}_i,\dots,\bar{\sigma}_j) = (j-i,\dots,2,1)\cdot(f(i)-\alpha+1,\dots,f(j)-\alpha+1)$ $\cdot(j-i,\dots,2,1)$, and $g = (g(1),\dots,g(i-1), g(i),\dots,g(j), g(j+1),\dots,g(m))$, where $g(s) = f(s)$ for any $s = 1,\dots,j-1$ and any $s = j+1,\dots,m$ and $g(k) = \sigma_k$ for any $k = i, i+1,\dots,j-1, j$, where $\sigma_k = \bar{\sigma}_k + \alpha - 1$. Then $\hat{\varkappa}(g) = \hat{\varkappa}(f)$.

Proof is analogous to the proof of Lemma 2.1.1.

2.2.2. Lemma. *The result of any perturbation of 1-singular join (3; 1)-configuration is a non-singular isotopy join (3; 1)-configuration.*

2.2.3. Theorem. *Two 1-singular isotopy join configurations of ≤ 6 lines of $\mathbb{R}P^3$ are rigidly isotopic if and only if they are p-equivalent.*

a) Any 1-singular isotopy join (3; 1)-configuration of degree 2 is rigidly isotopic to the join configuration $sjc(1,1)$.

b) Any 1-singular isotopy join (3; 1)-configuration of degree 3 is rigidly isotopic to the join configuration $sjc(1,1,2)$.

c) Any 1-singular isotopy join (3; 1)-configuration of degree 4 is rigidly isotopic to one of the following 3 pairwise non-isotopic join configurations: $sjc(1,1,2,3)$, $sjc(1,2,1,3)$, $sjc(3,2,1,1)$. Only $sjc(1,2,1,3)$ is a mirror configuration.

d) Any 1-singular isotopy join (3; 1)-configuration of degree 5 is rigidly isotopic to one of the following 8 pairwise non-isotopic join configurations:

1) $sjc(1,1,2,3,4)$, $sjc(1,2,1,3,4)$, $sjc(1,3,1,2,4)$, $sjc(1,1,3,4,2)$;

2) the mirror images of the configurations from 1): $sjc(4,3,2,1,1)$, $sjc(4,3,1,2,1)$, $sjc(4,2,1,3,1)$, $sjc(2,4,3,1,1)$.

e) Any 1-singular isotopy join (3; 1)-configuration of degree 6 is rigidly isotopic to one of the following 31 pairwise non-isotopic join configurations:

i) $sjc(1,1,2,3,4,5)$, $sjc(1,2,1,3,4,5)$, $sjc(1,1,2,3,5,4)$, $sjc(1,2,3,1,4,5)$, $sjc(1,1,3,4,5,2)$, $sjc(1,1,2,4,5,3)$, $sjc(1,3,1,2,4,5)$, $sjc(1,2,1,3,5,4)$, $sjc(1,3,4,1,5,2)$, $sjc(1,1,3,2,5,4)$, $sjc(1,2,4,1,3,5)$, $sjc(1,2,5,1,3,4)$, $sjc(1,3,1,4,2,5)$, $sjc(1,3,1,4,5,2)$;

ii) mirror configurations $sjc(1,1,3,5,2,4)$, $sjc(1,2,3,1,5,4)$, $sjc(1,2,4,1,5,3)$;

iii) the mirror images of the configurations from i): $sjc(5,4,3,2,1,1)$, $sjc(5,4,3,1,2,1)$, $sjc(4,5,3,2,1,1)$, $sjc(5,4,1,3,2,1)$, $sjc(2,5,4,3,1,1)$, $sjc(3,5,4,2,1,1)$,

$sjc(5,4,2,1,3,1)$, $sjc(4,5,3,1,2,1)$, $sjc(2,5,1,4,3,1)$, $sjc(4,5,2,3,1,1)$, $sjc(5,3,1,4,2,1)$, $sjc(4,3,1,5,2,1)$, $sjc(5,2,4,1,3,1)$, $sjc(2,5,4,1,3,1)$.

Proof of Theorem 2.2.3 is reduced to sorting all elements of $\hat{S}_m$ for $m \leq 6$. At the beginning, using Lemma 2.2.2, Remark 2.1.3, and Theorem 2.1.2, we split $\hat{S}_m (m \leq 6)$ into union of disjoint classes consisting of maps with p-equivalent corresponding 1-singular join configurations. Then, using Lemma 2.2.1, we prove that elements of each of these classes define rigidly isotopic 1-singular join configurations.

2.3. Adjacency graph of $SPC^m_{3,1}$ for $m \leq 5$.

2.3.1. Lemma. *Any 1-singular* (3; 1)*- configuration of degree* ≤ 4 *is an isotopy join.*

Proof. It is sufficient to prove the lemma for the case of 1-singular (3; 1)-configurations of degree 4. Let $K = \{K_1, K_2, K_3, K_4\}$ be a 1-singular (3; 1)- configuration. We assume that $K_3 \cap K_4 \neq \varnothing$. Let Q be the quadric containing disjoint lines K_1, K_2, and K_3. Up to a small rigid isotopy of K_4 it can be assumed to intersect Q in two different points A and B. Consider the generatrices L^1 and L^2 of Q which pass through points A and B respectively and belong to the family of generatrices dual to the family containing K_1, K_2, and K_3. It is clear that $K_i \cap L^j \neq \varnothing$ for any $i = 1,2,3,4$ and $j = 1,2$.

2.3.2. Lemma. *If there exists a line intersecting all lines of a given 1-singular (non-singular)* (3; 1)*-configuration* K*, then* K *is an isotopy join.*

Proof. (This proof is due to O. Ya. Viro). Let line L^1 intersect all lines of the given (3; 1)-configuration K. Consider a projective transformation of $\mathbb{R}P^3$ which preserves the orientation of $\mathbb{R}P^3$ and puts L^1 on the horizontal line at infinity. Let K' be the image of K under this transformation. It is easy to see that the affine parts of lines of K' lie on horizontal planes. Let line L^2 be transversal to the horizontal planes and, in the case when K is a 1-singular configuration, intersect the two lines of K' meeting each other. Obviously, every other line of K' can be translated (in its horizontal plane) so as to intersect L^2. After that we shall obtain a join configuration K'', which is evidently rigidly isotopic to the given configuration K.

2.3.3. Lemma. *Any 1-singular* (3; 1)*-configuration of degree* 5 *obtained by a perturbation of a framed configuration is an isotopy join.*

Proof is reduced to sorting all possible locations of the free line of a framed configuration and using Lemma 2.3.2.

2.3.4. Lemma. *Any 1-singular* (3; 1)*-configuration of degree* ≤ 6 *either is an isotopy join, or can be obtained by a perturbation of a framed configuration.*

Proof uses induction on the number of the lines of the configuration. Lemma 2.3.1 provides the basis of the induction. Assume that the statement is true for 1-singular (3; 1)-configurations of degree < 6.

Let $K = \{K_1, K_2, K_3, K_4, K_5, K_6\}$ be a 1-singular (3; 1)-configuration. Assume that $K_1 \cap K_3 \neq \varnothing$. If lines K_1 and K_3 of K are contiguous, then K is an isotopy join due to the inductive hypothesis and Lemmas 2.2.2 and 2.3.3. Let lines K_1 and K_3 be not contiguous. Then there exists a subconfiguration K'' of K which is a 1-singular mirror (3; 1)-configuration of degree 4. We can assume, without loss of generality, that

$K'' = \{K_1, K_2, K_3, K_4\}$. Let Q be the quadric containing the disjoint lines K_2, K_3, and K_4. Up to a small rigid isotopy of K_1 it can be assumed to intersect Q in two different points A and B. Let L^1 and L^2 be the two generatrices of Q which belong to the family dual to that containing K_2, K_3, and K_4 and pass through points A and B respectively. It is clear that $K_i \cap L^j \neq \varnothing$ for any $i = 1,2,3,4$ and $j = 1,2$. Let $K_1 \cap K_3 = A$. Consider a projective transformation of $\mathbb{R}P^3$ which preserves the orientation of $\mathbb{R}P^3$ and puts L^2 on the horizontal line of infinity. After that we shall have the following situation in $\mathbb{R}^3$ (canonically imbedded into $\mathbb{R}P^3$) : the images of lines K_1 and K_3 have a single common point, the images of the lines K_1, K_2, K_3, and K_4 lie on horizontal planes and intersect the image of L^1. Consider a sliding translation such that the image of L^1 is perpendicular to a horizontal plane. Let K_i' be the image of the affine part of K_i after the transformations given above, where $i = 1, 2, \ldots, 6$. It is easy to see that lines K_1' and K_3' have a single common point, lines K_1', K_2', K_3', and K_4' lie on horizontal planes and intersect a vertical line (i.e. a line perpendicular to a horizontal plane). We can assume, without loss of generality, that K_4' lies over K_2', and line K_2' lies over K_1' and K_3'. Consider the orthogonal projection of the affine configuration $K' = \{K_1', K_2', K_3', K_4', K_5', K_6'\}$ onto the plane perpendicular to K_4'. If this projection differs from the cases a) and b) below, then it is clear that configuration K' can be obtained by a perturbation of a framed configuration.

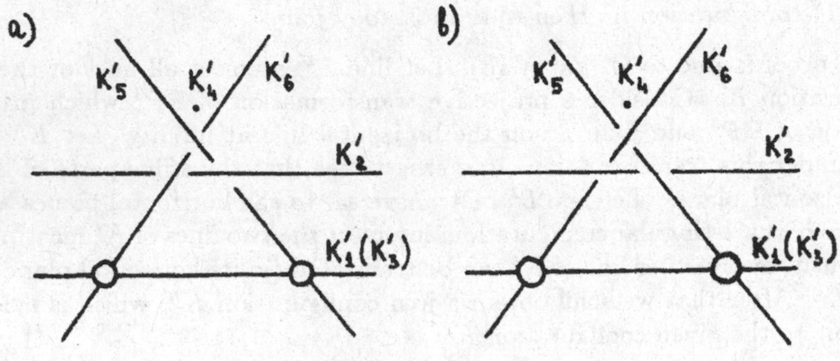

In cases a) and b) it can be proved, using Lemma 2.3.2 and sorting all possible locations of K_5' and K_6' in respect to K_1' and K_3', that K' is an isotopy join.

The next theorem is a consequence of Lemmas 2.3.4 and 2.3.3.

2.3.5. Theorem. *Any 1-singular (3; 1) - configuration of degree ≤ 5 is an isotopy join.*

2.3.6. Remark. Due to Theorem 2.3.5 and Lemma 2.2.2 any non-singular (3; 1) - configuraton of degree ≤ 5 is an isotopy join. Together with Remark 2.1.3 it yields the classification of non-singular (3; 1) - configurations of degree ≤ 5 up to rigid isotopy obtained by O.Ya.Viro in [6].

In what follows $[K]$ denotes the rigid isotopy type of configuration K. The following statements are consequences of Theorems 2.3.5 and 2.2.3.

2.3.7. Theorem. *Two 1-singular configurations of ≤ 5 lines of $\mathbb{R}P^3$ are rigidly isotopic if and only if they are p-equivalent. Any 1-singular (3; 1) - configuration of degree ≤ 5 is rigidly isotopic to one of the pairwise non-isotopic join configurations given in Theorem 2.2.3 a)-d).*

2.3.8. Theorem. *a) The adjacency graph of* $SPC^2_{3,1}$ *has one vertex and one edge-loop, which corresponds to the one-sided inner wall.*

b) The adjacency graph of $SPC^3_{3,1}$ *is shown on the diagram below.*

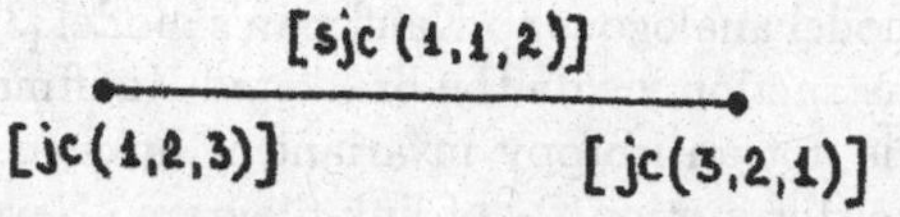

c) The adjacency graph of $SPC^4_{3,1}$ *is the graph presented below. The loop corresponds to the one-sided inner wall.*

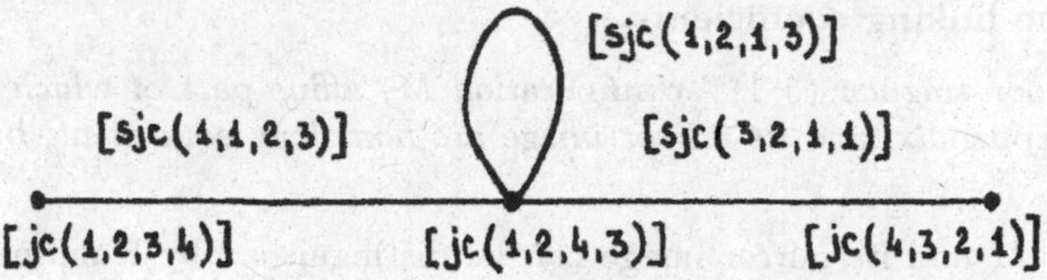

d) The adjacency graph of $SPC^5_{3,1}$ *is the following.*

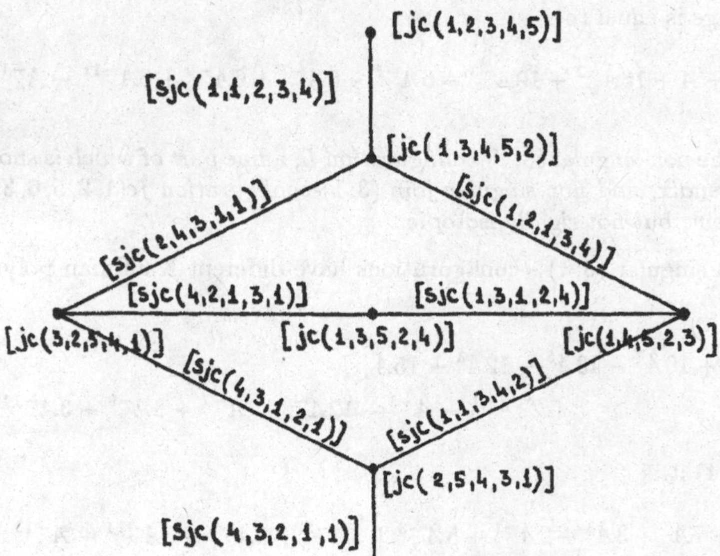

§ 3. Kauffman polynomial of non-singular configurations of lines of $\mathbb{R}P^3$

3.1. Polynomial invariant of non-singular (3; 1)**-configurations.** Yu. V. Drobotukhina in [1] defined an analogue of Jones polynomial for links in $\mathbb{R}P^3$. This polynomial was defined by means of the state model analogous to Kauffman's model [3] for the Jones polynomial of links in S^3. This construction yields the bracketed Kauffman polynomial of links in $\mathbb{R}P^3$. This polynimial is not an isotopy invariant of links in $\mathbb{R}P^3$, since it is not preserved under the Reidemeister motion Ω_1 of link diagram. Nevertheless, it is an rigid isotopy invariant of non-singular configurations of projective lines, since in the process of rigid isotopy this Reidemeister motion does not occur. This polynomial will be called the Kauffman polynomial of a non-singular configuration of lines of $\mathbb{R}P^3$.

3.2. Insufficiency of the linking coefficients.

3.2.1. Theorem. *The non-singular* (3; 1) *- configuration M, affine part of which is shown on diagram 1 in Appendix, and its mirror image are homology equivalent, but not rigidly isotopic.*

Proof. The configuration M and its mirror image can be distinguished by Kauffman polynomial, which for M is equal to

$$-A^{15}+6A^{11}+6A^{9}-5A^{7}-6A^{5}+10A^{3}+16A+A^{-1}-10A^{-3}+10A^{-7}+5A^{-9},$$

and for its mirror image is equal to

$$5A^{9}+10A^{7}-10A^{3}+A+16A^{-1}+10A^{-3}-6A^{-5}-5A^{-7}+6A^{-9}+6A^{-11}-A^{-15}.$$

3.2.2. Theorem. *The non-singular* (3; 1)*-configuration L, affine part of which is shown on diagram 2 in Appendix, and non-singular join* (3; 1)*-configuration* $jc(1,2,5,6,3,4)$ *are homology equivalent, but not rigidly isotopic.*

Proof. These two non-singular (3; 1) - configurations have different Kauffman polynomials: for L it is

$$A^{17}-5A^{13}+15A^{9}+10A^{7}-13A^{5}-12A^{3}+15A$$
$$+22A^{-1}-A^{-3}-12A^{-5}+A^{-7}+8A^{-9}+3A^{-11},$$

and for $jc(1,2,5,6,3,4)$ it is

$$A^{13}+A^{11}+4A^{7}+7A^{5}+3A^{3}+2A^{-1}+5A^{-3}+3A^{-5}+2A^{-9}+3A^{-11}+A^{-13}.$$

We denote the mirror images of the configurations L and M by L' and M' respectively.

3.2.3. Corollary. The non-singular (3; 1) - configurations M, L, M', and L' are not isotopy joins.

§ 4. Configurations of 6 lines of $\mathbb{R}P^3$

4.1. Singular framed configurations of 6 lines of $\mathbb{R}P^3$. Consider the projection of the skeleton of the affine part of a framed configuration onto a horizontal plane

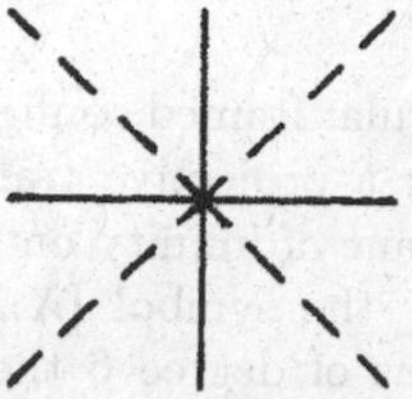

Here the projection of the frame in the horizontal plane $z = 1$ is shown by the continuous lines, the projection of the frame in the horizontal plane $z = -1$ is shown by the dash lines.

The frames of the skeleton divide the horizontal planes $z = 1$ and $z = -1$ into four open domains, which we denote as it is shown below

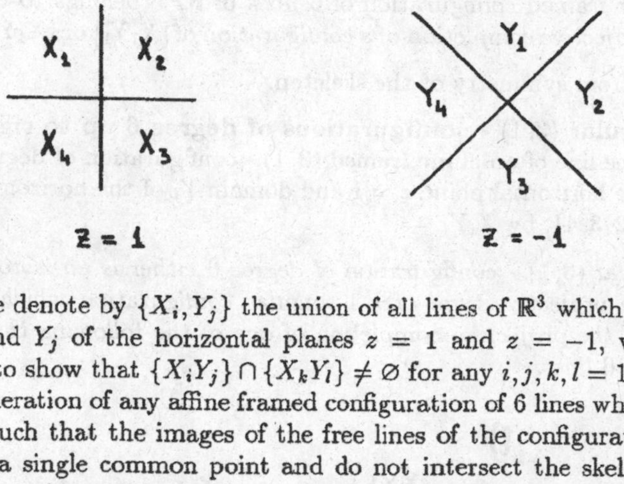

We denote by $\{X_i, Y_j\}$ the union of all lines of $\mathbb{R}^3$ which intersect the open domains X_i and Y_j of the horizontal planes $z = 1$ and $z = -1$, where $i, j = 1, 2, 3, 4$. It is easy to show that $\{X_i Y_j\} \cap \{X_k Y_l\} \neq \emptyset$ for any $i, j, k, l = 1, 2, 3, 4$. Thus there exists a degeneration of any affine framed configuration of 6 lines which keeps the skeleton fixed, and such that the images of the free lines of the configuration after this degeneration have a single common point and do not intersect the skeleton. The result of such a degeneration will be called an affine singular framed configuration of 6 lines. The free lines of an affine singular framed configuration are said to intersect in the middle part if their point of intersection lies between the horizontal planes $z = 1$ and $z = -1$.

Let $\mathcal{Z}_{-1,1} = \{(x, y, z) \in \mathbb{R}^3 | -1 < z < 1\}$, and let i, j, k, l be some fixed elements of the index set such that $\{X_i Y_j\} \cap \{X_k Y_l\} \cap \mathcal{Z}_{-1,1} \neq \emptyset$. Consider the space $\mathcal{X}$ consisting of ordered configurations (L_1, L_2) of two lines of $\mathbb{R}^3$ such that $L_1 \subset \{X_i Y_j\}$, $L_2 \subset \{X_k Y_l\}$, and L_1 and L_2 intersect at a point of $\mathcal{Z}_{-1,1}$.

4.1.1. Lemma. *Space $\mathcal{X}$ with the natural topology is arcwise connected.*

Proof. Let $\mathcal{B} = \{X_i Y_j\} \cap \{X_k Y_l\} \cap \mathcal{Z}_{-1,1}$. It is clear that $\mathcal{B}$ is an open convex subset of $\mathbb{R}^3$. Fix some point $b \in \mathcal{B}$. Let $\mathcal{F}_b$ be the set of all configurations of $\mathcal{X}$ the lines of which intersect at b. It is easy to see that the subset $\mathcal{F}_b$ of $\mathcal{X}$ is arcwise connected. Consider the map $p : \mathcal{X} \to \mathcal{B}$ which assigns to a configuration of $\mathcal{X}$ the point of intersection of its lines. It is not difficult to prove that the bundle $(\mathcal{X}, p, \mathcal{B})$ is topologically trivial. Since the fibre and the base of this bundle are arcwise connected, the total space $\mathcal{X}$ is also arcwise connected.

4.1.2. Lemma. *Any affine singular framed configuration of 6 lines is rigidly isotopic to an affine singular framed configuration of 6 lines the free lines of which intersect in the middle part.*

Proof. If the free lines of an affine singular framed configuration of 6 lines do not intersect in the middle part, then we consider a projective transformation which preserves the orientation of $\mathbb{R}P^3$ and moves the plane of infinity onto the horizontal plane $z = 0$.

Introduce the following notation: the symbol $[X_iY_j]$ denotes the set of all affine singular framed (3; 1) - configurations of degree 6 the free lines of which intersect in the middle part and one of them is contained in $\{X_iY_j\}$, where $i, j \in \{1, 2, 3, 4\}$. The projective (3; 1) - configuration of degree 6 will be called a singular framed configuration if it is the projective completion of an affine singular framed configuration of 6 lines.

Two projective configurations are said to belong to the same coarse projective type if either they belong to the same rigid isotopy type, or one of them is rigidly isotopic to the mirror image of the other.

4.1.3. Lemma. *Any singular framed configuration of 6 lines of $\mathbb{R}P^3$ belongs to the coarse projective type of the projective completion of a configuration of $[X_1Y_1]$ or $[X_1Y_2]$.*

Proof. It immediately follows from symmetry of the skeleton.

4.2. Classification of 1-singular (3; 1) - configurations of degree 6 up to rigid isotopy. We shall denote a free line of an affine framed (3; 1) - configuration of degree 6 intersecting domain X_i of the horizontal plane $z = 1$ and domain Y_j of the horizontal plane $z = -1$, where $i, j \in \{1, 2, 3, 4\}$, by X_iY_j.

4.2.1. Lemma. *Any 1-singular (3; 1) - configuration of degree 6 either is an isotopy join, or belongs to the coarse projective type of a 1-singular configuration which is obtained by a perturbation of the projective completion of one of the following three affine framed configurations of 6 lines:*

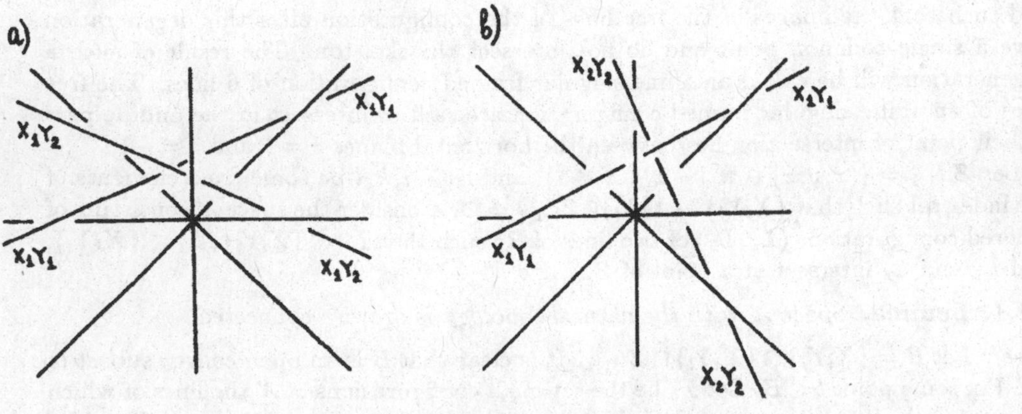

c)

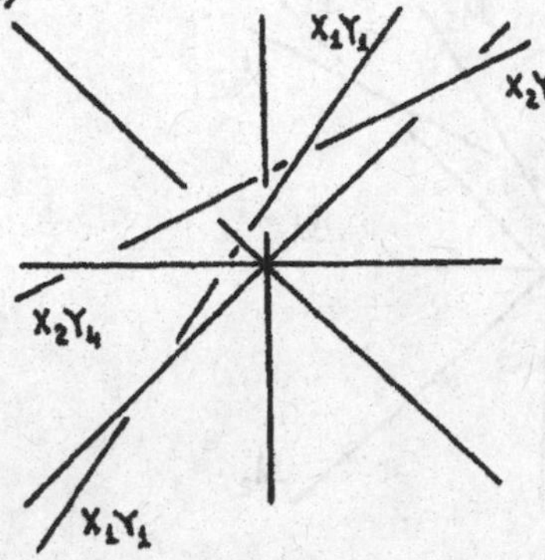

Proof. We sort all possible cases, using Lemmas 2.3.4, 4.1.1, 4.1.2, 4.1.3, 2.3.2. and 1.2.1.

4.2.2. Lemma. *Any 1-singular (3; 1) - configuration of degree 6 either is an isotopy join, or belongs to the coarse projective type of the projective completion of one of the following three affine configurations of 6 lines:*

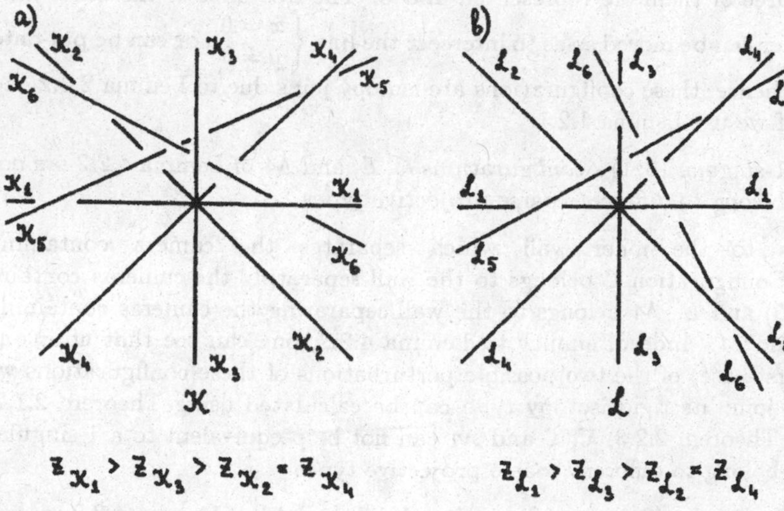

c)

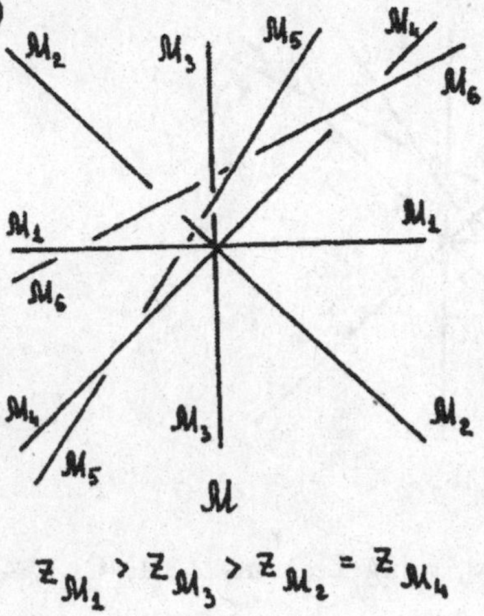

(The notation $z_{\mathcal{K}_1} > z_{\mathcal{K}_3} > z_{\mathcal{K}_2} = z_{\mathcal{K}_4}$ means that horizontal line $\mathcal{K}_1$ lies over horizontal line $\mathcal{K}_3$, line $\mathcal{K}_3$ lies over intersecting horizontal lines $\mathcal{K}_2$ and $\mathcal{K}_4$).

Proof. Perturbing the framed configurations of Lemma 4.2.1 we obtain twelve 1-singular configurations. Three of them are represented above. The free lines of the other nine configurations either can be moved so as to intersect the line $\begin{cases} x = 0 \\ y = 0 \end{cases}$, or can be put onto horizontal planes: hence, these configurations are isotopy joins due to Lemma 2.3.2. To complete the proof we use Lemma 1.2.1.

4.2.3. Lemma. *1-singular* (3; 1) *- configurations $\mathcal{K}$, $\mathcal{L}$, and $\mathcal{M}$ of Lemma 4.2.2 are not isotopy joins and belong to different coarse projective types.*

Proof. $\mathcal{K}$ belongs to the inner wall which separates the camera containing $jc(1,3,5,2,6,4)$. Configuration $\mathcal{L}$ belongs to the wall separating the cameras containing $jc(1,2,4,6,3,5)$ and L. $\mathcal{M}$ belongs to the wall separating the cameras containing $jc(1,3,5,2,6,4)$ and M. Indeed, similar to Lemma 4.2.2, one can see that after one (or both, in the first case) of the two possible perturbations of these configurations we obtain an isotopy join: its rigid isotopy type can be calculated using Theorem 2.1.2. Therefore, due to Theorem 2.2.3, $\mathcal{K}$, $\mathcal{L}$ and $\mathcal{M}$ can not be p-equivalent to a 1-singular isotopy join (3; 1)-belong to different coarse projective types.

4.2.4. Lemma. *1-singular* (3; 1) *- configurations $\mathcal{K}$, $\mathcal{L}$, and $\mathcal{M}$ of Lemma 4.2.2 are not mirror.*

Proof. $\mathcal{K}$ is not mirror, since its non-singular subconfiguration consisting of lines $\mathcal{K}_1$, $\mathcal{K}_3$, $\mathcal{K}_5$, $\mathcal{K}_6$, is not mirror. $\mathcal{M}$ is not mirror, since, due to Theorem 3.2.1, it is not p-equivalent to its mirror image. $\mathcal{L}$ is not mirror for the same reason.

As a consequence of Lemmas 4.2.2, 4.2.3, and 4.2.4 we obtain the following theorem.

4.2.5. Theorem. *Any 1-singular (3; 1) - configuration of degree 6 is either an isotopy join (see Theorem 2.2.3), or rigidly isotopic to one of the following six pairwise non-isotopic configurations:*

1) configuration $\mathcal{K}$,
2) configuration $\mathcal{L}$,
3) configuration $\mathcal{M}$,
4) the mirror image of $\mathcal{K}$,
5) the mirror image of $\mathcal{L}$,
6) the mirror image of $\mathcal{M}$.

4.2.6. Corollary. There exist exactly 37 rigid isotopy types of 1-singular (3; 1) - configurations of degree 6.

4.3. Classification of non-singular configurations of 6 lines of $\mathbb{R}P^3$ up to rigid isotopy.

4.3.1. Lemma. *Any non-singular (3; 1) - configuration of degree 6 either is an isotopy join, or belongs to the coarse projective type of the projective completion of one of the following two affine configurations:*

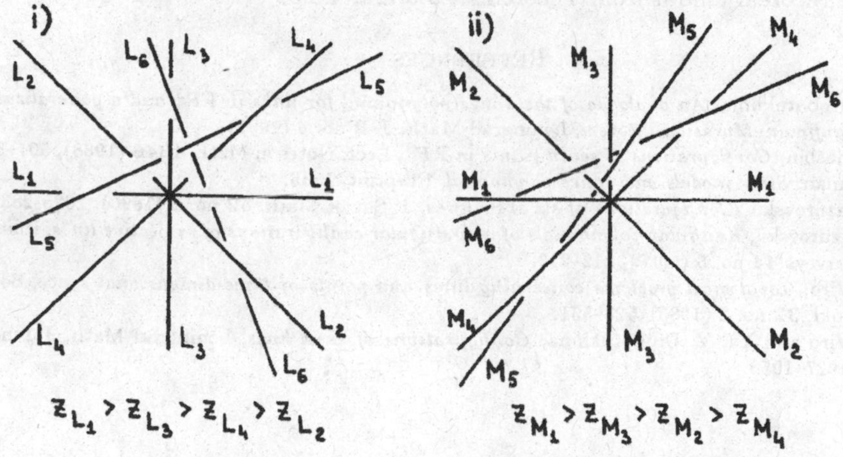

(It is easy to see that the configurations of i) and ii) are rigidly isotopic to L and M respectively (see, respectively, dia.2 and 1 in Appendix)).

Proof. Perturb the 1-singular (3; 1) - configurations of Lemma 4.2.2 and use Lemmas 2.3.2 and 1.2.1.

4.3.2. Lemma. *Non-singular (3; 1) - configurations M, L, and L' are not homology equivalent.*

As a consequence of Lemmas 4.3.1, 4.3.2, Theorems 3.2.1, and Corollary 3.2.3 we obtain the following theorem.

4.3.3. Theorem. *Any non-singular (3; 1) - configuration of degree 6 is either an isotopy join (see Theorem 2.1.2), or rigidly isotopic to one of the following four pairwise non-isotopic configurations:*

1) configuration L,
2) configuration M,
3) the mirror image of L,
4) the mirror image of M.

4.3.4. Corollary. There exist exactly 19 rigid isotopy types of non-singular (3; 1)-configurations of degree 6.

4.3.5. Theorem. *Non-singular configurations of 6 lines of $\mathbb{R}P^3$ are rigidly isotopic if and only if their Kauffman polynomials are equal.*

To prove this theorem it is sufficient to show that the configurations of each of the 19 rigid isotopy types of non-singular (3; 1) - configurations of degree 6 have different Kauffman polynomials. The complete list of these polynomials can be found in [5].

4.4. Adjacency graph of $SPC^6_{3,1}$. Denote the mirror images of the 1-singular (3; 1) - configurations $\mathcal{K}$, $\mathcal{L}$, and $\mathcal{M}$ by $\mathcal{K}'$, $\mathcal{L}'$, and $\mathcal{M}'$ respectively.

4.4.1. Theorem. *Adjacency graph of $SPC^6_{3,1}$ is shown on diagram 3 in Appendix. All loops of the graph correspond to one-sided unner walls.*

Proof. The theorem follows from Theorems 4.3.3 and 4.2.5.

References

[1] Yu. V. Drobotukhina, *An analogue of the Jones polynomial for links in $\mathbb{R}P^3$ and a generalization of the Kauffman–Murasugi theorem*, Leningrad Math. J. **2** no. 3 (1991).
[2] S. M. Finashin, *Configurations of seven points in $\mathbb{R}P^3$*, Lect. Notes in Math. **1346** (1988), 501–526.
[3] L. Kauffman, *State models and Jones polynomial*, Preprint, 1986.
[4] V. F. Mazurovskii, *Configurations of six skew lines*, J. Soviet Math. **52** no. 1 (1990), 2825-2832.
[5] V. F. Mazurovskii, *Kauffman polynomials of non-singular configurations of projective lines*, Russian Math. Surveys **44** no. 5 (1989), 212–213.
[6] O. Ya. Viro, *Topological problems concerning lines and points of three-dimensional space*, Soviet Math. Dokl. **32** no. 2 (1985), 528–531.
[7] O. Ya. Viro and Yu. V. Drobotukhina, *Configurations of skew lines*, Leningrad Math. J **1** no. 4 (1990), 1027–1050.

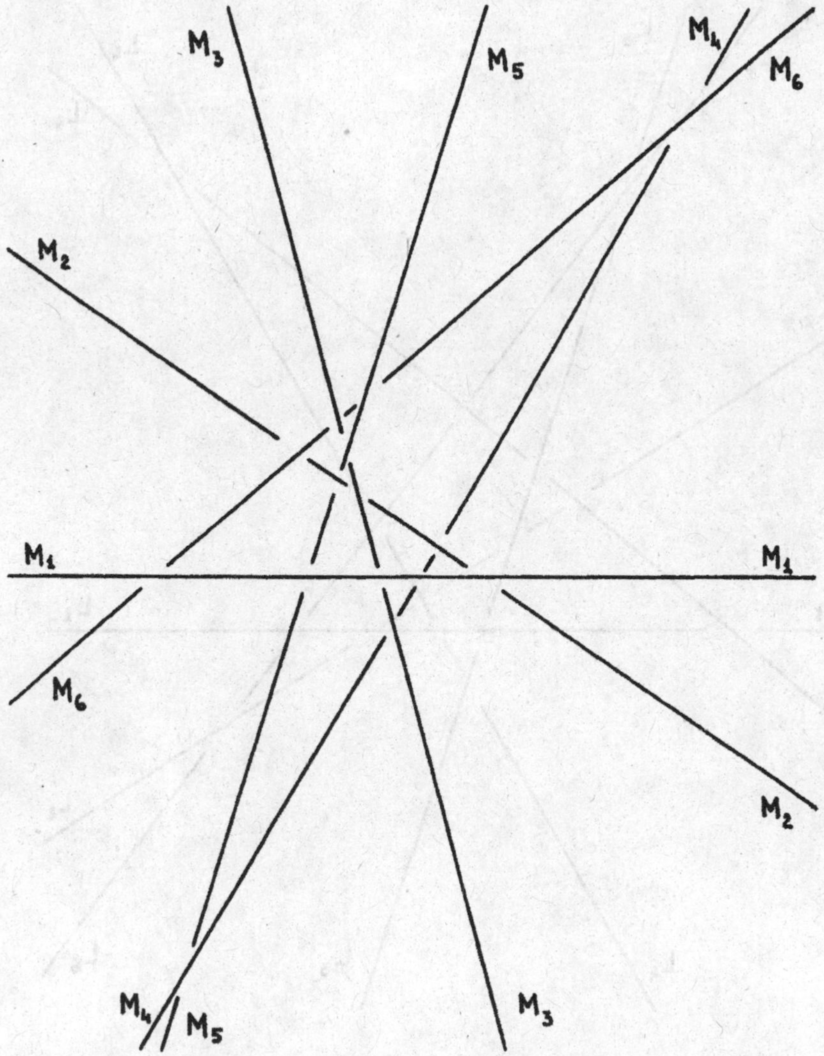

DIA. 1. Affine part of non-singular (3; 1)-configuration M.

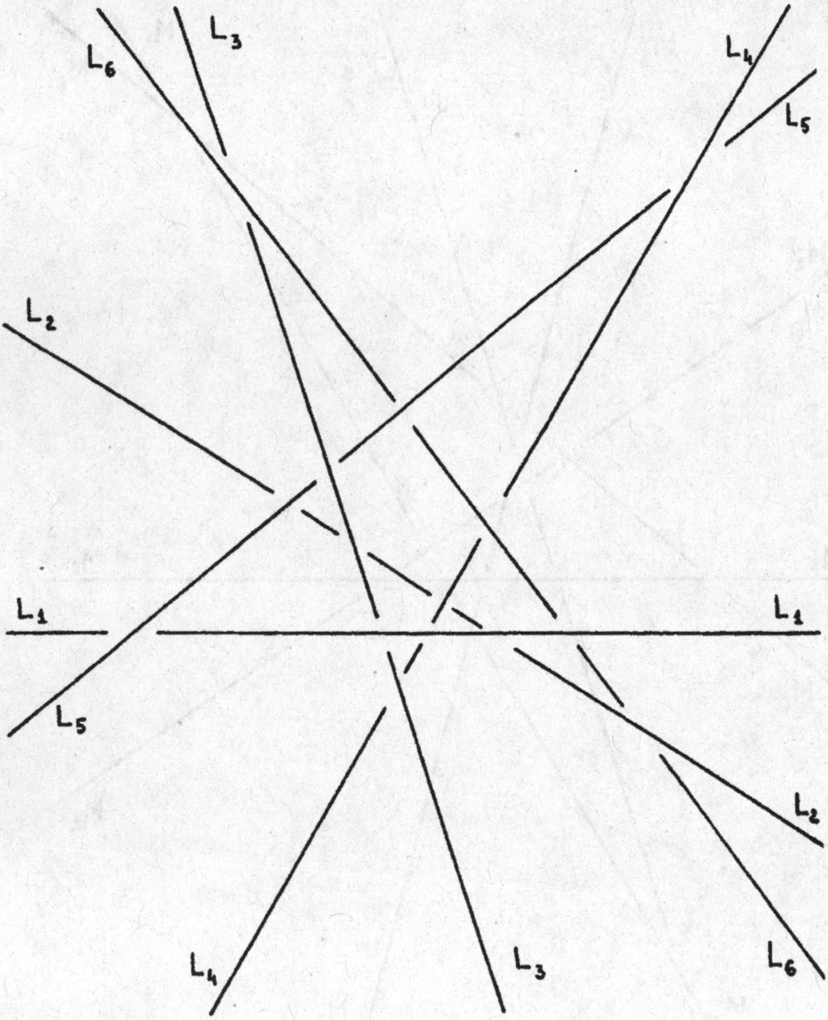

DIA. 2. Affine part of non-singular (3; 1)-configuration L.

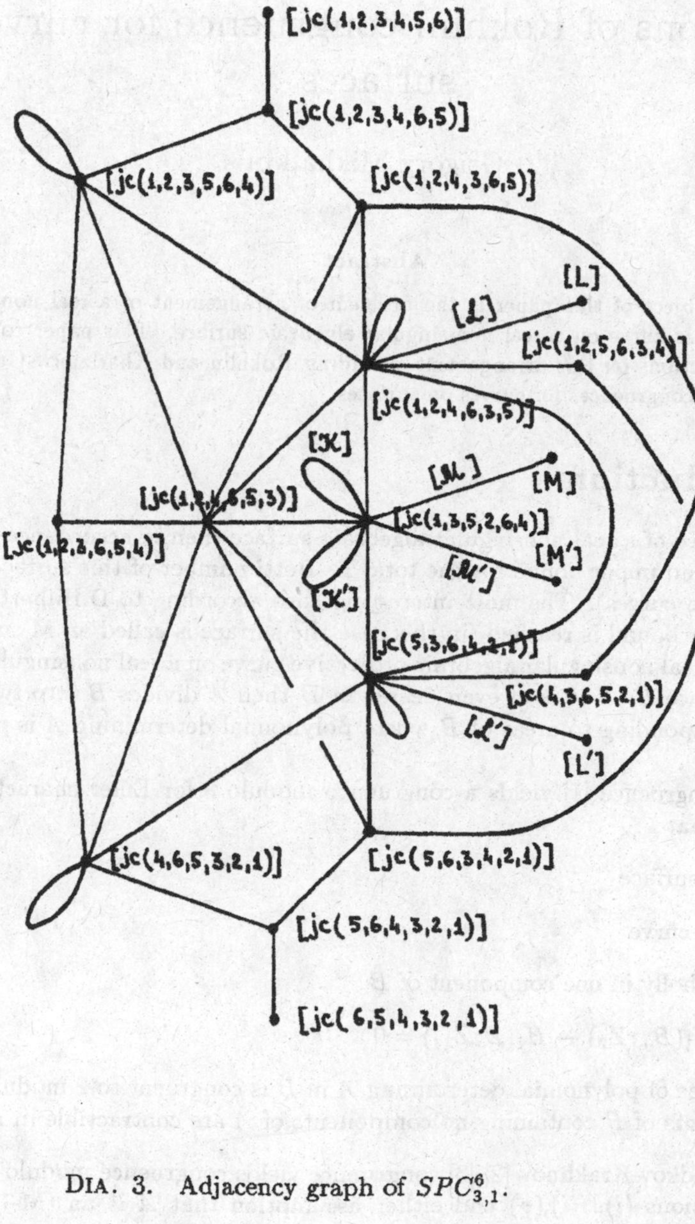

DIA. 3. Adjacency graph of $SPC^6_{3,1}$.

Extensions of Rokhlin congruence for curves on surfaces

Grigory Mikhalkin

Abstract

The subject of this paper is the problem of arrangement of a real nonsingular algebraic curve on a real non-singular algebraic surface. This paper contains new restrictions on this arrangement extending Rokhlin and Kharlamov-Gudkov-Krakhnov congruences for curves on surfaces.

1 Introduction

If we fix a degree of a real nonsingular algebraic surface then in accordance with Smith theory there is an upper bound for the total $\mathbf{Z}_2$-Betti number of this surface (the same thing applies to curves). The most interesting case according to D.Hilbert is the case when the upper bound is reached (in this case the surface is called an M-surface).

Let A be a real nonsingular algebraic projective curve on a real nonsingular algebraic projective surface B. If A is of even degree in B then A divides B into two parts B_+ and B_- (corresponding to areas of B where polynomial determining A is positive and negative).

Rokhlin congruence [1] yields a congruence modulo 8 for Euler characteristic χ of B_+ provided that

(i) B is an M-surface

(ii) A is an M-curve

(iii) B_+ lies wholly in one component of B

(iv) $rk(in_* : H_1(B_+; \mathbf{Z}_2) \to H_1(B; \mathbf{Z}_2)) = 0$

(v) if the degree of polynomial determining A in B is congruent to 2 modulo 4 then all components of B containing no components of A are contractible in P^q [1]

Kharlamov-Gudkov-Krakhnov [2],[3] congruence yields congruence modulo 8 for $\chi(B_+)$ under assumptions (i),(iii),(v) and either assumption that A is an (M-1)-curve and $rk(in_*) = 0$ or assumption that A is an M-curve and $rk(in_*) = 1$ and all components of A are $\mathbf{Z}_2$-homologically trivial. Recall that Rokhlin congruence for surfaces yields

[1] *Remark*. Paper [1] contains a mistake in the calculation of characteristic class of covering $Y \to CB$. It leads to the omission (after reformulation) of (v) in assumptions of congruence. The proof given in [1] really uses (v).

congruence modulo 16 for $\chi(B)$ so a congruence modulo 8 for $\chi(B_+)$ is equivalent to a congruence modulo 8 for $\chi(B_-)$.

One of the properties of M-surfaces (similar to the property of M-curves) remarked by V.I.Arnold [4] is that a real M-surface is a characteristic surface in its own complexification (for M-curves it means that a real M-curve divides its own complexification since a complex curve is orientable). We shall say that a real surface is of a characteristic type if it is a characteristic surface in its own complexification. Note that the notion of characteristic type of surfaces is analogous to the notion of type I of curves.

Consider at first the weakening of assumption (i) in Rokhlin and Kharlamov-Gudkov-Krakhnov congruences. Instead of (i) we can only assume that B is of characteristic type.

According to O.Ya.Viro [5] there are some extra structures on real surfaces of characteristic type ,namely, Pin_--structures and semiorientations or relative semiorientations (semiorientation is the orientation up to the reversing). In this paper we introduce another structure on surfaces of characteristic type — complex separation which is also determined by the arrangement of a real surface in its complexification. The complex separation is a natural separation of the set of components of a real surface of characteristic type into two subsets. Note that the set of semiorientations of a surface is an affine space over $\mathbf{Z}_2$-vector space of separations of this surface.

We use the complex separation to weaken assumption (iii). In theorem 1 instead of (iii) we assume only that B_+ lies in components of one class of complex separation.

The further extension ,theorem 2, can be applied not only for curves of even degree but also sometimes for curves of odd degree. Another weakening of assumptions in theorem 2 is that components of a curve are not necessarily $\mathbf{Z}_2$-homologically trivial.

These extensions can be applied to curves on quadrics and cubics. Theorem 1 together with an analogue of Arnold inequality for curves on cubics gives a complete system of restrictions for real schemes of flexible curves of degree 2 on cubics of characteristic type (see [6]). An application of theorem 2 to curves on an ellipsoid gives a complete system of restrictions for real schemes of flexible curves of degree 3 on an ellipsoid and reduces the problem of classification of real schemes of flexible curves of degree 5 on an ellipsoid to the problem of the existence of two real schemes (see [7]). An application of theorem 2 to curves on a hyperboloid extends Matsuoka congruences [8] for curves with odd branches on a hyperboloid (see [9]).

Applications of theorem 2 to empty curves on surfaces give restrictions for surfaces involving complex separation of surfaces. Restrictions for curves of even degree on surfaces can be obtained also by the application of these restrictions for surfaces to the 2-sheeted covering of surface branched along the curve , if we know the complex separation of this covering. This complex separation is determined by complex orientation of the curve. For example in this way one can obtain new congruences for complex orientations of curves on a hyperboloid.

These applications and further extension of Rokhlin congruence for curves on surfaces will be published separately in [9]. For example the assumption that the surface and the curve on the surface are complete intersections is quite unnecessary ,but this assumption simplifies definition of number c in formulations of theorems. The formulations of these results were announced in [7] as well as the formulations of results of the present paper.

The author is indebted to O.Ya.Viro for advices.

2 Notations and formulations of main theorems

Let the surface B be the transversal intersection of hypersurface in P^q defined by equations $P_j(x_0,\dots,x_q)=0, j=1,\dots,s-1$; $\mathbf{C}B$ and $\mathbf{R}B$ be sets of complex and real points of B; let A be a nonsingular curve on B defined by an equation $P_s(x_0,\dots,x_q)=0$,where P_j is a real homogenous polynomial of degree m_j $j=1,\dots,s$, $q=s+1$; $\mathbf{C}A$ and $\mathbf{R}A$ be sets of complex and real points of A. Let *conj* denote the involution of complex conjugation. Set c to be equal to $\frac{\prod_{j=1}^{s-1} m_j}{4}$.If m_s is even then denote $\{x \in \mathbf{R}B | \pm P_s(x) \geq 0\}$ by $B_\pm$ and set d to be equal to $rk(in_* : H_1(B_+;\mathbf{Z}_2) \to H_1(\mathbf{R}B;\mathbf{Z}_2))$.

A real algebraic variety is called an (M-j)-variety if its total $\mathbf{Z}_2$-Betti number is less by $2j$ then total $\mathbf{Z}_2$-Betti number of its complexification (Harnack-Smith inequality shows that $j \geq 0$). Let A be an (M-k)-curve. Let D_M be the operator of Poincaré duality of manifold M.We shall say that B is a surface of characteristic type if $[\mathbf{R}B] = D_{\mathbf{C}B} w_2(\mathbf{C}B) \in H_2(\mathbf{C}B;\mathbf{Z}_2)$ (as it is usual we denote by $[\mathbf{R}B]$ the element of $H_2(\mathbf{C}B;\mathbf{Z}_2)$ realized by $\mathbf{R}B$). We shall say that (B, A) is a pair of characteristic type if $[\mathbf{R}B] + [\mathbf{C}A] + D_{\mathbf{C}B}(w_2(\mathbf{C}B)) = 0 \in H_2(\mathbf{C}B;\mathbf{Z}_2)$.It is said that A is a curve of type I if $[\mathbf{R}A] = 0 \in H_1(\mathbf{C}A;\mathbf{Z}_2)$.It is said that A is of even(odd) degree if $[\mathbf{C}A] = 0 \in H_2(\mathbf{C}B;\mathbf{Z}_2)$ (otherwise).

As it is usual we denote by σ and χ the signature and the Euler characteristic.By $\beta(q)$ we mean Brown invariant of $\mathbf{Z}_4$-valued quadratic form q.

Theorem 1 *If B is a surface of characteristic type then there is defined a natural separation of surface $\mathbf{R}B$ into two closed surfaces B_1 and B_2 by the condition that $B_j, j=1,2$, is a characteristic surface in $\mathbf{C}B/conj$.Suppose that m_s is even,$B_+ \subset B_1$, every component of $\mathbf{R}A$ is $\mathbf{Z}_2$-homologous to zero in $\mathbf{R}B$ and if $m_s \equiv 2 \pmod 4$ then suppose besides that B_2 is contractible in $\mathbf{R}P^q$.*

- **a)** *If $d+k=0$ then* $\chi(B_+) \equiv c \pmod 8$.
- **b)** *If $d+k=1$ then* $\chi(B_+) \equiv c \pm 1 \pmod 8$.
- **c)** *If $d+k=2$ and $\chi(B_+) \equiv c+4 \pmod 8$ then A is of type I and B_+ is orientable.*
- **d)** *If A is of type I and B_+ is orientable then* $\chi(B_+) \equiv c \pmod 4$.

Theorem 2 *If (B, A) is a pair of characteristic type then there is defined a natural separation of surface $\mathbf{R}B \setminus \mathbf{R}A$ into two surfaces B_1 and B_2 with common boundary $\mathbf{R}A$ by the condition that $B_j \cup \mathbf{C}A/conj$ is a characteristic surface in $\mathbf{C}B/conj$, there is defined Guillou-Marin form q_j on $H_1(B_j \cup \mathbf{C}A/conj;\mathbf{Z}_2)$ and*

$$\chi(B_j) \equiv c + \frac{\chi(\mathbf{R}B) - \sigma(\mathbf{C}B)}{4} + \beta(q_j) \pmod 8.$$

3 Proof of theorem 2

Consider the Smith exact sequence of double branched covering $\pi : \mathbf{C}B \to \mathbf{C}B/conj$

$$\xrightarrow{\beta_3} H_3(\mathbf{C}B/conj, \mathbf{R}B;\mathbf{Z}_2) \xrightarrow{\gamma_3} H_2(\mathbf{R}B;\mathbf{Z}_2) \oplus H_2(\mathbf{C}B/conj, \mathbf{R}B;\mathbf{Z}_2) \xrightarrow{\alpha_3} H_2(\mathbf{C}B;\mathbf{Z}_2) \xrightarrow{\beta_2} .$$

Let ϕ denote the composite homomorphism

$$H_2(\mathbf{C}B/conj, \mathbf{R}B; \mathbf{Z}_2) \stackrel{0\oplus id}{\to} H_2(\mathbf{R}B; \mathbf{Z}_2) \oplus H_2(\mathbf{C}B/conj, \mathbf{R}B; \mathbf{Z}_2) \stackrel{\alpha_3}{\to} H_2(\mathbf{C}B; \mathbf{Z}_2).$$

Let j denote the inclusion map $(\mathbf{C}B/conj, \emptyset) \to (\mathbf{C}B/conj, \mathbf{R}B)$.Recall that $\phi_* \circ j_*$ is equal to Hopf homomorphism $\pi^!$. It is easy to deduce from the exactness of the Smith sequence that ϕ is a monomorphism. Indeed, $\pi_1(\mathbf{C}B) = 0$ hence $\pi_1(\mathbf{C}B/conj) = 0$ and $H_3(\mathbf{C}B/conj; \mathbf{Z}_2) = 0$. Therefore boundary homomorphism $\partial : H_3(\mathbf{C}B/conj, \mathbf{R}B; \mathbf{Z}_2) \to H_2(\mathbf{R}B; \mathbf{Z}_2)$ is a monomorphism.Therefore,since ∂ is the first component of γ_3, $Im\gamma_3 \cap (\{0\} \oplus H_2(\mathbf{C}B/conj, \mathbf{R}B; \mathbf{Z}_2)) = 0$ and ϕ is a monomorphism.

It is easy to check that

$$\pi^* w_2(\mathbf{C}B/conj) = D^{-1}_{\mathbf{C}B}[\mathbf{R}A] + w_2(\mathbf{C}B).$$

Thus $\pi^!(D_{\mathbf{C}B/conj} w_2(\mathbf{C}B/conj)) = [\mathbf{C}A]$, therefore,because of the injectivity of ϕ,we obtain that

$$j_* D_{\mathbf{C}B/conj} w_2(\mathbf{C}B/conj) = [\mathbf{C}A/conj, \mathbf{R}A] \in H_2(\mathbf{C}B/conj, \mathbf{R}B; \mathbf{Z}_2).$$

It means that there exists a compact surface $B_1 \subset \mathbf{R}B$ with boundary $\mathbf{R}A$ such that $B_1 \cup \mathbf{C}A/conj$ is a characteristic surface in $\mathbf{C}B/conj$. Surface $\mathbf{R}A$ is homologous to zero in $\mathbf{C}B/conj$ since $\mathbf{R}A$ is the set of branch points of π. Set B_2 to be equal to the closure of $(\mathbf{R}B \setminus B_1)$. Then $B_2 \cup \mathbf{C}A/conj$ is a characteristic surface $\mathbf{C}B/conj$, $B_1 \cup B_2 = \mathbf{R}B$,$B_1 \cap B_2 = \partial B_1 = \partial B_2 = \mathbf{R}A$.

Let us prove the uniqueness of pair $\{B_1, B_2\}$.It is sufficient to prove that the dimension of the kernel of inclusion homomorphism $H_2(\mathbf{R}B; \mathbf{Z}_2) \to H_2(\mathbf{C}B/conj; \mathbf{Z}_2)$ is equal to 1. This follows from the equality $\dim H_3(\mathbf{C}B/conj, \mathbf{R}B; \mathbf{Z}_2) = 1$ that can be deduced from the exactness of the Smith sequence.

We apply now Guillou-Marin congruence [10] to pair $(\mathbf{C}B/conj, B_j \cup \mathbf{C}A/conj), j = 1, 2$

$$\sigma(\mathbf{C}B/conj) \equiv [B_j \cup \mathbf{C}A/conj] \circ [B_j \cup \mathbf{C}A/conj] + 2\beta(q_j) \pmod{16}.$$

Hirzebruch index theorem gives an equality $\sigma(\mathbf{C}B/conj) = \frac{\sigma(\mathbf{C}B) - \chi(\mathbf{R}B)}{2}$.To finish the proof note that $[B_j \cup \mathbf{C}A/conj] \circ [B_j \cup \mathbf{C}A/conj] = 2c - 2\chi(B_j)$ (the calculation is similar to Marin calculation in [11]).

4 Proof of the theorem 1

Pair $(B, \emptyset)$ is of characteristic type since A is of even degree in B. Thus the first part of theorem 1 follows from theorem 2 — there exist a natural separation of B into two surfaces B_1 and B_2 such that B_1 and B_2 are characteristic surfaces in $\mathbf{C}B/conj$.

Let V denote $B_+ \cup \mathbf{C}A/conj$.Let W denote $V \cup B_1$. Recall that $B_+ \cap B_2 = \emptyset$ thus $V \cap B_2 = \emptyset$.

Lemma 1 $[V] = 0 \in H_2(\mathbf{C}B/conj; \mathbf{Z}_2)$

Proof. Since A is of even degree in B, there exists a 2-sheeted covering $p : Y \to \mathbf{C}B$ branched along $\mathbf{C}A$. Involution $conj$ can be lifted to involutions T_+ and $T_- : Y \to$

Y since $\mathbf{C}A$ is invariant under $conj$. It is easy to see using the straight algebraic construction of p that T_+ and T_- can be chosen in such a way that the set of fixed points of $T_\pm$ is $p^{-1}(B_\pm)$.

Consider the diagram

$$\begin{array}{ccc} Y & \xrightarrow{\ p\ } & \mathbf{C}B \\ \downarrow & & \downarrow \pi \\ Y/T_- & & \mathbf{C}B/conj. \end{array}$$

This diagram can be expanded to a commutative one by map $p' : Y/T_\mp \to \mathbf{C}B/conj$. It is easy to see that p' is a 2-sheeted covering branched along V. Therefore $[V] = 0 \in H_2(\mathbf{C}B/conj; \mathbf{Z}_2)$.

Using Lemma 1 we see that W is a characteristic surface in $\mathbf{C}B/conj$ as well as B_2.We apply Guillou-Marin congruence to these two surfaces :

$$\sigma(\mathbf{C}B/conj) \equiv [W] \circ [W] + 2\beta(q_W) \equiv 2c - 2\chi(B_+) - 2\chi(B_2) + 2\beta(q_W) \pmod{16}$$

$$\sigma(\mathbf{C}B/conj) \equiv [B_2] \circ [B_2] + 2\beta(q_{B_2}) \equiv -2\chi(B_2) + 2\beta(q_{B_2}) \pmod{16}$$

,where q_W and q_{B_2} are Guillou-Marin forms of W and B_2. Therefore

$$\chi(B_+) \equiv c + \beta(q_W) - \beta(q_{B_2}) \pmod{8}.$$

Lemma 2 $\forall x \in H_1(B_2; \mathbf{Z}_2), q_{B_2}(x) - q_W(x) = \begin{cases} 0 & \textit{if } x \textit{ is contractible in } \mathbf{R}P^q \\ \frac{m_4}{2} & \textit{if } x \textit{ is noncontractible in } \mathbf{R}P^q \end{cases}.$

Proof. It follows from the definition of Guillou-Marin form that values on x of q_{B_2} and q_W are differed by linking number of x and V in $\mathbf{C}B/conj$ that is equal to linking number of x and $\mathbf{C}A$ in $\mathbf{C}B$. The last linking number can be calculated from the straight construction of a 2-sheeted covering branched along $\mathbf{C}A$.

It was shown in [12] that Brown invariant of form q on the union of two surfaces with common boundary is equal to the sum of Brown invariants of restrictions of q on these surfaces in the case when q vanishes on the common boundary. Now theorem 1 follows from this additivity of Brown invariant and the classification of low-dimensional $\mathbf{Z}_4$-valued quadratic forms (see[12]). Indeed, since every component of $\mathbf{R}A$ is homologous to 0 in $\mathbf{R}B$, $\beta(q_W) = \beta(q_W|_{\mathbf{C}A/conj}) + \beta(q_W|_{B_+}) + \beta(q_W|_{B_2})$. Lemma 2 shows that under assumptions of theorem 1 $\beta(q_W|_{B_2}) = \beta(q_{B_2})$. To complete the proof note that ranks of intersection forms on $H_1(B_+; \mathbf{Z}_2)$ and $H_1(\mathbf{C}A/conj; \mathbf{Z}_2)$ are equal to d and k respectively.

References

[1] V.A.Rokhlin. Congruences modulo 16 in sixteenth Hilbert problem. Functsional'ni Analiz i ego Prilozheniya. 1972. Vol. 6(4). P. 58-64

[2] V.M.Kharlamov. New congruences for the Euler characteristic of real algebraic manifolds. Functsional'ni Analiz i ego Prilozheniya. 1973. Vol. 7(2). P. 74-78

[3] D.A.Gudkov and A.D.Krakhnov. On the periodicity of Euler characteristic of real algebraic (M-1)-manifolds. Functsional'ni Analiz i ego Prilozheniya. 1973. Vol. 7(2). P. 15-19

[4] V.I.Arnold. On the arrangement of the ovals of real plane curves, involutions of 4-dimensional smooth manofolds and the arithmetic of integral quadratic forms. Functsional'ni Analiz i ego Prilozheniya. 1971. Vol. 5(3). P.1-9

[5] O.Ya.Viro. The progress in topology of real algebraic varieties over last six years. Uspehi Mat. Nauk. 1986. Vol. 41(3). P. 45-67

[6] G.B.Mikhalkin. Real schemes of flexible M-curves of virtual degree 2 on cubics of type I rel. Diploma paper. Leningrad. 1991.

[7] G.B.Mikhalkin. Congruences for real aldebgaic curves on an ellipsoid. Zapiski Nauchnyh Seminarov LOMI. 1991. Vol. 192, Geometry and Topology 1

[8] S.Matsuoka. Congruences for M- and (M-1)-curves with odd branches on a hyperboloid. Preprint. 1990.

[9] G.B.Mikhalkin. The complex separation and extensions of Rokhlin congruence for curves and surfaces (to appear).

[10] L.Guillou and A.Marin. Une extension d'un theoreme de Rohlin sur la signature. C.R. Acad. Sci. Paris. 1977. Vol. 285. P. 95-97

[11] A.Marin. Quelques remarques sur les courbes algebriques planes reeles. Publ. Math. Univ. Paris VII. 1980. Vol.9. P.51-86.

[12] V.M.Kharlamov, O.Ya.Viro. Extensions of the Gudkov-Rokhlin congruence. Lect. Notes Math. 1988. Vol. 1346. P. 357-406

Complexité de la construction des strates à multiplicité constante d'un ensemble algébrique de $\mathbb{C}^n$

T.Mostowski E.Rannou

Résumé:

Nous décrivons dans cet article un algorithme qui construit la partition d'un ensemble algébrique de $\mathbb{C}^n$ en ensembles où la multiplicité est fixée. L'entrée de l'algorithme est constitué par un nombre fini de polynômes définissant l'ensemble algébrique. Chaque strate de la partition sera définie par l'algorithme à l'aide d'une formule du langage du premier ordre des corps algébriquement clos sans quantificateurs.

Cet algorithme est décrit par un réseau arithmétique, de complexité sequentielle polynomiale en la somme des degrés des polynômes d'entrée et simplement exponentielle en le nombre de variables.

Une conséquence immédiate de cet algorithme est la possibilité de construire des équations polynomiales définissant l'ensemble des points singuliers d'un ensemble algébrique de $\mathbb{C}^n$ avec une complexité séquentielle simplement exponentielle.

Le coeur de l'algorithme est constitué par une généralisation à plusieurs variables d'une méthode d'Hermite, qui permet de compter le nombre de points d'un ensemble algébrique de $\mathbb{C}^n$ de dimension zéro. Cette méthode est particulièrement simple à mettre en oeuvre.

Les auteurs sont très reconnaissants à Mme M-F. Roy qui, pendant plusieurs discussions, a suggéré le problème et les idées principales de la solution. Ils remercient le rapporteur pour ses critiques constructives très utiles.

1.Introduction:

1.1.Notation de base:

Soit $V \subset \mathbb{C}^n$ une ensemble algébrique défini par des polynômes $f_1, ..., f_s$ de $\mathbb{C}[X_1, ..., X_n]$.
Soit $D := \sum_{i=1}^{s} deg f_i$.

Nous noterons V^d la composante équidimensionnelle de dimension d de V.
Soit x un point de V.

Nous définirons la multiplicité $m_x(V)$ de V en x par:

$$m_x(V) := \sum_{d=0}^{n} m_x(V^d)$$

et, pour d=0,...,n nous poserons:

$$m_x(V^d) := \sharp L^d \cap V^d - \sharp H^d \cap V^d + 1$$

pour L^d un n-d-plan tel que $\sharp L^d \cap V^d$ soit fini et maximal, et H^d un $(n-d)$-plan passant par x tel que $\sharp H^d \cap V^d$ soit fini et maximal.
En effet, $m_x(V^d)$ est défini (dans [M], chapter 5, section 5A, par exemple) comme $\sharp L^d \cap V^d \cap B_\varepsilon$, où B_ε est une boule de centre x et de rayon ε, ε est assez petit, et L^d est un $(n-d)$-plan qui passe assez près de x et tel que $\sharp L^d \cap V^d$ soit fini et maximal. Remarquons que cette dernière condition est satisfaite si et seulement si L^d coupe V^d transversalement et $\overline{L}^d \cap \overline{V}^d \subset \mathbb{C}^n$, où $\overline{L}^d, \overline{V}^d$ sont les clôtures de L^d, V^d dans $\mathbb{CP}^{\kappa}$. Si H^d passe par x, alors $H^d \cap V^d = \{x, x_1, \ldots, x_k\}$ est maximal si et seulement si $\overline{L}^d \cap \overline{V}^d \subset \mathbb{C}^n$ et H^d coupe V^d transversalement en chaque x_i. Si B_ε, $B_{\varepsilon,i}$ sont les boules de centres x, x_i et de rayon ε, et si L^d est parallèle à H^d, coupe V^d transversalement et passe assez près de x, alors, pour chaque i, $L^d \cap V^d \cap B_{\varepsilon,i}$ contient exactement un point, et $(L^d \cap V^d) \setminus B_\varepsilon \subset \bigcup_i (L^d \cap V^d \cap B_{\varepsilon,i})$. On a donc

$$\sharp(L^d \cap V^d) = \sharp(L^d \cap V^d \cap B_\varepsilon) + k = m_x(V^d) + \sharp(H^d \cap V^d) - 1,$$

et il suffit de remarquer que $\sharp(L^d \cap V^d)$ ne dépend pas de L^d.
Nous appelerons la quantité $\sharp L^d \cap V^d$ degré de l'ensemble algébrique V^d que nous noterons $degV^d$.

Soit V_k (resp. V_k^d) l'ensemble des points de V (resp. de V^d) où la multiplicité de V (resp. V^d) vaut k.

1.2.Résultats:

Le but est de construire les ensembles $V_k = \{x \in V \mid m_x(V) = k\}$ avec une complexité séquentielle simplement exponentielle. Nous démontrerons donc le résultat suivant:

Théorème 1.2.1: *Soit V un ensemble algébrique de $\mathbb{C}^n$ défini par des polynômes $f_1, \ldots, f_s$ à coefficient dans un anneau $\mathbb{A}$. Soit $D := \sum_{i=1}^s deg f_i$. Soit $V_k = \{x \in V \mid m_x(V) = k\}$.*

Alors il existe un réseau arithmétique sur $\mathbb{A}$ qui construit avec une complexité séquentielle $D^{n^{O(1)}}$ un entier $k_{max} < D^{n^{O(1)}}$ et des formules du premier ordre du langage des corps algébriquement clos $\Phi_1, \ldots, \Phi_{k_{max}}$ tels que:
$V_k = \emptyset$ pour $k > k_{max}$
et
$V_k = \{x \in \mathbb{C}^n \mid \Phi_k\}$ sinon.

On en déduit le corollaire suivant:

Corollaire 1.2.2: *Soit V un ensemble algébrique de $\mathbb{C}^n$ défini par des polynômes $f_1, \ldots, f_s$ à coefficient dans un anneau $\mathbb{A}$. Soit $D := \sum_{i=1}^s deg f_i$.*

Alors il existe un réseau arithmétique sur $\mathbb{A}$ *qui construit avec une complexité séquentielle* $D^{n^{O(1)}}$ *des polynômes* $g_1, ..., g_t$ *de* $\mathbb{C}[X_1, ..., X_n]$ *définissant l'ensemble des points singuliers de* V *par:*

$$V_{sing} = \{g_1 = 0, ..., g_t = 0\}$$

Nous avons $V_{sing} = \overline{\cup_{k \geq 1} V_k}$. Pour obtenir les polynômes $g_1, ..., g_t$, il suffit d'utiliser le théorème 1.3.3.(ici $\overline{V_{sing}} = V_{sing}$).

Remarque: Le même résultat peut être obtenu sur un corps algébriquement clos de caractéristique 0.

1.3.Outils utilisés:

Nous utiliserons les résultats suivants pour construire l'algorithme annoncé.

Elimination "rapide" des quantificateurs:(Fitchas/Galligo/Morgenstern[FGM])

Théorème 1.3.1: *Soit* Φ *une formule prénexe du premier ordre du langage des corps algébriquement clos à* n *variables et à* r *blocs de quantificateurs definie par des polŷnomes à coefficient dans un anneau* $\mathbb{A}$. *Soit* D *la somme des degrés des polynômes apparaissant dans la formule* Φ.

Alors il existe un réseau arithmétique sur $\mathbb{A}$ *qui construit avec une complexité séquentielle* $D^{n^{O(r)}}$ *une formule* Ψ *sans quantificateurs équivalente à* Φ.

A priori l'élimination des quantificateurs est doublement exponentielle en n (cas où $r = n$)(voir [H]). Cependant dans de nombreuses situations significatives le nombre de blocs de quantificateurs est indépendant de n. Dans ce cas, l'éliminations "rapide" des quantificateurs permet d'obtenir des algorithmes simplements exponentiels.

Le calcul de la multiplicité d'une variété V en un point x se fera en sommant les multiplicités des composantes équidimensionnelles V^d de V en x. Aussi le résultat suivant nous sera d'une grande utilité:

Décomposition d'une variété V en composantes équidimensionnelles V^d (Giusti/Heintz[GH]):

Théorème 1.3.2: *Soit* V *un ensemble algébrique de* $\mathbb{C}^n$ *défini par des polynômes* $f_1, ..., f_s$. *Soit* $D := \sum_{i=1}^{s} deg f_i$.

Alors il existe un réseau arithmétique qui construit avec une complexité séquentielle $D^{O(n^2)}$ *des polynômes* $f_1^0, ..., f_{s_0}^0, ..., f_1^d, ..., f_{s_d}^d, ..., f_1^n, ..., f_{s_n}^n$ *définissant les composantes équidimensionnelles* V^d *de* V *et vérifiant:*

* $V^d = \{x \in \mathbb{C}^n \mid f_1^d = ... = f_{s_d}^d = 0\}$ *pour* $d = 0, ..., n$

* $\sum_{d=1}^{n} \sum_{i=1}^{s_d} deg f_i^d \leq D^{O(n^2)}$.

Cloture de Zariski d'un constructible (Heintz):

Théorème 1.3.3: *Soit Φ une formule sans quantificateur du langage des corps algébriquement clos définissant sur $\mathbb{C}^n$ un ensemble constructible V. Soit D la somme des degrés de tout les polynômes apparaissant dans Φ.*

Alors il existe un réseau arithmétique qui construit avec une complexité séquentielle $D^{n^{O(1)}}$ des polynômes $f_1, ..., f_s$ définissant la cloture de Zariski $\overline{V}$ de V:
$\overline{V} = \{x \in \mathbb{C}^n \mid f_1 = ... = f_s = 0\}$

Il semble qu'il n'y ait pas de référence précise pour ce résultat. En voici une démonstration, aimablement fournie par le rapporteur.
(i) Soit $V = \{f_1 = 0, ..., f_p = 0, g \neq 0\}$, d'après la définition de degré de [H], on a $\deg(\overline{V}) \leq D^n$.
(ii) En utilisant un raisonnement similaire à [H] Remark 4 + Prop. 3, on montre qu'il existe une famille de polynômes $g_1, ..., g_n$ telle que $\overline{V} = \{g_1 = 0, ..., g_n = 0\}$ et $\deg(g_i) \leq D^n$ (voir aussi [CGH] Prop 1.2).
(iii) Etant donné que $g.g_i(x) = 0$ $(1 \leq i \leq n)$ si $f_1(x) = ... = f_r(x) = 0$, on a $g.g_i \in Rad(f_1, ..., f_p)$, et , comme $\deg(g.g_i) \leq D^{n+1}$, $(g.g_i)^{D^{O(n^2)}} \in (f_1, ..., f_p)$ et il existe des représentations

$$(g.g_i)^{D^{O(n^2)}} = A_1 f_1 + ... + A_p f_p$$

avec $\deg(A_j \leq D^{O(n^2)}$ Voir [DFGS] Remark 7.
(iv) Soit B l'ensemble des polynômes $F \in k[X_1, ..., X_n]$ tels que $g^{D^{O(n^2)}} F = A_1 F_1 + ... + A_p F_p$, avec $\deg(A_i \leq D^{O(n^2)}$, $\deg(F) \leq D^{O(n^2)}$, B est un k-espace vectoriel de dimension $\leq D^{O(n^3)}$. On peut en calculer une base en résolvant un système d'équations où les inconnues sont les coefficients des polynômes $F, A_1, ..., A_p$. Si $H_1, ..., H_N$ en est une base, clairement $\overline{V} = \{H_1 = 0, ..., H_N = 0\}$.

2.Généralisation d'une méthode d'Hermite permettant de compter le cardinal d'une variété algébrique de dimension zéro:

2.1.Principe de la méthode:

La méthode consiste à généraliser au cas de plusieurs variables la méthode d'Hermite qui permet de compter le nombre de racines distinctes d'un polynôme d'une variable.(L'interprétation dans le cas réel clos de cette généralisation sera exposé par Pedersen et Roy dans Mega92 à paraître)

Soit $\mathbb{K}$ un corps algébriquement clos de caractéristique 0.

Soit $\mathfrak{p}$ un idéal de $\mathbb{K}[X_1, ..., X_n]$.
Supposons que l'ensemble $z(\mathfrak{p})$ des zéros de $\mathfrak{p}$ dans $\mathbb{K}^n$ soit de dimension zéro.

Posons $z(\mathfrak{p}) = \{\alpha_1, ..., \alpha_s\}$.

Notons A l'anneau de coordonnées suivant: $A = \mathbb{K}[X_1, ..., X_n]/\mathfrak{p}$.

Par hypothèse, A est noetherien et artinien, c'est un $\mathbb{K}$-espace vectoriel de dimension finie que nous pouvons écrire sous la forme:

$$A = \prod_{k=1}^{s} A_{\alpha_k}$$

où pour $k = 1, ..., s$, nous avons:

$A_{\alpha_k} = \mathbb{K}[X_1, ..., X_n]/\mathfrak{I}_{\alpha_k}$ qui vérifie $\sqrt{\mathfrak{I}_{\alpha_k}} = \mathfrak{m}_{\alpha_k}$ idéal maximal associé à l'ensemble algébrique $\{\alpha_k\}$.

Nous définissons la multiplicité n_{α_k} de $\mathfrak{p}$ au point α_k par:
$n_{\alpha_k} := dim_{\mathbb{K}} A_{\alpha_k} = dim_{\mathbb{K}} \mathbb{K}[X_1, ..., X_n]/\mathfrak{I}_{\alpha_k}$.

Soit $l_1, ..., l_q$ une base de A comme $\mathbb{K}$-espace vectoriel.
Soit $Q \in A$.

Considérons la forme bilinéaire symétrique de $A^q \times A^q$ dans $\mathbb{K}$ définie par:

$$\begin{aligned} B(\mathfrak{p}, Q)(a_1, ..., a_q) &:= \sum_{k=1}^{s} n_{\alpha_k} Q(\alpha_k)(a_1 l_1(\alpha_k) + ... + a_q l_q(\alpha_k))^2 \\ &= \sum_{i,j=1}^{q} (\sum_{k=1}^{s} n_{\alpha_k} Q(\alpha_k) l_i(\alpha_k) l_j(\alpha_k)) a_i a_j \end{aligned}$$

On a alors:

Proposition 2.1.1: *Le rang de la matrice de $B(\mathfrak{p}, Q)$ est égal au nombre de points de $z(\mathfrak{p})$ n'annulant pas Q.*

Preuve:

Il suffit de montrer que les formes linéaires $(a_1 l_1(\alpha_k) + ... + a_q l_q(\alpha_k))_{k=1,...,s}$ sont indépendantes.

Soient $\lambda_1, ..., \lambda_s \in \mathbb{K}$ tels que $\sum_{k=1}^{s} \lambda_k l_i(\alpha_k) = 0$ pour $i = 1, ..., q$.

Comme $(l_i)_{i=1,...,q}$ est une base de $A = \mathbb{K}[X_1, ..., X_n]/\mathfrak{p}$, la forme linéaire

$$\begin{array}{ccc} A = \mathbb{K}[X_1, ..., X_n]/\mathfrak{p} & \longrightarrow & \mathbb{K} \\ P & \longmapsto & \sum_{k=1}^{s} \lambda_k P(\alpha_k) \end{array}$$

est identiquement nulle.

Mais la forme linéaire

$$\begin{array}{ccc} \mathbb{K}[X_1, ..., X_n] & \longrightarrow & \mathbb{K} \\ P & \longmapsto & \sum_{k=1}^{s} \lambda_k P(\alpha_k) \end{array}$$

est identiquement nulle sur $\mathfrak{p}$, elle est donc identiquement nulle sur $\mathbb{K}[X_1, ..., X_n]$.

Soient $\mathfrak{I}_k$ l'idéal réduit associé à l'ensemble algébrique $\{\alpha_1, ..., \alpha_{k-1}, \alpha_{k+1}, ..., \alpha_s\}$ pour $k = 1, ..., s$.

Nous avons à cause du Nullstellensatz $\mathfrak{I}_k \neq \sqrt{\mathfrak{p}}$. Il existe donc un polynôme $P_k \in \mathfrak{I}_k \setminus \sqrt{\mathfrak{p}}$ pour chaque $k = 1, ..., s$.

Nous avons donc, pour chaque $k = 1, ..., s$:

$$\sum_{k'=1}^{s} \lambda_{k'} P_k(\alpha_{k'}) = \lambda_k P_k(\alpha_k) = 0$$

$$\Longrightarrow \lambda_k = 0 \text{ pour } k = 1, ..., s.$$

Par conséquent les formes linéaires $(a_1 l_1(\alpha_k) + ... + a_q l_q(\alpha_k))_{k=1,...,s}$ sont indépendantes.

Remarque:
Dans le cas où $\mathbb{K}$ est un corps réel clos, la signature de la matrice de $B(\mathfrak{p}, Q)$ est égale à la différence du nombre de zéros de $z(\mathfrak{p})$ rendant Q positif et du nombre de zéros de $z(\mathfrak{p})$ rendant Q négatif (voir [PRS]).

Il nous reste à exprimer la matrice de $B(\mathfrak{p}, Q)$ de manière à pouvoir effectuer son calcul de manière agréable. C'est l'objet du résultat suivant.

Proposition 2.1.2: *Soit $h \in A = \mathbb{K}[X_1, ..., X_n]/\mathfrak{p}$.*
Soit m_h l'endomorphisme suivant:

$$m_h : \begin{array}{ccc} A & \longrightarrow & A \\ P & \longmapsto & hP \end{array}$$

Alors:

les $(h(\alpha_k))_{k=1,...,s}$ sont les valeurs propres de m_h avec les multiplicités $(n_{\alpha_k})_{k=1,...,s}$.

Preuve:

Nous avons:

$$A = \prod_{k=1}^{s} A_{\alpha_k}$$

avec, pour $k = 1, ..., s$,

$$n_{\alpha_k} := dim_{\mathbb{K}} A_{\alpha_k} = dim_{\mathbb{K}} \mathbb{K}[X_1, ..., X_n]/\mathfrak{I}_{\alpha_k}$$

où $A_{\alpha_k} = \mathbb{K}[X_1, ..., X_n]/\mathfrak{I}_{\alpha_k}$ et $\sqrt{\mathfrak{I}_{\alpha_k}} = \mathfrak{m}_{\alpha_k}$ idéal maximal associé à l'ensemble algébrique $\{\alpha_k\}$.

Il suffit donc de montrer que les $(A_{\alpha_k})_{k=1,...,s}$ sont les sous-espaces propres associés au valeurs propres $(h(\alpha_k))_{k=1,...,s}$.

Comme $h - h(\alpha_k) \in \mathfrak{m}_{\alpha_k}$, il existe un entier n_k tel que $(h - h(\alpha_k))^{n_k} \in \mathfrak{I}_{\alpha_k}$.

Soit $P \in \mathbb{K}[X_1, ..., X_n]$. Soient Id l'identité de $\mathbb{K}[X_1, ..., X_n]$ et Id_A celle de A
Nous avons $(m_h - h(\alpha_k)Id)^{n_k} P \in \mathfrak{I}_{\alpha_k}$

Donc A_{α_k} est nilpotent par $m_h - h(\alpha_k)Id_A$.

On en déduit immédiatement

Corollaire 2.1.3: *$trace(m_h) = \sum_{k=1}^{s} n_{\alpha_k} h(\alpha_k)$.*

On peut donc écrire la forme bilinéaire $B(\mathfrak{p}, Q)$ de la manière suivante:
$B(\mathfrak{p}, Q)(a_1, ..., a_q) := \sum_{i,j=1}^{q}(\sum_{k=1}^{s} n_{\alpha_k} Q(\alpha_k) l_i(\alpha_k) l_j(\alpha_k)) a_i a_j$
$= \sum_{i,j=1}^{q} trace(m_{Ql_il_j}) a_i a_j$

2.2.Complexité de la méthode:

Pour mettre en œuvre cette méthode permettant de compter le nombre d'élément d'un ensemble algébrique de dimension zéro, nous avons besoin:
* d'une base $(l_1, ..., l_p)$ de $A = \mathbb{K}[X_1, ..., X_n]/\mathfrak{p}$
* d'effectuer des calculs dans A, c'est à dire modulo $\mathfrak{p}$.

Le calcul d'une base de Gröbner de $\mathfrak{p}$ remplit ces besoins. Tout les autres calculs sont ensuite des constructions d'algèbre linéaire pour lesquels nous disposons d'algorithmes polynomiaux, ou des réduction modulo la base de Gröbner de $\mathfrak{p}$. Les calculs de base de Gröbner ayant été particulièrement développés ces dernières années, l'implémentation de la méthode d'Hermite généralisée ne pose pas de problème majeur.

Pour construire notre algorithme de partition en strate de multiplicité donnée, nous aurons besoin d'une version paramétrée de la méthode d'Hermite généralisée. Il nous faudra construire une subdivision de l'ensemble des paramètres en sous-ensembles sur lesquels la base de Gröbner de $\mathfrak{p}$ prend une forme donnée. Il nous faudra en particulier connaître l'ensemble des paramètres où la dimension de l'ensemble algébrique $z(\mathfrak{p})$ n'est pas de dimension zéro.

Nous utiliserons le résultat suivant, qui résulte facilement de [DFGS]:

Proposition 2.2.1: *Soient $f_1, ..., f_t \in \mathbb{K}[Y_1, ..., Y_m, X_1, ..., X_n]$. Soient $y_1, ..., y_m$ des éléments de $\mathbb{K}$, $\mathfrak{p}_{y_1,...,y_m}$ désignera l'idéal de $\mathbb{K}[X_1, ..., X_n]$ engendré par $f_1, ..., f_t$ où les $Y_1, ..., Y_m$ ont été spécialisés en $y_1, ..., y_m$.*

Alors il existe un réseau arithmétique qui construit, ave un complexité $D^{(n+m)^{O(1)}}$, un entier $r \leq D^{(n+m)^{O(1)}}$ et une partition de $\mathbb{K}^n$ $(K_{-1}, K_0, K_1, ..., K_r)$ définie par:

$K_i = \{(y_1, ..., y_m) \in \mathbb{K}^m \mid dim\ z(\mathfrak{p}_{y_1,...,y_m}) = 0$ et $rang\ B(\mathfrak{p}_{y_1,...,y_m}, 1) = i\}$ pour $i \geq 0$
et
$K_{-1} = \{(y_1, ..., y_m) \in \mathbb{K}^m \mid dim\ z(\mathfrak{p}_{y_1,...,y_m}) > 0\}$
vérifiant:

$$K_r \neq \emptyset.$$

3.Démonstration du théorème principal:

Dans le théorème 1.2.1, nous voulons construire les ensembles

$$V_k = \{x \in V \mid m_x(V) = k\}$$

La multiplicité $m_x(V)$ de l'ensemble algébrique V au point x est définie par:

$$m_x(V) := \sum_{d=0}^{n} m_x(V^d)$$

où V^d désigne la composante équidimensionnelle de dimension d de V.
Pour d=0,...,n nous avons:

$$m_x(V^d) := degV^d - \sharp H^d \cap V^d + 1$$

pour H^d un n-d-plan passant par x tel que $\sharp H^d \cap V^d$ soit fini et maximal, $degV^d$ étant défini par la quantité $\sharp L^d \cap V^d$ pour L^d un n-d-plan tel que $\sharp L^d \cap V^d$ soit fini et maximal.

Dans un premier temps, nous décomposons V en variétés équidimensionnelles V^d à l'aide de l'algorithme de Giusti et de Heintz (Théorème 1.3.2).
Nous obtenons donc avec une complexité séquentielle $D^{O(n^2)}$ des polynômes: $f_1^0, ..., f_{s_0}^0, ..., f_1^d, ..., f_{s_d}^d, ..., f_1^n, ..., f_{s_n}^n$ vérifiant:

* $V^d = \{x \in \mathbb{C}^n \mid f_1^d = ... = f_{s_d}^d = 0\}$ pour $d = 0, ..., n$
* $\sum_{d=1}^{n} \sum_{i=1}^{s_d} degf_i^d \le D^{O(n^2)}$.

Pour chaque $d = 0, ..., n$,nous allons construire les ensembles $V_k^d = \{x \in V^d \mid m_x(V^d) = k\}$

Fixons $k \in \{0, ..., D^{n^{O(1)}}\}$.

* <u>Calculons tout d'abord le degré $degV^d$ de V^d:</u>

Soit $(a_{i,j})_{i,j} \in \mathbf{M}_{d,n+1}(\mathbb{C})$ l'ensemble des matrices à d lignes et $n+1$ colonnes à coefficients dans $\mathbb{C}$.
Soit $L_{(a_{i,j})_{i,j}}$ l'ensemble algébrique défini par les équations affines:

$$\begin{array}{ccccccc} a_{1,1}X_1 & + & ... & + & a_{1,n}X_n & = & a_{1,n+1} \\ ... & + & ... & + & ... & = & ... \\ a_{d,1}X_1 & + & ... & + & a_{d,n}X_n & = & a_{d,n+1} \end{array}$$

Soit $\mathfrak{p}_{(a_{i,j})_{i,j}}$ l'idéal engendré par:

$$\begin{array}{ccccccc} f_1^d & , & ... & , & f_{s_d}^d, & & \\ a_{1,1}X_1 & + & ... & + & a_{1,n}X_n & - & a_{1,n+1}, \\ ... & + & ... & + & ... & - & ..., \\ a_{d,1}X_1 & + & ... & + & a_{d,n}X_n & - & a_{d,n+1} \end{array}$$

En appliquant la version paramétrée de la méthode d'Hermite généralisée (proposition .2.1) à cette situation, nous obtenons avec une complexité $(D^{O(n^2)})^{(n+d(n+1))^{O(1)}} = D^{n^{O(1)}}$ un entier $r > 0$ tel que:

$$K_r = \{(a_{i,j})_{i,j} \in \mathbf{M}_{d,n+1}(\mathbb{C}) \mid dim\ z(\mathfrak{p}_{(a_{i,j})_{i,j}}) = 0 \text{ et } rang\ B(\mathfrak{p}_{(a_{i,j})_{i,j}}, 1) = r\} \neq \emptyset$$

et
$\{(a_{i,j})_{i,j} \in \mathbf{M}_{d,n+1}(\mathbb{C}) \mid dim\ z(\mathfrak{p}_{(a_{i,j})_{i,j}}) = 0 \text{ et } rang\ B(\mathfrak{p}_{(a_{i,j})_{i,j}}, 1) > r\} = \emptyset.$

Pour $(a_{i,j})_{i,j}$ fixé, $L_{(a_{i,j})_{i,j}}$ est soit vide, soit sous-espace affine de $\mathbb{C}^n$ de dimension supérieure ou égale à $n-d$.

Si $dim\ L_{(a_{i,j})_{i,j}} > n-d$, alors $L_{(a_{i,j})_{i,j}} \cap V^d$ est soit vide, soit infini.

Par conséquent $K_r \cap \{(a_{i,j})_{i,j} \in \mathbf{M}_{d,n+1}(\mathbb{C}) \mid dim\ \ L_{(a_{i,j})_{i,j}} > n-d\} = \emptyset$

En appliquant la proposition 2.1.1 de la méthode d'Hermite généralisée, nous avons:

$\{L\ n\text{-}d\text{-plan affine} \mid \sharp L \cap V^d = r\} \neq \emptyset$
et
$\{L\ n\text{-}d\text{-plan affine} \mid r < \sharp L \cap V^d < \infty\} = \emptyset.$

Par conséquent $degV^d = r$.

* <u>Calculons les ensembles $V_k^d = \{x \in V^d \mid m_x(V^d) = k\}$:</u>

Soit $(b_{i,j})_{i,j} \in \mathbf{M}_{d+1,n}(\mathbb{C})$.
Soit $H_{(b_{i,j})_{i,j}}$ l'ensemble algébrique défini par les équations affines:

$$\begin{array}{ccccccc} b_{1,1}(X_1 - b_{d+1,1}) & + & \dots & + & b_{1,n}(X_n - b_{d+1,n}) & = & 0 \\ \dots & + & \dots & + & \dots & = & \dots \\ b_{d,1}(X_1 - b_{d+1,1}) & + & \dots & + & b_{d,n}(X_n - b_{d+1,n}) & = & 0 \end{array}$$

Soit $\mathfrak{q}_{(b_{i,j})_{i,j}}$ l'idéal engendré par:

$$\begin{array}{ccccc} f_1^d & , & \dots & , & f_{s_d}^d, \\ b_{1,1}(X_1 - b_{d+1,1}) & + & \dots & + & b_{1,n}(X_n - b_{d+1,n}), \\ \dots & + & \dots & + & \dots, \\ b_{d,1}(X_1 - b_{d+1,1}) & + & \dots & + & b_{d,n}(X_n - b_{d+1,n}). \end{array}$$

En appliquant la version paramétrée de la méthode d'Hermite généralisée (proposition 2.2.1) à cette situation, nous obtenons avec une complexité $(D^{O(n^2)})^{(n+n(d+1))^{O(1)}} = D^{n^{O(1)}}$ un entier $r^d < D^{n^{O(1)}}$ et une partition de $\mathbf{M}_{d+1,n}(\mathbb{C})$ $(K_{-1}^d, K_0^d, K_1^d, \dots, K_{r^d}^d)$ vérifiant:

$K_i^d = \{(b_{i,j})_{i,j} \in \mathbf{M}_{d+1,n}(\mathbb{C}) \mid dim\ z(\mathfrak{q}_{(b_{i,j})_{i,j}}) = 0 \text{ et } rang\ B(\mathfrak{q}_{(b_{i,j})_{i,j}}, 1) = i\}$ pour $i \geq 0$
et
$K_{-1} = \{(b_{i,j})_{i,j} \in \mathbf{M}_{d+1,n}(\mathbb{C}) \mid dim\ z(\mathfrak{q}_{(b_{i,j})_{i,j}} > 0\}.$

En utilisant le même argument que pour le calcul du degré nous avons que V_k^d est l'ensemble des $x = (x_1, \dots, x_n) \in V^d$ qui vérifient :

$$\exists (b_{i,j})_{i,j} \in \mathbf{M}_{d+1,n}(\mathbb{C}) \text{ tel que: }$$
$$x_1 = b_{d+1,1}, ..., x_n = b_{d+1,n}$$
$$(b_{i,j})_{i,j} \in K^d_{degV^d+1-k}$$

et

$$\forall (c_{i,j})_{i,j} \in \mathbf{M}_{d+1,n}(\mathbb{C})$$
$$x_1 = b_{d+1,1}, ..., x_n = b_{d+1,n} \Longrightarrow (c_{i,j})_{i,j} \in K^d_{-1} \cup K^d_0 \cup ... \cup K^d_{degV^d+1-k}$$

Ceci constitue une formule du premier ordre du langage des corps algébriquement clos à $2n(d+1)$ variables, à 2 blocs de quantificateurs et de degré total au plus $D^{n^{O(1)}}.D^{n^{O(1)}} = D^{n^{O(1)}}$ (ou plus exactement 2 formules à $n(d+1)$ variables et 1 seul bloc).

Il suffit alors d'utiliser pour chaque $1 \leq k \leq r^d$ l'élimination "rapide" des quantificateurs (théorème 1.3.1) pour obtenir des formules sans quantificateurs définissant les V^d_k de degré total au plus $(D^{2(d+1)n^{O(1)}})^{(n^{O(2)})} = D^{n^{O(1)}}$.

Au total, pour chaque $0 \leq d \leq n$, avec une complexité $(n+1).D^{n^{O(1)}}.D^{n^{O(1)}} = D^{n^{O(1)}}$, nous obtenons une partition $(V^d_k)_{k=1,...,degV^d}$ de V^d.

Afin d'obtenir la partition $(V_k)_{k=1,...,k_{max}}$ de V, il suffit d'écrire les V_k sous la forme:

$$V_k = \bigcup_{\substack{(k_0,...,k_n)\in \mathbb{N}^n \\ k_0+...+k_n=k}} \bigcap_{d=0}^{n} V^d_{k_d}$$

Nous pouvons choisir pour k_{max} l'entier $degV = \sum_{d=0}^n degV^d$ $(k_{max} \leq D^{n^{O(1)}})$.

Il y a, au plus pour $k \leq k_{max}$ fixé, $k^n_{max} = D^{n^{O(1)}}$ n+1-uplet $(k_0, ..., k_n) \in \mathbb{N}^{n+1}$ vérifiant $k_0 + ... + k_n = k$.

Par conséquent chaque V_k est défini par une formule du premier ordre du langage des corps algébriquement clos sans quantificateur de degré total au plus $D^{n^{O(1)}}.(n+1).D^{n^{O(1)}} = D^{n^{O(1)}}$.

Références:

[CGH] L.Caniglia, A.Galligo, J.Heintz *Equations for the projective closure of an affine algebraic variety*, à paraître dans AAECC 7, Toulouse 1989

[DFGS] A.Dickenstein, N.Fitchas, M.Giusti, Sessa *The membership problem for unmixed polynomial ideals is solvable in single exponential time*, à paraître dans AAECC 7, Toulouse 1989

[FGM] N.Fitchas, A.Galligo, J.Morgenstern *Precise sequantial and parallel complexity bounds for the quantifier elimination in algebraically closed fields*, à paraître dans J. Pure Applied Algebra

[GH] M.Giusti, J.Heintz *Algorithmes "disons rapides" pour la décomposition d'une variété algébrique en composantes irréductibles et équidimensionnelles* , Effective Methods in Algebraic Geometry (MEGA 90), 169-191, Birkhauser (1991)

[H] J.Heintz *Definability and fast quantifier elimination in algebraically close fields* Theoret. Comp. Sci. 24 (1983) 239-277

[MR] T.Mostowski ,E.Rannou *Complexity of the computation of the canonical Whitney stratification of an algebraic set in* $\mathbb{C}^n$, AAECC 9 Springer Lecture Notes in Computer Science 539

[M] D.Mumford *Algebraic Geometry I: Complex Projective Varieties*, Springer 1976

[PRS] P. Pedersen, M-F. Roy, A. Szpiglas *Counting real zeros in the multivariate case*, soumis à MEGA 92

T. M. : University of Warsaw, Institute of Mathematics,
ul. Banacha 2, 00-913 Warszawae, Poland

E. R. : IRMAR (unité associée CNRS 305), Université de Rennes 1,
Campus de Beaulieu, 35042 Rennes Cedex, France

REAL PLANE ALGEBRAIC CURVES WITH MANY SINGULARITIES

E.I.Shustin
(Math. Dept., Samara State University, ul. Acad. Pavlova 1, 443011 Samara, USSR)

In this article we give the construction of real plane algebraic curves of a given degree with prescribed singularities. In particular, we construct real curves of degree m with $2m^2/9$ real cusps.

Introduction

Let $z \in \mathbb{C}P^2$ be an isolated singular point of the curve F, and $\mathcal{D} \subset \mathbb{C}P^2$ be the sufficiently small closed ball centred at z. The topological type of the pair $(\mathcal{D}, \mathcal{D} \cap F)$ we call the type of singular point $z \in F$ (or singularity). The topological type of triad $(\mathcal{D}; \mathcal{D} \cap F, \mathcal{D} \cap \mathbb{R}P^2)$ is called the real type of singular point $z \in \mathbb{R}P^2$ (or real singularity).

It is well-known [7] that for any integers d, m satisfying $0 \leq d \leq (m-1)(m-2)/2$ there is an irreducible curve of degree m in $\mathbb{C}P^2$ with d nodes. The same for real curves [2], although in this case the relation of numbers of knots (type $x^2-y^2=0$), of single points (type $x^2+y^2=0$) and of imaginary nodes is unknown in general (Viro [11] constructed curves with only single points). For other singularities there are results on complex curves: some constructions [3] with singularities number, which is a linear function of the curve degree, and upper bounds of singularities number by a quadratic function of the curve degree [5, 6].

Here we'll give the complete solution of the problem on nodal curves (sec. 2), the construction of real cuspidal curves (sec. 3), the construction of real curves with arbitrary singularities (sec 4). Our constructions are based on the theory of gluing algebraic curves and on the independence of singularities deformations.

From now on the term "curve" means the polynomial F and the locus of points defined by $F=0$.

We use below the joint numbering of formulae and statements.

1. Preliminaries

1.1. The gluing of singular curves. Further on we'll use the suggested in [8] modification of the Viro method [11-13]. Recall certain definitions (see [8, 11-13]).

Let z be an isolated singular point of a curve $F \subset \mathbb{C}P^2$. If z is zero-modal we put $b(z) = \mu(z) - 1$, where μ is a Milnor number. If the modality of z is positive we put $b(z)$ equal to the sum of orders of all infinitely near points in full resolution of z, except the nodal points of union of exceptional divisor with the proper transform of F.

Let $F(x,y)$ be a real polynomial with the non-degenerate Newton polygon Δ. The section F^ρ of $F(x,y)$ on edge $\rho \subset \Delta$ is the sum of items from F, corresponding to the edge ρ. If each section F^ρ hasn't any multiple factors (except x, y), we'll call $F(x,y)$ peripherally non-degenerate. Let $S(F)$ denote the singularities collection of the curve $F \setminus \{xy=0\}$.

Remind of the definition of the chart of a peripherally non-degenerate polynomial $F(x,y)$ [11-13]. Introduce the covering

$$E: \mathbb{C}^2 \to (\mathbb{C}\setminus 0)^2, \quad E(u,v) = (\exp u, \exp v).$$

Put $\mathbb{C}^2 = \mathbb{R}^2 \times (\mathbb{R}\sqrt{-1})^2$. There is a sufficiently large polygon $\Delta' \subset \mathbb{R}^2$ such that

(i) there is a homotety $h: \mathbb{R}^2 \to \mathbb{R}^2$, which transforms Δ' into Δ,

(ii) the pair $(\operatorname{Int}\widetilde{\Delta}, E^{-1}(F) \cap \operatorname{Int}\widetilde{\Delta})$, where $\widetilde{\Delta} = \Delta' \times (\mathbb{R}\sqrt{-1})^2$ is a strict deformation retract of the pair $(\mathbb{C}^2, E^{-1}(F))$,

(iii) $E^{-1}(F) \cap (V(\Delta') \times (\mathbb{R}\sqrt{-1})^2) = \emptyset$, where $V(\Delta')$ is a set of vertexes of Δ',

(iv) for each edge $\rho' \subset \Delta'$, the set $E^{-1}(F) \cap (\rho' \times (\mathbb{R}\sqrt{-1})^2)$ is, firstly, homeomorphic to a disjoint union of intervals, which correspond one-to-one to local branches of the curve F determined by a relevant edge $\rho \subset \Delta$, and, secondly, isotopic to $E^{-1}(F^\rho) \cap (\rho' \times (\mathbb{R}\sqrt{-1})^2)$ in $\rho' \times (\mathbb{R}\sqrt{-1})^2$.

Then there is a set $\widetilde{F} \subset \widetilde{\Delta}$ such that

(a) $\widetilde{F}$ is isotopic to $E^{-1}(F)$ in $(\Delta' \setminus V(\Delta')) \times (\mathbb{R}\sqrt{-1})^2$ by an isotopy, which is invariant with respect to the action of the complex conjugation and the automorphism group of the covering E,

(b) $\widetilde{F} \cap (\rho' \times (\mathbb{R}\sqrt{-1})^2) = E^{-1}(F^\rho) \cap (\rho' \times (\mathbb{R}\sqrt{-1})^2)$ for every edge $\rho' \subset \Delta'$.

At last, put

$$H: \widetilde{\Delta} = \Delta' \times (\mathbb{R}\sqrt{-1})^2 \longrightarrow \Delta \times (\mathbb{R}\sqrt{-1})^2, \quad H = (h, id)$$

$$\pi: \Delta \times (\mathbb{R}\sqrt{-1})^2 \to \mathbb{C}^2, \quad \pi(u_1, v_1, u_2\sqrt{-1}, v_2\sqrt{-1}) =$$
$$= (u_1 \exp(u_2\sqrt{-1}), v_1 \exp(v_2\sqrt{-1}))$$

The chart of the polynomial F is a pair

$$\langle F\rangle = (\hat{\Delta}, \hat{F}),$$

where $\hat{\Delta} = \pi(\Delta \times (\mathbb{R}\sqrt{-1})^2)$, $\hat{F} = \pi(H(\widetilde{F}))$.

Let

$$F_k(x,y) = \sum_{(i,j)\in\Delta_k} A_{ij}x^i y^j, \quad k=1,\dots,N, \tag{1.1.1}$$

be peripherally non-degenerate real polynomials with Newton polygons $\Delta_1,\dots,\Delta_N$. Assume

(i) $\Delta_1 \cup \dots \cup \Delta_N = \Delta$ is a convex polygon,

(ii) $\Delta_k \cap \Delta_l$ is empty, or a common vertex, or a common edge of Δ_k, Δ_l if $k \neq l$,

(iii) there is a convex continuous function $\Delta \to \mathbb{R}$, linear on any Δ_k, $1 \le k \le N$, and non-linear on any union $\Delta_k \cup \Delta_l$, $k \neq l$.

1.1.2. Remarks. We observe that the adjacency relation for $\hat{\Delta}_1,\dots,\hat{\Delta}_N$ is like one for $\Delta_1,\dots,\Delta_N$, and that $\hat{\Delta}_1 \cup \dots \cup \hat{\Delta}_N = \hat{\Delta} = \pi(\Delta \times (\mathbb{R}\sqrt{-1})^2)$. Also it follows from (1.1.1) that $F_k^\rho = F_l^\rho$, if $\rho = \Delta_k \cap \Delta_l$ is a common edge. In particular, $\hat{F}_k \cap \hat{\Delta}_k \cap \hat{\Delta}_l = \hat{F}_l \cap \hat{\Delta}_k \cap \hat{\Delta}_l$.

Define the gluing of charts of $F_1,\dots,F_N$ by

$$\langle F_1\rangle \# \dots \# \langle F_N\rangle = (\hat{\Delta}, \hat{F}_1 \cup \dots \cup \hat{F}_N).$$

It should be noted that the singularities set of $\langle F_1\rangle \# \dots \# \langle F_N\rangle$ is the disjoint union of $S(F_1),\dots, S(F_N)$.

1.1.3. Theorem [8]. If $S(F_1),\dots,S(F_N)$ consist of only nodes then there is a real polynomial $F(x,y)$ with the chart $\langle F_1\rangle \# \dots \# \langle F_N\rangle$ and Newton polygon Δ.

Now let Γ be an oriented graph of adjoining the polygons $\Delta_1,\dots,\Delta_N$ without cycles.

1.1.4. Theorem [8]. Assume that for any $k=1,\dots,N$ the curve $F_k \setminus \{xy=0\}$ is irreducible, and

$$\sum_{\sigma\in S(F_k)} b(\sigma) < \sum (\operatorname{card}(\gamma_k \cap \mathbb{Z}^2) - 1),$$

where γ_k runs through all edges of Δ_k, which don't correspond to arcs of Γ coming to Δ_k. Then there exists a real polynomial F with Newton polygon Δ and the chart $\langle F_1\rangle \# \dots \# \langle F_N\rangle$.

1.1.5. Remarks. (1) Proofs of theorems 1.1.3, 1.1.4 [8] imply that local branches of curve F with centres on coordinate axes correspond one-to-one to local branches of $F_1, \dots, F_N$ determined by edges of Δ. (2) In [8] it is shown that we can smooth an arbitrary subset of $S(F)$ in theorems 1.1.3, 1.1.4, while all other singularities are retained.

1.2. Singularities deformation. Let $\mathbb{R}P^n$, $n = m(m+3)/2$, be the space of real curves of degree m in plane. Let $F \in \mathbb{R}P^n$ be a reduced curve and $Sing(F) = S_1 \cup S_2$, where $S_1 \cap S_2 = \emptyset$ and S_1, S_2 are invariant with respect to complex conjugation $conj$. If $z = (0;0)$ in some affine coordinate system (x,y), then $F(x,y|z)$ denotes the polynomial, which defines the curve F in this affine plane. The minimal order K of a not vanishing K-jet of polynomial $F(x,y|z)$ is called the order $ord_F(z)$ of F at the point z. Put

$$\mathcal{D}(z) = \{\Phi \in \mathbb{R}P^n \mid ord_\Phi(z) \geq ord_F(z) - 1\}.$$

Let $T(z)$, $z \in Sing(F)$, mean the tangent cone at F to the germ of locus of curves $\Phi \in \mathbb{R}P^n$ with a singular point in some sufficiently small neighbourhood of z.

1.2.1. Lemma (see [9]). For any $z \in Sing(F)$,

$$T(z) = \{\Phi \in \mathbb{R}P^n \mid z \in \Phi\}.$$

1.2.2. Lemma (see [9]). Linear varieties

$$\mathcal{D}(z),\ z \in S_1/conj,\quad T(w),\ w \in S_2/conj,$$

intersect transversally in $\mathbb{R}P^n$.

Put $m_z = ord_F(z) - 2$, $z \in S_1$.

1.2.3. Corollary. For any polynomials $P_z(x,y)$, $z \in S_1$, with properties

(i) $\deg P_z \leq m_z$, $z \in S_1$;

(ii) $conj\, P_z = P_{conj\, z}$, $z \in S_1$;

(iii) P_z, $z \in S_1$, are sufficiently close to zero;

there exists the close to F curve $\Phi \in \mathbb{R}P^n$ satisfying the following conditions:

(1) m_z-jet of $\Phi(x,y|z)$ is equal to $P_z(x,y)$, $z \in S_1$,

(2) Φ is non-singular outside some neighbourhood of S_1.

1.2.4. Theorem [2]. If $F \in \mathbb{R}P^n$ is a nodal curve, then there exists the anyhow close to F curve $\Phi \in \mathbb{R}P^n$ with nodes in neighbourhoods of every point $z \in S_1$ and non-singular outside these neighbourhoods.

2. Construction of nodal curves

Here we consider curves with $S(F)$ consisting of only nodes. Any gluing of such curves is allowed by theorem 1.1.3.

The curve F with Newton polygon Δ and $S(F)$, consisting of a knots, b single points and $2c$ imaginary nodes, we'll denote as $F \in \Delta(a,b,c)$ or $F \in (a,b,c)$. If $F \in (a,b,c)$ is irreducible and has the degree m, then

$$a + b + 2c \le (m-1)(m-2)/2 \qquad (2.1)$$

according to the Plücker formula [14].

2.2. Theorem. For any non-negative integers a, b, c satisfying (2.1), there is an irreducible curve $F \in (a,b,c)$ of degree m.

Proof. According to theorem 1.2.4 it is enough to study the case $a + b + 2c = (m-1)(m-2)/2$. We'll use induction in m. All nodal curves of degree ≤ 4 are well-known [15], therefore we suppose $m \ge 5$.

1 step. Assume $b = 0$, $c \ge 2m - 7$. Then

$$a + 2(c - 2m + 7) = (m-5)(m-6)/2,$$

therefore there is a curve $\Phi \in (a, 0, c - 2m + 7)$ of degree $m - 4$. Using theorem 1.2.4 we obtain the desired curve F from curve $\Phi \cup C_2 \cup \mathrm{conj}\, C_2$, where C_2 is an imaginary conic with $C_2 \cap \mathbb{R}P^2 = \emptyset$.

2 step. Assume $b = 0$, $m - 3 \le c \le 2m - 8$. Then $a \ge 2$. Since

$$(a-1) + 2(c - m + 3) = (m-3)(m-4)/2,$$

there is a curve $\Phi \in (a-1, 0, c - m + 3)$ of degree $m - 2$. Also there is a real conic C_2 meeting Φ exactly at two real points, because $a - 1 \ge 1$. Using theorem 1.2.4 we smooth one of real points in $\Phi \cap C_2$ and obtain the desired curve $F \in (a, b, c)$.

3 step. Assume $b = 0$, $c \le m - 4$.

2.3. Lemma. For any $q \ge 0$ there exist curves $\mathcal{U}_q^-, \mathcal{U}_q^+ \in (q(q-1)/2, 0, 0)$ with Newton triangles $\{(0;1), (0;q+1), (q+1;0)\}$ and $\{(0;0), (0;q+1), (q;1)\}$ respectively.

Proof. We get the curves $\mathcal{U}_q^\pm$ by gluing the curve yF_q, where $F_q(x,y) = 1 + x\varphi(x) + y\psi(x,y)$ is a union of different real straight lines in general position, with curves

$G^{-}=y(1+x\varphi(x))+x^{q+1}$, $G^{+}=y(1+x\varphi(x))+1$ ■

Let $H=y+(x-1)(x-2)\cdot\ldots\cdot(x-m+2)$. There are real straight lines L_1, L_2 meeting H exactly at $2[c/2]$ and $2[(c+1)/2]$ imaginary points respectively. The gluing of $\mathcal{U}^{-}_{m-1}$ and H gives a curve $\Phi\in((m-3)(m-4)/2,0,0)$. Smoothing two real points in $\Phi\cap(L_1\cup L_2)$ we get the desired curve F from the curve $\Phi\cup L_1\cup L_2$.

4 step. Assume $b\geq 1$, $c\geq m-3$. Since

$$a+(b-1)+2(c-m+3)=(m-3)(m-4)/2,$$

there is a curve $\Phi\in(a,b-1,c-m+3)$ of degree $m-2$. Let L be a general imaginary straight line. Then we obtain the desired curve F from the curve $\Phi\cup L\cup \operatorname{conj} L$ by smoothing two conjugate points in $\Phi\cap(L\cup\operatorname{conj} L)$.

5 step. Assume $1\leq b\leq m-4$, $c\leq m-4$.

2.4. Lemma. For any triangle T with integral vertexes $(p_1;1),(p_2+p;2),(p_2-p;0)$ and $\operatorname{card}(\operatorname{Int} T\cap\mathbb{Z}^2)=q$, there is a curve $f\in T(0,q,0)$, and also, if q is even, a curve $g\in T(0,0,q/2)$.

Proof. A curve with Newton triangle T can be obtained from a curve with Newton triangle $\{(0;2),(q+1;1),(0;0)\}$ by birational transformation remaining $S(f)$ (or $S(g)$). Then f is an image of

$$y^2+2yP_{q+1}(x-2)+1=0,$$

where $P_{q+1}(x)=\cos((q+1)\arccos x)$ is the Chebyshev polynomial, and g is an image of

$$y^2+2yP_{q+1}(x\sqrt{-1})\sqrt{-1}-1=0 \quad ■$$

2.5. Remark. It should be noted that sections of polynomials f, g on edge $[(p_2+p;2),(p_2-p;0)]$ of Newton triangle T are products of real factors, linear in y .

Introduce the following polygons:

(i) quadrangle $\tau=\{(0;0),(0;2),(m-1;1),(m;0)\}$;

(ii) triangles $T^{+}_{q}=\{(0;m-q-2),(0;m-q),(q+1;m-q-1)$, $q=0,1,\ldots,m-4$;

(iii) triangles $T^{-}_{q}=\{(m-q-1;0),(m-q;q),(m-q-2;q+2)\}$, $q=0,1,\ldots,m-3$.

The corresponding curves from lemma 2.4 we denote by

$f(T_q^{\pm}),\ g(T_q^{\pm})$.

To get the desired curve F we have to glue curves $f(T_b^{+})$, $u_{b+1}^{-} y^{m-b-1}$ into a curve of type $(b(b-1)/2,\ b,\ 0)$ and then to complete the construction as in 3 step.

6 step. Assume $b \geq m-3,\ c \leq m-4$.

2.6. Lemma. For any non-negative integers α, β, γ satisfying

$$\alpha + \beta + 2\gamma = m-2 \qquad (m \geq 5)$$

there is a curve of type $\tau(\alpha, \beta, \gamma)$.

Proof. According to lemma 2.4 there is a curve $G \in$ $\in \tau(0, m-2, 0)$. The statement follows from the method to substitute (A) a single point for a knot, or (B) two single points for two imaginary nodes. Realize the Cremona transformation

$$x_0 = x_1' x_2', \quad x_1 = x_0' x_2', \quad x_2 = x_0' x_1', \tag{2.7}$$

supposing that $\mathrm{ord}_G (1:0:0) = m-2$, and $(0:1:0)$, $(0:0:1)$ are single points of G. According to [14] the proper image G^* of G has order $m-4$ at $(1:0:0)$, touches the line $x_0' = 0$ and meets the line $x_1' = 0$ at two imaginary points. We have to turn the line $x_1' = 0$ round the point $(1:0:0)$ so that a pair of imaginary intersection points with G^* substitutes for a pair of real these, or we have to substitute lines $x_1' = 0,\ x_2' = 0$ for a pair of imaginary conjugate lines through $(1:0:0)$. And then we realize the transformation

$$x_0' = x_1 x_2, \quad x_1' = x_0 x_2, \quad x_2' = x_0 x_1.$$

It is easy to see that the first operation is (A), the second operation is (B) ∎

Now we'll construct a curve $F \in (a, b, c)$ by the following algorithm. Put

$$b' = b, \quad c' = c, \quad F' = y^2 + yx^{m-1}, \quad q = m-3, \quad \varepsilon = -1.$$

The algorithm step is as follows: (1) if $q > \max\{b', 2c'\}$ or $q < 0$ then stop, (2) if q is even and $q \leq 2c'$ then glue F' with $g(T_q^{\varepsilon})$ and decrease c' by $q/2$, otherwise glue F' with $f(T_q^{\varepsilon})$ and decrease b' by q, (3) substitute F' for the obtained curve, ε for -1, and q for $q-1$, and then return to (1). After stopping algorithm we get:

If $q = -1$ then F is the gluing of F' and a curve $F'' \in \tau(a, b', c')$.

If $q > 0$, $b' + 2c' \leq m-2$ then F is the gluing

of F', the curve $u_{q+1}^{\varepsilon} y^{m-q-1}$ and a curve $F'' \in \tau(m-2-b'-2c', b', c')$.

Let $q > b'$, $q > 2c'$, $b' + 2c' > m-2$, $b' > (m-2)/2$, $\varepsilon = 1$. According to lemma 2.4 there is a curve $\Phi_1 \in (0, b', 0)$ with Newton triangle

$$\{(0; m-q-2), (2r; m-q), (q+1; m-q-1)\},$$

where $r = q - b'$. According to 2.5 the section of the polynomial Φ_1 on the edge $[(0; m-q-2), (2r; m-q)]$ has two real factors $y + sx^r$, $y + tx^r$. Therefore the curve

$$\Phi_2 = (y + s(x-1)\cdot\ldots\cdot(x-r))(y + t(x-1)\cdot\ldots\cdot(x-r))y^{m-q-2}$$

with Newton triangle $\{(0; m-q-2), (2r; m-q), (0; m-q)\}$ is of the type $(2r, 0, 0)$. Now put

$$\Phi_3 = x^{2r} y^{m-q-1}(uy(x-1)\cdot\ldots\cdot(x-q+2r)+1),$$

where $x^{2r} y^{m-q-1}(uyx^{q-2r}+1)$ is the section of the polynomial Φ_1 on the edge $[(2r; m-q), (q+1; m-q-1)]$. Gluing curves F', Φ_1, Φ_2, Φ_3 and a curve of type $\tau(m-2-2c', 0, c')$ we obtain a curve

$$F'' \in ((m-1)(m-2)/2 - q(q-1)/2, b, c)$$

with Newton polygon $\{(0;0), (0; m-q), (q; m-q), (m; 0)\}$. According to 1.1.5(1) a section of a polynomial F'' on the edge $[(0; m-q), (q; m-q)]$ is, evidently, a product of linear real factors. Hence a curve $F \in (a, b, c)$ can be obtained by gluing F'' with a relevant curve $F_q y^{m-q}$ from proof of lemma 2.3.

Analogously one has to construct the desired curve $F \in (a, b, c)$ in cases

$$q > b', \ q > 2c', \ b' + 2c' > m-2, \ b' > (m-2)/2, \ \varepsilon = -1$$

and

$$q > b', \ q > 2c', \ b' + 2c' > m-2, \ 2c' > (m-2)/2, \ \varepsilon = \pm 1.$$

3. Construction of cuspidal curves

The number K of ordinary cusps (i.e. of type $y^2 + x^3 = 0$) of a complex curve of degree m does not exceed $5m^2/16$ [5, 6]. On the other side there are curves of degree $m = 6q$ with $m^2/4$ cusps [6]. Namely, let $F(x_0, x_1, x_2)$ be the sextic curve dual to a non-singular cubic. It has 9 cusps. Then the curve $F(x_0^q, x_1^q, x_2^q)$ has $9q^2 = m^2/4$ cusps. However in this case at most m cusps are real, and also it is unknown what numbers $N \in (0; m^2/4)$ of cusps can be realized. We consider the problem to find the number

$\mathcal{K}(m) = \max \{ n \mid$ for each $N \in [0; n]$ there is a real curve of degree m with N real cusps as only its singularities $\}$.

It is well-known that $\mathcal{K}(1) = \mathcal{K}(2) = 0$, $\mathcal{K}(3) = 1$, $\mathcal{K}(4) = 3$, $\mathcal{K}(5) = 5$ [4].

3.1. <u>Theorem</u>. If $m \geq 6$ then

$$\mathcal{K}(m) \geq P(m) = 2[m/3]^2 - [m/3] \qquad (3.2)$$

Proof. We'll use below the construction 1.1.4, therefore, according to 1.1.5(2), it is enough to construct a curve of degree m with $P(m)$ real cusps.

3.3. <u>Lemma</u>. There exists a curve Φ with Newton quadrangle $\tau = \{(0;0), (0;3), (3;0), (3;3)\}$ and $S(\Phi)$ consisting of 4 real cusps.

Proof. Let C_4 be a quartic with 3 real cusps. Let the points $(0:1:0)$, $(0:0:1)$ lie on C_4, and the straight line $x_0 = 0$ touch C_4 at a real point. Then (see [14]) Φ is a transformation (2.7) image of C_4 ■

3.4. <u>Lemma</u>. There exists a real cuspidal cubic with Newton triangle $\{(0;0), (3;0), (0;3)\}$ and with prescribed sections on edges

$$[(0;0), (3;0)], \ [(0;0), (0;3)].$$

Proof. It is enough to show that anharmonic relation of origin and intersection points of C_3 on coordinate axes x and y are arbitrary. This can be got by means of a suitable choice of axes x and y ■

Suppose that $m = 3q$. The quadrangles

$$\tau_{ij} = \{(3i; 3j), (3i; 3j+3), (3i+3; 3j), (3i+3; 3j+3)\},$$
$$i \geq 0, \ j \geq 0, \ i+j \leq q-2,$$

and the triangles

$$T_i = \{(3i; m-3i-3), (3i; m-3i), (3i+3; m-3i-3)\}$$
$$i = 0, \dots, q-1,$$

form the regular subdivision of the triangle $\{(0;0), (m;0), (0;m)\}$. Orient all arcs of the adjoining graph Γ upwards or to the right. Define the following transformations of polynomials $\Psi(x,y)$ with Newton polygon τ :

$$R\psi(x,y)=x^3\psi(x^{-1},y),\quad H\psi(x,y)=y^3\psi(x,y^{-1})$$

Then the desired curve with $2q^2-q$ real cusps is a gluing of curves $x^{3i}y^{3j}R^iH^j\Phi(x,y),\ i\geq 0,\ j\geq 0,\ i+j\leq q-2$, and also suitable cuspidal cubics with Newton triangles T_i, $0\leq i\leq q-1$. Here conditions of theorem 1.1.4 are satisfied, because $b=1$ for any ordinary cusp.

If $m=3q+1$ or $3q+2$, it is necessary to add to the above gluing some curves ψ with $S(\psi)=\emptyset$ and Newton triangles $\{(0;m),(3i;3q-3i),(3i+3;3q-3i-3)\}$, $0\leq i\leq q-1$, $\{(0;m),(m;0),(3q;0)\}$.

4. Construction of curves with arbitrary singularities

4.1. Theorem. For any collection $\{S_1,\ldots,S_n\}$ of real singularities and for integer $m>0$ satisfying

$$m^2\geqslant\frac{4\sqrt{2}}{3}\sum_{i=1}^{n}(\mu(S_i)+3)^3, \tag{4.2}$$

where μ is a Milnor number, there exists a real irreducible curve of degree m with $S_1,\ldots,S_n$ as only its singularities.

4.3. Remarks. (1) The left hand side of (4.2) cannot be increased, because a singularities number of a degree m curve is at most $m^2/2$ [14]. The left hand sides of known estimates [3, 10] are linear functions of the curve degree. At the same time, probably, the singularities invariants in (4.2) can be decreased. (2) The getting of the type (4.2) estimate in the way of the formal comparison of a curve coefficients number with a number of conditions imposed by singularities is hardly possible, because it is very difficult to prove the system of these conditions gives a curve without additional degenerations.

Proof of theorem 4.1. If $F(x,y)=0$ defines the singularity S at $(0;0)$ then the equation $\Phi(x,y)=0$ with the same $(\mu(S)+1)$-jet defines the same singularity at $(0;0)$ (see [1]). Therefore, according to 1.2.3, it is enough to construct a reduced curve of degree m, satisfying (4.2), with points z_i of orders $\operatorname{ord}(z_i)\geqslant\mu(S_i)+3,\ 1\leq i\leq n$.

4.4. Lemma. For any integers $k_1\geqslant\ldots\geqslant k_n\geqslant 4$ there exists a reduced curve of degree

$$m\leq\sqrt{4\sqrt{2}/3\cdot(k_1^3+\ldots+k_n^3)} \tag{4.5}$$

with points z_i of orders

$$\mathrm{ord}(z_i) \geq K_i, \quad 1 \leq i \leq n. \tag{4.6}$$

Proof. Put

$$\pi(i) = (i-1)(i+2)/2, \quad i \geq 2.$$

Let $\pi(q) \leq n < \pi(q+1)$ and points $z_1, \dots, z_n$ be in general position in a plane. Note that $\pi(q+1)$ is a dimension of the space of the degree q curves. Since $n < \pi(q+1)$, there are different irreducible curves

$$\varphi_i^{(0)}, \quad 1 \leq i \leq K_{\pi(q)},$$

of degree q, which pass through $z_1, \dots, z_n$. For any $j = 1, \dots, q-2$, since $\pi(q-j+1)$ is a dimension of the space of the degree $(q-j)$ curves, there are different irreducible curves

$$\varphi_i^{(j)}, \quad 1 \leq i \leq K_{\pi(q-j)} - K_{\pi(q-j+1)},$$

of degree $q-j$, which pass through z_r, $1 \leq r \leq \pi(q-j+1)-1$. At last let

$$\varphi_i^{(q-1)}, \quad 1 \leq i \leq K_1 - K_{\pi(2)},$$

be different straight lines through z_1.

Consider the reduced curve $\varphi = \bigcup \varphi_i^{(j)}$. Since any point z_r, $\pi(q) \leq r \leq n$, belongs to $\varphi_i^{(0)}$, $1 \leq i \leq K_{\pi(q)}$, then

$$\mathrm{ord}_\varphi(z_r) \geq K_{\pi(q)} \geq K_r, \quad \pi(q) \leq r \leq n. \tag{4.7}$$

Analogously, any point z_r, $\pi(q-l) \leq r < \pi(q-l+1)$, $1 \leq l \leq q-2$, belongs to $\varphi_i^{(j)}$, $0 \leq j \leq l$, therefore

$$\mathrm{ord}_\varphi(z_r) \geq \mathrm{card}\{\varphi_i^{(j)} \mid 0 \leq j \leq l\} = K_{\pi(q-l)} \geq$$

$$\geq K_r, \quad \pi(q-l) \leq r < \pi(q-l+1). \tag{4.8}$$

At last

$$\mathrm{ord}_\varphi(z_1) \geq \mathrm{card}\{\varphi_i^{(j)} \mid 0 \leq j \leq q-1\} = K_1. \tag{4.9}$$

Inequalities (4.7), (4.8), (4.9) imply (4.6).

Now we'll give an upper bound for $\deg \varphi$:

$$m = \deg \varphi = K_{\pi(q)} \cdot q + (K_{\pi(q-1)} - K_{\pi(q)}) \cdot (q-1) + \dots +$$

$$+ (K_{\pi(2)} - K_{\pi(3)}) \cdot 2 + (K_1 - K_{\pi(2)}) =$$

$$= K_{\pi(q)} + K_{\pi(q-1)} + \ldots + K_{\pi(2)} + K_1 = \tau_q + \tau_{q-1} + \ldots + \tau_2 + \tau_1,$$

where $\tau_i = K_{\pi(i)}$, $2 \le i \le q$, $\tau_1 = K_1$. Since

$$K_r \geq K_{\pi(3)}, \quad 2 = \pi(2) < r \le \pi(3),$$

$$K_r \geq K_{\pi(i)}, \quad \pi(i-1) < r \le \pi(i), \quad 4 \le i \le q,$$

then

$$K_1^3 + \ldots + K_n^3 \geq K_1^3 + \ldots + K_{\pi(q)}^3 \geq$$

$$\geq K_1^3 + K_2^3 + \sum_{i=3}^{q} K_{\pi(i)}^3 (\pi(i) - \pi(i-1)) =$$

$$= K_1^3 + K_2^3 + \sum_{i=3}^{q} K_{\pi(i)}^3 \cdot i =$$

$$= \tau_1^3 + \tau_2^3 + 3\tau_3^3 + \ldots + q\tau_q^3 \overset{def}{=} M$$

The Minkovski inequality implies

$$m = \tau_1 + \ldots + \tau_q \le M^{1/3} (1 + 1 + 1/\sqrt{3} + \ldots +$$

$$+ 1/\sqrt{q})^{2/3} \le M^{1/3} (2\sqrt{q})^{2/3}. \tag{4.10}$$

Since $K_n \geq 4$ then

$$m^2 \geq \sum \mu(z_i) \geq \sum (K_i - 1)^2 \geq 9n \geq 9q^2/2.$$

Therefore (4.10) implies

$$m \le M^{1/3} (2\sqrt{q})^{2/3} \le M^{1/3} m^{1/3} (4\sqrt{2}/3)^{1/3},$$

what is equivalent to (4.5) ∎

Now for any degree satisfying (4.2), we have to add to the curve from lemma 4.4 a suitable set of straight lines in general position.

References

1. Arnol'd V.I., Gusein-zade S.M., Varchenko A.N. Singularities of differentiable maps, vol. 1. Boston, Basel, Stuttgart: Birkhäuser Verlag, 1985.

2. Brusotti L. Sulla "piccola variazione" di una curva piana algebrica reali. Rend. Rom. Ac. Lincei (5), 30, 375-379 (1921)

3. Gradolato M., Mezzetti E. Curves with nodes, cusps and ordinary triple points. Ann. Univ. Ferrara, sez. 7, vol. 31, 23-47

(1985)

4. Gudkov D.A. On the curve of 5th order with 5 cusps. Function. anal. Pril. 16, no. 3, 54-55 (1982) (Russian)

5. Hirzebruch F. Singularities of algebraic surfaces and characteristic numbers. Contemp. Math. 58, 141-155 (1986)

6. Ivinskis K. Normale Flächen und die Miyaoka-Kobayashi Ungleichung. Diplomarbeit, Bonn, 1985.

7. Severi F. Vorlesungen über algebraische Geometrie (Anhang F). Leipzig, Teubner, 1921.

8. Shustin E.I. Gluing of singular algebraic curves. In: Methods of Qualitative Theory, Gorky Univ. Press, Gorky, 1985, pp. 116-128 (Russian)

9. Shustin E.I. On manifolds of singular algebraic curves. Selecta Math. Sov. 10, no. 1, 27-37 (1991)

10. Vainstein A.D., Shapiro B.Z. Singularities of Hyperbolic Polynomials and Boundary of Hyperbolicity Domain. Uspekhi Math. Nauk 40, no. 5, 305 (1985) (Russian)

11. Viro O.Ya. Real varieties with prescribed topological properties. Doct. thesis, Leningrad Univ., Leningrad, 1983 (Russian)

12. Viro O.Ya. Gluing of algebraic hypersurfaces, smoothing of singularities and construction of curves. In: Proc. Leningrad Intern. Topological Conf., Leningrad, Nauka, 1983, pp. 149-197 (Russian)

13. Viro O.Ya. Gluing of plane real algebraic curves and construction of curves of degrees 6 and 7. In: Lect. Notes Math., vol. 1060, Springer, 1984, pp. 187-200.

14. Walker R. Algebraic curves. New York, Dover, 1950.

15. Zeuthen H.G. Sur les differentes formes des courbes planes du quadrieme ordre. Math. Annalen, 408-432 (1893)

Effective stratification of regular real algebraic varieties

Nicolai N. Vorobjov
St. Petersburg Dept. of Mathematical
Steklov Institute of Academy of Sci.
27, Fontanka, St. Petersburg, 191011 Russia

Abstract. An algorithm is proposed, producing a Whitney stratification for a real algebraic variety which is a union of transversally intersecting smooth varieties. The complexity of the algorithm and the estimates on the parameters of the produced strata are single exponential in the number of variables of the input polynomials.

Introduction.

Let us define a regular real algebraic variety to be the union $\bigcup_{1 \le i \le s} W_i$ of a finite number of smooth algebraic varieties $W_i \subset \mathbb{R}^n$ such that for any family of distinct indexes $j_1, \ldots, j_{s_1}$ $(s_1 \le s)$ if $\bigcap_{1 \le \ell \le s_1} W_{j_\ell} \ne \phi$ then this intersection is transversal.

In this paper an algorithm is presented for constructing in subexponential time (see below) a Whitney stratification (see e.g. [I]) of a regular real algebraic variety given by an arbitrary formula of the kind

$$\bigvee_{1 \le i \le t} \ \mathop{\&}_{1 \le j \le r} (f_{ij} = 0) \qquad \text{(I)}$$

where polynomial $f_{ij} \in \mathbb{Z}[X_1, \ldots, X_n]$.

Let us assume that the input formula (I) satisfies the following estimates : $\deg_{X_1, \ldots, X_n}(f_{ij}) < d$, the absolute value of every (integer) coefficient appearing in f_{ij} does not exceed 2^M (hence bit-length $\ell(f_{ij})$ of every coefficient does not exceed M) and $t+r = K$ for some natural d, M.

Theorem : There is an algorithm that produces for any formula of the form (I), defining a regular real algebraic variety $V \subset \mathbb{R}^n$ and satisfying the mentioned estimates, a Whitney stratification of V. This stratification is represented by a family of strata - semialgebraic sets $V_0 = \{\Theta_0\}, \ldots, V_m = \{\Theta_m\} \subset \mathbb{R}^n$ $(m \le n)$, where Θ_j $(0 \le j \le m)$ are quantifier-free formulas. Every Θ_j containes $(Kd)^{n^{O(1)}}$ atomic subformulas of the kind $(g \ge 0)$ where polynomial $g \in \mathbb{Z}[X_1, \ldots, X_n]$ satisfies the bounds :

$$\deg_{X_1,\dots,X_n}(g) < (Kd)^{n^{O(1)}}, \quad \ell(g) < M^{O(1)} (Kd)^{n^{O(1)}} .$$

The running time of the algorithm is polynomial in M, $(Kd)^{n^{O(1)}}$.

Let us admit that the natural idea of constructing a stratification by recursion (roughly speaking : defining firstly singular points with the help of a formula of the first-order theory of reals having small number of quantifiers, after that defining the singular points of the latter set and so on) will give the algorithm with the complexity double exponential in number of variables (more precisely - in the "depth" of the singularities). This idea can be applied to an arbitrary (not necessary regular) algebraic variety [9].

The set V admits a certain standard stratification. Namely, suppose that the stratification for the set $W^{(\ell)} = \bigcup_{1 \le i \le \ell} W_j$ $(1 \le \ell < s)$ is already defined. Then define the stratification for $W^{(\ell+1)} = \bigcup_{1 \le i \le \ell+1} W_j$ whose strata are :

1) the intersections of $W_{\ell+1}$ with the strata of $W^{(\ell)}$;
2) complements to the union of the strata defined by I) in strata of $W^{(\ell)}$;
3) complements to the union of the strata defined by I) in $W_{\ell+1}$.

Obviously this will be a Whitney stratification whose strata we shall denote by $V_0,\dots,V_m$ where $\dim(V_i) = i$ for $0 \le i \le m$.

In every component of the disjunction in formula (I) the system of equations is replaced by $f_i = 0$ where $f_i = \sum_{1 \le j \le r} f_{ij}^2$. Let $f = \prod_{1 \le i \le t} f_i$. It is easy to check that the variety $V = \{f=0\}$.

I. Infinitesimals.

Let F be any formally real field, denote by $\widetilde{F}$ its real closure (see e.g. [4]). Introduce an element $\varepsilon > 0$ infinitesimal relative to F (i.e. for every $0 < a \in F$ the inequality $\varepsilon < a$ is valid). The elements of $\widetilde{F(\varepsilon)}$ are Puiseux series in ε i.e. for $\alpha \in \widetilde{F(\varepsilon)}$ holds : $\alpha = \sum_{i \ge 0} \alpha_i \varepsilon^{\nu_i/\mu}$ where $0 \ne \alpha_i \in \widetilde{F}$ for all $i \ge 0$, integers $\nu_0 < \nu_1 < \dots$ increase and the natural number $\mu \ge 1$. If $\nu_0 < 0$, then $\alpha \in \widetilde{F(\varepsilon)}$ is infinitely large, if $\nu_0 > 0$ then α is infinitesimal relative to F. If $\nu_0 \ge 0$ then the standard part $st_\varepsilon(\alpha) \in \widetilde{F}$ is defined :

$$st_\varepsilon(\alpha) = \begin{cases} 0 & \text{if } \nu_0 > 0 \\ \alpha_0 & \text{if } \nu_0 = 0 \end{cases} .$$

The standard part of a vector $(\beta_1,...,\beta_n)$ is defined component-wise. By $st_{1,2}(.)$ we shall denote the superposition $st_{\varepsilon_1}(st_{\varepsilon_2}(.))$ in case the several extensions are involved.

The following "transfer principle" is valid : if $F_1 \subset F_2$ are two real closed fields, F_1 is a subfield of F_2 then every proposition expressed by a formula of first-order theory of real closed fields with coefficients (of the atomic polynomials) in F_1 is true over F_1 if and only if it is true over F_2.

Let $A \subset (\widetilde{F})^n$ be a semialgebraic set defined by a formula with coefficients in F. Then by $A^{(\varepsilon)}$ we shall denote the set defined in $(\widetilde{F(\varepsilon)})^n$ by the same formula. Analogous notation we shall use if several infinitesimals are involved.

2. Auxiliary propositions.

Introduce tree infinitesimals $0 < \varepsilon_2 < \varepsilon_1 < \varepsilon_0$ so that ε_i $(i = 0,1,2)$ is infinitesimal relative to the field $\mathbb{Q}(\varepsilon_0,...,\varepsilon_{i-1})$ and the variety $\mathcal{V} = \{f-\varepsilon_2=0\} \subset (\widetilde{\mathbb{Q}(\varepsilon_0,\varepsilon_1,\varepsilon_2)})^n$. Note that $\mathcal{V}$ is a smooth hypersurface. The main idea of the algorithm is that close to any point of p-dimension stratum V_p (of the standard stratification of V) the hypersurface $\mathcal{V}$ looks like a cylinder over $(n-p-1)$-dimension manifold having infinitely large absolute values of principal curvatures. This property one can express in the language of first-order theory of real closed fields and thus "define" the stratum V_p. The role of the base of the mentioned cylinder plays a section of $\mathcal{V}$ with a $(n-p)$-plane which is orthogonal to V_p at some point from V_p. The idea of studying a stratification of a zero-level set of a function via the behaviour of a close-level smooth set was introduced (in much more abstract setting)) in [10] (see also [11]).

Choose $x \in V_p$. Note that the intersection $M = V_p^{(\varepsilon_0,\varepsilon_1,\varepsilon_2)} \cap \mathcal{D}_x(\varepsilon_0)$ is connected since $V_p \cap \mathcal{D}_x(\delta)$ is connected for all sufficiently small $0 < \delta \in \mathbb{Q}$ and according to transfer principle. Here $\mathcal{D}_z(R)$ denotes an open ball of radius R about z. Let

$$\mathcal{M} = \bigcup_{y \in M} (\mathcal{V} \cap \mathcal{D}_y(\varepsilon_1)).$$

Let $L, L_1 \subset (\widetilde{\mathbb{Q}(\varepsilon_0,\varepsilon_1,\varepsilon_2)})^n$ be r-and m-planes correspondingly, $r \geq m > 0$. We shall say that L and L_1 are ε_0-collinear if for every unit vector $v_1 \in L_1$ there exists such a vector $v \in L$ that $v_1 = v + w$ where all components of w are infinitesimals relative to $\mathbb{Q}$.

Denote by P_z the $(n-p)$-plane orthogonal to M at the point $z \in M$ and by $\mathscr{L} \subset \mathscr{M}$ a subset of points of $\mathscr{M}$ at each of which the normal to $\mathscr{M}$ is not ε_0-collinear to $(n-p)$-plane P_x.

Let us remark that $\mathscr{L}$ is open in $\mathscr{V}$ in the topology with the base of all open disks in $\mathscr{V}$.

Note that the set $\mathscr{L}$ is not semialgebraic (its definition includes the indication on unspecified element infinitesimal relative to $\mathbb{Q}$). Nevertheless we shall consider the connected components of $\mathscr{L}$, namely, the linearly connected components. The subset $\mathscr{L}_0 \subset \mathscr{L}$ will be called linearly connected if for every pair of points $x,y \in \mathscr{L}_0$ there exists a connected semialgebraic curve belonging to $\mathscr{L}_0 \subset (\widetilde{\mathbb{Q}(\varepsilon_0,\varepsilon_1,\varepsilon_2)})^n$ and containing x, y.

Let us prove prove that $\mathscr{L}$ has a finite number of linearly connected components. Recall at first that if a semialgebraic set is defined by a quantifier-free formula of the first-order theory of real closed field with $\bar{K}$ atomic polynomials having $\bar{n}$ variables and degrees not exceeding $\bar{d}$ then the number of its connected components is less then a value of a certain function α in $\bar{K}$, $\bar{n}$, $\bar{d}$ (as such a function one can take $(\bar{K}\,\bar{d}\,)^{O(\bar{n})}$ but this is not essential here). Suppose that $\mathscr{L}$ has as infinite number of (linearly) connected components. Choose a finite number t of them. Then there exists an element $0 < \alpha \in \tilde{\mathbb{Q}}$ such that in each of the chosen components there is a point z for which the unit vector v_1 normal to $\mathscr{V}$ at z and every vector $v \in P_x$ the norm $|v-v_1| > \alpha$. Indeed, consider the function $\tau : \mathscr{L} \longrightarrow \tilde{\mathbb{Q}}$ which to every point $z \in \mathscr{L}$ puts in correspondence the element $0 < \beta \in \tilde{\mathbb{Q}}$ equal to the minimum of the value $st_{0,1,2}(|v'-v_1'|)$ where v_1' is the unit vector normal to $\mathscr{V}$ at z and v' runs through all vectors $v' \in P_x$. For every connected component $\mathscr{L}_0$ of $\mathscr{L}$ the image $\tau(\mathscr{L}_0) = (0,\lambda) \subset \tilde{\mathbb{Q}}$ where $1 \geq \lambda \in \mathbb{Q}$. Therefore for chosen components of $\mathscr{L}$ the intersection of τ-images is also an interval of the kind $(0,\lambda) \subset \tilde{\mathbb{Q}}$ thus for α one can take any element from $(0,\lambda)$.

Now consider a semialgebraic set Ω of all point z from $\mathscr{M}$ such that for a unit vector v_1 normal to $\mathscr{V}$ at z and every vector $v \in P_x$ the norm $|v-v_1| > \alpha$. The set Ω can be given by a formula of the first order theory with $\rho_1.k$ atomic polynomials of degrees less then $\rho_2.d$ and having $\rho_3.n$ variables for some natural numbers ρ_1,ρ_2,ρ_3. Therefore the number of the connected components of Ω is less then $\alpha(\rho_1 k,\rho_2 d,\rho_3 n)$. Obviously $\Omega \subset \mathscr{L}$. Thus the number of the connected components of Ω is not less then t. We can choose $t > \alpha(\rho_1 k,\rho_2 d,\rho_3 n)$- contradiction.

We shall call a subset $\mathcal{M}_1 \subset \mathcal{M}$ p-long if p is the maximal natural number such that for every point z from some p-dimension subset M_1 of M holds : $P_z \cap \mathcal{M}_1 \neq \phi$. It is clear that $\mathcal{M}_1$ is p-long iff $\dim(st_{1,2}(\mathcal{M}_1)) = p$.

Lemma 1. The set $\mathcal{L}$ is not p-long.

Proof. Suppose that $\mathcal{L}$ is p-long. Choose in $\mathcal{L}$ (correspondingly - in M) a p-dimensional smooth connected semialgebraic subset $\mathcal{L}_1$ (correspondingly - M_1) such that for every $z \in M_1$ the intersection $P_z \cap \mathcal{L}_1 \neq \phi$ consists of a unique point and this correspondence is bijective.

The sets $\mathcal{L}_1$ and M_1 can really be chosen. Indeed, since the number of the connected components of $\mathcal{L}$ is finite, among them there will be a p-long component $\mathcal{L}'$. Choose in the set $\mathcal{L}'$ which is open in $\mathcal{V}$ a p-dimension connected semialgebraic subset $\mathcal{L}''$ which is a smooth manifold. According to the definition of p-long set there exists a p-dimension open subset M'' of M such that for every $y \in M''$ the intersection $P_y \cap \mathcal{L}'' \neq \phi$. Besides, from the transfer principle follows that for all pairs $y_1, y_2 \in M'' \subset M$ holds : $P_{y_1} \cap P_{y_2} \cap \mathcal{M} = \phi$. Therefore we have the smooth map of smooth manifolds $\varphi_2 : \mathcal{L}'' \longrightarrow M''$, if $z \in \mathcal{L}'' \cap P_y$ then $\varphi_2(z) = y$. By Sard's theorem, there exists a connected p-dimension semialgebraic subset $M_1 \subset M''$ such that for every point $y \in M_1$ the fibre $\varphi_2^{-1}(y)$ consists of the noncritical points of the map φ_2 and thus - of finite number of points. Let $\mathcal{L}_1$ be such a smooth semialgebraic subset of $\varphi_2^{-1}(M_1)$ that $\varphi_2(\mathcal{L}_1) = M_1$ and the restriction of φ_2 on $\mathcal{L}_1$ is injective. Therefore we had constructed the sets $M_1, \mathcal{L}_1$ having the necessary properties.

Choose a point $x_0 \in M_1$ and denote by L the p-plane tangent to manifold M at x_0. Since the smooth manifolds V_p can be given by a formula of first-order theory of the field $\tilde{\mathbb{Q}}$ for every point $z \in M$ the intersection $P_z \cap L \neq \phi$ consists of the unique point and this correspondence $\varphi_1 : L \longrightarrow M$ is bijective.

Let $L_1 = \varphi_1^{-1}(M_1)$. Denote by $\psi_1 : L_1 \longrightarrow \widetilde{\mathbb{Q}(\varepsilon_0,\varepsilon_1,\varepsilon_2)}$ a smooth function taking the value of the euclidean distance between a point from L_1 and its image in M_1 in force of the smooth map φ_1. Denote by $\psi_2 : L_1 \longrightarrow \widetilde{\mathbb{Q}(\varepsilon_0,\varepsilon_1,\varepsilon_2)}$ a smooth function which for $y \in L_1$ is taking the value equal to the sum of $\psi_1(y)$ and the distance between $\varphi_1(y)$ and $\varphi_2^{-1}(\varphi_1(y))$.

Let us prove that there exists a point $a \in \mathcal{L}_1$ such that euclidean distance form a to M is an element from $\widetilde{\mathbb{Q}(\varepsilon_0)}$. This will contradict with the definition of $\mathcal{M}$.

According to Lagrange's theorem on finite increments, for every point $y \in L_1$ there exist points $\Theta_1 \in L_1$ and $\Theta_2 \in L_1$ such that

$$\psi_1(y) = <\mathrm{grad}_{\Theta_1}(\psi_1), (y-x_0)>,$$

$$\psi_2(y) - \psi_2(x_0) = <\mathrm{grad}_{\Theta_2}(\psi_2), (y-x_0)>,$$

therefore

$$\psi_2(y) - \psi_1(y) - \psi_2(x_0) = <\mathrm{grad}_{\Theta_2}(\psi_2) - \mathrm{grad}_{\Theta_1}(\psi_1), (y-x_0)>.$$

Choose a point $y_0 \in L_1$ such that $|y_0-x_0| = \varepsilon_0^\gamma$ for some $0 < \gamma \in \mathbb{Q}$, $\mathrm{grad}_{x_0}(\psi_2)$ is collinear to (y_0-x_0) and $\psi_2(y_0) \geq \psi_2(x_0)$. Then by definition of the set $\mathcal{L} \supset \mathcal{L}_1$ for corresponding Θ_2 the scalar product $<\mathrm{grad}_{\Theta_2}(\psi_2), (y_0-x_0)> \geq \alpha\, \varepsilon_0^\gamma$ where $0 < \alpha \in \mathbb{Q}$. Besides, $|\mathrm{grad}_{\Theta_1}(\psi_1)| < \varepsilon_0^\beta$ for a certain $0 < \beta \in \mathbb{Q}$ and thus $<\mathrm{grad}_{\Theta_1}(\psi_1), (y_0-x_0)>\ |\mathrm{grad}_{\Theta_1}(\psi_1)| \cdot |y_0-x_0| < \varepsilon_0^{\beta+\gamma}$. Let $a = \varphi_2^{-1}(\varphi_1(y_0))$. Then the distance from $a \in \mathcal{M}_1$ to M_1

$$\psi_2(y_0) - \psi_1(y_0) > \varepsilon_0^\gamma(\alpha - \varepsilon_0^\beta) + \psi_2(x_0) > \varepsilon_0^\gamma.$$

This contradicts the inclusion $st_{1,2}(a) \in M_1$ which is valid by definition of the set $\mathcal{M}_1 \subset \mathcal{M}$. The lemma is proved.

Let us remark that the lemma remaines true if we take instead of $\mathcal{V}$ an arbitrary smooth algebraic hypersurface the standard part st_2 of which coinsides with V. The property to be algebraic is important : consider for example a line on a plane for V and an infinitely close sinusoid having an infinitesimal period for the smooth hypersurface.

Denote by $\mathcal{G} : \mathcal{V} \longrightarrow S^{n-1}$ the Gauss map (see e.g. [5]) of $\mathcal{V}$, here

$$S^{n-1} = \{ x \in (\widetilde{\mathbb{Q}(\varepsilon_0,\varepsilon_1,\varepsilon_2)})^n : |x| = 1 \}.$$

Corollary. Fix a point $x \in V_p^{(\varepsilon_0)}$. For almost all point $y \in V_p^{(\varepsilon_0)} \cap \mathcal{D}_x(\varepsilon_0)$ (i.e. for all points with the exception, maybe, of the points of a semialgebraic subset of the dimension less then p) holds :

$$\dim(st_{0,1,2}(\mathcal{G}(\mathcal{V} \cap \mathcal{D}_y(\varepsilon_1)))) \leq n-p-1.$$

Moreover, the set $st_{0,1,2}(\mathcal{G}(\mathcal{V} \cap \mathcal{D}_y(\varepsilon_1)))$ belongs to the intersection of the $(n-1)$-sphere S^{n-1} with $(n-p)$-plane, passing through 0.

Proof. According to lemma I, $st_{1,2}(\mathcal{L}) \subset V_p^{(\varepsilon_0)}$ (where $\mathcal{L}$ appears from x as in lemma I) is a semialgebraic subset of the dimension less then p. Let the point y belong to the complement to $st_{1,2}(\mathcal{L})$ in $V_p^{(\varepsilon_0)}$. Then, by definition of $\mathcal{L}$, for any point $z \in (\mathcal{V} \cap \mathcal{D}_y(\varepsilon_1))$ the normal to $\mathcal{V}$ at this point is ε_0-collinear to $(n-p)$-plane which is passing through 0 and is collinear to P_x. The corollary is proved.

Consider the stratum V_p for a certain $0 \le p < m$ and all adjoining strata of greater dimensions $V_{p_1},\dots,V_{p_s}$ ($p < p_j \le m$, $1 \le j \le s$). Fix an arbitrary point $x \in V_p^{(\varepsilon_0)}$ satisfying the corollary to lemma I. According to the corollary, for every $1 \le j \le s$, for almost every point $y \in V_{p_j}^{(\varepsilon_0)} \cap \mathcal{D}_x(\varepsilon_0)$ holds :

$$\dim(st_{0,1,2}(\mathcal{I}(\mathcal{V} \cap \mathcal{D}_y(\varepsilon_1)))) \le n-p_j-1 \qquad (2)$$

Lemma 2. Let for almost every point $y \in V_{p_j}^{(\varepsilon_0)} \cap \mathcal{D}_x(\varepsilon_0)$ the equality take place in (2). Then

$$\dim(st_{0,1,2}(\mathcal{I}(\mathcal{V} \cap \mathcal{D}_x(\varepsilon_1)))) = n-p-1 .$$

Proof. According to the corollary to lemma I (being applied to V_p) the inequality holds :

$$\dim(st_{0,1,2}(\mathcal{I}(\mathcal{V} \cap \mathcal{D}_x(\varepsilon_1)))) = l \le n-p-1 .$$

Assume that the strict inequality takes place. Let $t_1 = \max\{l, n-p_1-1, \dots, n-p_s-1\}$. Then there exists an element $\delta : \varepsilon_0 \ge \delta > 0$ of the field $\widetilde{\mathbb{Q}(\varepsilon_0)}$ such that for all points $y \in V^{(\varepsilon_0)} \cap \mathcal{D}_x(\delta)$ the following holds :

$$\dim(st_{0,1,2}(\mathcal{I}(\mathcal{V} \cap \mathcal{D}_y(\varepsilon_1)))) \le t_1$$

(in the opposite case on can find $y_0 \in V^{(\varepsilon_0)} \cap \mathcal{D}_x(\delta_0)$, where δ_0 is infinitesimal relative to $\mathbb{Q}(\varepsilon_0)$ such that

$$\dim(st_{0,1,2}(\mathcal{I}(\mathcal{V} \cap \mathcal{D}_{y_0}(\delta_0)))) \ge t_1+1$$

therefore the last inequality is also valid if one changes δ_0 by ε_1 and y_0 by x).

Let $\mathcal{N} = \mathcal{V} \cap \mathcal{D}_x(\delta)$. Thus $\dim(st_{0,1,2}(\mathcal{I}(\mathcal{N}))) = t_1 \le n-p-2$.

There exists an element $0 < \nu \in \widetilde{\mathbb{Q}(\varepsilon_0)}$ infinitesimal relative to $\mathbb{Q}$ such that

$$\bigcup_{w \in (st_{0,1,2}(\mathcal{I}(\mathcal{N})))^{(\varepsilon_0,\varepsilon_1,\varepsilon_2)}} \mathcal{D}_w(\nu) \supset \mathcal{I}(\mathcal{N}) .$$

For each point $w \in (st_{0,1,2}(\mathcal{I}(\mathcal{N})))^{(\varepsilon_0,\varepsilon_1,\varepsilon_2)}$ denote by A_w the fibre $\mathcal{I}^{-1}(\mathcal{D}_w(\nu) \cap \mathcal{I}(\mathcal{N}))$. Obviously

$$\mathcal{N} = \bigcup_{w \in (st_{0,1,2}(\mathcal{I}(\mathcal{N})))^{(\varepsilon_0,\varepsilon_1,\varepsilon_2)}} A_w .$$

Since all normals to $\mathscr{V}$ at points of the set A_w are ε_0-collinear, every point from A_w is situated in an infinitesimal relative to $\mathbb{Q}$ distance from a certain (the same for all points) hyperplane, besides the normal to this hyperplane is ε_0-collinear to every normal to $\mathscr{V}$ at the points from A_w.

We shall say that a set $P \subset (\widetilde{\mathbb{Q}(\varepsilon_0,\varepsilon_1,\varepsilon_2)})^n$ at a point $z \in P$ has dimension r is for all sufficiently small elements $0 < \alpha \in \widetilde{\mathbb{Q}(\varepsilon_0,\varepsilon_1,\varepsilon_2)}$ the intersection $P \cap \mathscr{D}_z(\alpha)$ has dimension r.

Denote by T the subset of all points from the set $st_{0,1,2}(\mathscr{I}(\mathscr{N})) \subset S^{n-1}$ at which this set has dimension $t \leq t_1$.

Further we shall use considerations very similar to those which occur in the theory of the developable surfaces. One can prove that in fact every point from A_w where $w \in T^{(\varepsilon_0,\varepsilon_1,\varepsilon_2)}$ lies in infinitesimal relative to $\mathbb{Q}$ distance from a certain (the same for all points) $(n-t-1)$-plane, herewith the orthogonal $(t+1)$-plane is ε_0-collinear to each normal to $\mathscr{V}$ at the points from A_w. Indeed, consider $t+1$ points $w_j \in (st_{0,1,2}(\mathscr{I}(\mathscr{N})))^{(\varepsilon_0,\varepsilon_1,\varepsilon_2)}$ $(1 \leq t+1 \leq n-p-1)$ such that when $j_1 \neq j_2$, firstly $\mathscr{D}_{w_{j_1}}(v) \cap \mathscr{D}_{w_{j_2}}(v) = \phi$ and, secondly, the distance $|w_{j_1} - w_{j_2}|$ is infinitesimal relative to $\mathbb{Q}$. Besides, assume that the hyperplanes corresponding (see above) to the points w_j are in the general position. Then the fibres A_{w_j} over the points w_j which "do not deviate much" from the corresponding hyperplanes and from each other, also "do not deviate much" from the intersection of the hyperplanes i.e. from $(n-t-1)$-plane.

Every fibre A_w where $w \in T^{(\varepsilon_0,\varepsilon_1,\varepsilon_2)}$ can not be bounded by a set of points y such that

$$\dim(st_{0,1,2}(\mathscr{I}(\mathscr{V} \cap \mathscr{D}_y(\varepsilon_1)))) > t$$

since this would contradict the definition of T and is a $(n-t-1)$-long set.

Let us prove that T belongs to the union of finite family of $(t+1)$-planes passing through 0. Indeed, in the opposite case there exists a t-dimension subset $L \subset st_{0,1,2}(\mathscr{I}(\mathscr{N})))^{(\varepsilon_0,\varepsilon_1,\varepsilon_2)}$ such that none of its t-dimension subset $L_1 \subset L$ lies in $(t+1)$-plane passing through 0.

Consider two points w_1 and $w_2 \in L$ such that $(n-t-1)$-planes corresponding to fibres A_{w_1} and A_{w_2} are not ε_0-collinear. Every fibre A_{w_1} and A_{w_2} is an $(n-t-1)$-long set and an argumentation similar to the one in the proof of lemma I shows that there exists a point $\alpha \in A_{w_1} \cap \mathscr{D}_x(\delta)$ such that the nearest to α point $\beta \in A_{w_2}$ belongs to $\mathscr{D}_x(\delta)$ and the distance $|\alpha - \beta| > 0$ is not infinitesimal relative to $\mathbb{Q}(\varepsilon_0)$. It follows that the intersection $\mathscr{I}^{-1}(L) \cap \mathscr{D}_x(\delta)$ is a $(n-1)$-long set.

In case when there does not exist a stratum V_{p_j} of dimension $p_j > n-t-1$ which is adjacent to V_p this contradicts the corollary to lemma 1 and in the opposite case the inequality (2) (applied to V_{p_j}).

Therefore we have proved that $st_{0,1,2}(\mathcal{I}(\mathcal{N})) \subset S^{n-1}$ belongs to the union of a finite family of some planes (of maybe different dimensions) passing through 0, note also that for each pair $w_1, w_2 \in T^{(\varepsilon_0,\varepsilon_1,\varepsilon_2)}$ the $(n-t-1)$-planes corresponding to A_{w_1}, A_{w_2} are ε_0-collinear. Now let us prove that this family consists of not less then two members. Moreover, there exists a connected component $\mathcal{N}_1$ of $\mathcal{N}$ such that the same is true for $\mathcal{N}_1$.

Suppose the opposite : let it consist of unique $t = t_1$-dimensional member. Since $n-t-1 > p$, there exists a stratum V_{p_j} $(1 \le j \le s)$ such that the intersection $st_{1,2}(A_w) \cap V_{p_j}^{(\varepsilon_0)} \neq \phi$ for every $w \in T^{(\varepsilon_0,\varepsilon_1,\varepsilon_2)}$. Denote by B_w the $(n-t-1)$-plane corresponding to A_w. From the transversality condition it follows that B_w is ε_0-collinear to one of the connected components of $V_{p_j}^{(\varepsilon_0)} \cap \mathcal{D}_x(\delta)$. Consider two connected components $V' \subset V_{p_{j'}}^{(\varepsilon_0)} \cap \mathcal{D}_x(\delta)$, $V'' \subset V_{p_{j''}}^{(\varepsilon_0)} \cap \mathcal{D}_x(\delta)$ of some strata $V_{p_{j'}}^{(\varepsilon_0)}, V_{p_{j''}}^{(\varepsilon_0)}$ (possibly $j' = j''$) such that $V' \not\subset \bar{V}''$, $\bar{V}'' \not\subset V'$, $\bar{V}' \cap \bar{V}'' \neq \phi$ (here the bar denotes the closure in the topology with the base of all open disks). Choose a point $z \in V_p \cap \bar{V}' \cap \bar{V}'' \cap \widetilde{\mathbb{Q}^n}$ and consider a tangent $p_{j'}$-plane L' (correspondingly $p_{j''}$-plane L'') to $\bar{V}'$ (correspondingly - to $\bar{V}''$) at z. Note that B_w is defined over $\widetilde{\mathbb{Q}(\varepsilon_0)}$ while L', L''–over $\widetilde{\mathbb{Q}}$. Let us suppose that B_w is ε_0-collinear to both L' and L'' (we shall say that in this case B_w is ε_0-collinear to V' and V''). Then $st_0(B_w) \subset L' \cap L''$. Therefore $\min\{\dim(L'), \dim(L'')\} > \dim(L' \cap L'') \ge n-t-1$ and B_w is ε_0-collinear to $\bar{V}' \cap \bar{V}'' = \bar{V}^{(3)}$ where $V^{(3)}$ is the intersection of $\mathcal{D}_x(\delta)$ with a connected component of a certain stratum among $V_{p_j}^{(\varepsilon_0)}$. Repeating this argumentation if necessary we can assume that there does not exist another stratum among $V_{p_j}^{(\varepsilon_0)}$ $(1 \le j \le s)$ such that for an intersection $V^{(4)}$ of its connected component with $\mathcal{D}_x(\delta)$ holds : $V^{(4)} \not\subset \bar{V}^{(3)}$, $V^{(3)} \not\subset \bar{V}^{(4)}$ and B_w is ε_0-collinear to $V^{(3)}$ and $V^{(4)}$. The inequality $\dim(V_p) < n-t-1$ and the transversality condition imply that there exists a connected component of some stratum among $V_{p_j}^{(\varepsilon_0)}$ whose intersection with $\mathcal{D}_x(\delta)$ is transversal to $V^{(3)}$. From our assumption it follows that B_w is ε_0-collinear to this intersection. Thus there is a point $st_{0,1,2}(\mathcal{I}(\mathcal{N}_1))$ not belonging to the $(t+1)$-plane under consideration. This contradiction implies that there exist two points $w_1, w_2 \in st_{0,1,2}(\mathcal{I}(\mathcal{N}_1))$ such that fibres A_{w_1}, A_{w_2} intersect transversally at a point from $\mathcal{N}$. This contradicts the smoothness of $\mathcal{N}$. The lemma is proved.

Lemma 3. For almost every point $x \in V_p$ (with the exception of a semialgebraic subset of a dimension less then p) holds :

$$\dim(st_{1,2}(\mathcal{I}(\mathcal{V} \cap \mathcal{D}_x(\varepsilon_1)))) = \ell = n-p-1 .$$

Proof. From the transfer principle it follows that it is sufficient to prove the proposition for $V_p^{(\varepsilon_0)}$ instead of V_p and $st_{0,1,2}$ instead of $st_{1,2}$.

Proceed by induction. As a base of the induction consider the case when the stratum V_p is maximal i.e. it is not incident to any stratum of greater dimension. Then for almost every point $x \in V_p^{(\varepsilon_0)}$ the intersection $P_x \cap \mathcal{V}$ is obviously diffeomorphic to $(n-p-1)$-sphere, therefore $\ell \geq n-p-1$. On the other hand, according to the corollary to lemma I, $\ell \leq n-p-1$. Suppose that the lemma is proved for strata of the dimension greater then p. Consider non-maximal stratum $V_p^{(\varepsilon_0)}$ and apply to it lemma 2 (the hypothesis of lemma 2 is valid by the inductive hypothesis). The lemma is proved.

3. Representations of some semialgebraic sets.

Consider a semialgebraic set R given by a formula of first-order theory of $\widetilde{\mathbb{Q}}$. Let the dimension of R at every point (see the proof of lemma 2) be the same and equal to m. The fact that $z \in R$ is a smooth point in R (that is the fact that for any two pairs of distinct points in the neighbourhood of z in R, the lines passing through the first and second pair are almost collinear to the same m-plan) can be expressed by a first order formula. Note that in $R = V_p$ every point is smooth ; if V_{p+i} is incident to V_p then the point $z \in V_p \subset (V_{p+i} \cup V_p) = R$ is not smooth.

It is hence clear that the subset $Sm(R)$ of all smooth points of R is semialgebraic and can be written as a formula of first-order with a constant (i.e. not depending on R) number of quantifier alternations.

In the algorithm described below an important role is playing the set of the form :

$$K_p = st_{1,2}(\ \{x \in \mathscr{V} : \dim(st_{1,2}(\mathscr{I}(\mathscr{V} \cap \mathscr{D}_x(\varepsilon_1)))) = n-p-1\}\ ),$$

where the external st is taken for all points of the set for which it is defined.

Let us prove that K_p is a semialgebraic set which can be given by a formula with a constant number (not depending on $\mathscr{V}$) of quantifier alternations. Indeed, the image of a semialgebraic set under the Gauss map is obviously semialgebraic. The standard part of a semialgebraic set A in the representation of which atomic polynomials belong to $\mathbb{Q}[\varepsilon_1,\varepsilon_2]\,[X_1,\dots,X_n]$ is, as it was noted in [3], also semialgebraic (consider $\varepsilon_1,\varepsilon_2$ as new variables, then $st_{1,2}(A)$ coinsides with the intersection of n-planes $\{\varepsilon_1 = \varepsilon_2 = 0\}$ and the closure in euclidean topology of the set $A \cap \{\varepsilon_1 > 0\ \&\ \varepsilon_2 > 0\}$.

The proposition $\dim(.) = t$ can be written in the following way : there exists a linear transformation of coordinates $L : (X_1,\dots,X_n) \longmapsto (Y_1,\dots,Y_n)$ with the matrix

$$\begin{pmatrix} 1 & \lambda_2 & \dots & \lambda_n \\ 0 & 1 & \dots & 0 \\ \dots & \dots & \dots & \dots \\ 0 & 0 & 0 & 1 \end{pmatrix} \tag{3}$$

such that the projection of the set on the subspace of coordinates $Y_1,\dots,Y_t$ containes a t-ball and for every linear map of the kind (3) the projection on the $(t+1)$-subspace does not contain a $(t+1)$-ball.

4. The algorithm and its running time.

As an auxiliary procedures our algorithm essentially involves the effective algorithms for quantifier elimination in first-order theory of real closed fields proposed in [6] (see also [7]) and for finding all connected components of a semialgebraic set from [2] or [3].

The algorithm works recursively in p. Let $p = m = \dim(V)$. The algorithm defines the set $Sm(\{x \in V\ /\ \dim(V) \text{ in } x \text{ equals } m\})$ by a formula (with quantifiers) $\Theta_m^{(1)}$ of the first-order theory of $\widetilde{\mathbb{Q}(\varepsilon_1,\varepsilon_2)}$ as it was explaned at the beginning of section 3. After that using several times the quantifier elimination procedure from [6] the algorithm produces quantifier-free formula $\Theta_m^{(2)}$ which is equivalent to $\Theta_m^{(1)}$. Obviously $V_m = \{\Theta_m^{(2)}\}$.

Assume that strata $V_m,\dots,V_{p+1}$ are already constructed and given by formulas $\Theta_m,\dots,\Theta_{p+1}$ correspondingly. The algorithm defines the set $Sm(K_p)$ by a formula (with quantifiers) $\Theta_p^{(1)}$ as it was explaned in section 3, and then, with the help of procedure from [6] produces an equivalent quantifier-free formula

$\Theta_p^{(2)}$. Using [2] the algorithm finds all connected components of $\{\Theta_p^{(2)}\}$ and after that with the help of the procedure from [8] for solving systems of polynomial inequalities selects those which have empty intersections with every $V_m,\ldots,V_{p+1}$. Let the selected components be given by formulas $\Theta_p^{(2,1)},\ldots,\Theta_p^{(2,s_p)}$. For every $1 \leq i \leq s_p$ the algorithm writes a formula (with quantifiers) $\Theta_p^{(3,i)}$ defining the closure (in the topology in $(\widetilde{\mathbb{Q}(\varepsilon_1,\varepsilon_2)})^n$ with the base of all open balls) of the set $\Theta_p^{(2,i)}$ and then, eliminating quantifiers in $\Theta_p^{(3,i)}$ finds an equivalent quantifier-free formula $\Theta_p^{(4,i)}$.

Let $\Theta_p^{(5)} = \bigvee_{1 \leq i \leq s_p} \Theta_p^{(4,i)}$ and $\Theta_p^{(6)}$ be the formula defining $Sm(\{\Theta_p^{(5)}\})$. Eliminating quantifiers in $\Theta_p^{(6)}$ the algorithm obtaines the quantifier-free formula Θ_p.

Let us prove that $V_p = \{\Theta_p\}$. We must show that if $U_1,\ldots,U_t$ are all connected components of $Sm(K_p)$ which have empty intersections with each of the strata $V_m,\ldots,V_{p+1}$ and $V'_p = \bigcup_{1 \leq j \leq t} \overline{U}_j$ is the union of the closures of $U_1,\ldots,U_t$ then $V_p = Sm(V'_p)$. Let $x \in V_p$. Then either x satisfies the hypothesis of lemma 3 and therefore $x \in K_p$ by the definition of K_p or x belongs to a subset of V_p of a lower dimension and still $x \in K_p$, since K_p coincides with the standard part of a semialgebraic set and thus is closed in the topology with the base of all open balls. The point x may not belong to $Sm(K_p)$ but it lies in $\overline{U}_j$ for some value $1 \leq j \leq t$ since $V \setminus (V_m \cup \ldots \cup V_{p+1})$ is an algebraic variety and therefore every $U_j (1 \leq j \leq t)$ has a nonempty intersection with $\bigcup_{p+1 \leq i \leq m} V_i$ iff U_j belongs to this union. The inclusion $Sm(V'_p) \subset V_p$ can be proved by the reverse argumentation.

Let us estimate the running time of the algorithm and the parameters of the produced formulas Θ_p. Recall that the input variety is given by the formula (I) where $f_{ij} \in \mathbb{Z}[X_1,\ldots,X_n]$, $\deg_{X_1,\ldots,X_n}(f_{ij}) < d, \ell(f_{oj}) \leq M$ and the number of the atomic subformulas is K. The image $\mathcal{I}(\mathcal{V} \cap \mathcal{D}_x(\varepsilon_1))$ is defined by a formula $\Gamma^{(1)}$ in the prenex form with one block of existential quantifiers. The polynomials occuring in the quantifier-free part have total degrees less then Kd and bit-lengths of coefficients not exceeding $\log(K)M$. The application to $\Gamma^{(1)}$ of the procedure from [6] requires time $M^{0(1)} (Kd)^{n^{0(1)}}$ and the resulting quantifier-free formula $\Gamma^{(2)}$ has $(Kd)^{n^{0(1)}}$ atomic subformulas with polynomials of the degree $(Kd)^{n^{0(1)}}$ having coefficient lengths not exceeding $M^{0(1)} (Kd)^{n^{0(1)}}$.

The formula defining the set $st_{1,2}(\{\Gamma^{(2)}\})$ has two quantifier alternations, the result of quantifier elimination is a quantifier-free formula $\Gamma^{(3)}$ with the bounds of the same kind as in formula $\Gamma^{(2)}$.

The formula which defines the value of the dimension has the number of variables linearly depending on n and is a conjunction of two formulas in prenex forms the first of which has the prefix of existential quantifiers and the second - of universal quantifiers. As a result of quantifier elimination in both members of the conjunction a formula $\Gamma_p^{(4)}$ will be obtained having the bounds of the indicated type.

Elimnating both quantifier blocks in $st_{1,2}(\{\Gamma_p^{(4)}\})$ the algorithm obtaines a quantifier-free formula $\Gamma_p^{(5)}$ such that $K_p = \{\Gamma_p^{(5)}\}$ again having the estimates of the same kind as in $\Gamma^{(2)}$. The total running time for the constructing of $\Gamma_p^{(5)}$ will be $M^{0(1)}(Kd)^{n^{0(1)}}$.

The formula defining the set $Sm(\{\Gamma_p^{(5)}\})$ has a number of variables polynomially depending on n and one block of universal quantifiers. The elimination algorithm outputs quantifier-free formula $\Theta_p^{(2)}$ having analogous bounds on the parameteres in the similar time.

The algorithm constructs strata V_p by recursion beginning with $p = m$ (in this case $V_m = \{\Theta_m^{(2)}\}$). Assume that the formulas $\Theta_m = \Theta_m^{(2)}, \Theta_{m-1}, \ldots, \Theta_{p+1}$ defining strata $V_m, \ldots, V_{p+1}$ correspondingly are already obtained. The formula $\Theta_p^{(2,j)}$ $(1 \le j \le s_p)$ defining the connected component of $\{\Theta_p^{(2)}\}$ has according to [2,3] $(Kd)^{n^{0(1)}}$ atomic subformulas with polynomials of the degree $(Kd)^{n^{0(1)}}$ and coefficient lengths not exceeding $M^{0(1)}(Kd)^{n^{0(1)}}$. Besides, $s_p < (Kd)^{n^{0(1)}}$. The running time of the decomposition and of the selection of the components is $M^{0(1)}(Kd)^{n^{0(1)}}$ according to [2,3] and [8] (here we use the estimates on the parameters of formulas $\Theta_m, \ldots, \Theta_{p+1}$ which, as we shall see further, are of the same kind as for $\Theta_p^{(2)}$ and do not depend on p).

The formula $\Theta_p^{(3,j)}$ defining the closure of $\{\Theta_p^{(2,j)}\}$ $(1 \le j \le s_p)$ has three quantifier alternations and the number of variables linearly depending on n. The result of quantifier elimination in this formula will be the formula $\Theta_p^{(4,j)}$ having the bounds similar to just described. The formula $\Theta_p^{(5)}, \Theta_p^{(6)}$ obviously have the same type of bounds. Finally the formula Θ_p defining V_p has the similar estimates on the parameters and the total running time of the algorithm is $M^{0(1)}(Kd)^{n^{0(1)}}$.

The theorem is proved.

Remark. The problem of deciding for a given formula of the kind (1) whether is defines a regular algebraic variety does not seem to be more simple than the problem of stratification. However using quantifier elimination procedure from [6, 7] one can check whether the produced decomposition of the variety is a Whitney stratification and thus make sure that the algorithm had worked correctly for the given input.

Acknowledgement. I thank K. Bekka and M.-F. Roy for useful discussions and comments.

References

I. Goresky M., MacPherson R. : Stratified Morse Theory. Springer-Verlag, Berlin, 1988.

2. Canny J., Grigor'ev D. Yu., Vorobjov N.N. Jr. : Finding connected components of a semialgebraic set in subexponential time, 1990, to appear in AAECC.

3. Heintz J., Roy M.-F., Solerno P. : Description des composantes connexes d'un ensemble semi-algébrique en temps simplement exponentiel C.R. Acad. Sci. Paris 313,1991, p.167-170 .

4. Lang S. : Algebra. Addison-Wesley, New York, 1965.

5. Thorpe J.A. : Elementary Topics in Differential Geometry. Springer-Verlag, Berlin, 1979.

6. Heintz J., Roy M.-F., Solerno P. : Sur la complexité du principe de Tarski-Seidenberg, Bull. Soc. Math. France, 118, 1990, p. 101-126.

7. Renegar J. : On the computational complexity and geometry of the first order theory of the reals, Parts I, II, III, Tech. Report 856, Cornell University Ithaca, 1989.

8. Grigor'ev D.Yu., Vorobjov N.N. Jr. : Solving systems of polynomial inequalities in subexponential time. J. Symbolic Comp., 5, 1988, p. 37-64.

9. Mostowski T., Rannou E.: Canonical Whitney stratification of an algebraic set in $\mathbb{C}^n$. AAECC 1991, New Orleans.

10. Kashiwara M. : *B*-functions and holonomic systems. Inv. Math., 38, 1976, p. 33-53.

11. Henry J.-P., Merle M., Sabbah C. : Sur la condition de Thom stricte pour un morphisme analytique complexe. Ann. Scient. Ec. Norm. Sup., 4 série, 17, 1984, p. 227-268.

LIST OF PARTICIPANTS

ACQUISTAPACE Francesca	Univ. Pisa (Italy)
AKBULUT Selman	Michigan State Univ. (USA)
ALONSO Maria-Emilia	Univ. Complutense Madrid (Spain)
ANDRADAS Carlos	Univ. Complutense Madrid (Spain)
BECKER Eberhard	Univ. Dortmund (Germany)
BEKKA Karim	Univ. Rennes 1 (France)
BERR Ralph	Univ. Dortmund (Germany)
BIERSTONE Edward	Univ. Toronto (Canada)
BOCHNAK Jacek	Vrije Univ. Amsterdam (Netherlands)
BOROBIA Alberto	U.N.E.D., Madrid (Spain)
BOUZOUBAA Taoufik	Univ. Rennes 1 (France)
BRÖCKER Ludwig	Univ. Münster (Germany)
BUCHNER Michael	Univ. New Mexico, Albuquerque (USA)
CANNY John	I.C.S.I., Berkeley (USA)
CELLINI Paola	Univ. Pisa (Italy)
COSTE Michel	Univ. Rennes 1 (France)
CYGAN Ewa	Uniw. Jagielloński, Krakòw (Poland)
CYNK Slawomir	Uniw. Jagielloński, Krakòw (Poland)
DEDIEU Jean-Pierre	Univ. Paul Sabatier, Toulouse (France)
DEGTYAREV Alexander	Steklov Institute, Leningrad (USSR)
DE LA PUENTE Maria Jesus	Univ. Complutense Madrid (Spain)
DIOP Mahmadou	Univ. Rennes 1 (France)
DRUZKOWSKI Ludwik	Uniw. Jagielloński, Krakòw (Poland)
EFRAT Ido	Univ. Konstanz (Germany)
FERRAROTTI Massimo	Univ. Pisa (Italy)
FINASHIN S.	Leningrad Electrotechnical Institute (USSR)
FORTUNA Elisabetta	Univ. Pisa (Italy)
FRANÇOISE Jean-Pierre	Univ. Paris 6 (France)
GAMBOA Jose-Manuel	Univ. Complutense, Madrid (Spain)
GONDARD Danielle	Univ. Paris 6 (France)
GONZALEZ VEGA Laureano	Univ. Cantabria, Santander (Spain)
GUARALDO Francesco	Univ. degli Studi, Roma (Italy)
GUERGUEB Ahmed	Univ. Rennes 1 (France)
HAJTO Zbigniew	Uniw. Jagielloński, Krakòw (Poland)
HAROUNA Warou	Univ. Rennes 1 (France)
HEINZ Joos	Univ. Buenos Aires (Argentina)
HUBER Roland	Univ. Regensburg (Germany)
HUISMAN Johan	Vrije Univ., Amsterdam (Netherlands)
ISCHEBECK Friedrich	Univ. Münster (Germany)
ITENBERG Ilia	Leningrad State Univ. (USSR)
JAWORSKI Piotr	Univ. Warszaw (Poland)
JELONEK Zbigniew	Uniw. Jagielloński, Krakòw (Poland)

KING Henry	Univ. of Maryland (U.S.A.)
KNEBUSCH Manfred	Univ. Regensburg (Germany)
KORCHAGIN Anatolii	Nizhny Novgorod Univ. (USSR)
KRASIŃSKI Tadeusz	Univ. Lódź (Poland)
KRIEK Teresa	Univ. Buenos Aires (Argentina)
KURDIKA Krzysztof	Uniw. Jagielloński, Krakòw (Poland)
KWIECINSKI Michal	Univ. Lille 1 (France)
LAM Tsit Yuen	Univ. California, Berkeley (USA)
LE SAUX Frederic.	Univ. Rennes 1 (France)
LIGATSIKAS Zissis	Univ. Rennes 1 (France)
LOJASIEWICZ Stanislaw	Uniw. Jagielloński, Krakòw (Poland)
LOMBARDI Henri	Univ. Franche-Comté, Besançon (France)
MACRI Patrizia	Univ. degli Studi, Roma (Italy)
MAHE Louis	Univ. Rennes 1 (France)
MARIN Alexis	E.N.S. Lyon (France)
MARINARI Maria Grazia	Univ. Genova (Italy)
MARSHALL Murray	Univ. of Saskatchewan, Saskatoon (Canada)
MAZUROVSKII Vladimir	Ivanovo Civil Engineering Institute (USSR)
MEGUERDITCHIAN Ivan	Univ. Rennes 1 (France)
MIKHALKIN Grigory	Leningrad (USSR)
MILMAN Pierre	Univ. Toronto (Canada)
MIODEK Andrzej	Univ. Lódź (Poland)
MONTANA Jose Luis	Univ. Cantabria, Santander (Spain)
MOSTOWSKI Tadeusz	Univ. Warszaw (Poland)
NATANZON Serguey	Moskva (USSR)
NEUHAUS Ralph	Univ. Dortmund (Germany)
ORTEGA Jesus	Univ. Castilla-La-Mancha (Spain)
OTERO M.	Univ. Oxford (Great Britain)
PARDO VASALLO Luis Miguel	Univ. Cantabria, Santander (Spain)
PARIMALA	Tata Institute, Bombay (India)
PAUGAM Anne-Marie	Univ. Rennes 1 (France)
PAWLUCKI Wieslaw	Uniw. Jagielloński, Krakòw (Poland)
PECKER Daniel	Univ. Paris 6 (France)
PEDERSEN Paul	Courant Institute, New York (USA)
PFISTER Albrecht	Univ. Mainz (Germany)
POLOTOVSKII G.M.	Nizhny Novgorod Univ. (USSR)
PRESTEL Alexander	Univ. Konstanz (Germany)
RAMANAKORAISINA Rodolphe	Univ. Rennes 1 (France)
RANDRIAMAHALEO Solo	Univ. Fianarantsoa (Madagascar)
RANNOU Eric	Univ. Rennes 1 (France)
REZNICK Bruce	Univ. Illinois (USA)
RICHARDSON Donald	Univ. Bath (Great Britain)
ROBSON Robert O.	Oregon State Univ. (USA)
ROY Marie-Françoise	Univ. Rennes 1 (France)

RUIZ Jesus	Univ. Complutense, Madrid (Spain)
RUSEK Kamil	Uniw. Jagielloński, Krakòw (Poland)
SANDER Tomas	Univ. Dortmund (Germany)
SCHEIDERER Claus	Univ. Regensburg (Germany)
SCHMID Joachim	Univ. Konstanz (Germany)
SEPPALA Mikka	Univ. Helsinki (Finland)
SHUSTIN E.I.	Samara State Univ. (USSR)
SILHOL Robert	Univ. Montpellier 2 (France)
SIMON Odile	Univ. Rennes 1 (France)
SKIBINSKI Przemyslaw	Univ. Lódź (Poland)
SOLERNO Pablo	Univ. Buenos Aires (Argentina)
SPODZIEJA Stanislaw	Univ. Lódź (Poland)
STASICA Jacek	Uniw. Jagielloński, Krakòw (Poland)
STENGLE Gilbert	Lehigh Univ., Bethlehem (USA)
STES David	Univ. Gent (Belgium)
SZAFRANIEC Zbigniew	Univ. Gdansk (Poland)
TOGNOLI Alberto	Univ. Trento (Italy)
TOUGERON Jean-Claude	Univ. Rennes 1 (France)
TRAVERSO Carlo	Univ. Pisa (Italy)
TROTMAN David	Univ. Provence, Marseille (France)
VAN GEEL Jan	Univ. Gent (Belgium)
VIRO Oleg Ya.	Steklov Institute, Leningrad (USSR)
VOM HOFE Günter	Univ. Dortmund (Germany)
VOROBJOV Nicolai N.	Steklov Institute, Leningrad (USSR)
WÖRMANN Thorsten	Univ. Dortmund (Germany)
ZVONILOV V.I.	Syktyckar (USSR)

Printing: Weihert-Druck GmbH, Darmstadt
Binding: Buchbinderei Schäffer, Grünstadt